Microgrids

IEEE Press
445 Hoes Lane
Piscataway, NJ 08854

Microgrids

Dynamic Modeling, Stability and Control

Qobad Shafiee, Mobin Naderi, and Hassan Bevrani
University of Kurdistan, Iran

IEEE PRESS
WILEY

Published by John Wiley & Sons, Inc., Hoboken, New Jersey.
Published simultaneously in Canada.

Library of Congress Cataloging-in-Publication Data:

Names: Shafiee, Qobad, author. | Naderi, Mobin, author. | Bevrani, Hassan, author.
Title: Microgrids : dynamic modeling, stability and control / Qobad Shafiee, Mobin Naderi, Hassan Bevrani.
Description: Hoboken, New Jersey : Wiley, [2024] | Includes bibliographical references and index.
Identifiers: LCCN 2023040115 (print) | LCCN 2023040116 (ebook) | ISBN 9781119906209 (hardback) | ISBN 9781119906216 (adobe pdf) | ISBN 9781119906223 (epub)
Subjects: LCSH: Microgrids (Smart power grids)
Classification: LCC TK3105 .S43 2024 (print) | LCC TK3105 (ebook) | DDC 621.31–dc23/eng/20231107
LC record available at https://lccn.loc.gov/2023040115
LC ebook record available at https://lccn.loc.gov/2023040116

Cover Design: Wiley
Cover Image: © ArtemisDiana/Shutterstock.com

Set in 9.5/12.5pt STIXTwoText by Straive, Chennai, India

Dedicated to our beloved families.

Contents

About the Authors

Qobad Shafiee received a PhD degree in electrical engineering from the Department of Energy Technology, Aalborg University (Denmark), in 2014. He is currently an associate professor and co-leader of the Smart/Micro Grids Research Center at the University of Kurdistan (Sanandaj, Iran), where he was a lecturer from 2007 to 2011. In 2014, he was a visiting scholar at the Electrical Engineering Department, the University of Texas at Arlington, Arlington, TX, USA. He was a post-doctoral fellow and visiting professor at the Department of Energy Technology, Aalborg University in 2015 and 2017, respectively. He is a senior member of IEEE, associate editor of *IEEE Transactions on Power Electronics*, associate editor of *IEEE Transactions on Energy Conversion*, and associate editor of *e-Prime – Advances in Electrical Engineering, Electronics and Energy*. His research interests include dynamic modeling, stability, security, and control of renewable energy systems, power converter-dominated grids, and microgrids, and predictive/optimal control of modern power systems.

Mobin Naderi received the BSc, MSc, and PhD degrees in electrical engineering from Tabriz University, Tabriz, Iran; Iran University of Science and Technology, Tehran, Iran; and University of Kurdistan, Sanandaj, Iran in 2012, 2014, and 2019, respectively. He was a visiting PhD student at the Department of Energy Technology, Aalborg University, Aalborg, Denmark, in 2019, and a post-doctoral researcher at the University of Kurdistan from 2020 to 2022. He is currently a research fellow at the University of Sheffield, Sheffield, UK. His research interests include modeling, planning, stability, and control of autonomous/interconnected microgrids and hybrid energy storage systems as well as robust modeling and control methods.

Hassan Bevrani received a PhD degree in electrical engineering from Osaka University (Japan). Currently, he is an IEEE fellow, a full professor, and the program leader of the Smart/Micro Grids Research Center (SMGRC) at the University of Kurdistan. Over the years, he has worked as an experienced research fellow and visiting professor with Osaka University, Kumamoto University, Kyushu Institute of Technology, Doshisha University, Nagoya University (Japan), Queensland University of Technology (Australia), Centrale Lille (France), and Technical University of Berlin (Germany). He is the author/coauthor of 9 international books, 15 book chapters, and more than 500 journal/conference papers. His current research interests include microgrid dynamics and control, smart grid operation and control, and intelligent/robust control applications in the power electric industry.

Preface

The *microgrid* (MG) concept provides a quite appealing solution for overcoming the challenges of integrating renewable energy sources (RESs) and distributed energy resources (DERs) into the power grids. The MGs are small electrical distribution systems that interconnect multiple customers, DERs, and storage systems and are known as the main building blocks of the smart grids.

Like conventional power grids, the MGs have a hierarchical control structure with different operation layers. Advances in the *microgrids dynamic modeling, stability, and control* facilitate a higher integration of MGs into the conventional electrical grids. This issue not only covers their internal control performance and connection/disconnection procedures to limit the negative dynamic impacts on the main grid and other connected MGs but also includes the grid support functionalities to enhance the global operation of distribution networks.

The *microgrids dynamic modeling, stability, and control* book describe the most important issues on *individual MGs* and *interconnected MGs* (IMGs) modeling, stability, and control as well as new relevant perspectives and research outcomes. It covers different modeling and stability analysis methods together with various levels of MG control, including local (primary), secondary, central, and global controls.

The book provides a thorough understanding of the basic principles, e.g. MG/IMG structures, types, operating modes, dynamics, and control levels. The book presents several MG structure modeling, stability analysis methodologies, and fundamental control approaches that are mostly examined by real-time simulations and experimental studies. Advanced control synthesis approaches, virtual dynamic control, and cyber-security and cyber-challenges of MGs are also addressed in the book. A number of Matlab codes and simulation files are provided for the readers to regenerate and better understand the exercises and examples presented in the book. These codes allow readers to gain hands-on experience with implementing the modeling, stability analysis, and fundamental control techniques discussed. The book is organized into 11 chapters.

Chapter 1 addresses an introduction to the integration of MGs in modern power grids. Basic concepts, overall structure, main characteristics, and capabilities of MGs are briefly discussed. *Chapter 2* investigates the essential concepts and fundamentals of MG dynamic modeling. Different aspects including terminology, relevant concepts, classification, fundamental analysis tools, and corresponding requirements for small-signal modeling are studied. *Chapter 3* extends the knowledge learned from Chapter 2 to perform an overall representation structure of an MG and explains the process of overall modeling using several case studies.

Chapter 4 presents several small-signal stability analyses for ac and dc MGs using eigenvalue analysis, participation matrix, and sensitivity analysis. *Chapter 5* emphasizes the most important concepts and basics on the MG control. The chapter discusses a general overview of the main control layers and loops with some related examples. *Chapter 6* addresses the applications of advanced control methodologies, the virtual dynamic control concept for MGs, as well as challenges of MGs as cyber-physical systems. The topics are supplemented by several control synthesis examples.

Chapter 7 discusses architectures, definitions, benefits, opportunities, and challenges of IMGs. Then, the operation of IMGs is explained, and vacancies for future research are emphasized. *Chapter 8* addresses the dynamic modeling of IMGs for stability analysis and control synthesis, using a module-based modeling process. *Chapter 9* discusses the stability of the IMGs in both small-signal and large-signal perspectives. For this purpose, the applications of eigenvalues analysis and sensitivity analysis are presented. *Chapter 10* focuses on the control of IMGs with different structures. Frequency control, power sharing control, and emergency control for different architectures of IMGs are emphasized. The design examples are supplemented by simulations and real-time experimental studies. Finally, *Chapter 11* addresses the synchronization stability and control requirements in IMGs. The impacts of the secondary control loop, high loads, and physical constraints on the synchronization performance are discussed.

The book summarizes long-term academic/research outcomes and contributions influenced by the authors, practical experiences in dynamic modeling, stability analysis, and control of individual MGs and IMGs during the past two decades of professional activities in both academia and industry.

The book could be useful for engineers, researchers, and developers in power grid dynamic modeling, stability analysis, and control. Basic knowledge of electric circuits, power systems, power electronics, and linear control theory is preferable but not necessary. The book could be also useful as a supplementary text for university students in electrical engineering at both undergraduate and postgraduate levels in the standard courses of MG modeling and control, MG dynamics and stability, and smart grids.

September 2023

Qobad Shafiee, Mobin Naderi, and Hassan Bevrani
University of Kurdistan, Iran

Acknowledgments

Most contributions, outcomes, and insight presented in this book were achieved through long-term teaching and research conducted by the authors and their research center, Smart/Micro Grids Research Center (SMGRC) at the University of Kurdistan (Kurdistan, Iran), on the microgrids operation and control issues over the years. The given real-time simulation and tests in the present book have been mostly performed in the SMGRC. It is our pleasure to acknowledge the relevant received support and awards.

The authors also would like to thank their postgraduate students Yousef Khayat, Mehran Jami, Hamed Ahmadi, Abbas Ahmadi, Zeinab Karami, Babak Abdolmaleki, Kamran Moradi, Sahba Moslemimoghaddam, Zahra Shahbazi, Sirwan Shazdeh, and Sharara Rehimi for their contributions. Last but not least, the authors offer their deepest personal gratitude to their families for all their patience and help during the preparation of this book.

Qobad Shafiee, Mobin Naderi, and Hassan Bevrani
University of Kurdistan, Iran

Acronyms

AI	artificial intelligence
BESS	battery energy storage system
BTBC	back-to-back converter
CB	circuit breaker
CHP	combined heat and power
CN	communication network
CPL	constant power load
CRF	common reference frame
CSC	centralized secondary control
DER	distributed energy resource
DESC	decentralized secondary control
DLFM	dominated low-frequency mode
DNO	distribution network operator
DoS	denial of service
EA	Eigenvalue analysis
ESS	energy storage system
ETC	event-triggered control
EV	electric vehicle
FDI	false data injection
FLL	frequency-locked loop
FCS-MPC	finite control set model predictive control
GFL	grid-following
GFO	grid-forming
GSU	grid-supporting
GPS	global positioning system
HV	high voltage
IED	intelligent electronic devices
ID	interlinking device
IMG	interconnected microgrid
LC	load controller
LPF	low-pass filter
LQR	linear quadratic regulator
LQT	linear quadratic tracking
MC	micro source controller
MG	microgrid
MGCC	microgrid central controller

MOD	module-based
MOB	model-based
MPPT	maximum power point tracking
MV	medium voltage
PCC	point of common coupling
PF	participation factor
PI	proportional-integral
PLL	phase-locked loop
PR	proportional-resonant
PWM	pulse width modulation
QFT	quantitative feedback theory
RCT	robust control toolbox
RES	renewable energy source
RoCoF	rate of change of frequency
RoCoV	rate of change of voltage
SA	sensitivity analysis
SC	super capacitor
SG	synchronous generator
SISO	single-input single-output
SoC	state of charge
SS	static switch
SSR	state space representation
TDS	time-domain simulation
TF	transfer function
THD	total harmonic distortion
VDC	virtual dynamic control
VOC	virtual oscillator control
VSG	virtual synchronous generator
VSC	voltage source converter
WT	wind turbine

1

Introduction

1.1 Overview

This chapter provides an introduction and a general description of the present book, *Microgrids: Dynamic Modeling, Stability and Control*; and emphasizes its role in explaining the important relevant issues in a systematic and understandable way. *Microgrid dynamic modeling, stability, and control* addresses modeling methodologies and application of control theorems and relevant technologies to stability analysis and enhance the microgrid (MG) functions during real-world operations. The MGs' stability and control refer to keeping the desired performance and stabilizing of MGs following various disturbances, such as loss of distributed energy resources (DERs) and/or load units.

The capacity of installed inverter-based DERs and renewable energy sources (RESs) individually or through the MGs in power systems is rapidly growing, and a high penetration level is targeted for the next decade. Nowadays, increased needs for electrical energy and environmental concerns besides growing attempts to reduce dependency on fossil fuel resources have caused power grid industries to set an ambitious target on DERs/RESs. Most of the mentioned microsources connect to the power grid through the MGs as essential blocks of smart grids. The MGs can also operate independently as islanded small grids [1, 2].

MG control is becoming more significant today due to the changing structure, high penetration of RESs/DERs, environmental constraints, and increasing uncertainties of power grids. Moreover, modeling and stability analysis is a basic study, which is necessary for MGs to provide a secure operation. Both small-signal and dynamic stability analyses are important to fully identify the role of different MG subsystems in the overall MG stability [3]. Nevertheless, the stability studies, in turn, are based on dynamic modeling of MGs. In addition, many control methods use the dynamic model to design the controller. Therefore, the first step of stability analysis and model-based controller synthesis is finding an appropriate model for these systems. To go step by step with the stability analysis and the control of MGs, basic and new facilitating dynamic modeling methods of MGs are discussed in this book. The state space representations are obtained as small-signal models, and nonlinear models are represented to investigate large-signal disturbances.

The potential of much more flexibility has been revealed for MGs once the idea of interconnected microgrids (IMGs) was presented. IMGs provide a new operation mode in addition to the islanded and grid-connected modes, where the new MG provides higher capacity than the individual MGs [4]. Therefore, the flexibility of supplying consumers and optimized operating DERs/RESs increases. Moreover, reliability, resiliency, sustainability, and supply security improve with respect to the individual MGs. This new operating mode permits IMGs to exchange power to support the

Microgrids: Dynamic Modeling, Stability and Control, First Edition. Qobad Shafiee, Mobin Naderi, and Hassan Bevrani.

frequency/voltage of a critical MG in an emergency condition as well as supply power deficits in a planned manner [5]. IMGs are also able to be connected to the utility grid to profit from power selling and purchasing when it is required [6]. The idea of MGs interconnection can also be beneficial to divide an active distribution network into some financially independent MGs, while they are connected physically and have power exchanges. In such a new configuration and correlated electricity market, the energy consumption will be more economical. However, this interconnected system is very challenging in terms of modeling, stability assessment, and control architecture. The present book addresses these important issues in a systematic and understandable way.

1.2 Microgrid Concept and Capabilities

The MG concept provides a quite appealing solution for integrating DERs and RESs into power grids. The increasing interest in penetrating renewable energy power into the power grid among the MGs highlights the importance of these systems and addresses serious stability and control challenges in MGs design and operation.

The MGs are small electrical distribution systems that connect multiple customers to multiple DERs/RESs and storage systems. The MGs are typically characterized by multipurpose electrical power services to communities that are connected via low-voltage (LV) networks. A great interest is that these hybrid power grids have the potential to provide reliable power supply to remote communities where connection to transmission supply is unreasonable economically. In general, the MG concept assumes a cluster of loads and DERs/RESs operating as a controllable system that provides both power and heat to its local area. The benefits of MG, such as the enhancement of local reliability, the reduction of feeder losses, and the control of the local voltage, provide increased efficiency through the use of waste heat from combined heat and power (CHP) generators, the voltage sag correction or the provision of uninterruptible power supply functions [3].

With a high integration of DERs/RESs, various challenges in power grid reliability, stability, and efficiency are born. Recently, a complete view of the challenges appeared by the penetration of DERs/RESs and MGs is given by the researchers and engineers. The MG was introduced as a proper response to most challenges [2]. Indeed, studies show that MG is a solution to manage some new challenges and adds a valuable capability such as managing/creating desirable behaviors in modern power grids [7, 8]. In fact, in the energy production process, the behavior of MGs, their impact on the power grid, and the quality of produced energy can be comprehensively managed using appropriate MG controls [3, 9].

The MGs provide various capabilities, the most important of them are shown in Figure 1.1. Some of these capabilities are operating in different operation modes, simple circuitry and low-cost implementation, high efficiency and reliability, high flexibility and power quality, grid supporting and ancillary services, black start and expandable properties, fast response, and wide applications.

1.3 Microgrid Structure

The MGs comprise dispersed energy resources like wind turbines, photovoltaic (PV) panels, microgas turbines, and storage devices like flywheels, super-capacitors, fuel cells, and batteries, as well as controllable loads and electric vehicles (EVs). Generally, an MG is centrally controlled and managed by a microgrid central controller (MGCC) installed at the medium-voltage

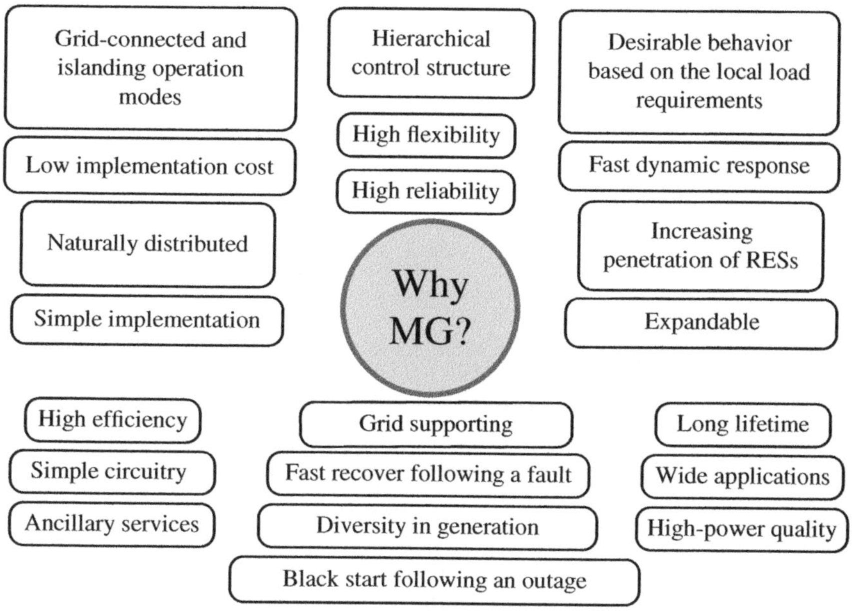

Figure 1.1 Microgrids capabilities.

(MV)/low-voltage (LV) substation [3, 10]. The MGCC includes several key functions, such as economic management and some overall control functionalities as the top point of the MG hierarchical control system. At the low hierarchical control level, load controllers (LCs) and microsource controllers (MCs) exchange information with the MGCC for managing the MG operation by providing set points to the LCs and MCs.

As mentioned, the MG is intended to operate in two different operating modes, i.e. the normal interconnected mode with a distribution network or other MGs, and the islanding operation mode. Most DERs/RESs installed in an MG are connected to the MG feeders and loads via power electronic interfaces. The MG is connected to the main grid (distribution system) or other MGs through a static switch (SS) or a circuit breaker (CB), at the point of common coupling (PCC). A general structure of an MG that can operate in both connected and islanded operation modes is shown in Figure 1.2.

In emergency conditions, e.g. following a serious fault in the main grid or other connected MGs, the MG can be disconnected from the main grid via SSs, as smoothly as possible. Before reconnection, the MGCC must implement the synchronization between the MG and the connected MGs/main grid.

The MG can be also islanded intentionally for specific reasons, even though there is no disturbance or serious fault in the main grid side. The balance between generation and demand of power is one of the most important requirements of MG management in islanded operation mode [3, 11].

The MG has a hierarchical control structure with different layers. The secure operation of MGs in different operation modes, as well as successful disconnection or reconnection processes, depend upon the MG control and protection systems. The controllers must guarantee seamless operation processes and safe working in the specified operating points. In an interconnected distribution network, each MG is locally controlled by the MGCC via its MCs and LCs. The LCs are installed at the controllable loads to provide load control capabilities, specifically for load shedding and demand

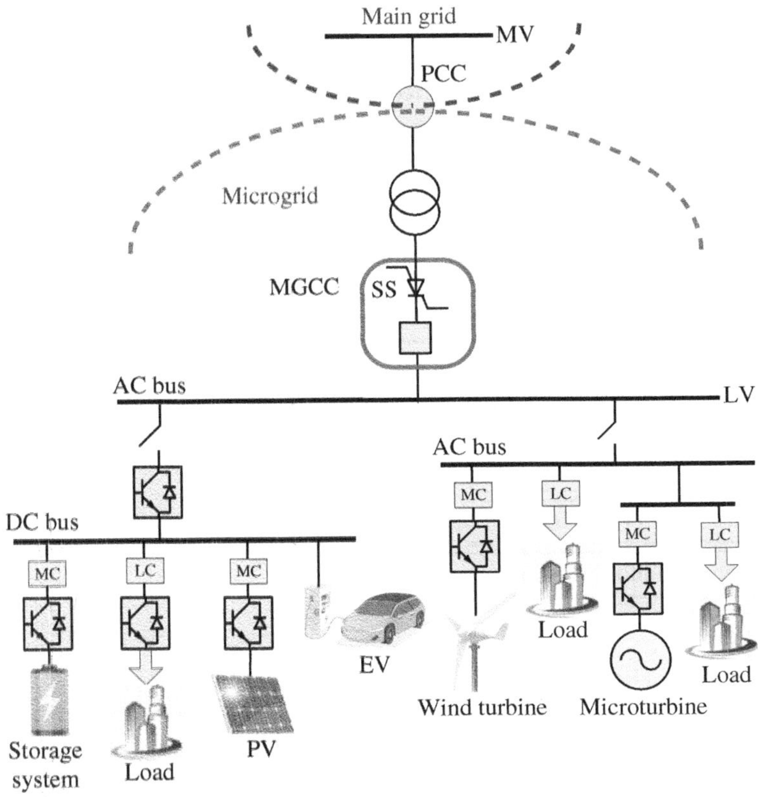

Figure 1.2 Microgrid structure.

response functionalities. For each MG, the MGCC interfaces with the distribution network operator (DNO) and the connected MGs. The DNO has the responsibility to manage the power exchange and operation of medium- and low-voltage areas in which more than one MG may exist [3].

The MG's control levels can be mainly classified into four control groups [3, 12]: local (primary), secondary, central, and global controls. The first three levels are associated with the operation of the MG itself, and the fourth level (global control) demonstrates the coordinated operation of the MG and neighbor MGs as well as the main grid. In contrast to the local control, operating without communication; secondary, central, and global controls may need communication channels. While the local controls are known as *decentralized* controllers, the other controllers may operate as *centralized* controllers.

The *local (primary) control* deals with initial primary control such as current and voltage control loops in the DERs/RESs. The *secondary control* ensures that the frequency and average voltage deviation of the MG are regulated toward zero after every change in load or supply. It is also responsible for local ancillary services. The *central control* covers all possible emergency control schemes and special protection plans to maintain the MG stability and availability in the face of contingencies. The emergency controls identify proper preventive and corrective measures that mitigate the effects of critical contingencies. The *global control* allows MG operation at an economic optimum and organizes the relation between an MG and the main grid as well as other connected MGs. Figure 1.3 represents the MG hierarchical control structure in an interconnected network of MGs with the main grid. These issues are well described in Chapter 5.

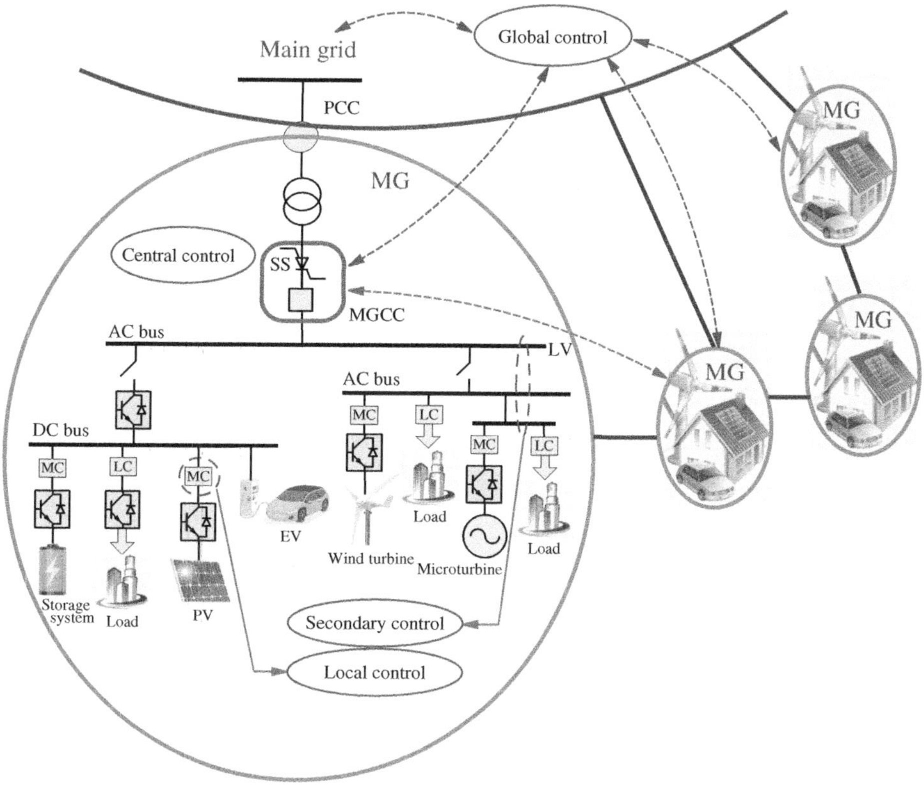

Figure 1.3 Microgrid with a hierarchical control structure in an interconnected distribution network.

1.4 Microgrids in the Future Smart Grids

The growing integration of MGs with power electronic-based DERs and loads has created new challenges in the operation and control of a modern power grid which is also known as smart grid. The conventional power system is challenged by environmental concerns, losses, and an increasing need for electric energy. In most countries, this system is in transition from a conventional hierarchical centralized structure with a one-way communication of power and data between generation and demand sides to a smart power grid with bidirectional exchange networks.

In a conventional power system, the electric network is not fully observable and controllable due to a limited number of sensors, flexible actuators, and appropriate control systems. Most of the operation and control processes, even the system restoration procedures, are accomplished manually. A smart grid makes a compromise between environmental needs, economic and efficiency issues, as well as system reliability. It is a fully sensor-enabled system with self-monitoring and self-healing capabilities.

A smart grid performs a network of DERs/RESs/MGs, with decentralized and distributed pervasive control and operation systems. In a smart grid, remote and automatic monitoring/control is dominant with a high diversity in both generation and demand. In addition to full generation/load control, emerging numerous DERs/RESs and MGs, as well as a wide network of monitoring units can be considered as new characteristics of smart grids.

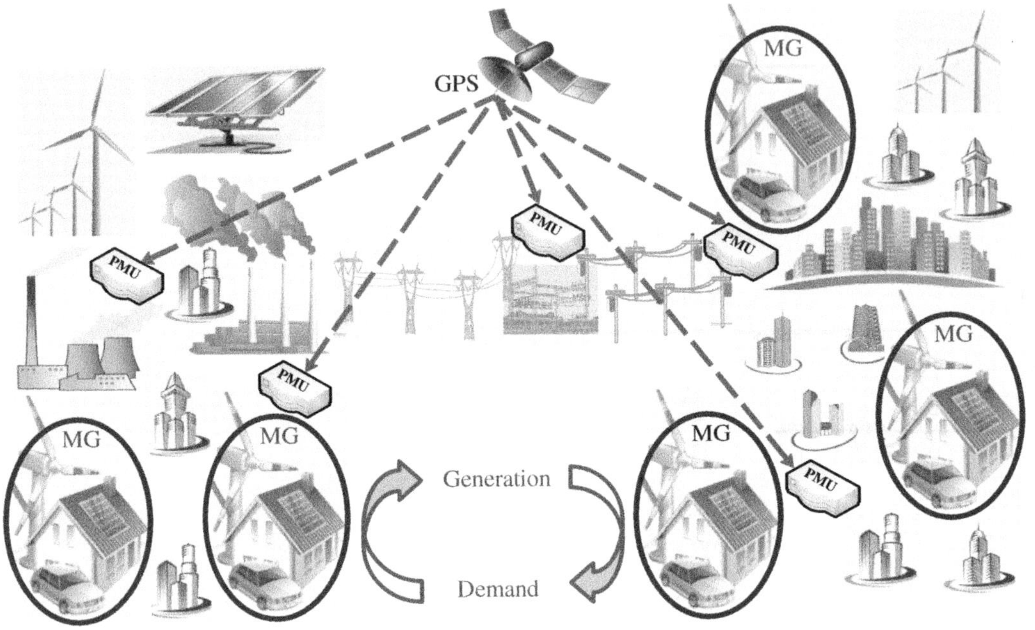

Figure 1.4 Conceptual representation of a smart grid.

The structure of a smart grid is conceptually shown in Figure 1.4. The system variables can be measured with distributed advanced sensors such as intelligent electronic devices (IEDs) and phasor measurement units (PMUs). These sensors can be synchronized by the global positioning system (GPS) via satellites [13].

The need for fast numerical calculating and data processing algorithms, facing a highly decentralized control structure with significant uncertainty, intermittent nature of RESs, reducing system inertia, as well as the necessity of updating grid codes and conventional dynamic analysis and control synthesis methods are the main challenges associated with the smart grid and MG integrated power systems. On the other hand, using smart wide-area monitoring and adaptive control systems, flexible demand response capabilities, regulation and ancillary services supports from the MGs, and constructive virtual dynamics emulation are some promising solutions in response to the mentioned challenges.

The main dynamic performance and stability concerns in a smart grid can be caused by the reduction of the system rotational inertia, as MGs with power converters gradually replace conventional power plants. Reducing rotational inertia in a smart grid can negatively affect the system response and may degrade the control capability and performance. This may lead to significant fluctuation in grid frequency, angle, and voltage, and even system instability [14].

Decreasing smart grid inertia due to the high integration of low-inertia DERs and MGs makes load-power balancing and frequency regulation extremely challenging. The intermittency of renewable power generation significantly magnifies this issue. Emulation of inertia and proper shaping of injected power from the controlled MGs and grid-connected converters [15] are promising solutions and may improve the grid robustness against various disturbances and reduce power fluctuations and parameter perturbations. Due to the fast response of MGs, this supplementary regulation power makes an effective impact in a short period.

Although MGs increase flexibility in the operation and regulation of power requirements, the high integration of RESs-based MGs in smart grids reduces system rotational inertia. The grid-connected MGs can receive the required operation/control set-points and references from the corresponding electric utility to produce the required regulation power support. These references are distributed between the existing MGs to determine the amount of contribution for each participant MG in the grid power regulation.

According to the initial objectives, an MG can operate in the grid-connected mode and supply a part of its load by the grid. However, in the case of any failure, which may result in a deviation of frequency and voltage of the main grid, the MG should disconnect from the grid and supply the loads autonomously by its microsources and existing energy storage systems (ESSs) which have been charged during the grid-connected mode. This characteristic will be served by modern MGs. In addition to the typical responsibility of converting the output power of resources to their own loads at a desired level, MGs control the voltage and frequency of DERs/RESs by adjusting the relevant set points. The enhancement of MGs' capabilities and control functionalities accelerates their integration into the smart grid as a service provider [16].

1.5 Microgrids-Integrated Power Grids

The capacity of installed MGs with converter-based DERs in modern power grids is rapidly growing, and a high integration level is expected for the next decade. A schematic diagram of an MG-integrated power grid with distributed PMUs is shown in Figure 1.5.

The increase of converter-based MGs in modern power grids has an important impact on CO_2 reduction and environmental pollution; however, this integration may have some undesirable impacts on smart grid dynamics, frequency, and voltage regulation.

The impacts of MGs on power grids' dynamic response, performance, and stability have been discussed in numerous reports over recent years. Whether a low inertia due to the high integration of the converter-based MGs can negatively affect the grid frequency dynamic performance and stability is investigated. Since the electric industry seeks to reliably integrate large amounts of MGs into the power grid in a secure environment, a considerable effort is needed to accommodate and effectively manage the installed distributed MGs.

A significant share of MGs increases the total system generation power, while mostly not contributing to the system rotational inertia, and requires the update of traditional control schemes because these well-known control schemes may fail in a modern power grid with lower inertia and faster dynamics.

For MG controllers' design in different control layers, usually reduced-order linearized models around the system operating points are used, and it is assumed that all MG parameters are known and time-invariant. These assumptions, however, are not valid in real MGs with dominated converter-based DERs. The main dynamic modes of the MG vary stochastically during the day because of the variation of load and system parameters, and a fixed linearized time-invariant model will not correctly represent the behavior of the MG.

Reduction in the cost of electricity transmission infrastructure from utilizing distributed energy resources has increased the integration of MGs. As mentioned, the modern MGs, in addition to participation in the electricity market, can offer some services such as improving inertia response and voltage/frequency regulation. These characteristics have brought the opportunity of employing MGs as ancillary service providers in a smart grid. With a proper design of MG control systems, they

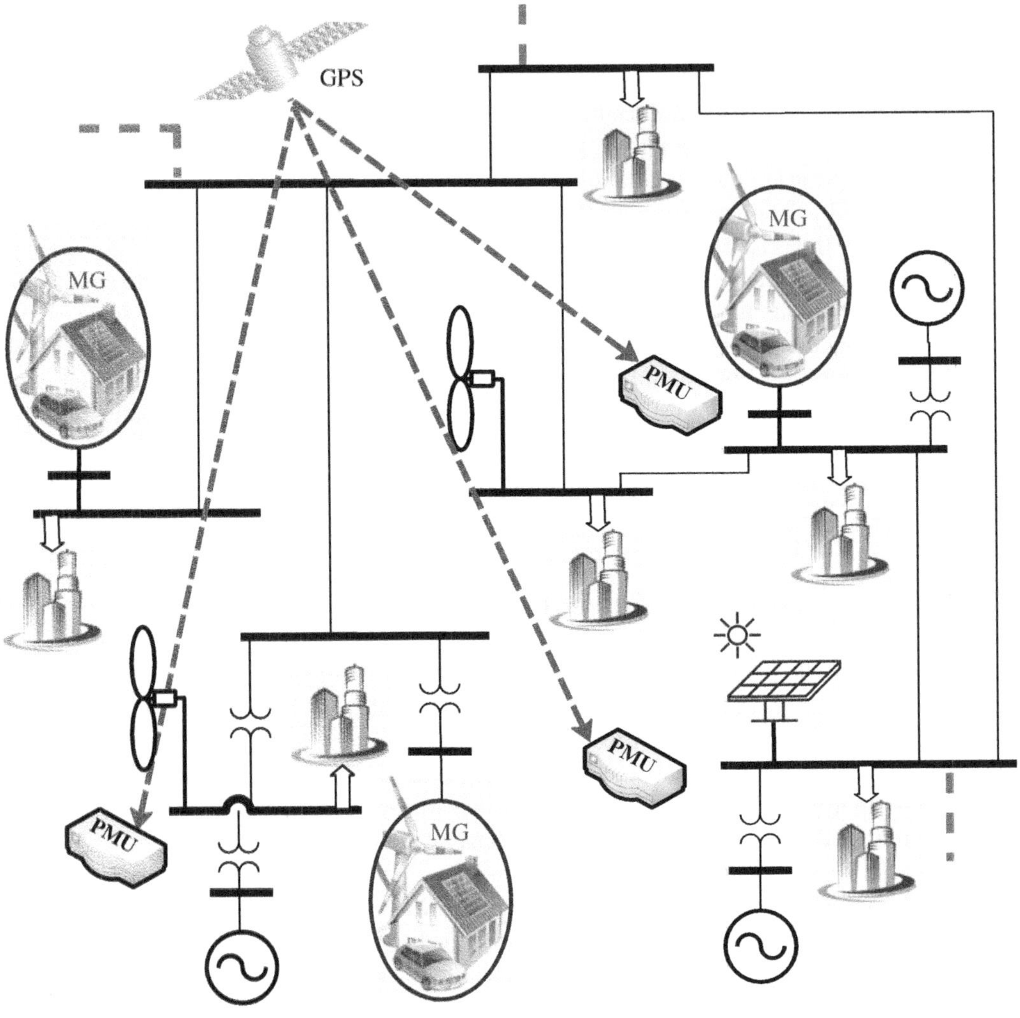

Figure 1.5 Microgrids-integrated power grid.

can contribute to power grid control and provide some services which are traditionally provided by conventional power plants.

1.6 Current Trends and Future Directions

Most current research activities in the field of MGs modeling, stability, and control rely on the assumption that they are working in nominal operating points. However, due to nonlinearities, and changes in system parameters and operation state, the MG dynamic response is becoming more changeable. Consequently, the MG dynamics characteristics may affect the operation of power grid relaying, power quality management and monitoring, and other grid operation and control issues.

To this end, proper MG dynamics estimation and modeling play an important role in modern power grids. It is also important to consider the concept of resilience in MG operation and control issues. This attempt aims to increase the resilience of MG control systems against cyber-attacks

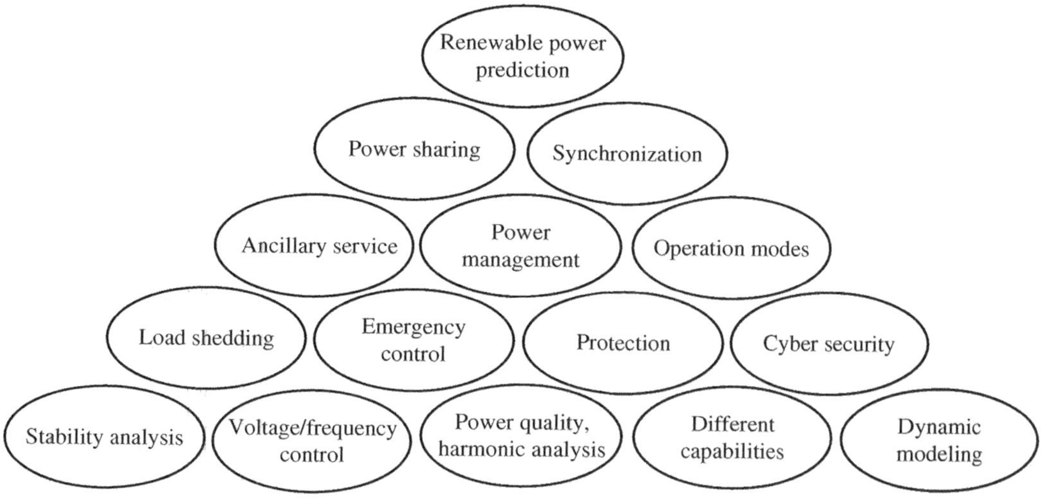

Figure 1.6 Most important topics in microgrids.

and the resulting manipulations. To facilitate a higher penetration level of DERs/RESs, it is also necessary to take benefit of the additional virtual inertia created by grid-forming converters individually or through the MGs for improving the grid dynamic performance and stability [16–18]. The most important MGs research trends and issues are presented in Figure 1.6. These topics include islanded and connected operation modes, active and reactive power sharing, synchronization methods, ancillary and regulatory services, dynamic modeling and parameter estimation, stability analysis, voltage and frequency control, load shedding and emergency control, cyber-security, renewable power prediction, protection schemes, power/energy management, power qualities and harmonic analysis, and various MGs capabilities.

The unpredictability of load is also an important challenge in MG control. Loads are becoming more and more erratic, especially in the distribution grids with electric transportation systems. These grids, without a global control system, can only rely on appropriate control of DERs set points for desirable dynamic shaping of output active and reactive power.

Recent works show the high capability of MGs as the main blocks of future smart grids to contribute to grid performance enhancement and control. The worldwide power grid utilities are going to revise their grid codes for this purpose [19, 20].

The ability of MGs to contribute to power reserve regulation is now required by the grid code of some utilities. Furthermore, if due to congestion management, sufficient inertia, and reserve provision, some MGs are disconnected, their energy can be used to produce upward energy reserves. The allocation of the ancillary reserve by the MGs in future smart communities will be different from the same method for the conventional plants, despite flexibility in control, since the RESs' outputs experience high intermittency. Here, some important directions in the application of MGs are briefly emphasized.

1.6.1 Dynamic Behavior of MGs and Their Impacts on Power Grids

Due to the different dynamic behavior of MGs compared to that of the conventional grids, the increasing share of MGs in modern power systems leads to a new classification in power system stability phenomena [21]. The high integration of the MGs provides a wider range of dynamic time scales in a modern power system. The time scale of MG control loops ranges from a few microseconds to tens of milliseconds, which belong to the wave and electromagnetic phenomena. Therefore, for analyzing the dynamic impacts of MGs on modern power grid stability and performance, there is a need to extend the bandwidth of the power system stability phenomena to cover this time scale. Stabilizing, protection, ride-through capability, and dynamic modeling are basic important technical issues in MGs-integrated power systems [3, 22, 23].

1.6.2 Microgrid-Based Ancillary Services

Contribution in the power grid demand response and inertia are key supportive services of the MGs in the smart grids. These abilities effectively support the power grid in all fields of voltage/frequency control, active/reactive control, reliability, security, and power quality. Although the contribution of the demand side to the power grid control support using the MGs is an important issue, it is still difficult to consider demand-side control support as a fixed participant in the overall power grid regulation requirement. Virtual inertia emulation requires the MG to be able to store or release an amount of energy depending on the grid frequency deviation from its nominal value, analogous to the inertia of a conventional generator [24]. This provides a promising solution to improve power grid stability and performance in the presence of high penetration of RESs.

1.6.3 Dynamic Modeling and Control

Measurement-based dynamics identification and modeling for adaptive control and online parameter tuning of the MGs is vital. The online tuning of MG set-points considering the unpredictable load changes can be quite challenging in operation and control. This emphasizes the significant role of data-driven modeling and control techniques for the distributed MGs in future relevant studies. A complete understanding of reliability considerations via effective modeling/aggregation techniques is vital to identifying a variety of ways that MGs in power grids can accommodate the large-scale integration of renewable power in the future. A proper dynamic modeling and aggregation of the MGs, for performance and stability studies, is a key issue to understanding the dynamic impact of distributed DERs/RESs and simulating their functions in a new environment [25, 26].

Effective control solutions for the MGs in a power grid with high integration of DERs/RESs, particularly in islanded grids due to their relatively low inertia, significant power fluctuation, and various uncertainties are needed [27]. In this direction, providing appropriate coordination between the MGs with conventional power plants and the large-scale ESSs is important. This direction also requires highlighting the cyber-security in the control and operation of MG systems.

1.7 The Book Content and Organization

The electric power industry can be significantly affected by the high integration of MGs. The essential knowledge about MG *modeling, stability, and control* helps engineers to design more efficient MGs in both grid-connected and islanding operation modes. They would be able to carry out systematic MG modeling, stability analysis, and control synthesis considering required constraints and physical limitations. In this book, these constraints and limitations are fully addressed. Moreover, some developed dynamic modeling methods are presented for IMGs, which facilitate the modeling process of such networked systems. Small-signal stability of IMGs is analyzed using eigenvalue and sensitivity analyses. Moreover, the transient stability is investigated using nonlinear models and time-domain simulations. Control of IMGs in both planned and emergency operations is taken into consideration. Finally, modeling, stability, and control of IMG synchronization are studied.

The book provides a comprehensive and strong handbook for students, researchers, and engineers working on modeling, stability, and control of MGs and IMGs; and it will also demonstrate a valid reference for further developments in the mentioned fields. The book tries to make a strong link between the fundamental concepts of MGs and advances in dynamic modeling, stability analysis, and control theory/techniques, together with practical engineering considerations.

The reader is motivated to be engaged with the content via numerous application examples of MG modeling, stability analysis, and controllers' design for single and interconnected MGs. This approach not only provides relevant readers with a better understanding of basic concepts but also reveals a systematic method of control synthesis in the real-world MGs. The content of the book is arranged and ordered so that anyone with a basic background in electrical engineering and mathematics can follow the chapters.

The book includes two parts, i.e. individual MGs (in six chapters) and interconnected MGs (in five chapters), which contain various issues of modeling, stability analysis, and control of MG/IMG. Most chapters are demonstrated by the outcomes of laboratory and real-time simulations. This book is organized into 11 chapters where their contents are briefly explained here.

Chapter 2 investigates the essential concepts and fundamentals of MG dynamic modeling. Firstly, terminology, concepts, and classification of dynamics and modeling of MGs are addressed. Fundamental analysis tools and corresponding requirements are studied including state-space modeling, module interconnection, detailed modeling, and simplification (order reduction) methods. Small-signal modeling of MG components is represented consisting of different types of power converters, loads, DERs as well as power networks and the corresponding considerations. Finally, small-signal modeling of MG controllers is considered including different converter control strategies in the primary control level of MG hierarchical control architecture, the secondary control level, and some considerations about higher control layers.

Chapter 3 clarifies the process of overall modeling of an MG using several case studies. The knowledge learned from Chapter 2 is extended to perform an overall representation structure of an MG. In this chapter, various cases of MG modeling are presented including small-signal modeling of a single grid-connected PV, a grid-connected ac MG with two grids following distributed energy resources, two autonomous ac MGs an autonomous dc MG, and large-signal (transient) modeling of an ac MG with different load types. Detailed modeling of an MG is emphasized, and a simplified frequency response model using low-order transfer functions is introduced.

Chapter 4 presents a classification of MGs' stability and the basic requirements for their stability analysis. Several small-signal stability analyses are provided using eigenvalue analysis, participation matrix, and sensitivity analysis to study different types of MGs and discuss their parameters and control gains. MGs under study consist of both ac and dc MGs. In stability analyses, the most important parameters of the MGs are investigated, e.g. droop characteristic gains, phase-locked loop (PLL) gains, coupling line parameters, virtual impedance gains, and secondary control gains. Finally, the chapter ends with a stabilization case for synchronverters, which are a type of virtual synchronous machines.

Chapter 5 emphasizes the most important concepts and basics of MG control. A general overview of the main control layers and loops is presented. The chapter classifies MG control strategies into four control levels: local (primary), secondary, central, and global, where the first three levels are associated with the operation of the MG itself, and the fourth level (global control) demonstrates the coordinated operation of the MG and neighbor MGs as well as the main grid. The mentioned control levels are discussed with some examples.

Chapter 6 addresses the applications of control methodologies for designing MG controllers. The topics are supplemented by several control synthesis examples.

Chapter 7 discusses the architectures, definitions, benefits, opportunities, and challenges of IMGs. In addition to giving a brief literature review on the relevant topics, the operation of IMGs is explained and vacancies for future research are emphasized.

Chapter 8 addresses the dynamic modeling of IMGs, which is realized in short-term studies in minute-timescale, rather than the static modeling in larger time slots. First, the module-based modeling process is discussed, and then the results are used to perform some comprehensive closed-loop models. The obtained IMG models are useful for stability analysis and parameter behavior identification. The accuracy of models is validated comparing real-time data, and finally, some reduced-order models are represented, which can be used in the controller synthesis.

Chapter 9 discusses the stability of the IMGs. Following a brief review, small-signal stability analysis is explained. Then, the applications of eigenvalues and sensitivity analysis in stability analysis of different IMG technologies are presented. To analyze the large-signal stability of IMGs, time-domain simulations are performed for two different case studies.

Chapter 10 focuses on the control of IMGs with different structures. Among the control loops, frequency control and power sharing control are emphasized. Moreover, both planned and emergency operations are represented, which focus on power exchange control among IMGs, and frequency/voltage support of a critical MG. The explained examples are supplemented by simulations and real-time experimental tests.

Finally, *Chapter 11* addresses the synchronization control requirements and stability analysis in IMGs. The impacts of the secondary control loop, high loads, and physical constraints on the synchronization performance and stability are discussed using examples and simulation results.

References

1 Robert Lasseter, Abbas Akhil, Chris Marnay, John Stephens, Jeff Dagle, Ross Guttromson, A Meliopoulos, Robert Yinger, and Joe Eto. White paper on integration of distributed energy resources. The CERTS microgrid concept. *Consortium for Electric Reliability Technology Solutions*, pages 1–27, 2002.

2 R H Lasseter and P Paigi. Microgrid: A conceptual solution. In *2004 IEEE 35th Annual Power Electronics Specialists Conference (IEEE Cat. No. 04CH37551)*, volume 6, pages 4285–4290, 2004. doi: 10.1109/PESC.2004.1354758.

3 Hassan Bevrani, Bruno François, and Toshifumi Ise. *Microgrid dynamics and control*. John Wiley & Sons, Hoboken, NJ, 2017.

4 Peng Zhang. *Networked microgrids*. Cambridge University Press, 2021.

5 Mobin Naderi, Yousef Khayat, Qobad Shafiee, Tomislav Dragicevic, Frede Blaabjerg, and Hassan Bevrani. An emergency active and reactive power exchange solution for interconnected microgrids. *IEEE J. Emerging Sel. Top. Power Electron.*, 9(5):5206–5218, 2019.

6 Mobin Naderi, Yousef Khayat, Qobad Shafiee, Frede Blaabjerg, and Hassan Bevrani. Dynamic modeling, stability analysis and control of inter- connected microgrids: A review. *Appl. Energy*, 334:120647, 2023.

7 Robert H Lasseter. Smart distribution: Coupled microgrids. *Proc. IEEE*, 99(6):1074–1082, 2011. doi: 10.1109/JPROC.2011.2114630.

8 Hêmin Golpîra, Arturo Román-Messina, and Hassan Bevrani. *Renewable integrated power system stability and control*. IEEE-Wiley, 2021.

9 Lilia Tightiz, Hyosik Yang, and Hassan Bevrani. An interoperable communication framework for grid frequency regulation support from microgrids. *Sensors (Basel, Switzerland)*, 21(13):4555, 2021.

10 Nikos Hatziargyriou, Hiroshi Asano, Reza Iravani, and Chris Marnay. An overview of ongoing research, development, and demonstration projects. *IEEE Power Energy Mag.*, 5(4):79–94, 2007.

11 Hassan Bevrani. *Robust power system frequency control*. Springer, 2014.

12 H Bevrani, M Watanabe, and Y Mitani. Microgrid controls. *Standard handbook for electrical engineers*, pages 160–176, 2012.

13 Hassan Bevrani, Masayuki Watanabe, and Yasunori Mitani. *Power system monitoring and control*. John Wiley & Sons, 2014.

14 Hassan Bevrani, Hêmin Golpîra, Arturo Román Messina, Nikos Hatziargyriou, Federico Milano, and Toshifumi Ise. Power system frequency control: An updated review of current solutions and new challenges. *Electr. Power Syst. Res.*, 194:107114, 2021. ISSN 0378-7796.

15 H Bevrani, T Kato, T Ise, and K Inoue. *Grid connected converters: Modeling, stability and control*. Elsevier Science, 2022. ISBN 9780323999021.

16 Hassan Bevrani, Masaki Imanaka, and Takeyoshi Kato. Grid dynamics shaping using flexible controlled power converters. *Energy Rep.*, 6:1490–1495, 2020. ISSN 2352-4847. doi: 10.1016/j.egyr.2020.10.065 URL https://www.sciencedirect.com/science/article/pii/S2352484720314189. 2020 The 7th International Conference on Power and Energy Systems Engineering.

17 Jia Liu, Yushi Miura, Hassan Bevrani, and Toshifumi Ise. Enhanced virtual synchronous generator control for parallel inverters in micro-grids. *IEEE Trans. Smart Grid*, 8(5):2268–2277, 2017. doi: 10.1109/TSG.2016.2521405.

18 Hassan Bevrani, Toshifumi Ise, and Yushi Miura. Virtual synchronous generators: A survey and new perspectives. *Int. J. Electr. Power Energy Syst.*, 54:244–254, 2014. ISSN 0142- 0615. doi: 10.1016/j.ijepes.2013.07.009 URL https://www.sciencedirect.com/science/article/pii/S0142061513003062.

19 IEEE standard for interconnecting distributed resources with electric power systems. *IEEE Std 1547-2003*, pages 1–28, 2003. doi: 10.1109/IEEESTD.2003.94285.

20 IEEE standard for interconnection and interoperability of distributed energy resources with associated electric power systems interfaces. *IEEE Std 1547-2018 (Revision of IEEE Std 1547-2003)*, pages 1–138, 2018. doi: 10.1109/IEEESTD.2018.8332112.

21 Nikos D Hatziargyriou, Jovica V Milanović, Claudia Rahmann, Venkataramana Ajjarapu, Claudio A Cañizares, István Erlich, David J Hill, Ian A Hiskens, I Kamwa, Bikash Chandra Pal, Pouyan Pourbeik, Juanita Gasca, Alexander M Stankovic, Thierry Van Cutsem, Vijay Vittal, and Costas D Vournas. Stability definitions and characterization of dynamic behavior in systems with high penetration of power electronic interfaced technologies. 2020.

22 Mobin Naderi, Qobad Shafiee, Hassan Bevrani, and Frede Blaabjerg. Low-frequency small-signal modeling of interconnected AC microgrids. *IEEE Trans. Power Syst.*, 36(4):2786–2797, 2021. doi: 10.1109/TPWRS.2020.3040804.

23 Mobin Naderi, Yousef Khayat, Qobad Shafiee, Tomislav Dragicevic, Hassan Bevrani, and Frede Blaabjerg. Comprehensive small-signal modeling and Prony analysis-based validation of synchronous interconnected microgrids. *Energy Rep.*, 7:6677–6689, 2021. ISSN 2352-4847.

24 Prabha Kundur, Neal J Balu, and Mark G Lauby. *Power system stability and control.* McGraw-Hill, New York, 1994.

25 Daniel E Olivares, Ali Mehrizi-Sani, Amir H Etemadi, Claudio A Cañizares, Reza Iravani, Mehrdad Kazerani, Amir H Hajimiragha, Oriol Gomis-Bellmunt, Maryam Saeedifard, Rodrigo Palma-Behnke, Guillermo A Jiménez-Estévez, and Nikos D Hatziargyriou. Trends in microgrid control. *IEEE Trans. Smart Grid*, 5(4):1905–1919, 2014. doi: 10.1109/TSG.2013.2295514.

26 Ritwik Majumder. Some aspects of stability in microgrids. *IEEE Trans. Power Syst.*, 28(3):3243–3252, 2013. doi: 10.1109/TPWRS.2012.2234146

27 IEEE Standards. IEEE standard for interconnecting distributed resources with electric power systems. https://fglongatt.org/OLD/Archivos/Archivos/SistGD/IEEE1547.pdf.

Part I

Individual Microgrids

2

Microgrid Dynamic Modeling: Concepts and Fundamentals

2.1 Introduction

Systems require to be modeled in order to study their behaviors in different time scales. In most cases, modeling is necessary to be done before implementing systems in order to show unknown challenges and represent solutions before making the system operation critical. Some disturbances and perturbations cannot be studied on the systems themselves due to some possible instabilities, failures, exceeding limits, and undesirable performances, which can damage systems. Therefore, systems' models are needed to investigate these disturbances. Moreover, most control design techniques are based on models. Power grids require various models to study their different aspects including planning, operation, security, stability, reliability, resilience, and transients. Each of these studies has its considerations and assumptions in modeling the power grids, which results in different long-term, medium-term, and short-term models. Fortunately, most of these studies and the corresponding models have been taken into account by software corporations, e.g. DIgSILENT Power Factory, ETAP, PSCAD/EMTDC, MATLAB, RTDS, OPAL-RT. These software usually provide the modules of power grids with their considerations regarding a certain study. However, there exist many studies that cannot be done using pre-defined modules in the introduced software, which should be fulfilled by using programming languages, e.g. C, C++, Python, and MATLAB.

Microgrids (MGs) represent small-scale power grids, which are implemented in low/medium voltages. Due to existing energy resources and distributed generations in MGs, some considerations of the generation level of traditional power systems should be taken into consideration. Moreover, the MG operation voltage, which is usually low or medium, and the fact that MGs can usually be connected to the distribution grids lead to paying attention to the distribution level of power systems and its considerations. Most studies of traditional power systems can be done for MGs, and the aforementioned software can generally be used for MG modeling and studies. Static long-term models are useful for planning, operation, and reliability analyses, and dynamic short-term models can be used for stability analysis and controller design/synthesis. Power electronics converters are widely used in all levels of modern power systems, and MGs are no exception. In fact, power electronics devices are essential and very applicable for MGs and should fundamentally be considered in MG modeling. In this book, dynamic modeling of different ac and dc MGs interfaced with power electronics converters in different operating modes, i.e. grid-connected, islanded, and interconnected modes, is taken into account.

Figure 2.1 shows a typical MG with different distributed energy resources (DERs), energy storage systems (ESSs), loads, and control layers. The MG can be connected/disconnected to/from the main

Microgrids: Dynamic Modeling, Stability and Control, First Edition. Qobad Shafiee, Mobin Naderi, and Hassan Bevrani.

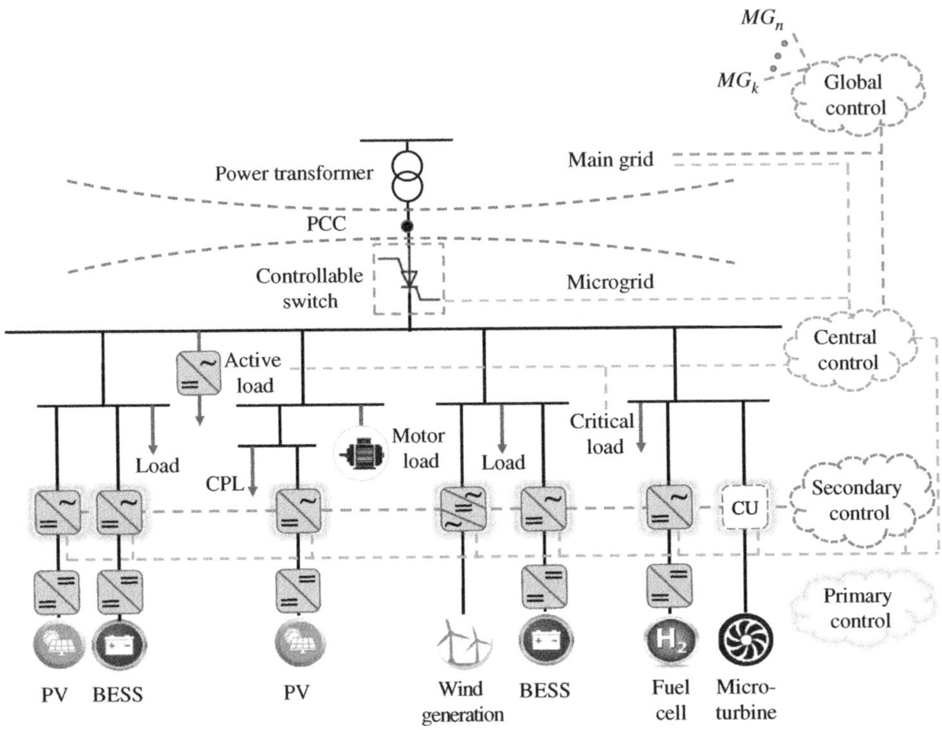

Figure 2.1 Schematic of a typical MG including different DERs, ESSs, loads, and control layers.

grid using a controllable switch. Different voltage levels at the point of common coupling (PCC) of MG and the main grid lead to the use of a transformer. DERs can include either non-renewable generation units such as micro-turbines and diesel generators or renewable energy resources, e.g. photovoltaic (PV) and wind generators. Due to the intermittency of renewable generations, ESSs are necessary for MGs like battery energy storage systems (BESSs), hydrogen storage fuel cells, flywheels, etc. MG loads can also be of different types including constant impedance loads, constant power loads (CPLs), motor loads, and active loads. DERs, ESSs, and active loads require different types of power converters, i.e. rectifiers (ac–dc converters), inverters (dc–ac converters), choppers (dc–dc converters), and cyclo-converters (ac–dc–ac converters). On the other hand, a hierarchical control architecture is required for a stable, reliable, flexible, and optimized operation of MGs (see Figure 1.3). Four control layers can be seen comprising primary, secondary, central, and global layers.

Modeling of different types of DERs and ESSs from several points of view has been investigated in the literature, which are generation-side modules of MGs. The detailed models of PV generation units [1–6], wind DERs [5, 7–10], micro-turbines [7], diesel generators [10–12], BESSs [2, 3, 6, 8, 10, 13], fuel cells [5], flywheels [14–18], pumped hydropower ESS [19], and supercapacitors [20–22] have been addressed for stability analysis, controller design, and stabilization. Furthermore, reduced-order and simplified models have been studied for specific objectives, e.g. frequency control [23–26]. On the other hand, some literature has focused on load types, their modeling, and their dynamics including motor loads [27–30], active dc/ac converter-interfaced loads [31] in ac MGs, and active dc/dc converter-interfaced loads in dc MGs [32, 33], CPLs [7, 8, 20, 33, 34], and constant impedance loads [35–37]. Some other literatures have considered simplified and ideal models for DERs, EESs, and loads and have focused on the modeling of grid-side power converters and their controllers, power networks, and interactions among these control loops and power grids [35, 38–41].

Although nonlinear models of MGs can easily be obtained and analyzed using the above-mentioned software and modern computers with appropriate computational capability, linear models and their small-signal analysis are still more applicable to analyze MG dynamics and stability due to more existing tools for small-signal analysis. On the other hand, in most studies, large-order nonlinear models are not necessary and reduced-order simplified models, focusing on specific dynamics and removing non-essential state variables, are more desirable.

In an order-reducing process, mostly, the fast dynamic modes are removed and the slow dominant modes are preserved using an appropriate order reduction technique, e.g. the balanced Residualization realization (singular perturbation theory) [36, 39–43]. Nevertheless, in [44], the balanced Truncation realization (regular perturbation theory) is used to achieve a second-order model of grid-feeding (GFE)/grid-following (GFL) converters. Furthermore, in [45], the aggregation methods and quasi-static equations are employed to map all similar modules in each ac MG, e.g. grid-forming (GFO) DERs have one equivalent module. Some other simplified models use measurement to focus on modeling just under-study variable(s) in MG, e.g. [46] finds improved load models in MGs using an adaptive machine learning-based approach.

The present chapter and Chapter 3 represent the dynamic modeling of MGs. In this chapter, basic concepts and fundamentals of MG dynamic modeling are provided. Firstly, terminology, concepts, and classification of dynamics and modeling of MGs are addressed. Fundamental analysis tools and corresponding requirements are studied including state-space modeling, module interconnection, detailed modeling, and simplification (order reduction) methods. Small-signal modeling of MG components is represented consisting of different types of power converters, loads, DERs, and ESSs as well as power network and the corresponding considerations. Finally, small-signal modeling of MG controllers is taken into account including different converter control strategies in the primary control level of MG hierarchical control architecture, the secondary control level, and some considerations about higher control layers. Therefore, all requirements and fundamentals of MG dynamic modeling are provided in this chapter in order to represent modeling applicable cases of ac and dc MGs in Chapter 3. Moreover, modeling interconnected microgrids (IMGs) is represented in Chapter 8, which is based on the contents of this chapter and Chapter 3. Note that stability analysis represented in Chapters 4 and 8 and some control designs in other chapters are done based on the small-signal models represented in Chapter 3 and using the fundamentals provided here.

Please note that source codes are provided for the modeling cases and examples within this chapter. This resource will be useful for readers and researchers seeking to implement the module models and examples outlined in this chapter, enabling them to adapt and apply these principles for different studies. Some codes are shown in the book, while some others are provided in the Wiley data center in [47].

2.2 Dynamics and Modeling

MG/IMG dynamics can be defined as changes and rate of changes of the phenomena related to MG/IMG control and operation. Accordingly, MG/IMG modeling is a mathematical equivalence in order to study the dynamics of a specific phenomenon or a group of phenomena in a range of frequencies shown in Figure 2.2. The phenomena can be classified into four groups in terms of change rate comprising very fast, fast/medium, slow, and very slow. Harmonic studies,

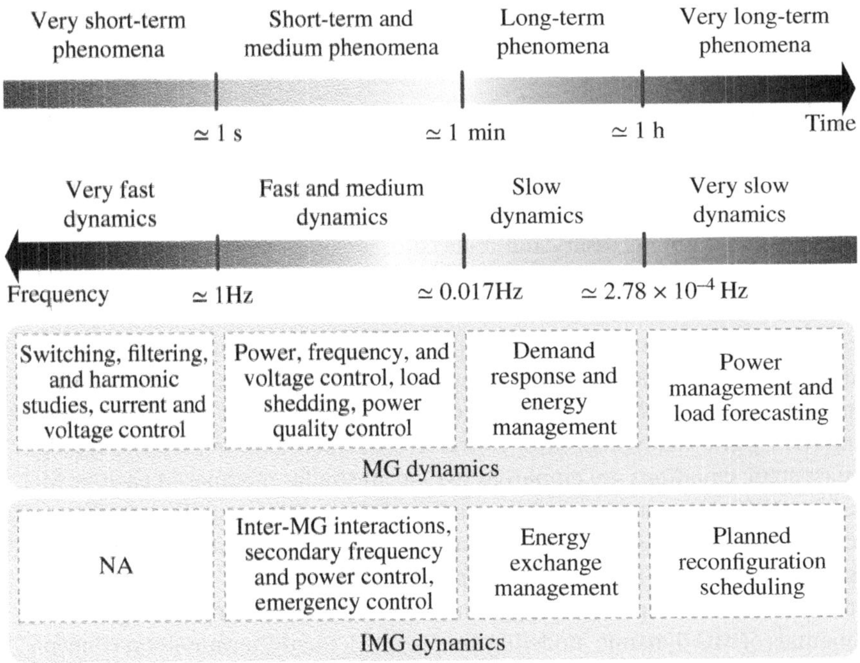

Figure 2.2 MG and IMG phenomena classification based on frequency ranges and correlated time durations.

voltage/frequency control, demand response, and power management are examples of such four groups, respectively [25].

For each frequency range, different types of modeling methods with their requirements are needed to model the corresponding dynamics, which in turn can be categorized into detailed/full-order and simplified/reduced-order modeling types. A detailed modeling method leads to a large range of frequencies, while simplified models focus on a special frequency range, e.g. low frequencies [26]. In Figure 2.3, two conceptual diagrams clearly show the differences between detailed and simplified modeling methods. Another classification presented in Figure 2.3 is based on the linear/nonlinear behavior of MG/IMG dynamics. Linear-dynamic modules can only be modeled using small-signal modeling methods including state-space representation and transfer function. Nevertheless, according to the requirement and application, both small-signal and large-signal modeling methods can be used for modeling nonlinear-dynamic modules. Common methods of large-signal modeling are nonlinear equations governing the module dynamics and block diagram simulations using modeling software, e.g. MATLAB and PSCAD/EMTDC.

2.3 Fundamental Analysis Tools and Requirements

Transfer function (TF) is a simple modeling method for low-order linear single-input single-output (SISO) systems, which can be extended as a TF matrix for multivariable systems. However, it is less flexible than state-space representation (SSR) and unsuitable for large-scale system modeling. Therefore, SSR is selected as the main modeling method in this book. Nevertheless, these two basic modeling tools can easily be transformed using MATLAB codes everywhere required, e.g. for robust H_∞ control design.

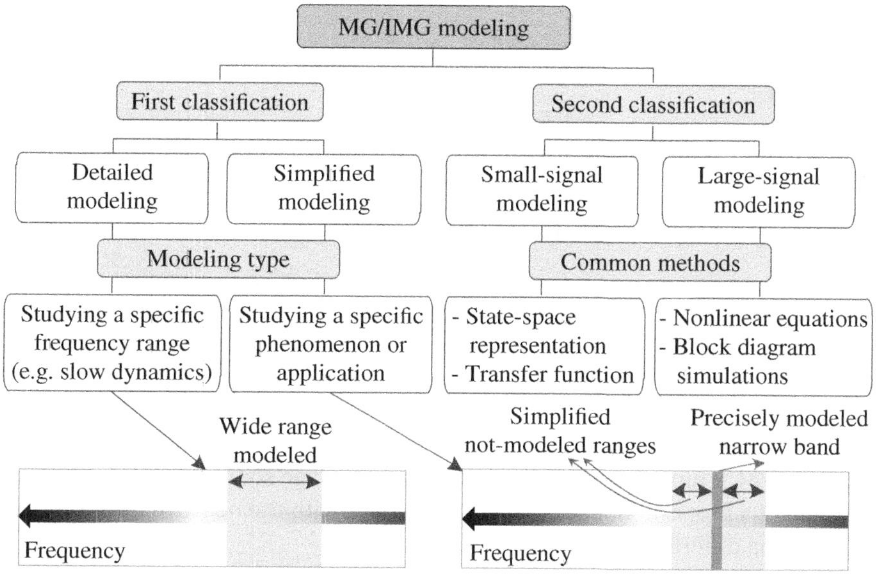

Figure 2.3 MG/IMG modeling classification including detailed, simplified, small-signal, and large-signal categories.

Block diagram is a common tool for modeling and is repeatedly used throughout the book, but it is not an independent analytical tool. It can be useful for software-defined modeling, and facilitating SSR and TF methods. Differential equations (DEs) are a more general modeling tool than TF and SSR because they can model both linear and nonlinear dynamics. Time domain simulations are also used for nonlinear modeling, which is based on block diagrams used in some software, e.g. MATLAB/Simulink and MATLAB/SimPowerSystems.

Static equations (SEs) are approximation of non-important dynamics to neglect them in model-based and module-based simplification methods. Static equations can also be called a truncation or regular perturbation method. Moreover, singular perturbation (SP) and aggregation methods are used as analytical modeling and simplification methods, which are taken into account in this section. Prony analysis (PA) is employed for model validation, while sensitivity analysis is utilized for stability investigations in Chapters 3 and 4, with their underlying principles explained here.

2.3.1 State-Space (Small-Signal) Modeling

Note that MGs can be considered multi-component systems, where their modeling procedure can generally be divided into two main steps. The first is obtaining an individual model of each component/module, and the second is to consider interconnections among the module models. In this book, the first step is done using SSR, and the second step is accomplished numerically using some MATLAB codes. In order to find an SSR, one or some steps are necessary to find relevant differential equations, transforming equations, and linearization.

2.3.1.1 Finding Differential Equations

In order to find an SSR for a module, the first step is finding mathematical equations. These equations are obtained based on the governing rules of the module. In an electrical circuit, the

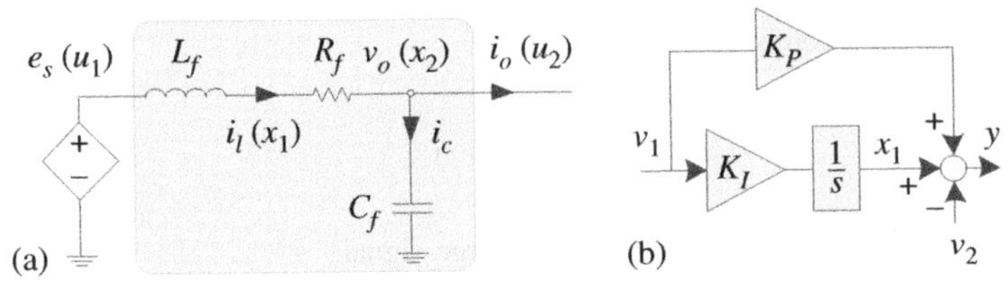

Figure 2.4 Basic examples: (a) single-line diagram of a three-phase LC filter and (b) block diagram of a typical PI controller.

basic rules are Kirchhoff voltage and current laws, and capacitor voltage and inductor current are usually considered as state variables. In control block diagrams, signal flow rules are governing, and, as a common rule, the output of each integrator is considered as a state variable [48].

Exercise 2.1 Figure 2.4(a) shows a single-line diagram of a three-phase LC filter connecting to a voltage source converter (VSC), modeled by an ideal dependent voltage source (e_s). Figure 2.4(b) shows the block diagram of a proportional-integral (PI) controller with the negative feed-forward signal v_2. Find the governing differential equations.

Solution

(a) Applying electrical circuit laws, the mathematical equations are calculated as:

$$e_{s,abc}(t) = L_f \frac{di_{l,abc}}{dt} + R_f i_{l,abc}(t) + v_{o,abc}(t), \tag{2.1a}$$

$$i_{c,abc}(t) = i_{l,abc}(t) - i_{o,abc}(t) = C_f \frac{dv_{o,abc}}{dt}, \tag{2.1b}$$

where χ_{abc} indicates a 3×1 vector including *abc* voltage/current components for $\forall \chi \in \{e, i, v\}$.

(b) In this case, one can find the governing equations using signal flow rules as follows:

$$\frac{dx_1}{dt} = K_I v_1, \tag{2.2a}$$

$$y = x_1 + K_P v_1 - v_2. \tag{2.2b}$$

2.3.1.2 Park and Clark Transformations

In three-phase systems, three-phase variables can be represented in three different reference frames for various applications. Figure 2.5 shows these reference frames including the *abc* stationary reference frame, $\alpha\beta\gamma$ stationary reference frame, and $dq0$ rotating reference frame. Clarke Transformation is used to transform *abc* variables into equivalent $\alpha\beta\gamma$ variables as follows:

$$\chi_{\alpha\beta\gamma}(t) = \frac{2}{3} \begin{bmatrix} 1 & -1/2 & -1/2 \\ 0 & \sqrt{3}/2 & -\sqrt{3}/2 \\ 1/2 & 1/2 & 1/2 \end{bmatrix} \chi_{abc}(t), \tag{2.3}$$

and Park Transformation is used to transform *abc* variables into equivalent $dq0$ variables as follows:

$$\chi_{dq0}(t) = \frac{2}{3} \begin{bmatrix} \cos(\theta) & \cos(\theta - 2\pi/3) & \cos(\theta + 2\pi/3) \\ -\sin(\theta) & -\sin(\theta - 2\pi/3) & -\sin(\theta + 2\pi/3) \\ 1/2 & 1/2 & 1/2 \end{bmatrix} \chi_{abc}(t), \tag{2.4}$$

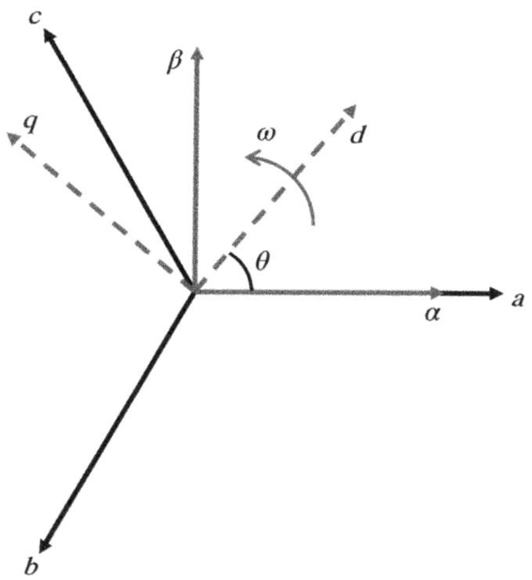

Figure 2.5 *abc* stationary, *αβγ* stationary, and *dq*0 rotating reference frames.

where $\theta = \omega t + \theta_0$ is the angular position, θ_0 is the initial angular position, and ω is the angular frequency of the rotating reference frame, which is usually selected the same as the frequency of sinusoidal *abc* voltages/currents to find constant *dq*0 components. In balanced power systems, γ and 0 components of $\alpha\beta\gamma$ and $dq0$ reference frames are fixed at zero, and the simplified transformation relationships can be obtained as follows:

$$\chi_{\alpha\beta}(t) = \frac{2}{3}(\, e^{j0} \;\; e^{j2\pi/3} \;\; e^{-j2\pi/3}\,)\chi_{abc}(t), \tag{2.5a}$$

$$\chi_{dq}(t) = (e^{-j\theta})\chi_{\alpha\beta}(t), \tag{2.5b}$$

where $\chi_{\alpha\beta}(t) = \chi_\alpha(t) + j\chi_\beta(t)$, and $\chi_{dq}(t) = \chi_d(t) + j\chi_q(t)$. Note that inverse Clarke and inverse Park transformations can also be used. In the case of transforming equations, which are commonly used, the constant coefficients are the same for all variables and can be removed. Therefore, equation transformation is an easy process.

Exercise 2.2 For the filter equations in Exercise 2.1, transform (2.1) to the equivalent equations in a rotating reference frame.

Solution

Note that we do this when we want to analyze stability or design a model-based controller in the rotating reference frame, especially in balanced power systems, where the 0 component equals zero. Hence, (2.1) can be transformed into the *dq* frame as follows:

$$e_{s,dq}(t) = jL_f\omega i_{l,dq}(t) + L_f\frac{di_{l,dq}}{dt} + R_f i_{l,dq}(t) + v_{o,dq}(t), \tag{2.6a}$$

$$i_{c,dq}(t) = i_{l,dq}(t) - i_{o,dq}(t) = jC_f\omega v_{o,dq}(t) + C_f\frac{dv_{o,dq}}{dt}. \tag{2.6b}$$

2.3.1.3 Linearization

Most of the modules of real MGs include nonlinear dynamics, e.g. nonlinear and constant-power loads, PV, and battery energy storages. Small-signal stability analysis methods and linear control theory cannot be applied to nonlinear models. Therefore, nonlinear equations should be linearized around an operating point in these cases. Consider a nonlinear equation in a general form as follows:

$$x = f(t_1, t_2, \dots, t_n), \tag{2.7}$$

where x is a dependent variable, f is a nonlinear function, t_1, t_2, to t_n are independent variables, t_{01}, t_{02}, to t_{0n} are the values of those variables at the operating point, and x_0 is the value of x at the operating point. The linearized relationship of (2.7) is a linear combination of the perturbed variables around the operating point, i.e. Δt_1 to Δt_n, which is given using the Taylor expansion as follows:

$$\Delta x = \left(\frac{\partial f}{\partial t_1}\bigg|_{x=x_0} \right) \Delta t_1 + \left(\frac{\partial f}{\partial t_2}\bigg|_{x=x_0} \right) \Delta t_2 + \cdots + \left(\frac{\partial f}{\partial t_n}\bigg|_{x=x_0} \right) \Delta t_n, \tag{2.8}$$

where $\partial f / \partial t_n$ is the partial derivative of f with respect to t_n, and $\Delta x = x - x_0$ represents the linearized variable around the operating point x_0.

Exercise 2.3 The terms $\omega i_{l,dq}(t)$ and $\omega v_{o,dq}(t)$ make (2.6) nonlinear. Find the linearized form of (2.6).

Solution

Considering the perturbed form of variables, and applying the Taylor expansion, the linearization results in the following relationships:

$$\Delta e_{s,dq}(t) = jL_f \omega_0 \Delta i_{l,dq}(t) + jL_f i_{l0,dq} \Delta\omega + L_f \frac{d\Delta i_{l,dq}}{dt} + R_f \Delta i_{l,dq}(t) + \Delta v_{o,dq}(t), \tag{2.9a}$$

$$\Delta i_{c,dq}(t) = \Delta i_{l,dq}(t) - \Delta i_{o,dq}(t) = jC_f \omega_0 \Delta v_{o,dq}(t) + jC_f v_{o0,dq} \Delta\omega + C_f \frac{d\Delta v_{o,dq}}{dt}. \tag{2.9b}$$

2.3.1.4 State-Space Representation

After finding linear/linearized single-order differential equations, one can transform them into two matrix equations, which are called SSR. In fact, each module of MGs represented by the corresponding equations, typically the kth module, can separately be modeled using an SSR as follows:

$$\dot{X}_M^k = A_M^k X_M^k + B_M^k U_M^k, \tag{2.10a}$$

$$Y_M^k = C_M^k X_M^k + D_M^k U_M^k, \tag{2.10b}$$

where $X_{M(n\times1)}^k$, $U_{M(m\times1)}^k$, and $Y_{M(l\times1)}^k$ are the state vector, input vector, and output vector of the kth module, respectively, and $A_{M(n\times n)}^k$, $B_{M(n\times m)}^k$, $C_{M(l\times n)}^k$, and $D_{M(l\times m)}^k$ are the state, input, output, and transfer matrices, respectively, which are scalar matrices. Input/output signal can be of an electrical, a control, a communication, or even an auxiliary mathematical type. In the process of module interconnection, the order of inputs and outputs within the U_M^k, and Y_M^k is very important. One can transform the SSR represented by (2.10) to an equivalent transfer function matrix as follows:

$$G_M^k(s) = C_M^k (sI - A_M^k)^{-1} B_M^k + D_M^k, \tag{2.11}$$

where $G_{M(l\times m)}^k(s)$ is the transfer function matrix, $I_{n\times n}$ is the identity matrix, and s is the complex frequency used in the Laplace operator.

Exercise 2.4 Consider the two systems introduced in Exercise 2.1 and shown in Figure 2.4. Find SSRs for both systems and consider the state variables as the output of the first system.

Solution

a) In this case, (2.9) includes four state variables, namely $\Delta i_{l,dq}(t)$ and $\Delta v_{o,dq}(t)$, and five input variables, i.e. $\Delta e_{s,dq}(t)$, $\Delta i_{o,dq}(t)$, $\Delta\omega$. Therefore, an SSR can be given by decoupling dq components in (2.9) and then rearranging the terms to be matched with (2.10) as follows:

$$\frac{d}{dt}\begin{bmatrix}\Delta i_{l,d}(t)\\ \Delta i_{l,q}(t)\\ \Delta v_{o,d}(t)\\ \Delta v_{o,q}(t)\end{bmatrix} = \overbrace{\begin{bmatrix}-R_f/L_f & \omega_0 & -1/L_f & 0\\ -\omega_0 & -R_f/L_f & 0 & -1/L_f\\ 1/C_f & 0 & 0 & \omega_0\\ 0 & 1/C_f & -\omega_0 & 0\end{bmatrix}}^{A_1}\overbrace{\begin{bmatrix}\Delta i_{l,d}(t)\\ \Delta i_{l,q}(t)\\ \Delta v_{o,d}(t)\\ \Delta v_{o,q}(t)\end{bmatrix}}^{X_1}$$

$$+\overbrace{\begin{bmatrix}1/L_f & 0 & 0 & 0 & i_{l0,q}\\ 0 & 1/L_f & 0 & 0 & -i_{l0,d}\\ 0 & 0 & -1/C_f & 0 & v_{o0,q}\\ 0 & 0 & 0 & -1/C_f & -v_{o0,d}\end{bmatrix}}^{B_1}\overbrace{\begin{bmatrix}\Delta e_{s,d}(t)\\ \Delta e_{s,q}(t)\\ \Delta i_{o,d}(t)\\ \Delta i_{o,q}(t)\\ \Delta\omega\end{bmatrix}}^{U_1},$$

$$Y_1 = \overbrace{I_{4\times 4}}^{C_1}\left[\Delta i_{l,d}(t)\ \Delta i_{l,q}(t)\ \Delta v_{o,d}(t)\ \Delta v_{o,q}(t)\right]^T$$
$$+\overbrace{0_{4\times 5}}^{D_1}\left[\Delta e_{s,d}(t)\ \Delta e_{s,q}(t)\ \Delta i_{o,d}(t)\ \Delta i_{o,q}(t)\ \Delta\omega\right]^T. \tag{2.12a}$$

b) In the PI controller, there are one state variable, two inputs, and a pre-defined output. An SSR is obtained as follows:

$$\dot{X}_2 = \overbrace{0}^{A_2}\cdot\overbrace{x_1}^{X_2} + \overbrace{[K_I\ 0]}^{B_2}\overbrace{[v_1\ v_2]^T}^{U_2}, \tag{2.13a}$$

$$Y_2 = \overbrace{1}^{C_2}\cdot x_1 + \overbrace{[K_P\ -1]}^{D_2}[v_1\ v_2]^T. \tag{2.13b}$$

2.3.1.5 Interconnecting Modules

It is obvious that finding SSRs is not the final step in modeling MGs, which are multi-module systems. A certain process is required to model interconnections among module SSRs including all electrical, control, communications, and logical/mathematical links. A routine method is substituting equations after finding individual models [1, 36, 39, 49], which results in analytical outputs. Nevertheless, in the case of large-scale MGs and IMGs, this method cannot succeed due to a large number of substitutions and complicated matrices obtained. In such cases, numerical methods using software are preferable. In this book, an interconnection method is used, which is based on Robust Control Toolbox (RCT) codes in MATLAB.

Simply, the numerical method based on RCT contains the following steps:

- The separated modeled modules are transformed to system matrix form using the *pck* function and are considered sub-systems using the *systemnames* function.
- All inputs to the modules are specified using *input_to* function.
- The desired inputs and outputs for the overall model of IMGs are specified by *inputvar* and *outputvar* functions.
- The overall model of IMGs is calculated by the *sysic* function.
- The obtained model is transformed to SSR by the *unpck* function to be used in eigenvalue analysis.

The details of the RCT functions in the general process of interconnecting the sub-systems are presented in [50].

Exercise 2.5 Consider the two systems introduced in Exercise 2.1, which are the main modules of a VSC shown in Figure 2.6(a). The voltage controller includes two PI controllers with mutual terms, the averaged model of the converter, and the output LC filter. Find a numerical SSR for this system considering the data shown in Table 2.1.

Solution

Three dynamic modules can be seen, including two PI controllers and the LC filter. The analytical models of these modules are represented in (2.12) and (2.13). The mutual terms should be

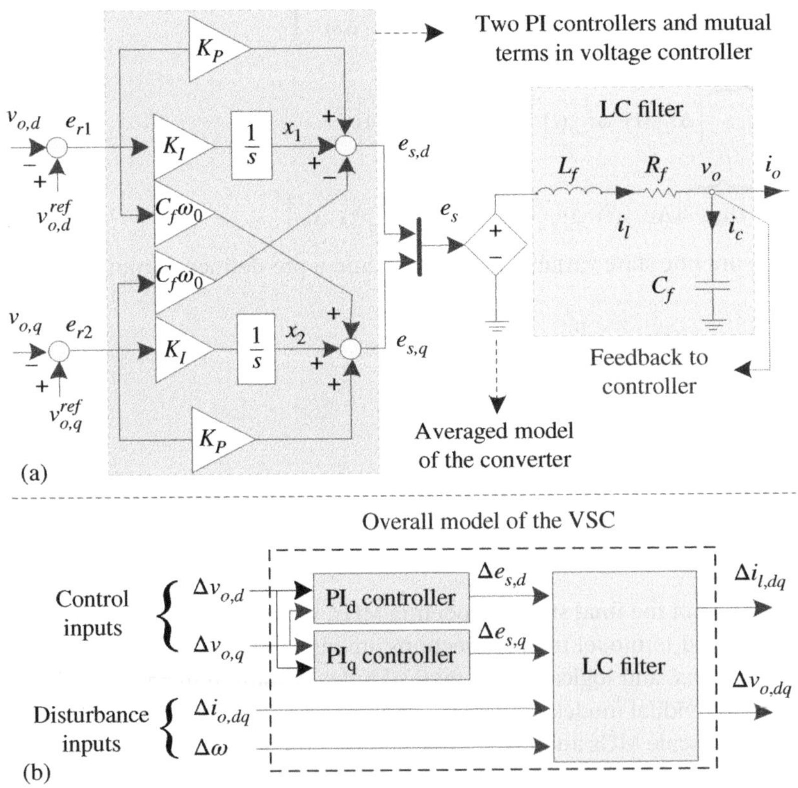

Figure 2.6 A VSC equipped with a voltage controller: (a) the block diagram including the voltage controller, averaging model of the converter, and LC filter and (b) the overall model and internal interconnections.

Table 2.1 Data of LC filter and PI controllers required for Exercise 2.5.

Parameter (unit)	Value	Parameter (unit)	Value	Variable	Initial value
$R_f(\Omega)$	0.1	ω_0 (rad/s)	314	i_{l0d} (A)	3
$C_f(\mu F)$	70	K_P	2	i_{l0q} (A)	−1
L_f (mH)	2	K_I	20	v_{o0d}, v_{o0q} (V)	325,0

considered as the second inputs of the PI controllers, where the gains $\pm C_f\omega_0$ appear in the input matrices. The independent voltage source is assumed as the input of the LC filter without any dynamics. The matrices are given as $C_{LC} = I_{4\times4}, D_{LC} = 0_{4\times5}$ and

$$\text{LC Filter: } A_{LC} = \begin{bmatrix} -50 & 314 & -500 & 0 \\ -314 & -50 & 0 & -500 \\ 14286 & 0 & 0 & 314 \\ 0 & 14286 & -314 & 0 \end{bmatrix}, \tag{2.14}$$

$$B_{LC} = \begin{bmatrix} 500 & 0 & 0 & 0 & -1 \\ 0 & 500 & 0 & 0 & -3 \\ 0 & 0 & -14286 & 0 & 0 \\ 0 & 0 & 0 & -14286 & -325 \end{bmatrix},$$

$$d-\text{PI Controller: } A_{PId} = 0, B_{PId} = \begin{bmatrix} 20 & 0 \end{bmatrix}, \quad C_{PId} = 1, \quad D_{PId} = \begin{bmatrix} 2 & -0.02 \end{bmatrix},$$

$$q-\text{PI Controller: } A_{PIq} = 0, B_{PIq} = \begin{bmatrix} 20 & 0 \end{bmatrix}, \quad C_{PIq} = 1, \quad D_{PIq} = \begin{bmatrix} 2 & +0.02 \end{bmatrix}.$$

Figure. 2.6(b) shows a block diagram of the VSC modules and their interconnections. One can apply these interconnections using *sysic* function and its requirements on the modules as shown in Table 2.2, where *G_LC*, *G_PI_d*, and *G_PI_q* are packed form of the module SSRs, and the outputs of the LC filter are considered as the output of the VSC model. Finally, the SSR matrices of the VSC are given as follows:

$$A_{VSC} = \begin{bmatrix} -50 & 314 & -500 & 0 & 500 & 0 \\ -314 & -50 & 0 & -500 & 0 & 500 \\ 14286 & 0 & 0 & 314 & 0 & 0 \\ 0 & 14286 & -314 & 0 & 0 & 0 \\ 0 & 0 & -20 & 0 & 0 & 0 \\ 0 & 0 & 0 & -20 & 0 & 0 \end{bmatrix}, \tag{2.15}$$

$$B_{VSC} = \begin{bmatrix} 0 & 0 & -1 \\ 0 & 0 & -3 \\ -14286 & 0 & 0 \\ 0 & -14286 & -325 \\ 0 & 0 & 0 \\ 0 & 0 & 0 \end{bmatrix}, \quad C_{VSC} = \begin{bmatrix} I_{4\times4} \\ 0_{4\times2} \end{bmatrix}^T, \quad D_{VSC} = 0_{4\times3}.$$

The MATLAB code for this exercise is available in [47].

Table 2.2 MATLAB code to specify interconnections among module models of the VSC introduced in Exercise 2.5.

```
systemnames =' G_LC G_PI_d G_PI_q';
sysoutname =' G_VSC';
inputvar =' [io{2}; w]';
input_to_G_LC =' [G_PI_d; G_PI_q; io; w]';
input_to_G_PI_d =' [-G_LC(3); -G_LC(4)]';
input_to_G_PI_q =' [-G_LC(4); -G_LC(3)]';
outputvar =' [G_LC]';
sysic;
[A_VSC, B_VSC, C_VSC, D_VSC] = unpck(G_VSC);
```

2.3.2 Detailed Modeling

In order to provide a detailed model of MGs, all subsystems/modules involved in a study should be separately modeled, and then their interconnections are considered including power lines and control signals. Control signals can be realized by physical cabling, communicating data, or a combination. The main modules constructing MGs are distribution lines, various load types, filters, power converters, various DERs, required control layers, and different controllers including power, current, voltage, and frequency controllers. In order to study small-signal stability and to design most model-based controllers, dynamic modeling of MG modules is often performed using SSR, which is also taken into account in this book. Then, the interconnections can be modeled precisely by considering the connections among the inputs and outputs of the module models. Although the basic analytical method of several equation substitutions is addressed in the literature [1, 35, 39, 49, 51, 52] for MG and IMG interconnection modeling, the analytical–numerical methods [37, 53] are preferred. Using such methods, which employ numerical calculations in some modeling steps, leads to reducing computational burdens of analytical modeling of numerous nested MG interconnections and consequently low computation/simulation time. Here, the numerical interconnection method, explained in Section 2.3.1.5, is used to consider interconnections in detailed modeling as well as simplified modeling represented in the next section.

Exercise 2.6 Note that Exercise 2.5 was an example for detailed modeling of a VSC, which is desirable for small-signal stability analysis of the dynamics slower than a few milliseconds and faster than several seconds. Find the eigenvalues of this model.

Solution

After finding the SSR model of the VSC using the analytical–numerical interconnection method, one can simply calculate the eigenvalues of this model using *eig* function in MATLAB as follows:

$$\lambda = -4.9 \pm j4943, -38.4 \pm j4315, -6.7 \pm j0.06. \tag{2.16}$$

2.3.3 Simplification Methods

Figure 2.7 shows a classification of simplification methods used in MG/IMG modeling. The simplified modeling methods can be organized into three clusters, i.e. model-based (MOB), module-based

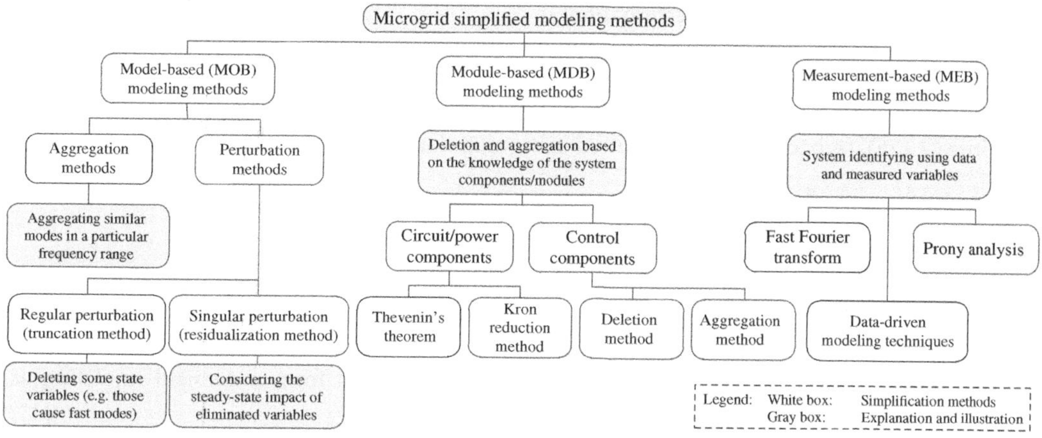

Figure 2.7 Classification of simplified modeling methods of MGs.

(MDB), and measurement-based (MEB) simplification techniques. In the first cluster, the detailed model should be obtained at first, which is a disadvantage, and then its order is reduced using aggregation and perturbation methods [54]. The perturbation methods including regular [44, 55] and singular types have often been used in the MG [39–43] and IMG [36, 56] dynamic modeling studies. MDB methods try to find the simplified model without calculating a detailed model. The dominant modules in the desired dynamics are preserved, and the others are removed according to the knowledge of the system [45, 57]. The MEB methods benefit from the measured data and system identification techniques to find the model with a minimum computational burden, e.g. fast Fourier transform and Prony analysis, which focus on modeling the desired variable(s), e.g. frequency. Despite being appropriate for large-scale systems, MEB methods are rarely used in, e.g. MG modeling [58] and load modeling in MGs [46].

According to dominant/non-dominant dynamic modes in the range of desired dynamics/frequencies, a simplified model can be obtained by dividing all system variables into preserved and removed state variables. In stability studies, slow dynamics are usually desirable to be preserved [26, 36, 39, 41]. To design controllers, the preserved frequency range is selected according to their bandwidth. The model (2.17) shows both group state variables and their interactions in a closed-loop form, i.e. without input/output. It can be decoupled using one of the simplification methods, i.e. MOB, MDB, or MEB.

$$\dot{x}_p = A_{pp}x_p + A_{pr}x_r, \tag{2.17a}$$

$$\dot{x}_r = A_{rp}x_p + A_{rr}x_r, \tag{2.17b}$$

where x_p and x_r represent the preserved and removed states, and A_{pp}, A_{pr}, A_{rp}, and A_{rr} are the sub-matrices of the main state matrix. The preserved states compose a reduced-order model.

2.3.3.1 Truncation (Regular Perturbation)

In the Truncation-based order reduction method, the undesirable state variables are completely removed without considering their impact on the preserved state variables. One can find the truncated model of (2.17) as follows:

$$\dot{x}_p = A_{pp}x_p. \tag{2.18}$$

2.3.3.2 Residualization (Singular Perturbation)

In the residualization method, the dynamics of undesirable state variables are removed. However, their impact on the steady state of preserved state variables is considered. This can be done by applying $\dot{x}_r = 0$, which means removing the corresponding dynamics, and then substituting the result into (2.17a). Hence, (2.17) is simplified using singular perturbation theory as follows:

$$\dot{x}_p = (A_{pp} - A_{pr}A_{rr}^{-1}A_{rp})x_p. \tag{2.19}$$

2.3.3.3 Aggregation

In the aggregation method, similar dynamic modes are aggregated to a lower number of dynamic modes. Let us explain the aggregation method for dominant low-frequency modes (DLFMs), which typically show the interactions among parallel VSCs in islanded MGs [39] in a complex conjugate form. Consider that k oscillatory DLFMs are obtained from a detailed modeling as $\left\{\lambda_{dlfm}^{1}, \dots, \lambda_{dlfm}^{2k}\right\}$, which can be represented by an SSR as follows:

$$\dot{x}_{dlfm} = A_{dlfm} \cdot x_{dlfm}, \tag{2.20}$$

where $x_{dlfm} \in R^{2k\times1}$ and $A_{dlfm} = diag\left\{\lambda_{dlfm}^{1}, \dots, \lambda_{dlfm}^{2k}\right\}$. The goal is to find the state matrix F of the aggregated system as

$$\dot{z}_{dlfm} = F \cdot z_{dlfm}, \tag{2.21}$$

where $z_{dlfm} \in R^{l\times1}$ is called the aggregation of x_{dlfm}, and they are correlated as $z_{dlfm}(t) = Cx_{dlfm}(t)$. $C \in R^{l\times2k}$, $(l < 2k)$ is a constant aggregation matrix of x_{dlfm}. Finally, F can be calculated as follows [54]

$$F = C \cdot A_{dlfm} \cdot C^{+}, \tag{2.22}$$

where C^{+} is the pseudo-inverse of the non-square matrix C, which can be calculated as $C^{+} = C^{T}(CC^{T})^{-1}$.

If all oscillatory DLFMs are interested in being aggregated in one oscillatory mode, z is selected as a 2×1 vector comprising a pair of complex conjugate eigenvalues. Thus, one can find a second-order transfer function as follows:

$$G_{Agg}(s) = \frac{1}{s^2 + 2\zeta\omega_0 s + \omega_0^2}, \tag{2.23}$$

where ζ and ω_0 are the damping ratio and undamped natural frequency of the aggregated model.

Exercise 2.7 The detailed model of the VSC, shown in Figure 2.6, has been represented in Exercise 2.5. (a) Find simplified second-order models of the VSC employing truncation, residualization, and aggregation methods. (b) Compare the simplified models using Bode diagrams and eigenvalues.

Solution

a) Codes to find simplified models according to truncation, residualization, and aggregation methods can be accessed in the data file [47]. In truncation and residualization methods, firstly, a balanced representation is applied to the detailed model (2.15) to rearrange the state variables from slow to fast dynamics. It can easily be done using the *sysbal* command. Then, the first two state variables are preserved. The state matrices of the different methods are given in Table 2.3. The results are the same for truncation and residualization methods due to using the balanced forms, i.e. $A_{pr} = A_{rp} = 0$ in (2.19). However, input and output matrices are different. All steps in Section 2.3.3.3 are required to find the aggregated model, which can be found in [47].

Table 2.3 State matrices of model-based simplification methods in Exercise 2.9.

Simplification method	Truncation	Residualization	Aggregation
State matrix	$\begin{bmatrix} -4.9 & 4943 \\ -4943 & -4.9 \end{bmatrix}$	$\begin{bmatrix} -4.9 & 4943 \\ -4943 & -4.9 \end{bmatrix}$	$\begin{bmatrix} -16.7 & 3086 \\ 3086 & -16.7 \end{bmatrix}$

b) The eigenvalues are shown in Figure 2.8(a), which also indicates the same eigenvalues of truncation and residualization methods. They are the same with the slowest dynamic modes of the detailed model. In order to have a more complete comparison, Figure 2.8(b) shows the bode diagrams of the detailed and simplified models. The residualization method yields best results in low frequencies, while it exhibits the poorest performance in high frequencies, where the truncation method excels. The aggregation method demonstrates a moderate response, outperforming the residualization method at higher frequencies but not surpassing the truncation method. It is important to note that the choice of reduced model order can significantly impact the results. Consequently, each method may find utility in specific system applications.

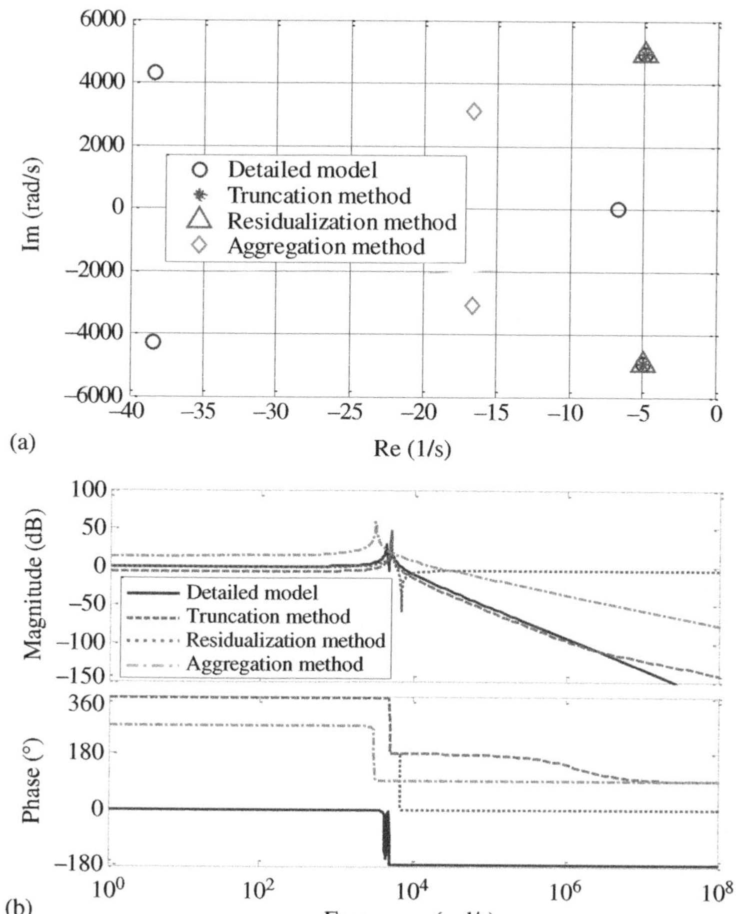

Figure 2.8 A comparison of model-based simplification methods: (a) eigenvalues and (b) Bode diagram of the detailed and simplified methods.

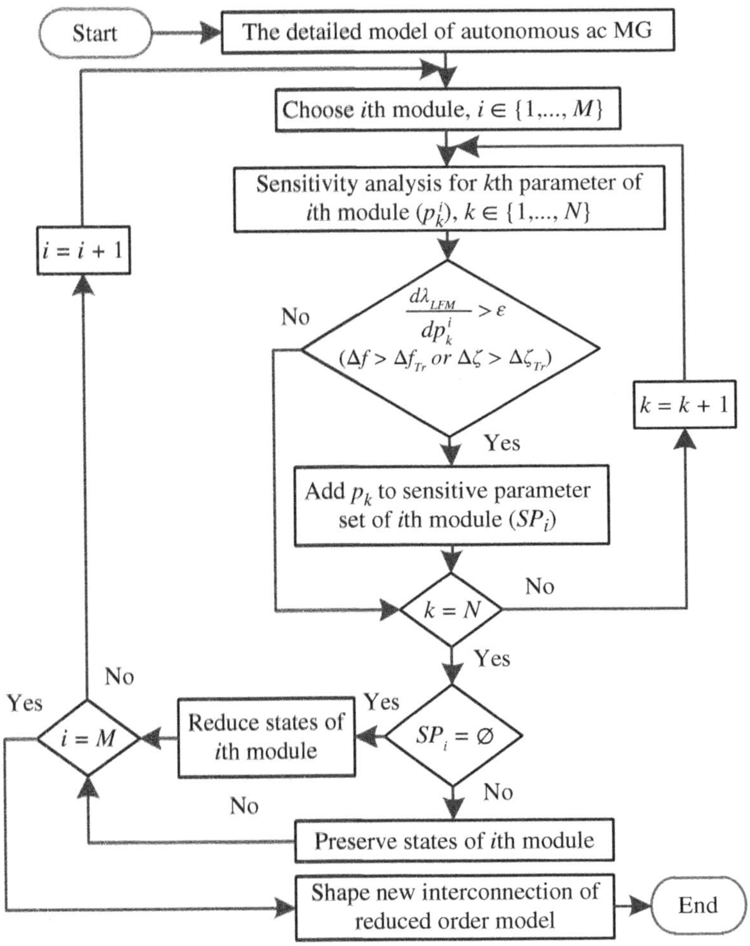

Figure 2.9 Sensitivity analysis-based simplification flowchart for a family of autonomous AC microgrids via changing the parameters in the analysis.

2.3.3.4 Sensitivity Analysis

Deletion or aggregation of modules can be done based on sensitivity analysis results in an MDB simplification method. Figure 2.9 shows how the sensitivity analysis is used for sifting out the significant modules for preserving the DLFMs from the insignificant ones as a sample study. The detailed model is the source model to be sure that all modules are checked. For each module, all its parameters are changed within acceptable ranges. The criterion for preserving a module is a minimum sensitivity (ε) of at least one of the DLFMs to at least one of its parameters:

$$d\lambda_{dlfm}/dp_k^i > \varepsilon, \tag{2.24}$$

where p_k^i is the kth parameter of the ith module. The variation of real value, imaginary value, damping ratio, or a combination of them for each eigenvalue can be selected to realize (2.24), i.e. recognizing the eigenvalue sensitivity to the parameter changes. Here, (2.24) is realized for oscillatory modes as follows:

$$\Delta f > \Delta f_{Tr}, \tag{2.25a}$$

$$\Delta\zeta > \Delta\zeta_{Tr}, \tag{2.25b}$$

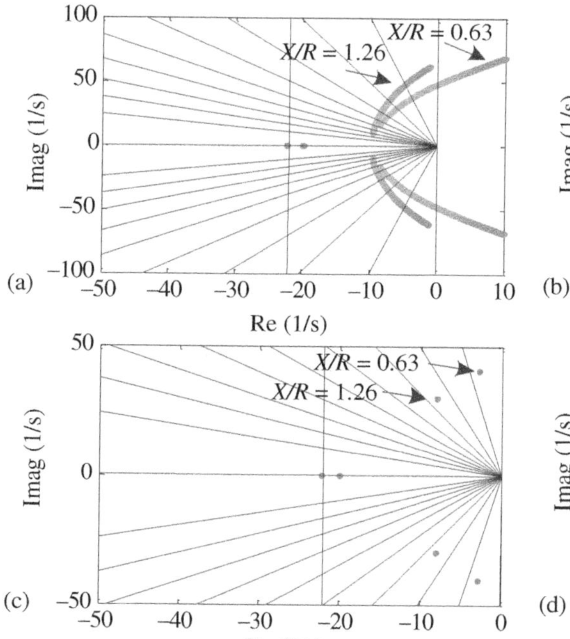

Figure 2.10 Sensitivity analysis results of dominant low-frequency modes typically for droop-based DER power controller parameters: (a) $0.001\% < m_p < 0.02\%$, (b) $0.01\% < n_q < 0.2\%$, and output filter parameters: (c) $0.3\ \text{mH} < L_f < 6.4\ \text{mH}$, and (d) $3\ \mu\text{F} < C_f < 60\ \mu\text{F}$.

where Δf and $\Delta\zeta$ are the frequency and damping ratio variations of the DLFM due to changing p_k^i. Δf_{Tr} and $\Delta\zeta_{Tr}$ are the threshold values specifying the ease/hardness of fulfilling (2.24). However, the absolute real value variation ($\Delta\sigma$) is considered for the non-oscillatory modes as follows:

$$\Delta\sigma > \Delta\sigma_{Tr}. \tag{2.26}$$

Exercise 2.8 Figure 2.10 shows the sensitivity analysis results for two modules of power controller (Figure 2.10(a),(b)) and output filter (Figure 2.10(c),(d)) of a GFO DER in an MG with two GFO DERs and one GFL DER, where the GFL DER is not connected. The studied MG is introduced in Section 3.3.3 and GFO and GFL control strategies as well as the corresponding DERs are modeled in Sections 2.5.1.1 and 2.5.1.2. Discuss the sensitivity of DLFMs to the module parameter changes.

Solution

The vertical line shows the margin between dominant and non-dominant modes. The radial lines determine margins for recognizing (2.25b). Obviously, the DLFMs are sensitive to the power controller parameters, i.e. $\omega - P$ droop gain (m_p) and $V - Q$ droop gain (n_q), but not to the DER output filter parameters, i.e. filter capacitance (C_f) and inductance (L_f). The sensitivity analysis done for two different X/R ratios of the DER coupling line, i.e. $X/R = 0.63$ and $X/R = 1.26$, shows different sensitivity, where a smaller X/R ratio causes more sensitive DLFMs.

2.3.4 Prony Analysis

In this section, the Prony method fundamentals as a signal analyzer tool are presented. This method is useful for MEB simplification and model validation methods. Let us consider a signal $y(t)$, which is sampled with the sampling time T, can be represented as follows:

$$y[k] = \sum_{i=1}^{L} C_i \mu_i^k, \tag{2.27}$$

where L is the total number of estimated damped sinusoid components and $k = 0, 1, \ldots, N-1$. Unknown parameters C_i and μ_i^k can be found by comparing (2.27) with the estimated signal $\hat{y}(t)$, which is constructed by the sum of complex damped sinusoids as follows [59]:

$$\hat{y}(t) = \sum_{i=1}^{L} A_i e^{\sigma_i(t)} \cos(2\pi f_i t + \phi_i), \tag{2.28}$$

where A_i, σ_i, f_i, and ϕ_i are the amplitude, damping coefficient, frequency, and phase of the ith component. Therefore, C_i and μ_i^k have the following relationships with the parameters of (2.28):

$$C_i = \frac{A_i}{2} e^{j\phi_i}, \tag{2.29a}$$

$$\mu_i^k = e^{(\sigma_i + j2\pi f_i)T}. \tag{2.29b}$$

The original Prony analysis computes C_i and μ_i^k in three following steps using the samples $y[k]$, such that $k = L, L+1, L+2, \ldots, N-1$ [59, 60].

Step 1: Predicting a linear model: A linear prediction model can be given by the observed data set as follows:

$$y[k] = a_1 y[k-1] + a_2 y[k-2] + \cdots + a_L y[k-L]. \tag{2.30}$$

In order to obtain a_i coefficients, (2.30) is written for all values of k to form an equation set, which is represented in a matrix form as follows:

$$\mathbf{A} = \mathbf{D}^{+}\mathbf{Y}, \tag{2.31}$$

where $\mathbf{D}^{+}$ means the pseudo-inverse of $\mathbf{D}$, $\mathbf{A} = [a_1 \ a_2 \ \cdots \ a_L]^T$, $\mathbf{Y} = [y[L] \ y[L+1] \ \cdots \ y[N-1]]^T$ and

$$\mathbf{D} = \begin{bmatrix} y[L-1] & y[L-2] & \cdots & y[0] \\ y[L] & y[L-1] & \cdots & y[1] \\ \vdots & \vdots & \vdots & \vdots \\ y[N-2] & y[N-3] & \cdots & y[N-L-1] \end{bmatrix}.$$

Step 2: Root calculation of the predicted model: The roots of the characteristic polynomial represented in (2.32), which is formed from the linear prediction coefficients a_i, can be found easily using many computation software e.g. MATLAB. In (2.32), $\hat{\mu}_i$ depicts the roots of the polynomial.

$$\mu^L - a_1\mu^{L-1} - \cdots - a_{L-1}\mu - a_L = (\mu - \hat{\mu}_1)(\mu - \hat{\mu}_2)\ldots(\mu - \hat{\mu}_L).$$

Step 3: Coefficient calculation: C_i coefficients are calculated by substituting $\hat{\mu}_i$ in (2.27) as follows:

$$\mathbf{C} = \mathbf{U}^{+}\mathbf{Y}, \tag{2.32}$$

where $\mathbf{C} = [\, C_1 \ C_2 \ \cdots \ C_L \,]^T$, and $\mathbf{U}^+$ means the pseudo-inverse of $\mathbf{U}$, where $\mathbf{U}$ is as follows:

$$\mathbf{U} = \begin{bmatrix} 1 & 1 & \cdots & 1 \\ \hat{\mu}_1 & \hat{\mu}_2 & \cdots & \hat{\mu}_L \\ \vdots & \vdots & \vdots & \vdots \\ \hat{\mu}_1^{N-1} & \hat{\mu}_2^{N-1} & \cdots & \hat{\mu}_L^{N-1} \end{bmatrix}.$$

Finally, A_i, σ_i, f_i, and ϕ_i can be calculated only by decoupling the real and imaginary components of both sides of (2.29a) and (2.29b).

2.3.5 Large-Signal Modeling

Differential equations are the main tool for large-signal modeling, which can represent the nonlinearities well. General rules to find differential equations are represented in Section 2.3.1.1. Their examples can also be found. Furthermore, nonlinear SSRs can be used [38] to model nonlinear systems, which are generally given as follows:

$$\dot{X}_M^k = A_M^k(X_M^k)X_M^k + B_M^k U_M^k, \tag{2.33a}$$

$$Y_M^k = C_M^k X_M^k + D_M^k U_M^k, \tag{2.33b}$$

where at least, the state matrix is a function of the state vector.

Graphical tools, block diagrams, and time-domain simulations are complementary tools for nonlinear modeling based on differential equations. Different graphical tools are used to simplify the display and analysis of nonlinearities, e.g. characteristics curves and phase-portrait diagrams. Block diagrams are also used to facilitate understanding the relations among different modules of a nonlinear model and to show some nonlinear functions easier than their equations, e.g. limiters, rate limiters, saturations, and dead bands. Time-domain simulation is the most powerful non-analytical nonlinear modeling tool, which can numerically consider a large number of nonlinear dynamics and can show their behaviors using time responses. Throughout this book, time-domain simulations using MATLAB/Simulink and MATLAB/SimPowerSystems are used for different objectives including modeling validation, dynamics studies, stability analysis, and controller test. In addition to offline time-domain simulations, real-time simulations have the advantages of fast simulations and being real-time, which are done in this book using an OPAL-RT digital simulator.

In order to see an example of nonlinear modeling see Section 2.6, and to compare linear and nonlinear models see Exercise 2.10.

2.4 Small-Signal Modeling of Microgrid Components

Figure 2.1 shows a typical MG with its essential components and modules. In this section, the SSR of common modules and components of MGs is represented individually in each subsection including different types of power electronics converters, MG loads, distributed energy resources, and energy storage systems. Moreover, the SSR of common control strategies required for the operation of MGs is given with different DERs and converters. Considerations of each control layer of the MG hierarchical control are also represented in this section.

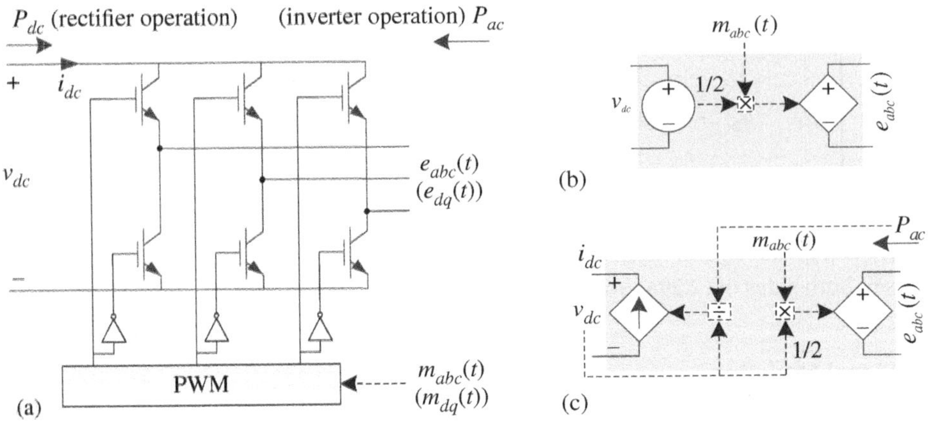

Figure 2.11 Schematics of dc–ac and ac–dc converters: (a) topology of a three-phase six-switch converter, (b) averaged model of dc–ac converter with constant dc voltage (inverter operation), and (c) general averaged model for both rectifier and inverter operations (P_{ac} direction shows rectifier operation).

2.4.1 DC–AC Converter (Inverter)

There exist different types and structures of dc–ac converters including different numbers and types of power-electronic switches and various connection topologies. Here a common structure used in ac MG applications is considered for modeling. In order to model other types of inverters, readers can follow basic references to power-electronic converter modeling [61, 62].

Figure 2.11(a) shows a schematic of the standard topology of a three-phase six-switch converter, which can be used in both inverter and rectifier operations. In the inverter operation as VSC, the dc link voltage can be provided by a primary source with a permanent dc voltage. Therefore, in some cases, the dc link voltage can ideally be considered constant. Figure 2.11(b) shows a block diagram of an averaged model of these three-phase dc–ac converters with three arms and six switches. Switching and pulse width modulation (PWM) dynamics are neglected because they are too fast to be considered in the interested dynamics here, i.e. milliseconds to a few seconds. Therefore, dc–ac converter model does not include any dynamics, i.e.,

$$e_{abc}(t) = \frac{1}{2}V_{dc}m_{abc}(t), \tag{2.34}$$

where three-phase output voltages $e_{abc}(t)$ are obtained as a multiplication of half of dc voltage and three-phase modulation signals $m_{abc}(t)$. In fact, $m_{abc}(t)$ is obtained from the inverse Park transformation of dq components of innermost controller outputs.

The assumption of constant dc link voltage can be eliminated to find a more general and non-ideal model of inverters. This model is required when the dynamics of the dc link, the primary source, or the dc voltage controller should be modeled. The modeling process is the same for the rectifier operation modeling, which is represented in the next section.

2.4.2 AC–DC Converter (Rectifier)

Figure 2.11(c) shows an averaged model of a three-phase controlled rectifier with six switches shown in Figure 2.11(a). Note that the power direction in rectifiers is reversed with respect to inverters. However, this fact cannot be observed in the modeling process. The dc side can be modeled as a current source dependent on the ac power drained by the ac side of the rectifier from an ac source.

This model is static as follows:

$$i_{dc}(t) = P_{ac}(t)/v_{dc}(t), \tag{2.35a}$$

$$e_{dq}(t) = \frac{1}{2}V_{dc}m_{dq}(t), \tag{2.35b}$$

where P_{ac} can be calculated for three-phase systems from the ac voltage and currents in the rotating reference frame as follows:

$$P_{ac}(t) = \frac{3}{2}(e_d i_d + e_q i_q). \tag{2.36}$$

Note that (2.35b) shows the ac voltage side model of the rectifier. In fact, it models the ac side of the converter during the rectifying mode. Note that (2.35) is a nonlinear model, which can be linearized as follows:

$$\begin{aligned} \Delta i_{dc} &= T_{ac/dc}\left[\Delta v_{dc}\ \ \Delta i_{dq}\ \ \Delta e_{dq}\right]^T, \\ \Delta e_{dq} &= T_{dc/ac}\left[\Delta v_{dc}\ \ \Delta m_{dq}\right]^T, \end{aligned} \tag{2.37}$$

where

$$T_{ac/dc} = \frac{3}{2v_{dc0}}\left[-\frac{2}{3}I_{dc0}\ \ e_{d0}\ \ e_{q0}\ \ i_{d0}\ \ i_{q0}\right], \tag{2.38}$$

$$T_{dc/ac} = \frac{1}{2}\begin{bmatrix} m_{d0} & v_{dc0} & 0 \\ m_{q0} & 0 & v_{dc0} \end{bmatrix}. \tag{2.39}$$

One can augment both $T_{ac/dc}$ and $T_{dc/ac}$ in one matrix, namely G_{rec}, for $U_{rec} = [v_{dc}\ m_{dq}\ i_{dq}\ e_{dq}]^T$ and $Y_{rec} = [i_{dc}\ e_{dq}]^T$ as follows:

$$Y_{rec} = G_{rec}U_{rec}, \tag{2.40}$$

where G_{rec} can be used as the averaged model of both the rectifier and the inverter as follows:

$$G_{rec} = \begin{bmatrix} -\frac{I_{dc0}}{v_{dc0}} & 0 & 0 & \frac{3e_{d0}}{2v_{dc0}} & \frac{3e_{q0}}{2v_{dc0}} & \frac{3i_{d0}}{2v_{dc0}} & \frac{3i_{q0}}{2v_{dc0}} \\ 0.5m_{d0} & 0.5v_{dc0} & 0 & 0 & 0 & 0 & 0 \\ 0.5m_{q0} & 0 & 0.5v_{dc0} & 0 & 0 & 0 & 0 \end{bmatrix}. \tag{2.41}$$

2.4.3 DC–DC Converter (Chopper)

Three basic types of dc–dc converters include buck, boost, and buck–boost converters [62]. Here, boost and buck converters are modeled, and buck–boost converter model can be performed using the rules represented here. Figure 2.12(a) shows a schematic of a typical boost converter as a case to find a dynamic model including both switching devices and filtering elements. The modeling method is generally similar for other types of dc–dc converters. Two operation modes include continuous conduction mode (CCM) and discontinuous conduction mode (DCM), where the conduction refers to the inductor current. In both modes, the switch can be closed ($S = 1$, and $D_i = 0$) or open ($S = 0$ and $D_i = 1$). For each situation, the equivalent circuit can be obtained, then the SSRs can be given as follows:

$$s = 0 : \dot{X}_{ddc}(t) = A_{s=0}X_{ddc}(t) + B_{ddc}\tilde{U}_{ddc}(t), \tag{2.42a}$$

$$s = 1 : \dot{X}_{ddc}(t) = A_{s=1}X_{ddc}(t) + B_{ddc}\tilde{U}_{ddc}(t), \tag{2.42b}$$

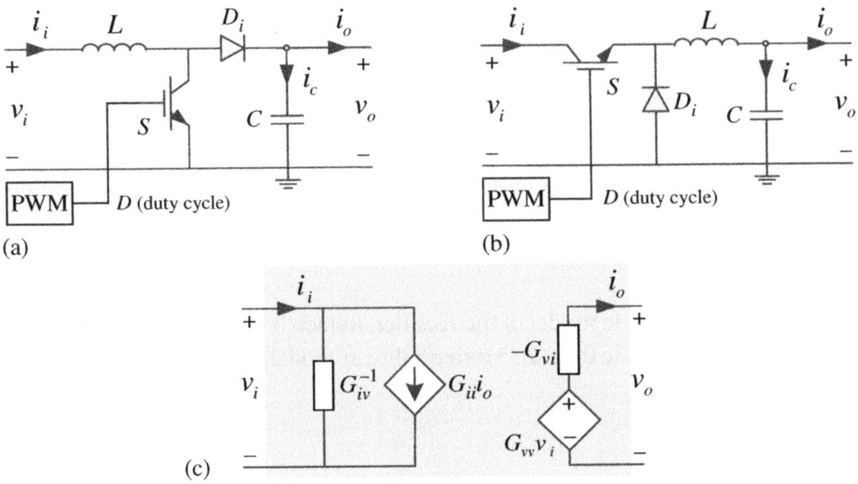

Figure 2.12 Circuit schematic and averaged model of dc–dc converters: (a) dc–dc boost converter, (b) dc–dc buck converter, and (c) general averaged model of both dc–dc boost and buck converters.

where $X_{ddc}(t) = [i_l(t)\ v_o(t)]^T$, $\tilde{U}_{ddc}(t) = [v_i(t)\ i_o(t)]^T$, and

$$A_{s=0} = \begin{bmatrix} 0 & -1/L \\ 1/C & 0 \end{bmatrix}, \quad A_{s=1} = \begin{bmatrix} 0 & 0 \\ 0 & 0 \end{bmatrix}, \quad B_{ddc} = \begin{bmatrix} 1/L & 0 \\ 0 & -1/C \end{bmatrix}.$$

One can rewrite (2.42) as follows:

$$\dot{X}_{ddc}(t) = [(1-s)A_{s=0} + sA_{s=1}]X_{ddc}(t) + B_{ddc}\tilde{U}_{ddc}(t). \tag{2.43}$$

Finally, averaging the model (2.43) during (or a coefficient of) the switching time [63, 64] results in an averaged model as follows:

$$\dot{X}_{ddc}(t) = (1 - D_{ch}(t))A_{s=0}X_{ddc}(t) + B_{ddc}\tilde{U}_{ddc}(t), \tag{2.44}$$

where $D_{ch}(t)$ is the duty cycle of the switch used in the chopper. The equation (2.44) is nonlinear due to the multiplication of $D(t)$ and $X_{ddc}(t)$. After linearization, (2.44) is changed as follows:

$$\Delta\dot{X}_{ddc}(t) = (1 - D_{ch,0}(t))A_{s=0}\Delta X_{ddc}(t) + B_{ddc}\Delta\tilde{U}_{ddc}(t) - A_{s=0}X_{ddc,0}\Delta D_{ch}(t). \tag{2.45}$$

Finally, considering $\Delta U_{ddc}(t) = [\Delta\tilde{U}_{ddc}(t)\ \Delta D_{ch}]^T$, the SSR model of the boost converter is obtained as follows:

$$\Delta\dot{X}_{ddc}(t) = A_{ddc}\Delta X_{ddc}(t) + B_{ddc}\Delta U_{ddc}(t), \tag{2.46a}$$

$$Y_{ddc}(t) = \Delta X_{ddc}(t), \tag{2.46b}$$

and the corresponding matrices are given as follows:

$$A_{ddc} = \begin{bmatrix} 0 & (D_{ch,0}-1)/L \\ (1-D_{ch,0})/C & 0 \end{bmatrix}, \quad B_{ddc} = \begin{bmatrix} 1/L & 0 & v_{o,0}/L \\ 0 & -1/C & -i_{l,0}/C \end{bmatrix}, \tag{2.47}$$

where 0 in subscripts denotes the value of the variable in the operating point.

A similar procedure can be performed to find an SSR for the dc–dc buck converter shown in Figure 2.12(b) with a similar state and input vectors as $X_{ddc}(t)$ and $U_{ddc}(t)$. A similar procedure to

Table 2.4 The transfer functions of the impedance-based two-port model of dc–dc converters shown in Figure 2.12(c).

Converter type	G_{ii}	G_{iv}	G_{vv}	G_{vi}
Boost	$\frac{1-D}{LCs^2+(1-D)^2}$	$\frac{CS}{LCs^2+(1-D)^2}$	$\frac{1-D}{LCs^2+(1-D)^2}$	$\frac{-LS}{LCs^2+(1-D)^2}$
Buck	$\frac{1}{LCs^2+1}$	$\frac{CDs}{LCs^2+1}$	$\frac{D}{LCs^2+1}$	$\frac{-Ls}{LCs^2+1}$

find the equivalent circuits for $s = 0$ and $s = 1$ results in the same SSR model as represented in (2.46), but with different matrices as follows:

$$A_{ddc} = \begin{bmatrix} 0 & -1/L \\ 1/C & 0 \end{bmatrix}, \quad B_{ddc} = \begin{bmatrix} D_{ch,0}/L & 0 & v_{i,0}/L \\ 0 & -1/C & 0 \end{bmatrix}. \tag{2.48}$$

Since the SSR shown in (2.46) is the same for both dc–dc converter types, their averaged models have also the same form shown in Figure 2.12(c). The averaged model is a two-port with parameters G_{ii}, G_{iv}, G_{vv}, and G_{vi}, which can be obtained in the s domain by inverting the SSR (2.46) before linearization to the equivalent transfer function matrix. In fact, the SSR model can also be represented as an impedance-based transfer function in the frequency domain. The corresponding two-port parameters (transfer functions) are listed in Table 2.4 for the buck and boost converters. The equivalent impedance-based transfer function can also be calculated for the linearized SSR, i.e. (2.46). Note that models of power converters represented here are widely used in modeling different renewable energy resources and energy storage systems.

2.4.4 LC Filter

LC or LCL filters are generally used in a three-phase form at the output of inverters, i.e. universal bridges in ac MGs. They can also be used in a single-phase form in single-phase ac and dc MGs. Figure 2.4 presents the single-line diagram of LC filters in Exercise 2.1. The details of modeling of three-phase LC filters are expressed in Exercises 2.1, to 2.4 including mathematical formulation and numerical calculation. Furthermore, the SSR of the three-phase LC filter can be found in (2.12). The process is not repeated here.

In order to model a single-phase LC filter, the process is similar except for using the Park transformation. In fact, in single-phase LC filter modeling, the step of transforming equations from the *abc* reference frame to the *dq0* reference frame should be eliminated. Furthermore, one can easily model an LCL filter, whether three-phase or single-phase, just by considering two (one in single-phase) more state variables (variable) for the current of the second inductor.

2.4.5 Power Network

AC MGs are usually operated in low-voltage and medium-voltage configurations, where short and medium overhead lines and underground cables are used. Therefore, the power grid of MGs can be considered as a combination of resistive–inductive (RL) lines. Although the X/R ratio is small in low/medium-voltage grids/MGs, both inductance and resistance components should be considered. However, the capacitance component is not necessary due to the small voltage level. Each RL line can easily be modeled like an LC filter. In fact, a three-phase series RL branch should be

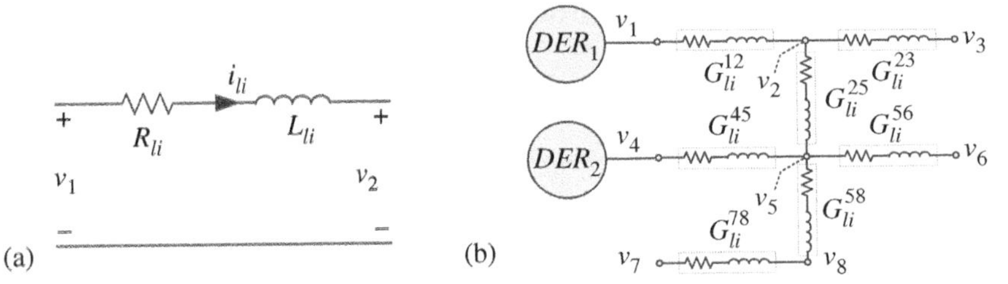

Figure 2.13 Power network modeling: (a) single-line diagram of RL lines and (b) a typical MG power network to specify interconnections.

considered, whose single-line diagram is shown in Figure 2.13(a). After Park transformation, linearization, and decoupling of real and imaginary components, the linear differential equation of each line of the power network is given as follows:

$$\begin{aligned}\Delta\dot{i}_{li,d} &= -(R_{li}/L_{li})\Delta i_{li,d} + \omega_0 \Delta i_{li,q} + (1/L_{li})\Delta v_{1,d} - (1/L_{li})\Delta v_{2,d} + i_{li,q0}\Delta\omega_{com},\\ \Delta\dot{i}_{li,q} &= -\omega_0 \Delta i_{li,d} - (R_{li}/L_{li})\Delta i_{li,q} + (1/L_{li})\Delta v_{1,q} - (1/L_{li})\Delta v_{2,q} - i_{od0}\Delta\omega^{n}_{com},\end{aligned} \tag{2.49}$$

where i_{li}, R_{li}, and L_{li} are the current, resistance, and inductance of the RL line, and v_1 and v_2 are the voltages of the sending and receiving ends of the line. $\Delta\omega_{com}$ is the perturbed frequency of the MG common reference frame (CRF), which will be explained later. Finally, an SSR is given as follows:

$$\begin{aligned}\dot{X}_{li} &= A_{li}X_{li} + B_{li}U_{li} + B_{li,\omega}U_{li,\omega},\\ Y_{li} &= C_{li}X_{li},\end{aligned} \tag{2.50}$$

where $X_{li} = Y_{li} = \Delta i_{li,dq}$, $U_{li} = [\Delta v_{1,dq}\ \Delta v_{2,dq}]^T$, $U_{li,\omega} = \Delta\omega_{com}$, and matrices can be found comparing (2.49) and (2.50). Note that modeling a power network, including a number of lines, is done by interconnecting modules, where they are the lines SSR models.

2.4.5.1 Virtual Resistor Calculation

According to Figures 2.13 and 2.14, MG power network and load models are related to the node voltages. In fact, the grid-side node voltages $v_{n,dq}$ are unknown disturbance inputs in the line and load models, which do not have any correlated equations. This inequality in unknowns and equations causes a lack of solving the set of existing differential equations. Hence, a virtual resistor r_v is assumed at nodes with unknown voltages, which results in complementary relationships for the set of equations. After adding the virtual resistor and applying KCL at node n, the node voltage is obtained as:

$$V_{n,dq} = \sum_{j=1}^{S} R_V i^{j}_{in,dq} - \sum_{l=1}^{M} R_V i^{l}_{out,dq}, \tag{2.51}$$

where $R_V = r_V I_2$ and $i^{j}_{in,dq}$ is the input current of the jth line to the node, and $i^{l}_{out,dq}$ is the output current of the lth line from the node. Note that the virtual resistor must be selected large enough because its current is zero.

If the MG network includes N nodes, L lines, and an arbitrary topology, node voltage vector $V_{N\times 1}$ can be given based on line current vector $I_{L\times 1}$ and virtual resistors as follows:

$$V_{N\times 1} = G_{N\times L} I_{L\times 1}, \tag{2.52}$$

where each row of V (I) is a couple of dq voltage (current) components, i.e. $v_{n,dq}$ for the nth node and $i_{nl,dq}$ for the lth line connected to the nth node. Each g_{nl} is the array of the nth row and the lth column of $G_{N\times L}$, and can be calculated as follows:

$$g_{nl} = \begin{cases} R_V, & \text{if the } l\text{th line current enters to the } n\text{th node.} \\ -R_v, & \text{if the } l\text{th line current exits from the } n\text{th node.} \\ 0_{2\times 2}, & \text{if the } l\text{th line and } n\text{th node do not have a direct connection.} \end{cases} \tag{2.53}$$

DER filter nodes should not be included in (2.52), because they are known based on the given converter voltage. Note that both voltage source and current source converters are usually operated based on voltage-controlled methods, and this fact is also considered in this book. Hence, the converter output voltage and the filter voltage are known. Another important point is a correct test after calculating (2.52) to validate it. The correct test is a zero-sum of arrays of each column of the $G_{N\times L}$ except columns belonging to DER coupling line currents. In these columns, the sum should equal 1 if the current direction is considered from DERs to the network.

Exercise 2.9 Consider the typical network shown in Figure 2.13(b), where the line data is assumed as $R_{li}^{number} = (number/100)R_{\Omega/\text{km}}$, and $L_{li}^{number} = (number/100)L_{\text{H/km}}$, where $R_{\Omega/\text{km}} = 0.164\,\Omega/\text{km}$, and $L_{\text{H/km}} = 0.26\,\text{mH/km}$. G_{li}^{number} denotes a line of the grid including the aforementioned resistance and inductance. Find an SSR for the typical network, and consider node voltage equations in the SSR with $r_v = 10000$. Neglect frequency variations in this example.

Solution

For each line shown in Figure 2.13(b), (2.50) is obtained and the matrices are available [47]. Therefore, the packed forms of the line SSRs exist separately. Before applying RCT functions, the static model of node voltages should be calculated according to (2.52) and (2.53). v_1 and v_4 are DER node voltages, which are not considered in the static model. Therefore, the static model is given as follows:

$$\begin{bmatrix} v_2 \\ v_3 \\ v_5 \\ v_6 \\ v_7 \\ v_8 \end{bmatrix} = r_v \overbrace{\begin{bmatrix} I_2 & -I_2 & -I_2 & 0_2 & 0_2 & 0_2 & 0_2 \\ 0_2 & I_2 & 0_2 & 0_2 & 0_2 & 0_2 & 0_2 \\ 0_2 & 0_2 & I_2 & I_2 & -I_2 & -I_2 & 0_2 \\ 0_2 & 0_2 & 0_2 & 0_2 & I_2 & 0_2 & 0_2 \\ 0_2 & 0_2 & 0_2 & 0_2 & 0_2 & 0_2 & -I_2 \\ 0_2 & 0_2 & 0_2 & 0_2 & 0_2 & I_2 & I_2 \end{bmatrix}}^{G_{netw}} \begin{bmatrix} i_{12} \\ i_{23} \\ i_{25} \\ i_{45} \\ i_{56} \\ i_{58} \\ i_{78} \end{bmatrix}. \tag{2.54}$$

In order to obtain the SSR using the RCT function, the packed form variables of network lines are shown in Figure 2.13(b) and the packed form variable of unknown network voltages is as G_{netw}. The interconnections should be specified as shown in Table 2.5. The complete modeling process and output matrices can be achieved using the corresponding MATLAB code in [47].

2.4.6 Loads

Various load types may be supplied in an MG, which leads to the appearance of different dynamics. Electrical loads can be classified into three groups, i.e. constant impedance loads, constant current loads, and constant power loads (CPLs). Constant impedance loads are a combination of passive elements, which are also called passive loads, e.g. light bulbs, heaters, and direct-connected induction motors. Constant current/power loads are generally a combination of passive elements

Table 2.5 MATLAB code to specify interconnections among module models of the distribution network (the complete version is available in [47]).

```
systemnames =' G_li12 G_li23 G_li25 G_li45 G_li56 G_li58 G_li78 G_netw';
sysoutname =' G_network';
inputvar =' [v12; v42]';
input_to_G_li12 =' [v1; G_netw(1 : 2)]';
input_to_G_li23 =' [G_netw(1 : 2); G_netw(3 : 4)]';
input_to_G_li25 =' [G_netw(1 : 2); G_netw(5 : 6)]';
input_to_G_li45 =' [v4; G_netw(5 : 6)]';
input_to_G_li56 =' [G_netw(5 : 6); G_netw(7 : 8)]';
input_to_G_li58 =' [G_netw(5 : 6); G_netw(11 : 12)]';
input_to_G_li78 =' [G_netw(9 : 10); G_netw(11 : 12)]';
input_to_G_netw =' [G_li12; G_li23; G_li25; G_li45; G_li56; G_li58; G_li78]';
outputvar =' [G_netw]';
sysic;
[A_ntw, B_ntw, C_ntw, D_ntw] = unpck(G_network);
```

and other components like electronic and power-electronic devices, which inherently or compulsorily using controllers leads to a constant current/power. Controlled constant current/power loads are also called active loads [31]. If there are both current and voltage control loops in the cascade, power is controlled [65]. However, CPLs can also be obtained using direct power control methods [66, 67]. DC constant-torque motors are examples of constant current loads. Power-controlled rectifier-inverter-interfaced induction motors and dc loads interfaced with a voltage or current-controlled rectifier are examples of CPLs.

Static models of loads are usually used for planning, power flow, reliability, and load estimation studies [68]. Nevertheless, they are not desirable for stability and control analyses of MGs, where dynamic models are used. Most constant impedance loads can be modeled by a simplified dynamic model including a series RL branch, e.g. heating loads and small uncontrolled induction motors. Furthermore, a simple model of CPLs can also be obtained using the simplified series RL (or static R) branch, which is useful in many stability and control studies [69]. However, detailed dynamic models should be used for some active loads or large uncontrolled motors, where they have a considerable impact on the MG performance and stability.

2.4.6.1 Constant RL Impedance Load

Figure 2.14(a) shows a series RL branch as a single-line diagram of three-phase constant impedance loads for their simplified modeling. Obviously, modeling of constant impedance RL loads is similar to RL line modeling due to the same structures, while variables are only different. Hence, an SSR can be represented as follows:

$$\dot{X}_{lo} = A_{lo}X_{lo} + B_{lo}U_{lo} + B_{lo,\omega}U_{lo,\omega}, \tag{2.55a}$$

$$Y_{lo} = C_{lo}X_{lo}, \tag{2.55b}$$

where $X_{lo} = Y_{lo} = \Delta i_{lo,dq}$, $U_{lo} = \Delta v_{lo,dq}$, and $U_{lo,\omega} = \Delta\omega_{com}$.

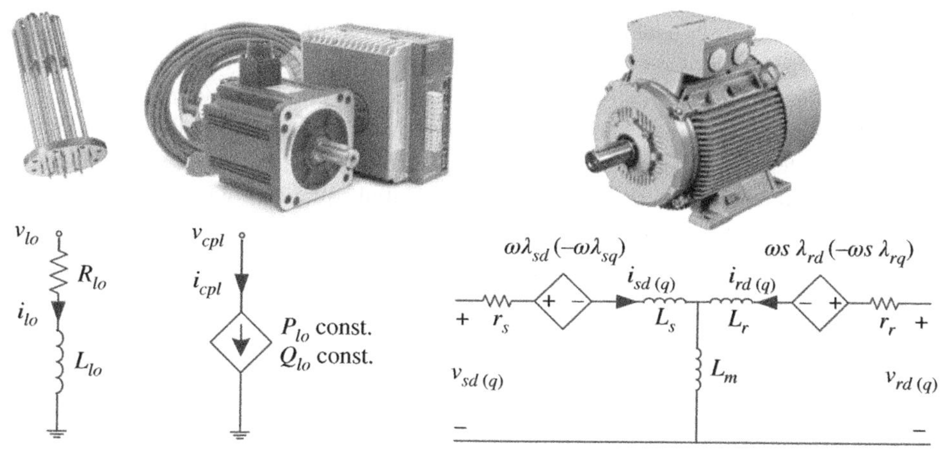

Figure 2.14 Block diagrams for modeling different load types: (a) RL modeling of constant impedance loads, (b) simplified modeling of constant power loads, and (c) detailed modeling of induction motors.

2.4.6.2 Constant Power Load (CPL)

Consider the RL branch shown in Figure 2.14(b) as a single-phase load with variable resistance and inductance. The governing differential equation is given neglecting the rate of change of inductance as follows:

$$v_{lo}(t) = R_{lo}(t)i_{lo}(t) + L_{lo}(t)\frac{di_{lo}}{dt}. \tag{2.56}$$

From the viewpoint of consumption power, constant active and reactive powers, $P_{const.}$ and $Q_{const.}$, can be considered for the CPL, which are related to the aforementioned resistance and inductance as follows:

$$P_{const.} = R_{lo}(t)i_{lo}^2(t), \tag{2.57a}$$

$$Q_{const.} = L_{lo}(t)\omega i_{lo}^2(t). \tag{2.57b}$$

Substituting (2.57a) and (2.57b) in (2.56), then linearizing, the dynamic of the CPL, shown in Figure 2.14(b) as a dependent current source, is given as follows:

$$\frac{d\Delta i_{cpl}}{dt} = \left(\frac{\omega P_{const.}}{Q_{const.}}\right)\Delta i_{cpl} + \left(\frac{\omega i_{cpl,0}^2}{Q_{const.}}\right)\Delta v_{cpl}. \tag{2.58}$$

In most studies of dc MGs, the inductive part of the load is not considered, which can be done by applying $\frac{d\Delta i_{cpl}}{dt} = 0$ in (2.58) as follows:

$$\Delta i_{cpl} = -\left(\frac{P_{const.}}{v_{cpl,0}^2}\right)\Delta v_{cpl}, \tag{2.59}$$

where it shows a negative resistance for CPLs, i.e. $(-v_{cpl,0}^2/P_{const.})$, which results in a poor damping or even instability [70, 71].

2.4.6.3 Motor Load

Figure 2.14(c) shows a dynamic model of a typical symmetrical induction motor in the rotating reference frame [27]. The model is a combination of passive elements and two voltage sources, which presents the induced speed voltages of the motor. The rotor end is shown open. In fact, it depends on the rotor type. For squirrel-cage rotors, the rotor output terminal should be considered as a short-circuit. For outer resistor-connected rotors, the corresponding resistance should be considered at the rotor output terminal in the model. The differential equations of this detailed model can be obtained as follows:

$$\frac{di_{sd}}{dt} = k_1\left(V_{sd} - \left(r_s i_{sd} - \omega_{com} L_s i_{sq} - \omega_{com} L_m i_{rq}\right)\right) + k_1 k_2\left(V_{rd} - \left(r_r i_{rd} - s\omega_{com} L_r i_{rq} - s\omega_{com} L_m i_{sq}\right)\right), \tag{2.60a}$$

$$\frac{di_{sq}}{dt} = k_1\left(V_{sq} - \left(r_s i_{sq} + \omega_{com} L_s i_{sd} + \omega_{com} L_m i_{rd}\right)\right) + k_1 k_2\left(V_{rq} - \left(r_s i_{rq} + s\omega_{com} L_{rr} i_{rd} + s\omega_{com} L_m i_{sd}\right)\right), \tag{2.60b}$$

$$\frac{di_{rd}}{dt} = -k_2\frac{di_{sd}}{dt} + k_3\left(V_{rd} - \left(r_r i_{rd} - s\omega_{com} L_r i_{rq} - s\omega_{com} L_m i_{sq}\right)\right), \tag{2.60c}$$

$$\frac{di_{rq}}{dt} = -k_2\frac{di_{sq}}{dt} + k_3\left(V_{rq} - \left(r_s i_{rq} + s\omega_{com} L_r i_{rd} + s\omega_{com} L_m i_{sd}\right)\right), \tag{2.60d}$$

$$\frac{s}{dt} = (1/J\omega_{com})(T_{L,im} - T_{e,im}), \tag{2.60e}$$

where $k_1 = 1/(L_s - L_m^2/L_r)$, $k_2 = L_m/L_r$, $k_3 = 1/L_r$. r_s, L_s, r_r, L_r, and L_m, are stator resistance/inductance, rotor resistance/inductance, and magnetizing inductance of the induction motor. P is the number of poles and J is the moment of inertia of motor-load coupling. T_L is the load torque, and T_e is the electrical torque of the induction motor calculated as follows:

$$T_e = \left(\frac{3}{2}\right)\left(\frac{P}{2}\right) L_m \left(i_{sq} i_{rd} - i_{sd} i_{rq}\right). \tag{2.61}$$

Flux linkage components, shown in Figure 2.14(c), are functions of the inductance and current products, which are considered in (2.60) and can be found in chapter 6 of [27]. One can find an SSR of the detailed model of the induction motor after linearizing (2.60) and considering state and input vectors as $X_{im} = \left[\Delta i_{s,dq}\ \Delta i_{r,dq}\ s\right]^T$, and $U_{im} = \left[\Delta v_{s,dq}\ \Delta v_{r,dq}\ \Delta\omega_{com}\ \Delta T_L\right]^T$. More details on the modeling of motor loads can be found in [27].

2.4.6.4 Active Load

Figure 2.15 shows an active load including a dc load, a rectifier, a dc voltage control, a current controller, a phase-locked loop (PLL), an LC filter, and a coupling RL line. Figure 2.16 shows the dc voltage and current controllers. Simple PI controllers are considered for current and voltage controllers [31]. The studied active load consists of a number of modules, which are previously modeled, and the corresponding SSRs are obtained except the PLL. The dc voltage PI controller is shown in Figure 2.4(b), but without considering the feed-forward input, and assuming the first input v_1 as the perturbed form of the negative dc voltage feedback of the rectifier, i.e. $-\Delta v_{dc}$. The structure of the dq current controllers is similar to that of dq voltage controllers in Figure 2.6(a) including two PI controllers and the mutual terms. However, the mutual terms, gains, and signals are different, which should be considered in finding vectors and matrices of the SSR. Therefore, modeling of the current controller can easily be done similarly to (2.14) with different values and

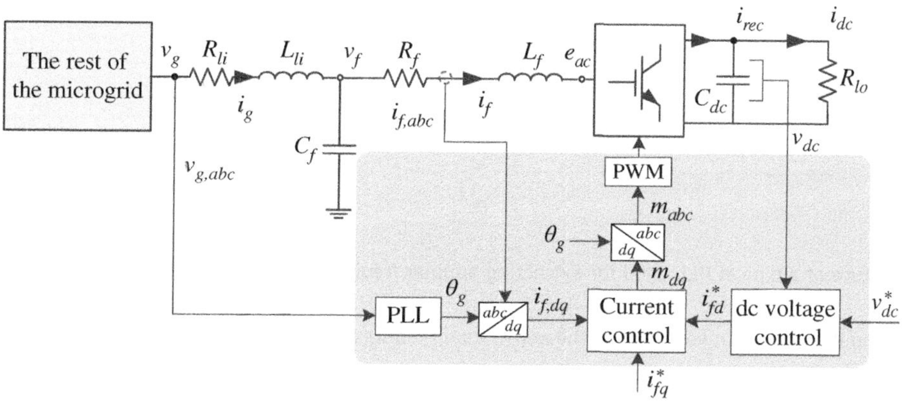

Figure 2.15 Block diagram of an active load connected to an MG including a rectifier controlled by a dc voltage controller and a current controller.

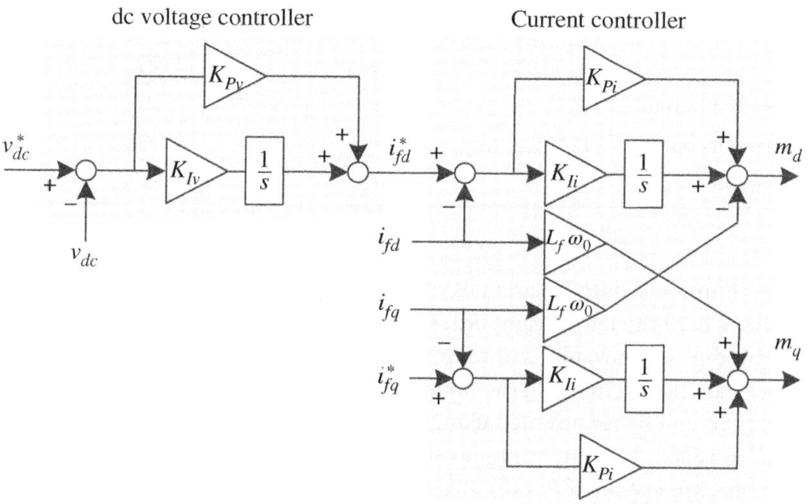

Figure 2.16 Block diagram of dc voltage and current PI controllers of the active load shown in Figure 2.15.

variables. The ac side of the rectifier comprises an LC filter and a coupling RL line, and their modeling is addressed in Sections 2.4.4 and 2.4.5. The dc load is a constant resistance load paralleled with a capacitor to regulate the dc voltage. Its modeling is easy according to the rules in Section 2.4.6.1 and Exercise 2.1. The PLL is the only module that has not been modeled already. Figure 2.17 shows a schematic of a simple-structure PLL including two dynamics of the integrators used for the PI controller and phase angle producer [72]. Hence, an SSR can be given as follows:

$$\begin{aligned} \dot{X}_{pll} &= A_{pll}X_{pll} + B_{pll}U_{pll} + B_{pll,com}U_{pll,com}, \\ Y_{pll} &= C_{pll}X_{pll} + D_{pll}U_{pll}, \end{aligned} \tag{2.62}$$

where $U_{pll} = \Delta v_{g,dq}$ $U_{pll,com} = \Delta\omega_{com}$, $Y_{pll} = [\Delta\omega_g\ \Delta\delta_g]$, $C_{pll} = I_{2\times 2}$ and the rest of matrices are as follows:

$$A_{pll} = \begin{bmatrix} 0 & 0 \\ 1 & 0 \end{bmatrix}, \quad B_{pll} = \begin{bmatrix} 0 & K_{I,pll} \\ 0 & K_{p,pll} \end{bmatrix}, \quad B_{pll,com} = \begin{bmatrix} 0 \\ -1 \end{bmatrix}, \quad D_{pll} = \begin{bmatrix} 0 & K_{p,pll} \\ 0 & 0 \end{bmatrix}.$$

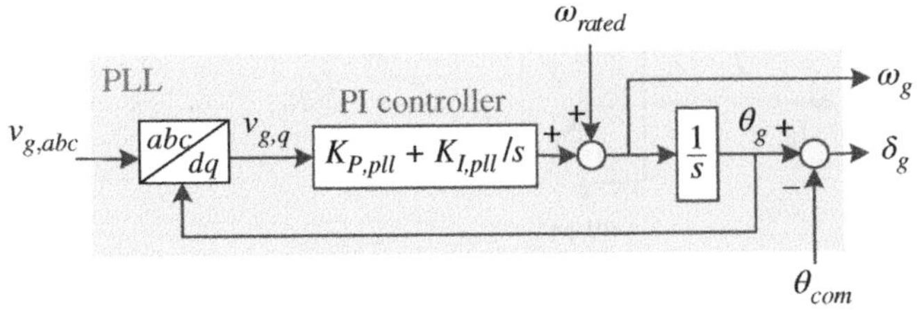

Figure 2.17 Structure of a typical PLL used for extracting angular frequency and correlated phase angle.

Table 2.6 State, input, and output variables of the active load modules shown in Figure 2.15.

Module	State variables (X)	Input variables (U)	Output variables (Y)	Packed form (G)
RL line	$i_{g,dq}$	$v_{g,dq}$, $v_{f,dq}$	$= X$	G_{li}
LC filter	$i_{f,dq}$, $v_{f,dq}$	$i_{g,dq}$, e_{dq}	$= X$	G_{lc}
Rectifier	—	v_{dc}, m_{dq}, $i_{f,dq}$, e_{dq}	i_{rec}, e_{dq}	G_{rec}
dc load	v_{dc}	i_{rec}	$= X$	G_{dc}
dc voltage controller	Integrator output	v_{dc}	i^*_{fd}	$G_{v,dc}$
Current controller	Integrators output	$i^*_{f,dq}$, $i_{f,dq}$	m_{dq}	G_{cc}
PLL	Integrator output	$v_{g,dq}$, ω_{com}	θ_g	G_{pll}

Table 2.6 shows the state, input, and output variables of the active load modules as well as the packed form variable of SSRs, where the left-to-right order shows the exact order of the equivalent column vectors. Since all three-phase variables are represented in their *dq* form, Park and Park inverse transformation blocks are not included in the table. In order to find a unified model of the active load, the interconnections can be represented using RCT functions as shown in Table 2.7.

Finally, *A_AL* in Table 2.7 consists of 12 state variables shown in the second column of Table 2.6. The output and input variables are specified in *outputvar* and *inputvar*. The complete MATLAB code for the modeling process of the sample active load depicted in Figure 2.15 is available in [47] to facilitate further practice and model development.

2.4.7 Energy Resources and Storages

In this section, common structures of a wind generation unit, a PV generation unit, a battery energy storage system, a hybrid PV-battery system, and a super-capacitor (SC) energy storage are modeled using block diagrams, SSRs, and interconnecting modules.

2.4.7.1 Wind Generation Unit

The main components of a wind generation unit consist of a wind turbine, a gearbox, an electrical generator, power electronics interfaces, filters, and control loops. In each component, different structures and several types can be used. Generators can be dc generators, ac synchronous generators, e.g. permanent-magnet synchronous generators, or ac induction generators, e.g. doubly fed induction generators and squirrel cage induction generators [73]. Power electronics converters may be used for all or a portion of the wind generation. Moreover, several topologies can be considered

Table 2.7 MATLAB code to specify interconnections among module models of the active load.

```
systemnames =' G_li G_lc G_rec G_dc G_vdc G_cc G_pll';
sysoutname =' G_AL';
inputvar =' [v_g2; v_dc_ref; i_fq_ref; w_com]';
input_to_G_li =' [V_g; G_lc(3 : 4)]';
input_to_G_lc =' [G_li; G_rec(2 : 3)]';
input_to_G_rec =' [G_dc; G_cc; G_lc(1 : 2); G_rec(2 : 3)]';
input_to_G_dc =' [G_rec(1)]';
input_to_G_vdc =' [v_dc_ref; G_dc]';
input_to_G_cc =' [G_vdc; i_fq_ref; G_lc(1 : 2)]';
input_to_G_PLL =' [V_g; w_com]';
outputvar =' [G_rec(1); G_dc]';
sysic;
[A_AL, B_AL, C_AL, D_AL] = unpck(G_AL);
```

according to rotor axis orientation and speed, control objectives, and drive trains [73]. In this book, the goal is not to study these types, topologies, and structures to compare. However, the objective is to apply the modeling method to the common components of MGs. Hence, a typical structure of wind generation units is considered in Figure 2.18, where most main components are included in a general form.

The main input is wind features like wind speed v_{wind}, which leads to generating mechanical torque T_{wt}. Then, the torque is changed by the gearbox, transformed to the electrical torque T_g, and injected into the grid as the active power P_{grid}. On the other hand, the maximum power point (MPP) is tracked using the wind features and maximum power point tracking (MPPT) procedure. In renewable generations like wind generation units, the control objective is usually power/current control, which is done in Figure 2.18 by the equivalent torque control on the rectifier, i.e. applying m_{rec}, and the grid-injected power/current control on the inverter, i.e. applying m_{inv}. Moreover, a dc voltage control should be cascaded with the inverter current control to stabilize the dc voltage v_{dc} and in turn to stabilize all power exchanges [25].

Most modules shown in Figure 2.18 are previously modeled including the rectifier, the inverter, the LC filter, the current and dc voltage controllers, and the PLL. However, other modules should be modeled before applying interconnections to find the overall model.

Wind Turbine The produced torque T_{wt} and angular frequency ω_{wt} of the wind turbine are related through the swing equation as follows:

$$J_{sh}\frac{d\omega_{wt}}{dt} = T_{wt} - T_{gb} - D_{f,sh}\omega_{wt}, \tag{2.63}$$

where J_{sh} and $D_{f,sh}$ are the moment of inertia and friction coefficient of the turbine shaft, T_{gb} is the gearbox torque, and T_{wt} can be calculated from the wind and turbine features as follows:

$$T_{wt}(t) = \frac{1}{2}C_T(\lambda, \beta)\rho S_b R_b v_{wind}^2(t), \tag{2.64}$$

where C_T is the torque coefficient, which is a nonlinear function of the blade pitch angle β and the tip speed ratio λ [74]. ρ is the air density, and R_b and S_b are the blade radius and the area swept by

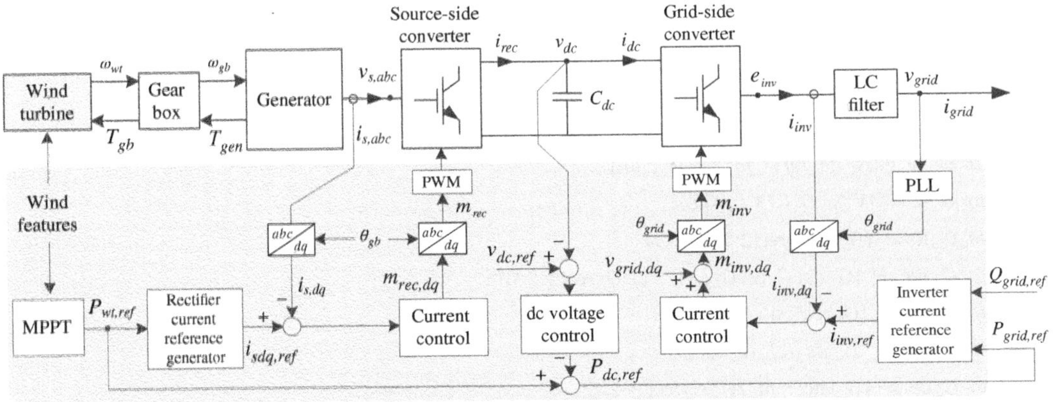

Figure 2.18 Block diagram of a wind generation unit including mechanical, electrical, and control subsystems.

the blades. Here, a constant pitch angle and a constant operating point are considered for the wind turbine. After linearization, an SSR is given as follows:

$$\Delta\dot{\omega}_{wt}(t) = (D_{f,sh}/J_{sh})\Delta\omega_{wt}(t) + \left[K_{wt}/J_{sh} \quad -1/J_{sh}\right] U_{wt}(t), \tag{2.65a}$$

$$Y = \Delta\omega_{wt}(t), \tag{2.65b}$$

where $U_{wt}(t) = [\Delta v_{wind} \ \Delta T_{gb}]^T$, and $K_{wt} = C_T(\lambda_0, \beta_0)\rho_0 S_b R_b v_{wind,0}$.

Gear Box The torque and the angular frequency can be increased/decreased with the inverse gain m_{gb} from the input terminal to the output terminal of the gear box as follows:

$$\omega_{gen}(t) = m_{gb}\omega_{wt}(t), \tag{2.66a}$$

$$T_{gb}(t) = m_{gb}T_{gen}(t), \tag{2.66b}$$

where ω_{gen} and T_{gen} are the output angular frequency and input torque to/from the generator, and ω_{wt} and T_{gb} are the input angular frequency and output torque from/to the wind turbine.

Synchronous Generator Considering a fixed rotor flux such as permanent magnet synchronous generators, and representing the dynamics in the rotating reference frame, one can show the governing equations on the generator as:

$$\frac{di_{sd}}{dt} = -\frac{R_s}{L_s}i_{sd}(t) + \omega_{gen}(t)i_{sq}(t) + \frac{1}{L_s}v_{sd}(t), \tag{2.67a}$$

$$\frac{di_{sq}}{dt} = -\omega_{gen}(t)i_{sd}(t) - \frac{R_s}{L_s}i_{sq}(t) + \frac{1}{L_s}v_{sq}(t), \tag{2.67b}$$

$$T_{gen}(t) = \frac{3}{2}p\phi_{rd}i_{sq}, \tag{2.67c}$$

where R_s and L_s are the stator resistance and inductance, i_s and v_s are the stator current and terminal voltage, p is the number of generator pole pairs, and ϕ_{rd} is the constant rotor flux. Therefore, an SSR can be obtained after linearizing (2.67) as follows:

$$\dot{X}_{gen}(t) = A_{gen}X_{gen}(t) + B_{gen,s}U_{gen,s}(t) + B_{gen,\omega}U_{gen,\omega}(t), \tag{2.68a}$$

$$Y_{gen}(t) = C_{gen}X_{gen}(t), \tag{2.68b}$$

where $X_{gen} = \Delta i_{s,dq}$, $U_{gen,s} = \Delta v_{s,dq}$, $U_{gen,\omega} = \Delta\omega_{gen}$, $Y_{gen} = [\Delta i_{s,dq}\ \Delta T_{gen}]^T$, and the matrices are as follows:

$$A_{gen} = \begin{bmatrix} -R_s/L_s & \omega_{gen,0} \\ -\omega_{gen,0} & -R_s/L_s \end{bmatrix}, \quad B_{gen,s} = \begin{bmatrix} 1/L_s & 0 \\ 0 & 1/L_s \end{bmatrix}, \quad B_{gen,\omega} = \begin{bmatrix} i_{sq0} \\ -i_{sd0} \end{bmatrix},$$

$$C_{gen} = \begin{bmatrix} 1 & 0 \\ 0 & 1 \\ 0 & 3p\phi_{rd}/2 \end{bmatrix}. \tag{2.69}$$

DC Link The capacitor voltage v_{dc} dynamics can be expressed in an SSR as follows:

$$\dot{v}_{dc} = (0)v_{dc} + \begin{bmatrix}1/C_{dc} & -1/C_{dc}\end{bmatrix} \begin{bmatrix}i_{rec} & i_{dc}\end{bmatrix}^T, \tag{2.70a}$$

$$y_{dc} = v_{dc}, \tag{2.70b}$$

where C_{dc} is the dc link capacitor, and i_{rec} and i_{dc} are the input and output currents to/from the dc link, which are the inputs of the SSR.

Rectifier Current Reference Generator Note that the MPPT block is considered as a static block with a constant output $P_{wt,ref}$ in the studied time spans. The equivalent torque reference $T_{gb,ref}$ can easily be found as:

$$T_{gb,ref} = P_{wt,ref}/\omega_{wt}, \tag{2.71}$$

where ω_{wt} should be measured from the turbine rotor to have a precise torque reference based on the operating angular frequency. However, it can be obtained according to the nominal angular frequency $\omega_{wt,n}$ without measurement of ω_{wt}. In order to find the reference of the rectifier current, which equals the stator current, the generator-side torque reference $T_{gen,ref}$ is required. It is given as follows:

$$T_{gen,ref} = T_{gb,ref}/m_{gb}. \tag{2.72}$$

Therefore, reference of the quadrature current component can be achieved as:

$$i_{sq,ref} = 2T_{gen,ref}/(3p\phi_{rd}), \tag{2.73}$$

and the direct current component reference can be assumed as zero, i.e. $i_{sd,ref} = 0$. Note that $\Delta i_{sdq,ref} = 0_{2\times1}$ if $P_{wt,ref}$ is considered constant. In other words, the rectifier's current reference generator does not participate in the small-signal model under the mentioned assumptions. The dc voltage reference $v_{dc,ref}$ has the same situation in modeling the dc voltage controller.

Inverter's Current Reference Generator Since the inverter's current reference generator has an input from the dc voltage control output, i.e. $P_{dc,ref}$, it has a considerable role in the small-signal modeling, unlike the rectifier's current reference generator. The inverter's current references are given as follows [61]:

$$i_{inv,ref} = (2/3V_n) \begin{bmatrix} 1 & 0 \\ 0 & -1 \end{bmatrix} \begin{bmatrix}P_{grid,ref} & Q_{grid,ref}\end{bmatrix}^T, \tag{2.74}$$

where V_n is the nominal peak voltage, $Q_{grid,ref}$ is the arbitrary reference of the output reactive power and $P_{grid,ref}$ is the output active power obtained by $P_{grid,ref} = P_{wt,ref} - P_{dc,ref}$. However, the references must be selected such that the apparent power is smaller than the inverter power rating.

Table 2.8 State, input, and output variables of the wind generation unit modules.

Module	State variables (X)	Input variables (U)	Output variables (Y)	Packed form (G)
Wind turbine	ω_{wt}	v_{wind}, T_{gb}	$= X$	G_{wt}
Gearbox	—	ω_{wt}, T_{gen}	ω_{gen}, T_{gb}	G_{gb}
Generator	$i_{s,dq}$	$v_{s,dq}, \omega_{gen}$	$i_{s,dq}, T_{gen}$	G_{gen}
Rectifier	—	$v_{dc}, m_{rec,dq}, i_{s,dq}, v_{s,dq}$	$i_{rec}, v_{s,dq}$	G_{rec}
Dc link	v_{dc}	i_{rec}, i_{dc}	$= X$	G_{dc}
Inverter	—	$v_{dc}, m_{inv,dq}, i_{inv,dq}, e_{inv,dq}$	$i_{dc}, e_{inv,dq}$	G_{inv}
Filter	$i_{inv,dq}, v_{grid,dq}$	$e_{inv,dq}, i_{grid,dq}$	$= X$	G_{lc}
Rectifier control	Integrator outputs	$i_{sdq,ref}, i_{s,dq}$	$m_{rec,dq}$	G_{rc}
Inverter's current ref. generator	—	$P_{wt,ref}, P_{dc,ref}, Q_{grid,ref}$	$i_{inv,ref}$	G_{ig}
Inverter control	Integrator outputs	$i_{inv,ref}, i_{inv,dq}, v_{grid,dq}$	$m_{inv,dq}$	G_{ic}
Voltage control	Integrator output	$v_{dc,ref}, v_{dc}$	$P_{dc,ref}$	$G_{v,dc}$
PLL	Integrator outputs	$v_{grid,dq}, \omega_{com}$	θ_{grid}	G_{pll}

Overall Model of the Wind Generation Unit Table 2.8 shows the state, input, and output variables of all modules of the wind turbine, where the left-to-right order shows the exact order of the equivalent column vectors. Moreover, the packed SSR variables can be seen in the last column. Hence, an overall model of the wind generation unit can be obtained using the RCT functions as shown in Table 2.9.

The complete small-signal modeling of the wind generation unit in MATLAB is provided in [47], where the SSR of all modules and interconnections among them are represented. This small-signal model is closed-loop, which is useful for small-signal stability analysis. It can also be changed to an open-loop model to design each controller.

2.4.7.2 Photovoltaic Generation Unit

The diversity mentioned for wind generation systems also exists for PV generation units. Nevertheless, a general form of PV generation is considered to show how it can be modeled using the analytical–numerical method. Figure 2.19 shows the PV generation unit including a PV module, a dc–dc chopper, a dc link, an inverter, LC filters, and the basic control loops. In PV generation units, MPPT can also be used, which works based on the solar irradiance data. This data is also used for modeling the PV module. In general, four control loops are used to regulate PV output voltage v_{pv}, to track the MPP by regulating PV output current i_f, to stabilize dc link voltage v_{dc}, and to control output ac current. The PV voltage and current controllers are cascaded and their actuator is the dc–dc chopper. The dc voltage and output inverter current controllers are also cascaded. The PV voltage and current controllers as well as the dc voltage controller are of SISO PI controllers with voltage/current feed-forward (see Figure 2.4(b)), and the output inverter current controller is of two-axis current controllers with mutual terms and grid voltage feed-forward (see Figure 2.6(a)). The module models are available in the previous sections except for the models of the PV module, the single-phase LC filter, and virtual resistors r_{vr1} and r_{vr2}.

Table 2.9 MATLAB code to specify interconnections among module models of the wind generation unit.

```
systemnames =' G_wt G_gb G_gen G_rec G_dc G_inv G_lc G_rc G_ig G_ic G_vdc G_pll';
sysoutname =' G_WGU';
inputvar =' [v_wind; P_wt_ref; i_sdq_ref{2}; v_dc_ref; Q_grid_ref; i_grid_dq{2}; w_com]';
input_to_G_wt =' [v_wind; G_gb(2)]';
input_to_G_gb =' [G_wt; G_gen(3)]';
input_to_G_gen =' [G_rec(2 : 3); G_gb(1)]';
input_to_G_rec =' [G_dc; G_rc; G_gen(1 : 2); G_rec(2 : 3)]';
input_to_G_dc =' [G_rec(1); G_inv(1)]';
input_to_G_inv =' [G_dc; G_ic; G_lc(1 : 2); G_inv(2 : 3)]';
input_to_G_lc =' [G_inv(2 : 3); i_grid_dq]';
input_to_G_rc =' [i_sdq_ref; G_gen(1 : 2)]';
input_to_G_ig =' [P_wt_ref - G_vdc; Q_grid_ref]';
input_to_G_ic =' [G_ig; G_lc]';
input_to_G_vdc =' [v_dc_ref; G_dc]';
input_to_G_pll =' [G_lc(3 : 4); w_com]';
outputvar =' [G_lc]';
sysic;
[A_WGU, B_WGU, C_WGU, D_WGU] = unpck(G_WGU);
```

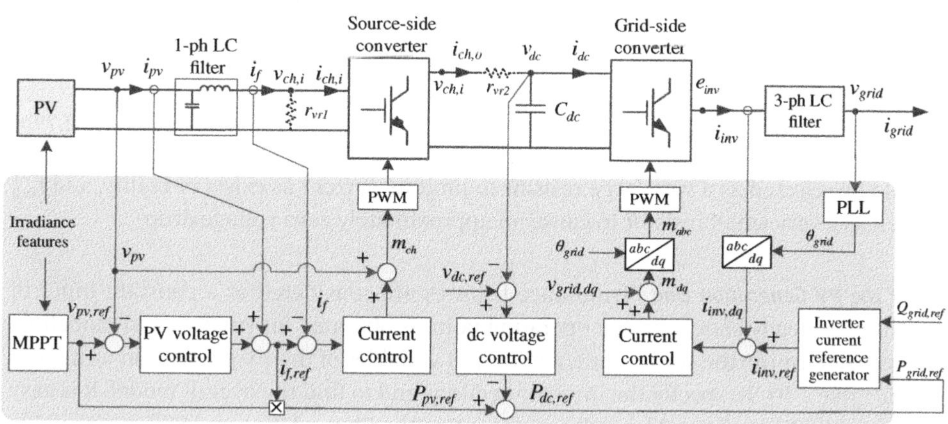

Figure 2.19 Block diagram of a PV generation unit including dc and ac parts, interface converters, and the basic control loops.

PV Modeling The PV can be considered as a PV array including N_P parallel strings and N_c panels in each string. The PV current i_{pv} can be represented based on the PV terminal voltage v_{pv}, the solar irradiance features, and the solar module data as follows:

$$i_{pv}(t) = N_P I_{ph} - N_P I_{rs}\left(e^{\frac{q v_{pv}(t)}{k T_T A_{IF} N_S N_C}} - 1\right), \tag{2.75}$$

where I_{ph} is the short-circuit current of each string (A), I_{rs} is the diode inverse saturation current (A), q is electron charge (1.6×10^{-19} C), k is the Boltzmann constant (1.38×10^{-23} (J/K)), T_T is the p-n junction temperature (°K), A_{IF} is the diode factor, N_s is the number of series cells in each panel. After linearization, the following static model can be obtained for the PV array:

$$\Delta i_{pv}(t) = G_{pv} \Delta v_{pv}(t), \tag{2.76}$$

where

$$G_{pv} = -\frac{qN_P I_{rs}}{kT_T A_{IF} N_s N_c} e^{\frac{qv_{pv0}}{kT_T A_{IF} N_S N_C}}. \tag{2.77}$$

Single-Phase LC Filter The single-phase dc LC filter modeling is similar to the three-phase ac LC filter modeling except in the Park transformation. Furthermore, there are just two state variables, i.e. i_f and v_{pv}. Therefore, the state and input matrices can be found as follows:

$$A_{fpv} = \begin{bmatrix} 0 & -1/c_{fpv} \\ 1/L_{fpv} & 0 \end{bmatrix}, \quad B_{fpv} = \begin{bmatrix} 1/c_{fpv} & 0 \\ 0 & -1/L_{fpv} \end{bmatrix}. \tag{2.78}$$

Virtual Resistors Inputs and outputs of two correlated modules are matched when the input of each module can be obtained from the output of another module. However, they are not matched when a common variable of them is the input of both modules. There are two examples in the PV generation unit. The voltage between the PV LC filter and the boost chopper, namely $v_{ch,i}$ is a common input of both modules. On the dc link side, a similar situation exists for the current $i_{ch,o}$, which is the common input of the boost chopper and the dc link capacitor.

The solution for modeling these modules is to use a virtual resistor as explained in Section 2.4.5.1. Here, for the two unmatched couples, two virtual resistors are considered as follows:

$$v_{ch,i} = \begin{bmatrix} r_{vr1} & -r_{vr1} \end{bmatrix} \begin{bmatrix} i_f & i_{ch,i} \end{bmatrix}^T, \tag{2.79a}$$

$$i_{ch,o} = \begin{bmatrix} \frac{1}{r_{vr2}} & -\frac{1}{r_{vr2}} \end{bmatrix} \begin{bmatrix} v_{ch,o} & v_{dc} \end{bmatrix}^T, \tag{2.79b}$$

where r_{vr1} should be selected as a very large resistor to limit its current as exists in reality, and r_{vr2} should be chosen as a very small resistor to cause an approximately zero voltage drop.

Overall Model of the PV Generation Unit Irradiance features are considered as a constant input of the system during the steady state, and the operating point is obtained for small-signal modeling. Therefore, one can determine the state, input, and output variables of the PV generation unit modules as shown in Table 2.10. To specify the interconnections and to find the overall model, it is easy to use Table 2.10 and RCT functions. The codes representing the PV modeling and SSR are available in [47] including individual module modeling, and interconnections of modules.

2.4.7.3 Battery

The ESSs are important players in modern power systems and microgrids to improve the intermittent output power of RESs, represent ancillary services in different layers of the power system, and undertake a backup role, e.g. in electric vehicle charging stations. Although many types of ESSs include flywheels, fuel cells, hydrogen storages, several battery storages, super-capacitor, and super-conducting magnetic energy storage, here a BESS and an SC are just considered to be modeled, which are applicable and enough to show the modeling method. The modeling approach

Table 2.10 State, input, and output variables of the PV generation unit modules.

Module	State variables (X)	Input variables (U)	Output variables (Y)	Packed form (G)
PV array	—	v_{pv}	i_{pv}	G_{pv}
PV LC filter	i_f, v_{pv}	i_{pv}, $v_{ch,i}$	$= X$	G_{fpv}
Vir. resistor 1	—	i_f, $i_{ch,i}$	$v_{ch,i}$	G_{vr1}
Boost chopper	$i_{ch,i}$, $v_{ch,o}$	$i_{ch,i}$, $v_{ch,o}$, D_{ch}	$= X$	G_{ch}
Vir. resistor 2	—	$v_{ch,o}$, v_{dc}	$i_{ch,o}$	G_{vr2}
Dc link	v_{dc}	$i_{ch,o}$, i_{dc}	$= X$	G_{dc}
Inverter	—	v_{dc}, $m_{inv,dq}$, $i_{inv,dq}$, $e_{inv,dq}$	i_{dc}, $e_{inv,dq}$	G_{inv}
Filter	$i_{inv,dq}$, $v_{grid,dq}$	$e_{inv,dq}$, $i_{grid,dq}$	$= X$	G_{lc}
PV voltage control	Integrator output	$v_{pv,ref}$, v_{pv}, i_{pv}	$i_{f,ref}$	G_{pvc}
Chopper current control	Integrator outputs	$i_{f,ref}$, i_f, v_{pv}	D_{ch}	G_{chc}
Inverter current ref. generator	—	$Q_{grid,ref}$, $P_{pv,ref}$, $P_{dc,ref}$	$i_{inv,ref}$	G_{ig}
Inverter control	Integrator outputs	$i_{inv,ref}$, $i_{inv,dq}$, $v_{grid,dq}$	$m_{inv,dq}$	G_{ic}
Voltage control	Integrator output	$v_{dc,ref}$, v_{dc}	$P_{dc,ref}$	$G_{v,dc}$
PLL	Integrator outputs	$v_{grid,dq}$, ω_{com}	θ_g	G_{pll}

can be applied to other ESSs, similarly. Modeling of different ESSs and their challenges have been represented in [75].

Figure 2.20 shows the block diagram of a hybrid PV-battery generation unit, where the sources are connected to the dc link. The light-gray parts are modeled and discussed in Section 2.4.7.2 including the PV module, its dc–dc converter, and the cascaded controllers as well as dc link, ac side, VSC, and the corresponding controllers. The BESS consists of a battery, a bi-directional dc–dc converter, and a current controller to regulate output current/power. The current controller can be considered as a PI controller, which its modeling has already been presented.

Battery Modeling A BESS requires a bi-directional power converter to have the ability to be charged and discharged [76–78]. This storage system has several constraints such as minimum/maximum state of charge (SoC), minimum/maximum rate of charge/discharge, and state of health [79–81]. However, they can usually be considered using nonlinear components which are not of interest in linear modeling. Figure 2.21 shows a generic model of a battery considering the dynamic of the SoC, which in turn is a function of the output current i_{bat} as follows:

$$SoC = SoC_0 - \frac{x_{bat}}{Q}, \tag{2.80}$$

where Q is the battery capacity, SoC_0 is the initial state of battery charge, and x_{bat} is the actual battery charge, which is the battery model state variable. The internal battery voltage E_{bat} is obtained as a function of the SoC as follows [13, 82]:

$$E_{bat} = E_{bat,0} - \frac{K}{SoC(t)} + Ae^{-BQ(1-SoC(t))}, \tag{2.81}$$

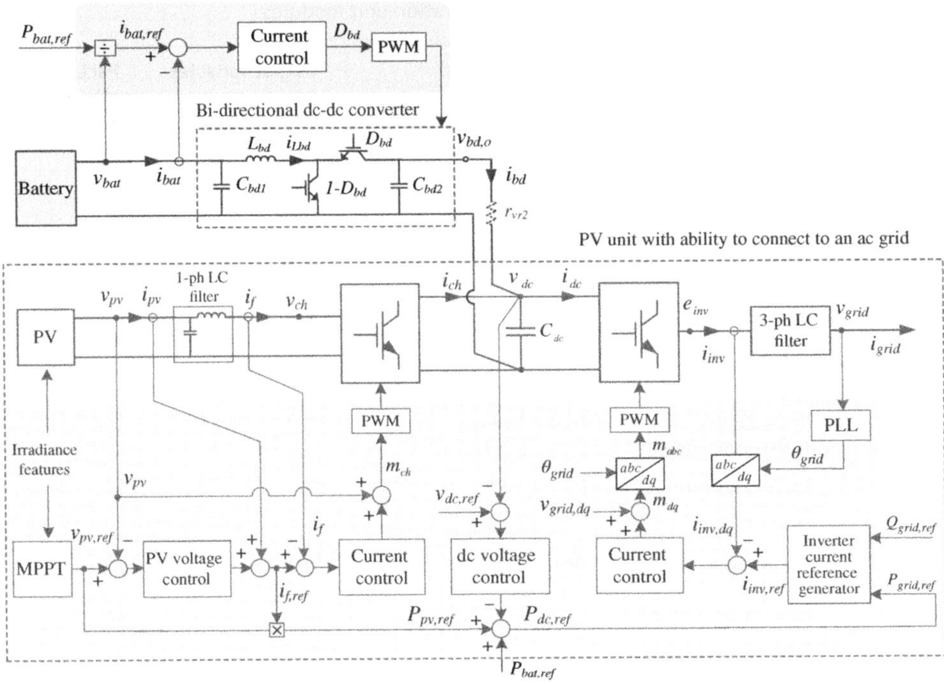

Figure 2.20 Block diagram of a hybrid PV-battery generation unit with a focus on the battery, its bi-directional converter, and the controller.

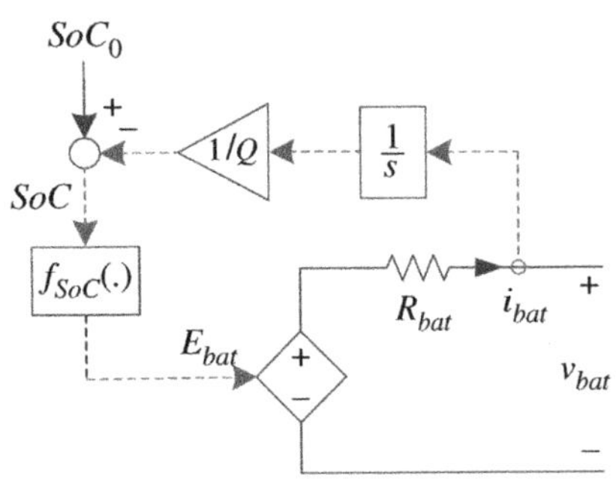

Figure 2.21 Schematic of a generic battery model.

where $E_{bat,0}$ is the battery constant voltage, K is the polarization voltage, and A and B are the parameters determining the battery charge and discharge characteristics. One can find an SSR for the battery as follows:

$$\dot{x}_{bat} = a_{bat}x_{bat} + b_{bat}u_{bat}, \tag{2.82a}$$

$$y_{bat} = a_{bat}x_{bat} + b_{bat}u_{bat}, \tag{2.82b}$$

where $u_{bat} = \Delta v_{bat}, y_{bat} = \Delta i_{bat}, b_{bat} = -1/R_{bat}$ and

$$a_{bat} = \frac{-1}{QR_{bat}}\left(\frac{K}{SoC_0^2} + ABQe^{-BQ(1-SoC_0)}\right), \tag{2.83}$$

Bi-directional dc–dc Converter Modeling As shown in Figure 2.20, the bi-directional dc–dc converter comprises two controllable static switches, two capacitors to reduce voltage ripples on both sides and an inductor to reduce current ripples. Therefore, considering three state variables, assuming D_{bd} as the duty cycle of the horizontal switch, and complementary considering $(1 - D_{bd})$ as the duty cycle of the vertical switch, one can find an SSR for the converter. Note that a procedure similar to that shown in Section 2.4.3 is required. In fact, two SSRs, i.e. SSR_D and SSR_{1-D}, are obtained for the periods of D_{bd} and $(1 - D_{bd})$ and then they are averaged as follows:

$$SSR_{bd} = D_{bd} \cdot SSR_D + (1 - D_{bd}) \cdot SSR_{1-D}. \tag{2.84}$$

Hence, the SSR_{bd} is given as follows:

$$\dot{X}_{bd} = \left[D_{bd}A_{bd1} + (1 - D_{bd})A_{bd2}\right] X_{bd} + \tilde{B}_{bd}\tilde{U}_{bd}, \tag{2.85a}$$

$$Y_{bd} = X_{bd}, \tag{2.85b}$$

where $X_{bd} = [v_{bat}\ v_{dc}\ i_{Lbd}]^T$ and $\tilde{U}_{bd} = [i_{bat}\ i_{bd}]^T$. Note that (2.85) is a nonlinear SSR due to the multiplication terms of $D_{bd}(t)$ and $X_{bd}(t)$. Hence, one can find the linearized SSR as follows:

$$\begin{aligned} \dot{X}_{bd} &= \left[D_{bd,0}A_{bd1} + (1 - D_{bd,0})A_{bd2}\right] X_{bd} + \left[B_{bd}\ (A_{bd1} - A_{bd2})X_{bd,0}\right] U_{bd}, \\ Y_{bd} &= X_{bd}, \end{aligned} \tag{2.86}$$

where $U_{bd} = [\tilde{U}_{bd}\ D_{bd}]^T$, and

$$A_{bd1} = \begin{bmatrix} 0 & 0 & \frac{-1}{C_{bd1}} \\ 0 & 0 & \frac{-1}{C_{bd2}} \\ \frac{1}{L_{bd}} & \frac{-1}{L_{bd}} & 0 \end{bmatrix}, \quad A_{bd2} = \begin{bmatrix} 0 & 0 & \frac{-1}{C_{bd1}} \\ 0 & 0 & 0 \\ \frac{1}{L_{bd}} & 0 & 0 \end{bmatrix}, \quad B_{bd} = \begin{bmatrix} \frac{1}{C_{bd1}} & 0 \\ 0 & \frac{-1}{C_{bd2}} \\ 0 & 0 \end{bmatrix}.$$

Current Reference Generator Modeling As shown in Figure 2.20, it is obvious that the current reference $i_{bat,ref}$ is obtained through a nonlinear relationship, which requires a linearization. Hence, a static equation representing the model of the current reference generator is given as follows:

$$\Delta i_{bat,ref} = -\frac{P_{bat,ref}}{v_{bat,0}^2} \Delta v_{bat}. \tag{2.87}$$

Overall PV-Battery Model State, input, and output variables of a number of modules of the hybrid PV-battery generation unit have already been shown in Table 2.10. Furthermore, Table 2.11 shows the input, output, and state variables of the individual battery components including the battery, the bi-directional dc–dc converter, the virtual resistor, the current reference generator, and the current controller. Therefore, one can specify the interconnections and find the overall model, using the data represented in Tables 2.10 and 2.11 and RCT functions in MATLAB. The codes representing the overall PV-battery generation unit SSR are available in [47].

2.4.7.4 Super-Capacitor

Figure 2.22 shows an SC ESS including an SC, a bi-directional dc–dc converter without the capacitors shown in Figure 2.20, a voltage controller, and a current controller. Note that the SC voltage is charged using the grid energy and the voltage controller, is in charge of this action after each discharging. In fact, during sudden changes in the dc grid, the SC tries to inject instantaneous powers very fast. After the transient disturbance, the voltage controller immediately tries to charge the SC. The current controller is also used to regulate the output current within its limits. Voltage and current controllers are already modeled.

Table 2.11 State, input, and output variables of the battery modules.

Module	State variables (X)	Input variables (U)	Output variables (Y)	Packed form (G)
Battery	x_{bat}	v_{bat}	i_{bat}	G_{bat}
Bi-directional converter	v_{bat}, v_{dc}, i_{Lbd}	i_{bat}, i_{bd}, D_{bd}	$= X$	G_{bd}
Virtual resistor 3	—	$v_{bd,o}$, v_{dc}	i_{bd}	G_{vr3}
Current control	Integrator output	$i_{bat,ref}$, i_{bat}	D_{bd}	G_{bdc}
Current ref. generator	—	v_{bat}	$i_{bat,ref}$	G_{bdrg}

Bi-directional dc–dc Converter The bi-directional dc–dc boost converter is also modeled in Section 2.4.7.3, where there were two input and output regulator capacitors. The modeling process for the bi-directional converter without these capacitors is not different and will be left for readers as an exercise. The model can be obtained as follows:

$$\frac{d\Delta i_{bd}}{dt} = 0\Delta i_{bd} + \begin{bmatrix} \frac{D_{bd0}}{L_{bd}} & \frac{-1}{L_{bd}} & \frac{v_{bd1,0}}{L_{bd}} \end{bmatrix} U_{bd}, \tag{2.88a}$$

$$Y_{bd} = \Delta i_{bd}, \tag{2.88b}$$

where $U_{bd} = [\Delta v_{bd1}\ \Delta v_{bd2}\ \Delta D_{bd}]^T$.

Super-Capacitor The SC is not modeled previously in this book. Since it is a capacitor with an internal resistance, its modeling is too easy. The governing equations are given as follows:

$$\frac{d\Delta v_{sc}}{dt} = 0\Delta v_{sc} + \frac{1}{C_{sc}}\Delta i_{bd}, \tag{2.89a}$$

$$\Delta v_{bd2} = \Delta v_{sc} + R_{sc}\Delta i_{bd}. \tag{2.89b}$$

Considering the state, input, and output variables as $X_{sc} = \Delta v_{sc}$, $U_{sc} = \Delta i_{bd}$, and $Y_{sc} = \Delta v_{bd2}$, respectively, one can easily find the SSR as follows:

$$\dot{X}_{sc} = a_{sc}X_{sc} + b_{sc}U_{sc}, \tag{2.90a}$$

$$Y_{sc} = c_{sc}X_{sc} + d_{sc}U_{sc}, \tag{2.90b}$$

where $a_{sc} = 0$, $b_{sc} = 1/C_{sc}$, $c_{sc} = 1$, and $d_{sc} = R_{sc}$.

The overall model of the SC ESS shown in Figure 2.22 including four modules, i.e. the SC, the bi-directional dc–dc converter, the voltage controller, and the current controller, is calculated using RCT functions and can be found in [47].

Exercise 2.10 As mentioned, the small-signal model of the SC ESS is available in [47]. In [47], the nonlinear model simulated in MATLAB/SimPowerSystems is also provided. The parameters and operating point are considered the same for both models. An ideal voltage source 700 V series with a resistance of 2 Ω is connected to v_{bd1}. The initial voltage of the SC is considered 400 V, where the voltage controller aims to reach 500 V. (a) Show the SC voltage of the nonlinear model. (b) Compare the dynamics of linear and nonlinear models through the SC current.

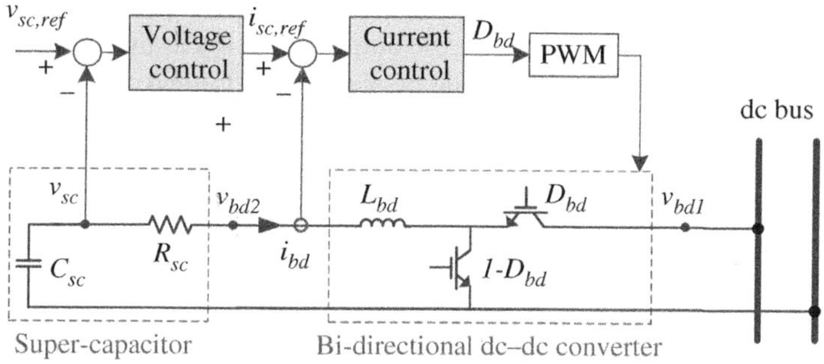

Figure 2.22 Block diagram of a super-capacitor energy storage including a super-capacitor and a bi-directional dc–dc converter with its controller.

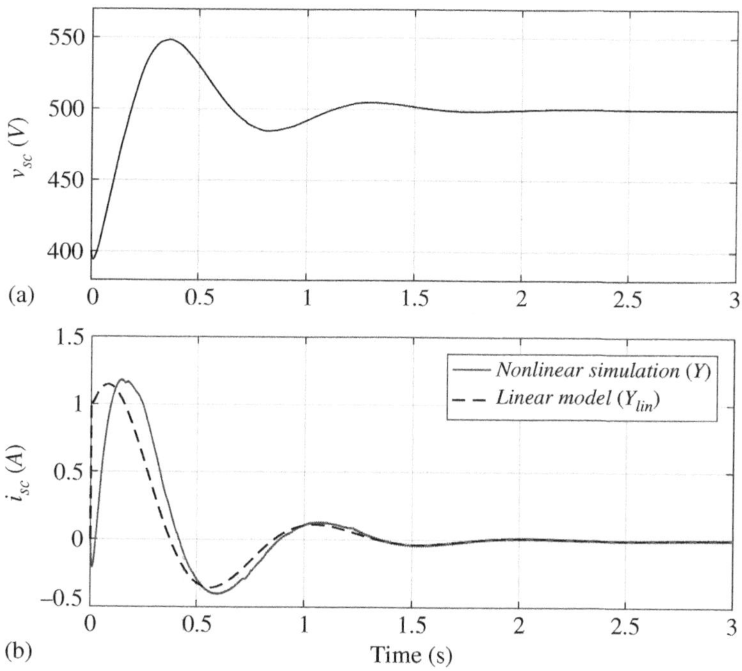

Figure 2.23 A comparison of linear model and nonlinear simulation of the super-capacitor shown in Figure 2.22: (a) the SC voltage of the nonlinear simulation and (b) linear and nonlinear SC currents.

Solution

a) Figure 2.23(a) shows the SC voltage, where the reference voltage, i.e. 500 V, is tracked from the initial voltage of 400 V. In fact, the SC is more charged in this study. Note that the control gains are selected based on trial and error. Therefore, a better response can be achieved by tuning the gains or designing an advanced controller.

b) The outputs of the linear and nonlinear models cannot be compared directly with each other. In fact, the input and output of the nonlinear model are generally U and Y, and those of the linear model are $\Delta U = U - U_0$ and $\Delta Y = Y - Y_0$. In order to compare the models, three basic steps are required. Firstly, ΔX should be obtained for the linear model according to the nonlinear model simulation. Here, $\Delta U = v_{sc,ref} - v_{sc,0} = 100\,V$. Next, the required initial values of the linear model should be obtained from the simulated nonlinear model. These values can be checked by the linear model code addressed in [47]. Finally, the output of the linear model, Y_{lin}, which can be compared with the nonlinear output, Y, should be obtained as follows:

$$Y_{lin} = Y_0 + \Delta Y.$$

In the case under study, $Y_0 = i_{sc,0} = 0$, due to the steady state of the SC before applying the $v_{sc,ref}$ at $t = 0$. Figure 2.23(b) shows the comparison between the outputs of the linear and nonlinear model. Generally, the linear model has been able to indicate the dynamics of the nonlinear models. A more precise comparison can be performed using indices, e.g. overshoot percent and settling time. A significant difference is immediately observed after $t = 0$ s, which is due to the nonlinearity of the system.

2.5 Small-Signal Modeling of Microgrid Controllers

2.5.1 Primary Control Strategies

Control aspects of MGs are fully illustrated in Chapter 5 including MG control architectures, control layers, and DER control strategies. Here, the four-level hierarchical control strategy is assumed consisting of primary, secondary, central, and global control layers [25]. The control objectives in each layer are related to the MG operating mode. Nonetheless, the main objectives of these layers (from the primary layer to the global layer) are frequency/voltage stabilization and preliminary power sharing, frequency/voltage restoration and power sharing improvement, power management and intra-MG coordination, and finally energy management and ultra-MG coordinations, respectively.

On the other hand, DERs can be mainly categorized into GFO and GFE/GFL types [83]. GFO DERs support frequency and voltage controls. GFL DERs aim to inject a certain power to the grid/MG, and unlike GFOs are not able to contribute to voltage/frequency regulation. In fact, a compromise is made between these two control goals using control gains. Generally, GFOs can be considered as voltage sources, and GFLs can be assumed as current sources.

From the viewpoint of operating modes, DERs have some considerations. In the grid-connected mode, grid-side power plants provide necessary voltage and frequency control and stabilization. However, in the grid-isolated operating modes, i.e. autonomous and interconnected modes, at least one GFO is required to play the role of control and stabilizing voltage/frequency, which can be called the leader GFO. In such a case, other GFOs can be operated as follower GFOs, where they receive their voltage/frequency references from the leader GFO. In small scales of power generation in MGs, GFOs are usually used just in the grid-isolated modes in order to prevent voltage/frequency control interferences with the main grid's huge power plants. In contrast, GFLs can be used in both autonomous and grid-connected modes for the sake of supplying active and reactive powers. Although GFLs can be operated in the grid-connected mode unconditionally, they cannot be used alone in the autonomous modes. While GFOs are flexible to operate in both modes through the tuning of compromise gains, they exhibit greater flexibility when employed for supplementary control services, such as enhancing inertia response and mitigating power fluctuations. Virtual synchronous generators (VSGs) can be considered as a type of GFOs, which are taken into account in this book as a solution for providing virtual inertia and virtual dynamics.

In this section, different DER control strategies are modeled in the primary and secondary control layers. The central and global control layers consist of very slow control actions in a few to several minutes, usually longer than 15 min. These actions can be considered as scheduled set points for the lower control layers. Therefore, the central and global control layers usually are not considered in dynamic modeling, except in a few cases like synchronization.

Figure 2.24 shows the details of the studied MG control architecture including inner, primary, and secondary control layers in which control loops of leader and follower GFO DERs as well as GFL DERs are represented. Inner and primary control units are similar for both leader and follower DERs. Inner voltage and current controllers regulate the voltage and current, respectively, and provide the PWM unit with the inverter voltage reference to work as a VSC. The primary control unit consists of $\omega - P$ and $v - Q$ droop characteristics, where GFO frequency and voltage amplitude references are provided. The secondary control is realized using the pinning consensus-based architecture in which selecting neighbor GFOs is different for leader and follower GFO DERs [84–86]. MGs are generally considered to have a single-leader GFO and multiple follower GFOs. Only the leader GFOs have rated voltage and frequency values. They have two fundamental duties. The first is communicating the rated voltage and frequency instantaneously with its own follower GFOs inside the MG such that $a_{mi} = 1$ to achieve stable intra-MG power sharing and restore voltage and frequency to the rated values. The second is communicating power, frequency, and voltage data with other leader GFOs in the neighbor MGs such that $a_{mi} = 1$ to control multi-MGs power sharing after the interconnection. There is another communication among neighbor follower GFOs ($a_{mj} = 1$) to improve power sharing and voltage/frequency restoration in the islanded MG mode. Note that the secondary control architecture of the leader GFOs is the same as follower GFOs in the islanded mode.

To control GFL DERs, represented as GFL_k in Figure 2.24, voltage control and droop characteristics are eliminated from the primary level due to a lack of frequency and voltage control in these DERs. There is an inverter current reference generator similar to the cases of inverters in wind and PV generation units to transform the active and reactive power references communicated from the central control and in turn from a higher-level controller, e.g. the distributed network operator (DNO), into the direct and quadrature components of the current reference. Furthermore, a PLL is used to synchronize the GFL unit to the grid. Modeling the control strategies is done in the next sections. The readers are also encouraged to study Chapter 5 for more details on the control of MGs.

2.5.1.1 Grid-Forming Strategy

The main modules of leader and follower GFOs are shown in Figure 2.24, which are the same for both GFO types in inner and primary control levels. Figure 2.25 shows details of the modules of primary and inner control loops of GFO DERs. Please note that the *dq* voltage PI control (see Figure 2.6) and the *dq* current PI control (the same structure with different mutual terms and gains as explained in Section 2.4.6.4) have already been modeled. The virtual impedance and the entire primary control layer should be modeled.

Virtual Impedance Here, a virtual inductive reactance $L_{vi}\omega_{rated}$ is considered, which acts as a mutual voltage drop. This module does not include any dynamics and can be modeled as follows:

$$\begin{bmatrix} v^m_{vi,d} \\ v^m_{vi,q} \end{bmatrix} = \begin{bmatrix} 0 & L_{vi}\omega_{rated} \\ -L_{vi}\omega_{rated} & 0 \end{bmatrix} \begin{bmatrix} i^m_{od} \\ i^m_{oq} \end{bmatrix}. \tag{2.91}$$

Primary Control Layer (Droop Control) As shown in Figure 2.25, the power controller of each GFO DER consists of a power calculator, two low-pass filters (LPFs), and $\omega - P$ and $V_d - Q$ droop characteristics. The droop characteristics are proportional controllers. Therefore, in order to find the small-signal model of the power controller, it is sufficient to consider the dynamics of LPFs

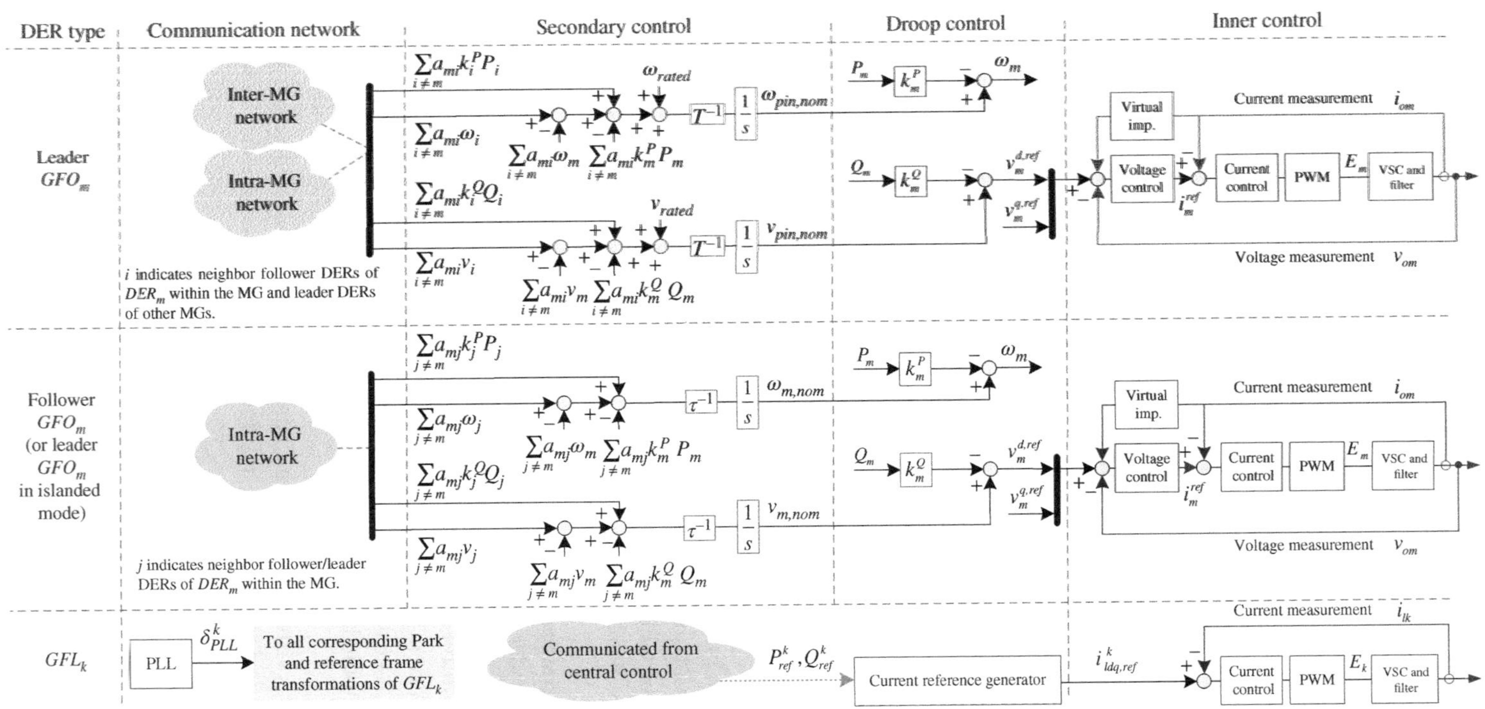

Figure 2.24 Main control functions of MGs including inner, droop and secondary controller loops and communication during islanded and interconnected modes.

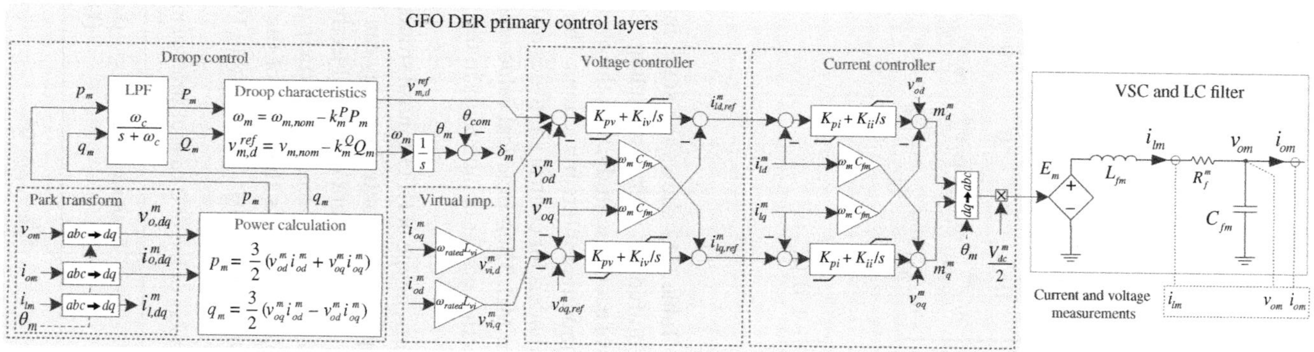

Figure 2.25 Details of the modules of droop and inner control loops of GFO DERs.

and the integrator of the local voltage phase (θ_m) producer. According to the reason presented in Section 3.2.1, the angle difference of the GFO DER$_m$'s reference frame from the CRF (δ_m) is considered as a state variable instead of the θ_m which can be calculated as follows:

$$\delta_m = \theta_m - \theta^n_{com} = \int (\omega_m - \omega^n_{com})dt, \tag{2.92}$$

where ω_m is the GFO DER$_m$ frequency and θ_{com} is the voltage phase of the DER$_1$, which the MG's CRF is based on it. Consequently, an SSR is deduced as:

$$\begin{aligned} \dot{X}^m_{pc} &= A^m_{pc}X^m_{pc} + B^m_{pc}U^m_{pc} + B^m_{com}\Delta\omega_{com}, \\ Y^m_{pc} &= C^m_{pc}X^m_{pc}, \end{aligned} \tag{2.93}$$

where $X^m_{pc} = [\Delta\delta_m \ \Delta P_m \ \Delta Q_m]^T$, $U^m_{pc} = [v^m_{o,dq} \ i^m_{o,dq}]^T$, $Y^m_{pc} = [\Delta\omega_m \ \Delta v^m_{dq,ref} \ \Delta\delta_m]^T$, and

$$A^m_{pc} = \begin{bmatrix} 0 & -k^P_m & 0 \\ 0 & -\omega_c & 0 \\ 0 & 0 & -\omega_c \end{bmatrix}, \quad B^m_{pc} = \frac{3}{2}\begin{bmatrix} 0 & 0 & 0 & 0 \\ \omega_c i^m_{od0} & \omega_c i^m_{oq0} & \omega_c v^m_{od0} & \omega_c v^m_{oq0} \\ -\omega_c i^m_{oq0} & \omega_c i^m_{od0} & \omega_c v^m_{oq0} & -\omega_c v^m_{od0} \end{bmatrix},$$

$$C^m_{pc} = \begin{bmatrix} 0 & -k^P_m & 0 \\ 0 & 0 & -k^Q_m \\ 0 & 0 & 0 \\ 1 & 0 & 0 \end{bmatrix}, \quad B^m_{com} = \begin{bmatrix} -1 & 0 & 0 \end{bmatrix}^T.$$

Overall Model of GFO Control Strategy Table 2.12 shows the state, input, and output variables of the GFO control strategy modules required for overall modeling. The packed form of the SSRs should be finalized for the parameters of each case study, then the overall model can be obtained by applying RCT functions to specify the modules' interconnections. The MATLAB codes are available in [47]. Please note that all MATLAB models for control strategies result in open controller models unlinked to a system, like a DER in an MG. Their practical value emerges when you consider the required connections with their respective power components in the system.

Grid-Supporting Strategy In some reports, a type of GFO control strategy with specific functionality is called grid-supporting (GSU) strategy. A GSU control strategy can be implemented as a current source or a voltage source. When the power/current control objective has a dominant weighting coefficient, the GSU is modeled by a current source. The participation of this GSU type in the voltage/frequency control is not zero, but it is less than 50% of the entire control effort. On the other hand, the voltage/frequency control objective dominates in the voltage source GSU, i.e. a

Table 2.12 State, input, and output variables of the GFO control strategy modules.

Module	State variables (X)	Input variables (U)	Output variables (Y)	Packed form (G)
Primary control	δ_m, P_m, Q_m	$v^m_{o,dq}, i^m_{o,dq}, \omega_{com}$	$\omega_m, v^m_{dq,ref}, \delta_m$	G_{pc}
Voltage control	Integrator outputs	$v^m_{dq,ref}, v^m_{o,dq}, v^m_{vi,dq}$	$i^m_{ldq,ref}$	G_{vc}
Current control	Integrator outputs	$i^m_{ldq,ref}, i^m_{l,dq}, v^m_{o,dq}$	m^m_{dq}	G_{cc}
Virtual impedance	—	$i^m_{o,dq}$	$v^m_{vi,dq}$	G_{vi}

major control effort is dedicated to voltage/frequency control, and less than 50% of the actuator control effort is considered for the power/current control objective. In this section, a current source GSU is studied.

The details of the current source GSU strategy are shown in Figure 2.26, where similar to the GFL strategy, there are three main modules including a current controller, a current reference generator, and a PLL. The current controller structure is fully the same as the GFO and GFL current controllers. The PLL is already modeled in (2.62). Nevertheless, the current reference generator differs from that of the GFL DER due to the existence of $P - \omega$ and $Q - V$ droop characteristics as follows:

$$P^*_{ref} = P_{ref} + k_\omega(\omega_{rated} - \omega_{gs}), \tag{2.94a}$$

$$Q^*_{ref} = Q_{ref} + k_v(v_{rated} - v_{gs}), \tag{2.94b}$$

where P^*_{ref} and Q^*_{ref} are the final references and P_{ref} and Q_{ref} are the scheduled references of active and reactive powers. k_ω and k_v are the droop gains and can be obtained as follows:

$$k_\omega = \frac{\Delta P_{max}}{\Delta \omega_{max}}, \tag{2.95a}$$

$$k_v = \frac{\Delta Q_{max}}{\Delta v_{max}}, \tag{2.95b}$$

where ΔP_{max} (ΔQ_{max}) is the maximum desirable variation of the active (reactive) power within power limitations, and $\Delta \omega_{max}$ (Δv_{max}) is the maximum acceptable variation of the frequency (voltage amplitude). Therefore, compromises are between frequency and active power tracking, and voltage amplitude and reactive power tracking. $\Delta \omega_{max}$ and Δv_{max} are the tuning gains of the trade-off. Considering ω_{rated} and v_{rated} as constant inputs, one can find the model of GSU's current reference generator as follows:

$$i^{gs}_{ldq,ref} = \left(\frac{2}{3v_{rated}}\right)\begin{bmatrix} 1 & -k_\omega & 0 & 0 \\ 0 & 0 & -1 & k_v \end{bmatrix}\begin{bmatrix} P_{ref} & \omega_{gs} & Q_{ref} & v^{gs}_{od} \end{bmatrix}^T. \tag{2.96}$$

Table 2.13 shows the state, input, and output variables of the GSU control strategy modules needed for overall modeling. The packed form of the SSRs and static models should be finalized for the parameters of each case study. The last column of Table 2.13 shows variables selected for the packed form of module models. The overall model can be obtained by specifying the interconnections among module models by applying RCT functions. The model is available in [47].

Virtual Synchronous Generator Strategy VSG is another type of the GFO control strategy, which is an idea to virtually increase the inertia of modern power converter-dominated power systems for the sake of frequency stability improvement and power oscillation damping. The basic VSG topology is to implement the swing equation instead of $\omega - P$ droop characteristic to add dynamics to frequency and control its changes and rate of changes [87]. However, this control strategy has

Table 2.13 State, input, and output variables of the current-source GSU control strategy modules.

Module	State variables (X)	Input variables (U)	Output variables (Y)	Packed form (G)
PLL	Integrator outputs	$v^{gs}_{o,dq}$, ω_{com}	ω_{gs}, δ_{gs}	G_{pll}
Current reference generator	—	P^k_{ref}, ω_{gs}, Q^k_{ref}, v^{gs}_{od}	$i^{gs}_{ldq,ref}$	$G_{cr,gs}$
Current control	Integrator outputs	$i^{gs}_{ldq,ref}$, $i^{gs}_{l,dq}$, $v^{gs}_{o,dq}$	m^{gs}_{dq}	G_{cc}

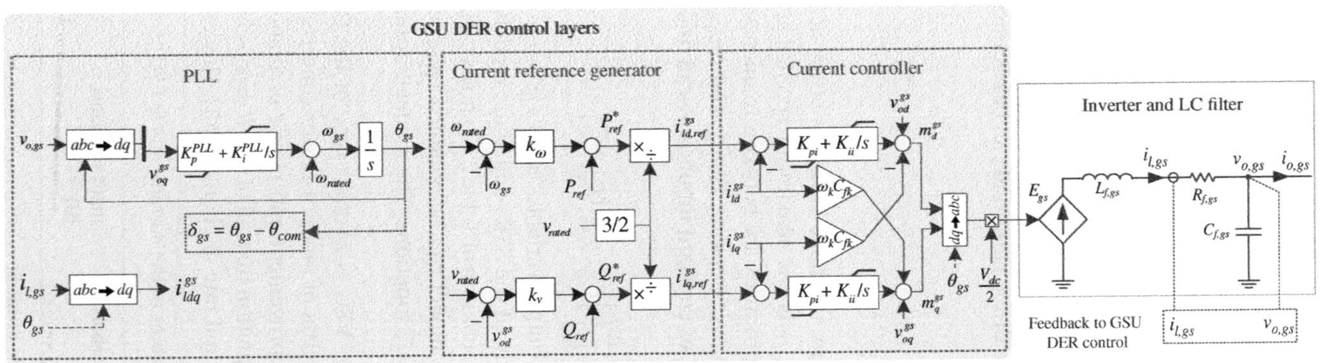

Figure 2.26 Details of the modules of control loops of a current-source GSU DER.

been modified for ancillary services and applying different control strategies and methods [88–90]. Here a basic VSG strategy is considered including the fundamental blocks of inertial frequency control [88].

Figure 2.27(a) shows the basic topology of VSGs, where three highlighted modules are added instead of $\omega - P$ droop characteristic, and other modules are preserved with respect to the GFO control strategy. The different modules of VSGs are a PLL, a virtual governor, and a virtual swing equation, and their details are shown in Figure 2.27(b). The PLL is used to provide a regulation for the internal VSG frequency ω_{vsg} compared with the real frequency ω_{pll}. Note that the PLL structure is completely similar to the previous PLL structures. Hence, its SSR can be found in (2.62). The virtual governor is a $P - \omega$ droop characteristic with the active power reference, which is given as follows:

$$\Delta P_{vsg} = P_{ref} + k_{\omega}(\omega_{rated} - \omega_{vsg}) - P_{vsg}, \tag{2.97}$$

where ΔP_{vsg} is the difference between measured VSG output active power P_{vsg} and the final active power reference P^*_{ref}. Note that the $V - Q$ droop characteristic is the same as that shown in Figure 2.25 and the model can be given as:

$$v^{vsg}_{od,ref} = v_{rated} - k_q Q_{vsg}. \tag{2.98}$$

The virtual swing equation includes an integrator, a virtual inertia coefficient J, and a virtual damping coefficient k_D. Considering the integrator producing the voltage phase angle θ_{vsg} from ω_{vsg} as a part of the virtual swing equation module, one can find an SSR of this module as follows:

$$\begin{aligned} \dot{X}_{vse} &= A_{vse}X_{vse} + B_{vse}U_{vse} + B_{com}\Delta\omega_{com}, \\ Y_{vse} &= C_{vse}X_{vse}, \end{aligned} \tag{2.99}$$

where $X_{vse} = [\theta_{vsg}\ \omega_{vsg}]^T$, $U_{vse} = [\Delta P_{vsg}\ \omega_{pll}]^T$, $Y_{vse} = X_{vse}$, and the matrices are given as follows:

$$A_{vse} = \begin{bmatrix} 0 & 1 \\ 0 & \frac{-k_D}{J\omega_{rated}} \end{bmatrix}, \quad B_{vse} = \begin{bmatrix} 0 & 0 \\ \frac{1}{J\omega_{rated}} & \frac{k_D}{J\omega_{rated}} \end{bmatrix}, \quad B_{com} = \begin{bmatrix} -1 \\ 0 \end{bmatrix}.$$

Other modules of the VSG control strategy have already been modeled including the power calculation block, LPF, virtual impedance, voltage control, and current controller. Note that the power calculation and LPF perform the module of primary control of GFO strategy, which requires to be independently modeled in VSG modeling. Finally, the state, input, and output variables, along with the packed form variables of the VSG module models are listed in Table 2.14. These variables can be utilized with MATLAB's RCT functions to define the interconnections among the VSG modules and, in turn, to derive the comprehensive model of the fundamental VSG control strategy. This model is available in [47].

2.5.1.2 Grid-Following Strategy

The details of the control strategy of GFL DERs can be seen in Figure 2.28, in which there are three main modules including a current controller, a current reference generator, and a PLL. The current controller structure is identical to the GFO current controller shown in Figure 2.25. The current reference generator is also the same as the inverter current reference generators of wind and PV generation units modeled by (2.74). Another module is the PLL already modeled in (2.62). Table 2.15 shows the state, input, and output variables of the GFL control strategy modules needed for overall modeling. The packed form of the SSRs should be finalized for the parameters of each case study, which are considered to be obtained in Table 2.15. Finally, the overall model can be

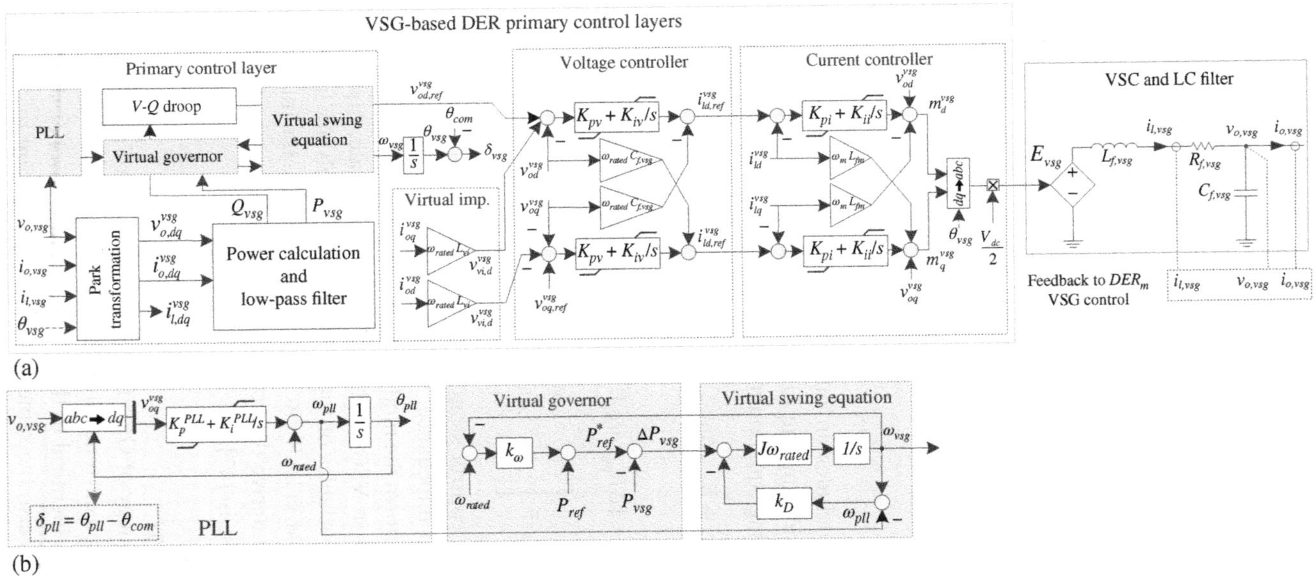

Figure 2.27 Details of the modules of (a) primary control loops and (b) PLL, virtual governor, and virtual swing equation of VSG DERs.

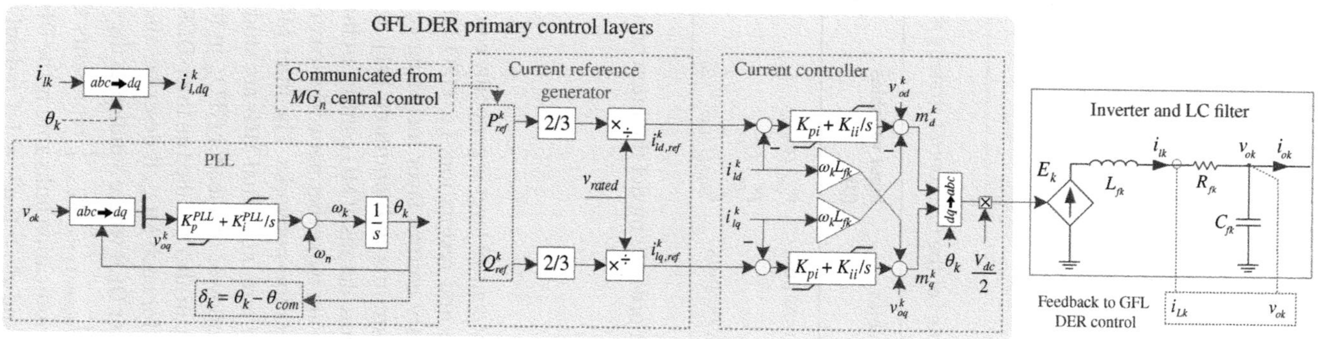

Figure 2.28 Details of the modules of primary control loops of GFL DERs.

Table 2.14 State, input, and output variables of the VSG control strategy modules.

Module	State variables (X)	Input variables (U)	Output variables (Y)	Packed form (G)
PLL	Integrator outputs	$v_{o,dq}^{vsg}, \omega_{com}$	$\omega_{pll}, \delta_{pll}$	G_{pll}
Virtual governor	—	$P_{ref}, P_{vsg}, \omega_{vsg}$	ΔP_{vsg}	G_{vg}
Virtual swing equation	$\delta_{vsg}, \omega_{vsg}$	$\Delta P_{vsg}, \omega_{pll}, \omega_{com}$	$= X$	G_{vse}
V–Q droop characteristic	—	Q_{vsg}	$v_{od,ref}^{vsg}$	G_{vq}
Power calc. and LPF	P_{vsg}, Q_{vsg}	$v_{o,dq}^{vsg}, i_{o,dq}^{vsg}$	$= X$	G_{pq}
Voltage control	Integrator outputs	$v_{odq,ref}^{vsg}, v_{o,dq}^{vsg}, v_{vi,dq}^{vsg}$	$i_{ldq,ref}^{vsg}$	G_{vc}
Current control	Integrator outputs	$i_{ldq,ref}^{vsg}, i_{l,dq}^{vsg}, v_{o,dq}^{vsg}$	m_{dq}^{vsg}	G_{cc}
Virtual impedance	—	$i_{o,dq}^{vsg}$	$v_{vi,dq}^{vsg}$	G_{vi}

Table 2.15 State, input, and output variables of the GFL control strategy modules.

Module	State variables (X)	Input variables (U)	Output variables (Y)	Packed form (G)
PLL	Integrator outputs	$v_{o,dq}^{k}, \omega_{com}$	ω_k, δ_k	G_{pll}
Current reference generator	—	P_{ref}^{k}, Q_{ref}^{k}	$i_{ldq,ref}^{k}$	G_{cr}
Current control	Integrator outputs	$i_{ldq,ref}^{k}, i_{l,dq}^{k}, v_{o,dq}^{k}$	m_{dq}^{k}	G_{cc}

obtained by applying RCT functions in order to model the interconnections among module SSRs. The overall model is available in [47].

2.5.2 Secondary Control

As shown in Figure 2.24, the secondary control layer is generally considered for participating GFO DERs in frequency/voltage regulation and restoration, i.e. GFO DERs. Although all three fundamental architectures of the secondary control layer including centralized, decentralized, and distributed may be used with their constraints and conditions [91], here a simple distributed architecture is assumed for modeling. For more details of MG control architectures, see Chapter 5 as well as [91]. According to Figure 2.24, there are two cyber networks to communicate required comprising inter-MG and intra-MG communication networks. The inter-MG cyber network is not taken into account in this part of the book, where individual MGs are studied. Therefore, the secondary control should only be modeled for a follower or a leader GFO_m in the islanded mode.

The intra-MG cyber network, which can also be called an inter-DER cyber network (only for GFO DERs) can be defined as $\chi_n(\vartheta_n, \varepsilon_n, A_n)$ for each MG_n. It describes the communication interactions among $S-1$ follower DERs, and a leader DER where S indicates the total number of DERs of MG_n. The node set $\vartheta_n = \left\{\vartheta_{n,1}, \dots, \vartheta_{n,pln-1}, \vartheta_{n,pin+1}, \dots, \vartheta_{n,S_n}\right\}$ represents the follower DERs, the set of edges $\varepsilon_n \subseteq \vartheta_n \times \vartheta_n$, expresses the communication links among the DERs, and $A_n = (a_{mj}^n)_{(S_n-1)\times(S_n-1)}$ is an adjacency matrix, where $a_{mm}^n = 0$ and $a_{mj}^n \geq 0$. $a_{mj}^n = 1$ if and only if $(\vartheta_{n,m}, \vartheta_{n,j}) \in \varepsilon_n$. The neighbor set of MG_n can be expressed by $N_{n,m} = \left\{\vartheta_{n,j} \in \vartheta_n \left|(\vartheta_{n,m}, \vartheta_{n,j}) \in \varepsilon_n\right.\right\}$. The access of each DER to the MG_n leader values is specified by the leader-adjacency matrix $B_n = diag\{a_{10}^n, \dots, a_{m0}^n, \dots, a_{S0}^n\}$, where $a_{m0}^n = 1$ if the leader DER values are available for the $DER_{n,m}$, otherwise, $a_{m0}^n = 0$.

In the third row of Figure 2.24 from the title row, one can define the two integrator outputs as the state variables, and find an SSR for the secondary controller of each GFO DER$_m$ as follows:

$$\dot{X}^m_{sc} = A^m_{sc}X^m_{sc} + B^m_{sc1}U^m_{sc1} + B^m_{sc2}U^m_{sc2} + B^m_{sc3}U^m_{sc3},$$
$$Y^m_{sc} = C^m_{sc}X^m_{sc}, \tag{2.100}$$

where $Y^m_{sc} = X^m_{sc}$ and equal $[\omega^{nom}_{pin}\ v^{nom}_{pin}]^T$ for leader DERs, or $[\omega_{m,nom}\ v_{m,nom}]^T$ for follower DERs. Input vectors are generally three sets of variables from different modules. In leader/follower DER$_m$, U^m_{sc1} is the set of frequency, voltage amplitude, active power, and reactive power of DER$_m$, which is a 4×1 vector. The second input U^m_{sc2} is the set of frequency, voltage amplitude, active power, and reactive power of all adjacent DERs of DER$_m$, which is a $4\gamma \times 1$, where γ is the number of adjacent DERs. U^m_{sc3} is only for the leader DER, which is the set of ω_{rated} and v_{rated}. These inputs can be constant values or variables from a higher control level. For both leader and follower DERs, $A^m_{sc} = 0_{2\times2}$ and $C^m_{sc} = I_{2\times2}$. However, the size of the transfer matrix is different for leader/pinning and follower DERs, i.e. $D^{pin}_{sc} = 0_{2\times4(\gamma+1)+2}$ is the transfer matrix of the leader DER and $D^{fol}_{sc} = 0_{2\times4(\gamma+1)}$ is the transfer matrix for follower DERs. Obviously, the size and content of the input matrices are strongly related to the inter-DER communication network configuration. More communications among adjacent DERs result in more inputs in the SSR of the DER secondary control.

Exercise 2.11 Consider a typical MG including two GFO DERs, where the DER$_1$ is the leader DER and the DER$_2$ is the only follower DER. Then, find the SSR matrices for both DERs.

Solution

The state matrices are always $0_{2\times2}$ for all DERs, i.e. $A^1_{sc} = A^2_{sc} = 0_{2\times2}$. The output matrices are also the same for all DERs in each configuration, i.e. $C^1_{sc} = C^2_{sc} = I_{2\times2}$. For each DER, there is one adjacent DER, i.e. $\gamma = 1$. Therefore, transfer matrices are as $D^{pin}_{sc} = 0_{2\times10}$ and $D^{pin}_{sc} = 0_{2\times8}$. In order to find input matrices, one should use the points presented in this section as well as Figure 2.24 with a focus on the secondary control components and signal flow. The input matrices of the leader DER (DER$_1$) are obtained as follows:

$$B^{DER_1}_{SC1} = \begin{bmatrix} -T^{-1} & 0 & -T^{-1}k^P_1 & 0 \\ 0 & -T^{-1} & 0 & -T^{-1}k^P_1 \end{bmatrix}, \tag{2.101a}$$

$$B^{DER_1}_{SC2} = \begin{bmatrix} T^{-1} & 0 & T^{-1}k^P_1 & 0 \\ 0 & T^{-1} & 0 & T^{-1}k^P_1 \end{bmatrix}, \tag{2.101b}$$

$$B^{DER_1}_{SC3} = \begin{bmatrix} 1 & 0 \\ 0 & 1 \end{bmatrix}, \tag{2.101c}$$

and those of the follower DER (DER$_2$) are calculated as follows:

$$B^{DER_2}_{SC1} = \begin{bmatrix} -\tau^{-1} & 0 & -\tau^{-1}k^P_1 & 0 \\ 0 & -\tau^{-1} & 0 & -\tau^{-1}k^P_1 \end{bmatrix}, \tag{2.102a}$$

$$B^{DER_2}_{SC2} = \begin{bmatrix} \tau^{-1} & 0 & \tau^{-1}k^P_1 & 0 \\ 0 & \tau^{-1} & 0 & \tau^{-1}k^P_1 \end{bmatrix}, \tag{2.102b}$$

$$B^{DER_2}_{SC3} = \begin{bmatrix} 0 & 0 \\ 0 & 0 \end{bmatrix}. \tag{2.102c}$$

2.5.3 Higher Control Levels

As explained earlier, long-term actions such as energy management, demand response, and power scheduling are usually done in central and global control layers. These actions have very slow dynamics, which are not in the frequency range of slow and relatively fast phenomena studied here for dynamic modeling. However, there are some fast control actions such as emergency control and synchronization, which will be discussed in the corresponding chapters.

2.6 Large-Signal Modeling: An Example

The nonlinearity of some MG modules is too extensive to be linearized with an acceptable precision. In such cases, large-signal modeling should be performed to find accurate models. In order to follow fundamentals and basic classifications of large-signal/transient modeling, see Section 2.3.5. Here, an example of large-signal modeling is represented about Synchronverters.

The Synchronverter control strategy is one of the methods employed to produce the virtual inertia for inverter-interfaced DERs. Generally, Synchronverter can be categorized as a type of VSG. Modeling Synchronverters is similar to modeling VSGs represented in Section 2.5.1.1. However, nonlinearities are highlighted in modeling Synchronverters. As an example of large-signal/nonlinear modeling of MG components, Synchronverter modeling is represented in this section.

2.6.1 Governing Equations on Synchronverter

Figure 2.29 shows the basic Synchronverter topology as presented in [92]. In fact, the output voltage e implied to the PWM is modified such that the bus voltage V_{bus} and grid frequency ω are stabilized and certain amounts of active/reactive power P_{ref}/Q_{ref} are produced. The output filter of the Synchronverter can be modeled using KVL as follows:

$$v_s = -R_f i_f - L_f \frac{di_f}{dt} + e_s, \tag{2.103}$$

where $v_s = [v_{sa}\ v_{sb}\ v_{sc}]^T$, $e_s = [e_{sa}\ e_{sb}\ e_{sc}]^T$, $i_f = [i_{fa}\ i_{fb}\ i_{fc}]^T$, and R_f and L_f are, respectively, the resistance and inductance of the output filter, which emulate the virtual stator winding. v_s is the vector of real phase terminal voltages and e_s is the virtual back electromotive force (EMF) due to the virtual rotor movement made by PWM, and is given by:

$$e_s = -\frac{d}{dt}(M_f i_r \tilde{\cos}\,\theta_s), \tag{2.104}$$

where $M_f > 0$ is the mutual inductance, i_r is the virtual rotor current that is considered constant and wavy line above $\cos(\theta)$ means vector of balanced cosine waves in the following form

$$\tilde{\cos}\,\theta = [\cos\theta\ \ \cos(\theta - 2\pi/3)\ \ \cos(\theta + 2\pi/3)]^T. \tag{2.105}$$

Hence, the virtual EMF can be calculated as follows

$$e_s = \psi_f \dot{\theta}_s \sin\theta_s, \tag{2.106}$$

where $\psi_f = M_f I_r$ is the amplitude of the virtual rotor flux, $\tilde{\sin}\theta_s$ mentions a vector of balanced sine waves like (2.105), and the virtual angular speed ω_s and the voltage phase angle θ_s are correlated as:

$$\dot{\theta}_s = \omega_s. \tag{2.107}$$

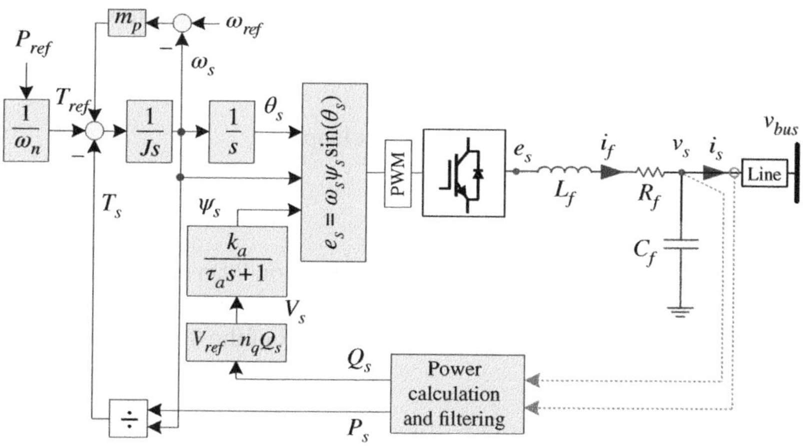

Figure 2.29 A Synchronverter topology for primary control of an inverter-based distributed generation unit within an ac MG.

According to assumptions given in [92], (2.106) is the most important part of the electrical Synchronverter side. On the other hand, the virtual mechanical side of the Synchronverter can be represented as follows:

$$\dot{\omega}_s = \frac{1}{J}(T_{ref} - T_s + m_p(\omega_{ref} - \omega_s)), \tag{2.108}$$

where J is the virtual rotor moment of inertia, and m_p is the damping coefficient, which is also the $P - \omega$ droop gain. T_{ref} is the virtual mechanical torque or the reference torque, which equals to P_{ref}/ω_n, ω_n is the nominal angular frequency, and T_s is the virtual electromagnetic torque or Synchronverter torque given by:

$$T_s = \frac{3}{2}\psi_f I \cos\delta_s, \tag{2.109}$$

where I is the amplitude of output inverter current (i_f), $\delta_s = \theta_s - \varphi_f$, and φ_f is the current phase angle. Although the active and reactive powers are calculated from the output measured voltage v_s and current i_s and then are filtered, they can be obtained internally using the control variables:

$$P_e = \frac{3}{2}\omega\psi_f I \cos\delta_s, \tag{2.110a}$$

$$Q_e = \frac{3}{2}\omega\psi_f I \sin\delta_s. \tag{2.110b}$$

The voltage amplitude control loop, which can be called the virtual rotor flux control loop, is formed from two blocks including a $V - Q$ droop and a flux control transfer function emulating the flux decay equation of synchronous generators. The $V - Q$ droop is realized as

$$V_s = V_{ref} - n_q Q_s, \tag{2.111}$$

where V_{ref} is the reference voltage amplitude and n_q is the $V - Q$ droop gain. It is easy to calculate both m_p and n_q according to the maximum allowable frequency/voltage deviation ($\Delta\omega_{max}$, ΔV_{max}) and rated active/reactive power (P_n, Q_n) as

$$m_p = P_n/\Delta\omega_{max}, \tag{2.112a}$$

$$n_q = \Delta V_{max}/Q_n. \tag{2.112b}$$

Finally, as shown in Figure 2.29, the flux control transfer function is determined as follows:

$$G_{flux}(s) = \frac{\psi_s(s)}{V_s(s)} = \frac{k_a}{\tau_a s + 1}, \tag{2.113}$$

where k_a and τ_a are the control gain and time constant of the flux loop.

2.6.2 Nonlinear State-Space Representation

By taking (2.107) and (2.108) into account, finding the differential equation form of (2.113), taking a derivative of (2.109) to find the torque dynamics, and considering the perturbed state vector of the Synchronverter as $X = [\Delta\delta_s\ \Delta\omega_s\ \Delta\psi_s\ \Delta T_s]^T$, one can express the SSR of the Synchronverter as follows

$$\dot{x}_1 = x_2, \tag{2.114a}$$

$$\dot{x}_2 = a_1 x_2 + a_2 x_4, \tag{2.114b}$$

$$\dot{x}_3 = a_3 x_3 + a_4 u, \tag{2.114c}$$

$$\dot{x}_4 = a_5 u I \cos x_1 + a_3 x_4 + a_6 \dot{I} x_3 \cos x_1 - a_6 i x_3 \sin x_1 + d, \tag{2.114d}$$

where the assumption $\Delta\dot{\delta} \simeq \Delta\dot{\theta}$ can be found in [93], and $a_1 = -m_p/j, a_2 = -1/j, a_3 = -1/\tau_a, a_4 = k_a/\tau_a, a_5 = 3k_a/2\tau_a$, and $a_6 = 3/2$. The control input u is considered to be added to the input of the virtual flux controller like a power system stabilizer, which applies a correction term to the voltage loop in order to stabilize the frequency. Note that u appears in (2.114d) by taking the derivative of (2.109) and substituting (2.114c) in it. Follow Section 4.5.2 for an application of using the nonlinear Synchronverter model in stability analysis.

2.7 Summary

This chapter provides a comprehensive guide to the dynamic modeling of the main components of MGs, starting from the very basic steps to the full requirements. Although the focus is on small-signal modeling, which is essential for stability analysis, sensitivity analysis, and model-based control synthesis, some points on nonlinear modeling are also covered. Furthermore, various detailed and simplified modeling methods are presented that are useful for modeling different dynamic and control studies of MGs. By gathering all the requirements for small-signal modeling and providing useful exercises, this chapter enables researchers to form their own models for their case studies independently without relying on other references.

The modeling steps and relevant MATLAB codes are provided for a large number of MG components, including power converters, electrical loads, renewable energy resources, energy storage systems, and control schemes. Additionally, a simple method for interconnecting several sub-models is presented, which is based on the Robust Control Toolbox functions in MATLAB. The main requirement of this method is to calculate a small-signal model of each subsystem using a state-space representation, which is explained in detail in this chapter for different electrical and control subsystems. This method significantly reduces the computational burden of small-signal modeling and enables the modeling of complex MGs with several components, as presented in Chapter 3. The same method is used in Chapter 3 to calculate small-signal models of different ac and dc MGs.

References

1 Zhuoli Zhao, Ping Yang, Yuewu Wang, Zhirong Xu, and Josep M Guerrero. Dynamic characteristics analysis and stabilization of PV-based multiple microgrid clusters. *IEEE Trans. Smart Grid*, 10(1):805–818, 2017.

2 Konstantinos F Krommydas and Antonio T Alexandridis. Modular control design and stability analysis of isolated PV-source/battery-storage distributed generation systems. *IEEE J. Emerging Sel. Top. Circuits Syst.*, 5(3):372–382, 2015.

3 Xiaoling Xiong and Yuchen Yang. A photovoltaic-based DC microgrid system: Analysis, design and experimental results. *Electronics*, 9(6):941, 2020.

4 Yanghong Xia, Wei Wei, Miao Yu, and Peng Wang. Stability analysis of PV generators with consideration of P&O-based power control. *IEEE Trans. Ind. Electron.*, 66(8):6483–6492, 2018.

5 Hamid Reza Baghaee, Mojtaba Mirsalim, Gevork B Gharehpetian, and Heidar Ali Talebi. Decentralized sliding mode control of WG/PV/FC microgrids under unbalanced and nonlinear load conditions for on-and off-grid modes. *IEEE Syst. J.*, 12(4):3108–3119, 2017.

6 Jipeng Gu, Luyao Xie, Xiaodong Yang, Youbing Zhang, Licheng Wang, Qianru Wang, Weijie Zhang, and Binjie Wang. Fuzzy piecewise coordinated control and stability analysis of the photovoltaic-storage direct current microgrid. *IET Renew. Power Gener.*, 16(15):3311–3324, 2022.

7 Allahyar Akhbari and Mohsen Rahimi. Control and stability analysis of DFIG wind system at the load following mode in a DC microgrid comprising wind and microturbine sources and constant power loads. *Int. J. Electr. Power Energy Syst.*, 117:105622, 2020.

8 Sajad Ghadiriyan and Mohsen Rahimi. Mathematical representation, stability analysis and performance improvement of DC microgrid system comprising hybrid wind/battery sources and CPLs. *IET Gener. Transm. Distrib.*, 13(10):1845–1855, 2019.

9 Mohammadreza F M Arani and Yasser Abdel-Rady I Mohamed. Analysis and impacts of implementing droop control in DFIG-based wind turbines on microgrid/weak-grid stability. *IEEE Trans. Power Syst.*, 30(1):385–396, 2014.

10 Xisheng Tang, Wei Deng, and Zhiping Qi. Investigation of the dynamic stability of microgrid. *IEEE Trans. Power Syst.*, 29(2):698–706, 2013.

11 Asit Mohanty, Meera Viswavandya, Prakash K Ray, and Sandipan Patra. Stability analysis and reactive power compensation issue in a microgrid with a DFIG based WECS. *Int. J. Electr. Power Energy Syst.*, 62:753–762, 2014.

12 Zhixin Miao, Alexander Domijan, and Lingling Fan. Investigation of microgrids with both inverter interfaced and direct AC-connected distributed energy resources. *IEEE Trans. Power Delivery*, 26 (3):1634–1642, 2011.

13 Mostafa Farrokhabadi, Sebastian König, Claudio A Cañizares, Kankar Bhattacharya, and Thomas Leibfried. Battery energy storage system models for microgrid stability analysis and dynamic simulation. *IEEE Trans. Power Syst.*, 33(2):2301–2312, 2017.

14 Gastón Orlando Suvire, Marcelo Gustavo Molina, and Pedro Enrique Mercado. Improving the integration of wind power generation into AC microgrids using flywheel energy storage. *IEEE Trans. Smart Grid*, 3(4):1945–1954, 2012.

15 Pranav Pattabi, Eman Hammad, Abdallah Farraj, and Deepa Kundur. Simplified implementation and control of a flywheel energy system for microgrid applications. In *2017 IEEE Global Conference on Signal and Information Processing (GlobalSIP)*, pages 1105–1109. IEEE, 2017.

16 Abid Soomro, Mustafa E Amiryar, Keith R Pullen, and Daniel Nankoo. Comparison of performance and controlling schemes of synchronous and induction machines used in flywheel energy storage systems. *Energy Procedia*, 151:100–110, 2018.

17 Sudipta Ghosh and Sukumar Kamalasadan. An integrated dynamic modeling and adaptive controller approach for flywheel augmented DFIG based wind system. *IEEE Trans. Power Syst.*, 32(3):2161–2171, 2016.

18 Aws Saleh, Abdalkarim Awad, and Wasel Ghanem. Modeling, control, and simulation of a new topology of flywheel energy storage systems in microgrids. *IEEE Access*, 7:160363–160376, 2019.

19 Beibei Xu, Diyi Chen, M Venkateshkumar, Yu Xiao, Yan Yue, Yanqiu Xing, and Peiquan Li. Modeling a pumped storage hydropower integrated to a hybrid power system with solar-wind power and its stability analysis. *Appl. Energy*, 248:446–462, 2019.

20 Xiaoyong Chang, Yongli Li, Xuan Li, and Xiaolong Chen. An active damping method based on a supercapacitor energy storage system to overcome the destabilizing effect of instantaneous constant power loads in DC microgrids. *IEEE Trans. Energy Convers.*, 32(1):36–47, 2016.

21 Mehran Jami, Qobad Shafiee, Mehrdad Gholami, and Hassan Bevrani. Control of a super-capacitor energy storage system to mimic inertia and transient response improvement of a direct current micro-grid. *J. Energy Storage*, 32:101788, 2020.

22 C R Arunkumar, Udaya Bhasker Manthati, and Punna Srinivas. Accurate modelling and analysis of battery–supercapacitor hybrid energy storage system in DC microgrid systems. *Energy Syst.*, 13:1055–1073, 2021.

23 Hassan Bevrani, Mohammad Ramin Feizi, Sirwan Ataee. Robust frequency control in an islanded microgrid: H_∞ and μ-synthesis approaches. *IEEE Trans. Smart Grid*, 7(2):706–717, 2015.

24 Abdolwahhab Fathi, Qobad Shafiee, and Hassan Bevrani. Robust frequency control of microgrids using an extended virtual synchronous generator. *IEEE Trans. Power Syst.*, 33(6):6289–6297, 2018. doi: 10.1109/TPWRS.2018.2850880.

25 Hassan Bevrani, Bruno François, and Toshifumi Ise. *Microgrid dynamics and control*. John Wiley & Sons, Hoboken, NJ, 2017.

26 Mobin Naderi, Qobad Shafiee, Hassan Bevrani, and Frede Blaabjerg. Low-frequency small-signal modeling of interconnected AC microgrids. *IEEE Trans. Power Syst.*, 36(4):2786–2797, 2020.

27 Paul C Krause, Oleg Wasynczuk, Scott D Sudhoff, and Steven D Pekarek. *Analysis of electric machinery and drive systems*, volume 75. John Wiley & Sons, 2013.

28 Alireza Kahrobaeian and Yasser Abdel-Rady I Mohamed. Analysis and mitigation of low-frequency instabilities in autonomous medium-voltage converter-based microgrids with dynamic loads. *IEEE Trans. Ind. Electron.*, 61(4):1643–1658, 2013.

29 Francis K Tuffner, Kevin P Schneider, Jacob Hansen, and Marcelo A Elizondo. Modeling load dynamics to support resiliency-based operations in low-inertia microgrids. *IEEE Trans. Smart Grid*, 10(3):2726–2737, 2018.

30 Juan Diego Rios Penaloza, James Amankwah Adu, Alberto Borghetti, Fabio Napolitano, Fabio Tossani, and Carlo Alberto Nucci. Influence of load dynamic response on the stability of microgrids during islanding transition. *Electr. Power Syst. Res.*, 190:106607, 2021.

31 Nathaniel Bottrell, Milan Prodanovic, and Timothy C Green. Dynamic stability of a microgrid with an active load. *IEEE Trans. Power Electron.*, 28(11):5107–5119, 2013.

32 Ling-Ling Fan, Vahidreza Nasirian, Hamidreza Modares, Frank L Lewis, Yong-Duan Song, and Ali Davoudi. Game-theoretic control of active loads in DC microgrids. *IEEE Trans. Energy Convers.*, 31(3):882–895, 2016.

33 Omid Lorzadeh, Iman Lorzadeh, Mohsen N Soltani, Amin Hajizadeh, and Kasper Jessen. Active load stabilization of cascaded systems in DC microgrids using adaptive parallel-virtual-resistance control scheme. *IEEE J. Emerging Sel. Top. Power Electron.*, 1, 2021. doi: 10.1109/JESTPE.2021.3129456.

34 Jiawei Chen and Jie Chen. Stability analysis and parameters optimization of islanded microgrid with both ideal and dynamic constant power loads. *IEEE Trans. Ind. Electron.*, 65(4):3263–3274, 2017.

35 Nagaraju Pogaku, Milan Prodanovic, and Timothy C Green. Modeling, analysis and testing of autonomous operation of an inverter-based microgrid. *IEEE Trans. Power Electron.*, 22(2):613–625, 2007.

36 Iraklis P Nikolakakos, Hatem H Zeineldin, Mohamed Shawky El-Moursi, and Nikos D Hatziargyriou. Stability evaluation of interconnected multi-inverter microgrids through critical clusters. *IEEE Trans. Power Syst.*, 31(4):3060–3072, 2016.

37 Mobin Naderi, Yousef Khayat, Qobad Shafiee, Tomislav Dragicevic, Hassan Bevrani, and Frede Blaabjerg. Interconnected autonomous AC microgrids via back-to-back converters—Part I: Small-signal modeling. *IEEE Trans. Power Electron.*, 35(5):4728–4740, 2019.

38 Mahmoud Kabalan, Pritpal Singh, and Dagmar Niebur. A design and optimization tool for inverter-based microgrids using large-signal nonlinear analysis. *IEEE Trans. Smart Grid*, 10(4):4566–4576, 2018.

39 Iraklis P Nikolakakos, H H Zeineldin, Mohamed Shawky El-Moursi, and James L Kirtley. Reduced-order model for inter-inverter oscillations in islanded droop-controlled microgrids. *IEEE Trans. Smart Grid*, 9(5):4953–4963, 2017.

40 Ling Luo and Sairaj V Dhople. Spatiotemporal model reduction of inverter-based islanded microgrids. *IEEE Trans. Energy Convers.*, 29(4):823–832, 2014.

41 Petr Vorobev, Po-Hsu Huang, Mohamed Al Hosani, James L Kirtley, and Konstantin Turitsyn. High-fidelity model order reduction for microgrids stability assessment. *IEEE Trans. Power Syst.*, 33(1):874–887, 2017.

42 Valerio Mariani, Francesco Vasca, Juan C Vásquez, and Josep M Guerrero. Model order reductions for stability analysis of islanded microgrids with droop control. *IEEE Trans. Ind. Electron.*, 62(7):4344–4354, 2014.

43 Md Rasheduzzaman, Jacob A Mueller, and Jonathan W Kimball. Reduced-order small-signal model of microgrid systems. *IEEE Trans. Sustainable Energy*, 6(4):1292–1305, 2015.

44 Aswad Adib, Fariba Fateh, Mohammad B Shadmand, and Behrooz Mirafzal. A reduced-order technique for stability investigation of voltage source inverters. In *IEEE Energy Conversion Congress and Exposition (ECCE'2018)*, pages 5351–5356, 2018.

45 Zhikang Shuai, Yelun Peng, Xuan Liu, Zuyi Li, Josep M Guerrero, and Z John Shen. Dynamic equivalent modeling for multi-microgrid based on structure preservation method. *IEEE Trans. Smart Grid*, 2018. doi: 10.1109/TSG.2018.2844107

46 L Chávarro-Barrera, S Pérez-Londoño, and Juan Mora-Flórez. An adaptive approach for dynamic load modeling in microgrids. *IEEE Trans. Smart Grid*, 12(4):2834–2843, 2021.

47 Matlab codes for the microgrids dynamic modeling, stability and control book. www.wiley.com/go/shafiee/microgrids. Accessed: 2023-07-07.

48 Katsuhiko Ogata. *Modern control engineering*, volume 5. Prentice Hall, Upper Saddle River, NJ, 2010.

49 Jinghan He, Xiaoyu Wu, Xiangyu Wu, Yin Xu, and Josep M Guerrero. Small-signal stability analysis and optimal parameters design of microgrid clusters. *IEEE Access*, 7:36896–36909, 2019.

50 Gary J Balas, John C Doyle, Keith Glover, Andy Packard, and Roy Smith. μ-analysis and synthesis toolbox: For use with Matlab. The MathWorks Inc., 1994.

51 Xiaoyu Wu, Yin Xu, Xiangyu Wu, Jinghan He, Josep M Guerrero, Chen-Ching Liu, Kevin P Schneider, and Dan T Ton. A two-layer distributed control method for islanded networked microgrid systems. *IEEE Trans. Smart Grid*, 11(2):942–957, 2019.

52 Mohammad S Golsorkhi, David John Hill, and Hamid Reza Karshenas. Distributed voltage control and power management of networked microgrids. *IEEE J. Emerging Sel. Top. Power Electron.*, 6(4):1892–1902, 2018.

53 Mobin Naderi, Yousef Khayat, Qobad Shafiee, Tomislav Dragicevic, Hassan Bevrani, and Frede Blaabjerg. Comprehensive small-signal modeling and Prony analysis-based validation of synchronous interconnected microgrids. *Energy Rep.*, 7:6677–6689, 2021.

54 Mohammad Jamshidi. *Large-scale systems: Modeling, control, and fuzzy logic*. Prentice-Hall, Inc., 1996.

55 Rui Wang, Qiuye Sun, Pengfei Tu, Jianfang Xiao, Yonghao Gui, and Peng Wang. Reduced-order aggregate model for large-scale converters with inhomogeneous initial conditions in DC microgrids. *IEEE Trans. Energy Convers.*, 36(3):2473–2484, 2021.

56 Chengshan Wang, Hao Yu, Peng Li, Jianzhong Wu, and Chengdi Ding. Model order reduction for transient simulation of active distribution networks. *IET Gener. Transm. Distrib.*, 9(5):457–467, 2015.

57 Yun Zhang, Le Xie, and Qifeng Ding. Interactive control of coupled microgrids for guaranteed system-wide small signal stability. *IEEE Trans. Smart Grid*, 7(2):1088–1096, 2016.

58 Mehdi Monadi, Hossein Hooshyar, Luigi Vanfretti, Farhan Mahmood, Jose Ignacio Candela, and Pedro Rodriguez. Measurement-based network clustering for active distribution systems. *IEEE Trans. Smart Grid*, 10(6):6714–6723, 2019.

59 Panagiotis N Papadopoulos, Theofilos A Papadopoulos, Paul Crolla, Andrew J Roscoe, Grigoris K Papagiannis, and Graeme M Burt. Measurement-based analysis of the dynamic performance of microgrids using system identification techniques. *IET Gener. Transm. Distrib.*, 9(1):90–103, 2014.

60 Tomasz Zieliński and Krzysztof Duda. Frequency and damping estimation methods-an overview. *Metrol. Meas. Syst.*, 18(4):505–528, 2011.

61 Amirnaser Yazdani and Reza Iravani. *Voltage-sourced converters in power systems*, volume 34. Wiley Online Library, Hoboken, NJ, 2010.

62 Muhammad H Rashid. *Power electronics handbook*. Butterworth-heinemann, 2017.

63 Peter Azer and Ali Emadi. Generalized state space average model for multi-phase interleaved buck, boost and buck-boost DC–DC converters: Transient, steady-state and switching dynamics. *IEEE Access*, 8:77735–77745, 2020.

64 C A Nwosu and M Eng. State-space averaged modeling of a nonideal boost converter. *Pacific J. Sci. Technol.*, 9(2):302–308, 2008.

65 Xiongfei Wang, Josep M Guerrero, Frede Blaabjerg, and Zhe Chen. A review of power electronics based microgrids. *Int. J. Power Electron.*, 12(1):181–192, 2012.

66 Nishit Tiwary, Venkata Ramana Naik N, Anup Kumar Panda, Rajesh Kumar Lenka, and Ankireddy Narendra. Sliding mode and current observer-based direct power control of dual active bridge converter with constant power load. *Int. Trans. Electr. Energy Syst.*, 31(5):e12879, 2021.

67 Mariusz Malinowski, Marek Jasinski, and Marian P Kazmierkowski. Simple direct power control of three-phase pwm rectifier using space-vector modulation (DPC-SVM). *IEEE Trans. Ind. Electron.*, 51(2):447–454, 2004.

68 Prabha Kundur, Neal J Balu, and Mark G Lauby. *Power system stability and control*. McGraw-Hill, New York, 1994.

69 Zeinab Karami, Qobad Shafiee, Yousef Khayat, Meysam Yaribeygi, Tomislav Dragičević, and Hassan Bevrani. Decentralized model predictive control of DC microgrids with constant power load. *IEEE J. Emerg. Sel. Top. Power Electron.*, 9(1):451–460, 2019. doi: 10.1109/JESTPE.2019.2957231

70 Alexis Kwasinski and Chimaobi N Onwuchekwa. Dynamic behavior and stabilization of DC microgrids with instantaneous constant-power loads. *IEEE Trans. Power Electron.*, 26(3):822–834, 2010.

71 Suresh Singh, Aditya R Gautam, and Deepak Fulwani. Constant power loads and their effects in DC distributed power systems: A review. *Renewable Sustainable Energy Rev.*, 72:407–421, 2017.

72 Saeed Golestan, Josep M Guerrero, and Juan C Vasquez. Three-phase PLLs: A review of recent advances. *IEEE Trans. Power Electron.*, 32(3):1894–1907, 2016. doi: 10.1109/TPEL.2016.2565642

73 Dehong Xu, Frede Blaabjerg, Wenjie Chen, and Nan Zhu. *Advanced control of doubly fed induction generator for wind power systems*. John Wiley & Sons, 2018.

74 Tony Burton, Nick Jenkins, David Sharpe, and Ervin Bossanyi. *Wind energy handbook*. John Wiley & Sons, 2011.

75 Hengzhao Yang. A review of supercapacitor-based energy storage systems for microgrid applications. In *2018 IEEE Power & Energy Society General Meeting (PESGM)*, pages 1–5. IEEE, 2018.

76 Shigenori Inoue and Hirofumi Akagi. A bidirectional DC–DC converter for an energy storage system with galvanic isolation. *IEEE Trans. Power Electron.*, 22(6):2299–2306, 2007.

77 Xiaokang Xu, Martin Bishop, Donna G Oikarinen, and Chen Hao. Application and modeling of battery energy storage in power systems. *CSEE J. Power Energy Syst.*, 2(3):82–90, 2016.

78 Fabian Calero, Claudio A Cañizares, and Kankar Bhattacharya. Dynamic modeling of battery energy storage and applications in transmission systems. *IEEE Trans. Smart Grid*, 12(1):589–598, 2020.

79 Andrew J Hutchinson and Daniel T Gladwin. Verification and analysis of a battery energy storage system model. *Energy Rep.*, 8:41–47, 2022.

80 S Castano, L Gauchia, F. Voncila, and J Sanz. Dynamical modeling procedure of a Li-ion battery pack suitable for real-time applications. *Energy Convers. Manage.*, 92:396–405, 2015.

81 Antonio Ernesto Sarasua, Marcelo Gustavo Molina, and Pedro Enrique Mercado. Dynamic modelling of advanced battery energy storage system for grid-tied AC microgrid applications. Energy Storage: Technology *and* Applications, 2013.

82 Olivier Tremblay, Louis-A Dessaint, and Abdel-Illah Dekkiche. A generic battery model for the dynamic simulation of hybrid electric vehicles. In *2007 IEEE Vehicle Power and Propulsion Conference*, pages 284–289. IEEE, 2007.

83 Joan Rocabert, Alvaro Luna, Frede Blaabjerg, and Pedro Rodriguez. Control of power converters in AC microgrids. *IEEE Trans. Power Electron.*, 27(11):4734–4749, 2012.

84 Jingang Lai, Xiaoqing Lu, Xinghuo Yu, and Antonello Monti. Cluster-oriented distributed cooperative control for multiple AC microgrids. *IEEE Trans. Ind. Informat.*, 15(11):5906–5918, 2019.

85 Shengxuan Weng, Yusheng Xue, Jianbo Luo, and Yanman Li. Distributed secondary control for islanded microgrids cluster based on hybrid-triggered mechanisms. *Processes*, 8(3):370, 2020.

86 Xiaoqing Lu, Jingang Lai, and Xinghuo Yu. A novel secondary power management strategy for multiple AC microgrids with cluster-oriented two-layer cooperative framework. *IEEE Trans. Ind. Informat.*, 17(2):1483–1495, 2020.

87 Hassan Bevrani, Toshifumi Ise, and Yushi Miura. Virtual synchronous generators: A survey and new perspectives. *Int. J. Electr. Power Energy Syst.*, 54:244–254, 2014. ISSN 0142-0615. doi: 10.1016/j.ijepes.2013.07.009 URL https://www.sciencedirect.com/science/article/pii/S0142061513003062.

88 Jia Liu, Yushi Miura, Hassan Bevrani, and Toshifumi Ise. Enhanced virtual synchronous generator control for parallel inverters in microgrids. *IEEE Trans. Smart Grid*, 8(5):2268–2277, 2016.

89 Amin Karimi, Yousef Khayat, Mobin Naderi, Tomislav Dragičević, Rahmatollah Mirzaei, Frede Blaabjerg, and Hassan Bevrani. Inertia response improvement in AC microgrids: A fuzzy-based virtual synchronous generator control. *IEEE Trans. Power Electron.*, 35(4):4321–4331, 2019.

90 Vincenzo Mallemaci, Fabio Mandrile, Sandro Rubino, Andrea Mazza, Enrico Carpaneto, and Radu Bojoi. A comprehensive comparison of virtual synchronous generators with focus on virtual inertia and frequency regulation. *Electr. Power Syst. Res.*, 201:107516, 2021.

91 Yousef Khayat, Qobad Shafiee, Rasool Heydari, Mobin Naderi, Tomislav Dragičević, John W Simpson-Porco, Florian Dörfler, Mohammad Fathi, Frede Blaabjerg, Josep M Guerrero, et al. On the secondary control architectures of AC microgrids: An overview. *IEEE Trans. Power Electron.*, 35(6):6482–6500, 2019. doi: 10.1109/TPEL.2019.2951694.

92 Qing-Chang Zhong and George Weiss. Synchronverters: Inverters that mimic synchronous generators. *IEEE Trans. Ind. Electron.*, 58(4):1259–1267, 2010.

93 Seyed Mahdi Ashabani and Yasser Abdel-Rady I Mohamed. A flexible control strategy for grid-connected and islanded microgrids with enhanced stability using nonlinear microgrid stabilizer. *IEEE Trans. Smart Grid*, 3(3):1291–1301, 2012.

3

Microgrid Dynamic Modeling: Overall Modeling and Case Studies

3.1 Introduction

Microgrids (MGs) can be either small-scale power grids with few modules or medium-scale power grids with a large number of modules and units. They can also be operated at low or medium voltages, as well as be isolated or grid-connected. MG dynamic modeling contains two main steps. The first step is finding the model of each module of the MG. This is done in Chapter 2 using state-space representation (SSR). The second step is to realize interconnections among modules, which is explained in Section 2.3.1.5 and is called the interconnection method. The interconnections include electrical connections, control signals, or communication data transfers. Note that most MG modules, e.g. wind generation units or active loads, include several submodules, where their modeling will be completed after applying the interconnection method. Therefore, the interconnection method makes the process of modeling easy even for very large-scale MGs with many modules. Especially, this step is done numerically using MATLAB coding without the hardness of several substitutions of equations.

In the literature, the modeling of ac MGs with entirely inverter-based distributed energy resources (DERs) is undertaken to address their dynamic challenges. Both detailed models [1, 2] and simplified models [3–10] have been investigated. Detailed models are usually used to study unknown dynamics caused by interactions of MG modules, e.g. droop-based primary control and DER coupling lines [4]. The simplification procedure can be done in two steps: simplifying the electrical network and reducing the order of control modules individually. The first step can be accomplished using the Kron reduction method [3–5]. However, non-simplified networks [6–8] and simple networks with a single point of common coupling (PCC) [9, 10] are also investigated in the literature. In addition, a reduced-order model based on the modified Krylov subspace method is developed for active distribution networks [11]. Regarding the second step, several reduced-order models have been presented for inverters consisting of eighth-order [6], fifth-order [3], third-order [3, 7, 10, 12], second-order [4], and single-order [3, 9]. In [3, 4, 9, 13], the coupling line is included in the inverter model as a significant module in forming the dominant modes.

Dynamic modeling of dc MGs has also been investigated for different objectives. To analyze the asymptotic stability of dc MGs and verify the stability of a designed controller, small-signal modeling is employed [14–17]. Challenges of constant power loads (CPLs) in dc MGs are broadly investigated [18–22]. Different dc–dc power converters required in dc MGs are modeled, including buck converters, boost converters, buck–boost converters, bidirectional converters, and some other

Microgrids: Dynamic Modeling, Stability and Control, First Edition. Qobad Shafiee, Mobin Naderi, and Hassan Bevrani.

special structures of power converters [21–25]. The dynamic impact of different DERs and energy storage systems (ESSs) on the stability of dc MGs is studied [26–28]. Some literature has focused on modeling of communication links and corresponding control loops to study the corresponding challenges, e.g. delay, loss of data, and cyberattacks [29–33]. Simplified models are also investigated for dc MGs using balanced truncation [24] and a module-based method [17].

In Chapter 2, basic concepts, classifications, fundamentals, and important considerations for modeling MG modules were presented. Moreover, modeling energy resources, energy storage systems, and different control loops and strategies is provided in which module interconnections are taken into account and their details are represented for potential readers who want to implement models in the MATLAB/Editor environment. Here, some complementary considerations are presented for the overall modeling of MGs, and then various cases of MG modeling are provided, including small-signal modeling of a single grid-connected photovoltaic (PV), a grid-connected ac MG with two grid-following (GFL) DERs, two autonomous ac MGs, an autonomous dc MG, and a large-signal (transient) modeling of an ac MG with different load types. Note that these case studies will be considered for MG stability analysis and control studies in the next chapters.

3.2 Overall Microgrid Dynamic Modeling

By modeling all or most possible components of ac or dc MGs in the previous sections, modeling of each type or each structure of MGs can easily be done. The input and output of each module are specified, and their interconnection should be based on the MG structure in each case. It is enough to apply the interconnections among all module models, i.e. SSRs or static gain matrices, using RCT functions in MATLAB. In overall MG modeling, if the first step is assumed to be finding individual SSRs/models for all modules, and the final step is assumed to be specifying the interconnections using RCT functions, there is one more step between them, i.e. common reference frame (CRF) considerations.

3.2.1 Common Reference Frame

Though the frequencies are equal in stablesteady-state operation due to the electrical connection and the $\omega - P$ droop characteristic performance, their dynamics can be different. The DER frequency references is produced independently in power controllers of grid-forming (GFO) DERs. The frequency of GFL DERs is also obtained through a phase-locked loop (PLL) process. Therefore, corresponding variables are obtained in different and independent reference frames, and a CRF should be considered to model DER frequency interactions [1, 13, 34]. For this purpose, the following five steps are necessary.

i) One of the DER reference frames should be considered as the CRF. In this book, the reference frame of the DER_1 is considered as the CRF.
ii) $\Delta\delta$ is determined as a state variable of each DER, which is realized in the power (primary) controller of GFO units and in the PLL of GFL units. This state shows the synchronism dynamics of DERs in each individual ac MG.
iii) ω_{com} is used in the modeling of passive modules without self-produced frequency such as loads and power networks. Moreover, ω_{com} is considered as an input for all modules that have the frequency reference generation unit or frequency measurement to transfer individual frequencies and voltage phase angles to the CRF.

iv) The output variables of each individual reference frame to the modules, which are stated in the CRF, should be transformed to the CRF and vice versa. For instance, the $v^{m}_{o,dq}$ is stated in the individual m'th reference frame as an input to the power network, which is stated in the CRF. Hence, the $v^{m}_{o,dq}$ should be transformed to the CRF. Similarly, the inverse transformation is needed for $i^{m}_{o,dq}$. Both the transformation and its inverse in a perturbed form are as follows:

$$\Delta v_{DQ} = T_s \cdot \Delta v_{dq} + T_{\delta 1} \cdot \Delta\delta, \tag{3.1a}$$

$$\Delta v_{dq} = T_s^{-1} \cdot \Delta v_{DQ} + T_{\delta 2} \cdot \Delta\delta, \tag{3.1b}$$

where the v_{DQ} and v_{dq} express variables in CRF and individual frames, respectively, $\Delta\delta$ is the angle difference between each individual reference frame and the CRF of the MG, and

$$T_s = \begin{bmatrix} \cos\delta_0 & -\sin\delta_0 \\ \sin\delta_0 & \cos\delta_0 \end{bmatrix},$$

$$T_{\delta 1} = \begin{bmatrix} -(v_{d0}\sin\delta_0 + v_{q0}\cos\delta_0) \\ v_{d0}\cos\delta_0 - v_{q0}\sin\delta_0 \end{bmatrix}, T_{\delta 2} = \begin{bmatrix} -v_{d0}\sin\delta_0 + v_{q0}\cos\delta_0 \\ -(v_{d0}\cos\delta_0 + v_{q0}\sin\delta_0) \end{bmatrix}.$$

3.2.2 Microgrid General State-Space Model

An intensive generalized model can be given for each MG with any number of DERs, lines, and loads as an SSR model as follows:

$$\begin{aligned} \dot{X}_{MG} &= A_{MG}X_{MG} + B_{MG}U_{MG}, \\ Y_{MG} &= C_{MG}X_{MG} + D_{MG}U_{MG}, \end{aligned} \tag{3.2}$$

where X_{MG} is the state vector, which can be organized in any order of the modules' state variables. U_{MG} is the input vector. Input variables can be selected from the module inputs or can be new inputs required for the overall MG model. The input vector can be a null vector in the cases of stability analysis, where a closed-loop model is needed including both control and power parts of the MG. Nevertheless, in the case of using the MG model for designing a controller, an open-loop model is required, where the inputs should be selected as the control signals. The same considerations can be expressed for the output vector Y_{MG}. In the analytical–numerical interconnection modeling method based on robust control toolbox (RCT) functions, the matrices can be obtained just numerically for the case studies.

3.2.3 Grid Model

The grid model can be considered as a module in the modeling of grid-connected MGs. The main grid can be modeled using a three-phase voltage source series with a resistive–inductive (RL) line. Two different situations can be assumed for the voltage source component depending on the MG rating with respect to the grid rating. When a small-scale MG is connected to the main grid, where its rating is much smaller than the grid capacity, the grid can be modeled by an ideal voltage source series with the RL line, because low penetration of MGs may only cause a voltage drop at the point of connection. MGs in low penetrations cannot have a considerable impact on the voltage and frequency of the grid, except a voltage drop modeled by the RL line. When DERs and ESSs of the MG have a considerable capacity compared to the rated grid power, an ideal voltage source cannot show the interactions among the MG and the grid considering voltage and frequency disturbances.

Therefore, a nonideal voltage source should be assumed in such cases. If the grid generally consists of traditional rotary power plants, it can be simplified with the dynamics of an equivalent synchronous generator such as the case shown at the beginning of Section 2.4.7.1 for the wind generator. If the grid is a composition of rotary and low-inertia generation units, a GFO strategy, or its special subgroups, i.e. a grid-supporting (GSU), or a virtual synchronous generator (VSG) strategy is suitable to model the grid with a large value of the active and reactive rated powers, i.e. very small droop gains showing a stiff grid with respect to the MG. In the case of considering VSG, an inertia value can be considered for the grid model. Therefore, depending on the case study, an appropriate model of the main grid should be selected.

3.3 Small-Signal Modeling of DC and AC Microgrids

In this section, some basic case studies for detailed modeling of MGs are represented. The first case is a grid-connected PV DER in which dc side dynamics are considered. The second case is a grid-connected MG including two GFL DERs focusing only on the ac side dynamics. A grid-isolated MG is modeled in the third case, and the modeling of a dc MG is presented in the fourth case.

3.3.1 A Grid-Connected PV

Figure 3.1 shows a schematic of the grid-connected PV including a PV array, a dc link, an inverter and its control unit, an LC filter, a coupling line to the PCC, a constant impedance load, the upstream grid, and its feeder. The PV setup to the LC filter has been modeled previously in Section 2.4.7.2. However, here, dc–dc chopper and its control unit are not considered. The model can easily be reconfigured with a lower number of components. Other modules are also modeled previously. The only node voltage to be modeled using the virtual resistor concept is the PCC voltage. The details of node voltage modeling can be found in Section 2.4.5.1. The interconnections among the module models of the grid-connected PV case study are specified using RCT functions in MATLAB, which can be found in [35]. Note that the obtained model is a closed-loop model to include both the power system and the control system in the model. The obtained close-loop model can be given as follows:

$$\frac{d}{dt}X_{gc-pv} = A_{gc-pv}X_{gc-pv}, \tag{3.3}$$

where the state vector X_{gc-pv} includes 16 state variables, which can be shown as follows:

$$X_{gc-pv} = \left[X_{lc}\ X_{li}\ X_{cc}\ X_{pll}\ X_{dcs}\ X_{vdc}\ X_{li,grid}\ X_{lo}\right], \tag{3.4}$$

where X_{lc} includes four state variables, namely direct and quadrature components of the filter current and voltage, X_{li} and $X_{li,grid}$ are direct and quadrature components of PV coupling line current and grid line current, respectively, X_{cc} and X_{pll} are the state variables of the current controller and the PLL, X_{dcs} is the state variable of the dc side, i.e. the dc link voltage, X_{vdc} is the state variable of the dc voltage controller, and X_{lo} is the direct and quadrature components of the load current.

Figure 3.2 shows some results of the small-signal model of the grid-connected PV unit. Eigenvalues of the system can be seen in Figure 3.2(a), where they are dispersed in a large range of real and imaginary values. It is due to obtaining the detailed model and existing various modules with different dynamics, e.g. PLL is a module with slow dynamics, while generally network lines and loads are considered as modules with fast dynamics. A more complete analysis of eigenvalues is

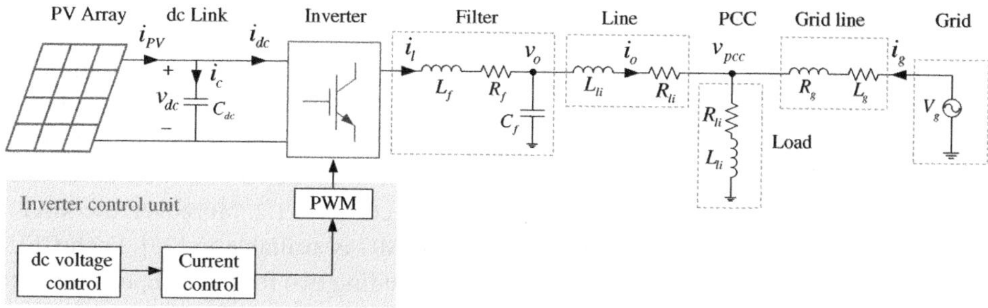

Figure 3.1 A schematic of the grid-connected PV case study.

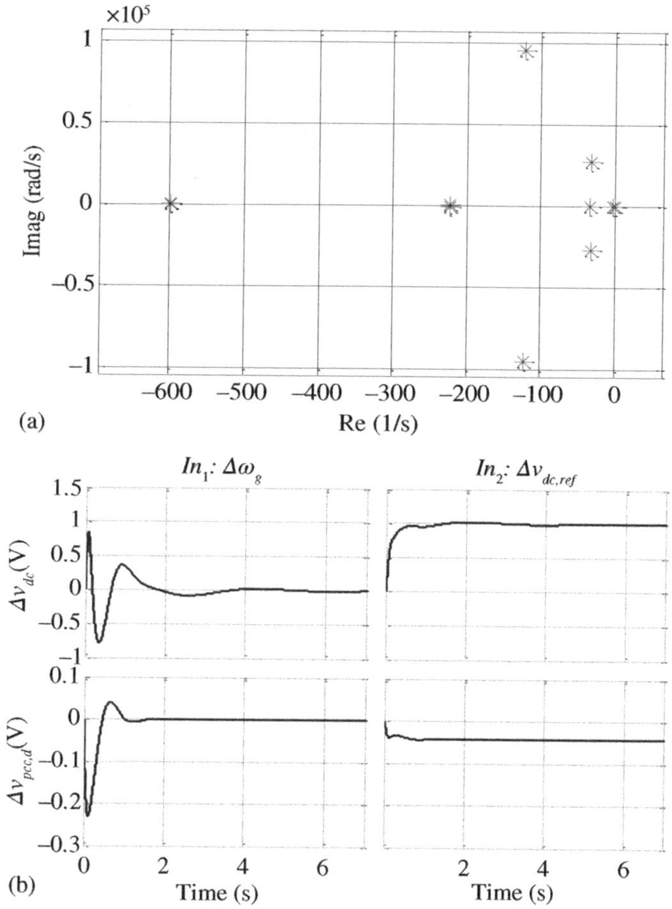

Figure 3.2 Results of modeling the grid-connected PV system: (a) eigenvalues on the s plane, (b) step response to the input changes, i.e. dc voltage reference and grid frequency.

done in Chapter 4. Figure 3.2(b) shows step responses for two important inputs of the systems, i.e. grid frequency $\Delta\omega_g$ and dc voltage reference $v_{dc,ref}$. The first is a disturbance input from the grid and the second is a disturbance input from the maximum power point tracking (MPPT). Fluctuations and variations of the two important outputs, i.e. dc and ac voltage amplitude v_{dc} and $v_{pcc,d}$, can be seen.

3.3.2 Grid-Connected AC Microgrids

Figure 3.3 shows the studied grid-connected ac MG, where it consists of two GFL DERs, an integrated constant RL load, and the model of the upstream grid. The grid model includes an ideal voltage source and an RL branch modeling the low-voltage cables and the medium-voltage feeder referred to as the low-voltage side. All modules have already been modeled and their SSRs can be found in Chapter 2, e.g. GFL control strategy modeling in Section 2.5.1.2. Moreover, the MATLAB code for small-signal modeling of this grid-connected ac MG is available in [35]. Each GFL DER to the end of its coupling line forms 10 state variables including two for the PLL, two for the current/power controller, four for the LC filter, and two for the coupling line. The grid model comprises 2 state variables for the RL line, and the RL load has two state variables. Therefore, the studied grid-connected MG consists of 24 state variables, and consequently, its SSR has 24 eigenvalues.

Figure 3.4(a) shows all 24 eigenvalues, which are spread in a large range of real and imaginary values due to existing and modeling different modules with several various bandwidths. Although most of the dynamic modes are so fast, a slower set of them is shown in Figure 3.4(b), where they are placed in a smaller range of real values. This set includes eight eigenvalues (four complex conjugate dynamic modes) with very large imaginary values. However, their real value is still large enough such that their large oscillations are damped very fast. The dominant eigenvalues are shown in Figure 3.4(c), which are the slowest dynamic modes and the most important eigenvalues in the stability analysis. Although all the dominant eigenvalues can be critical and sensitive to parameter changes, a set of them in the vicinity of the imaginary axis is the most critical eigenvalues, which will be studied in Chapter 4 with more details of the stability aspects.

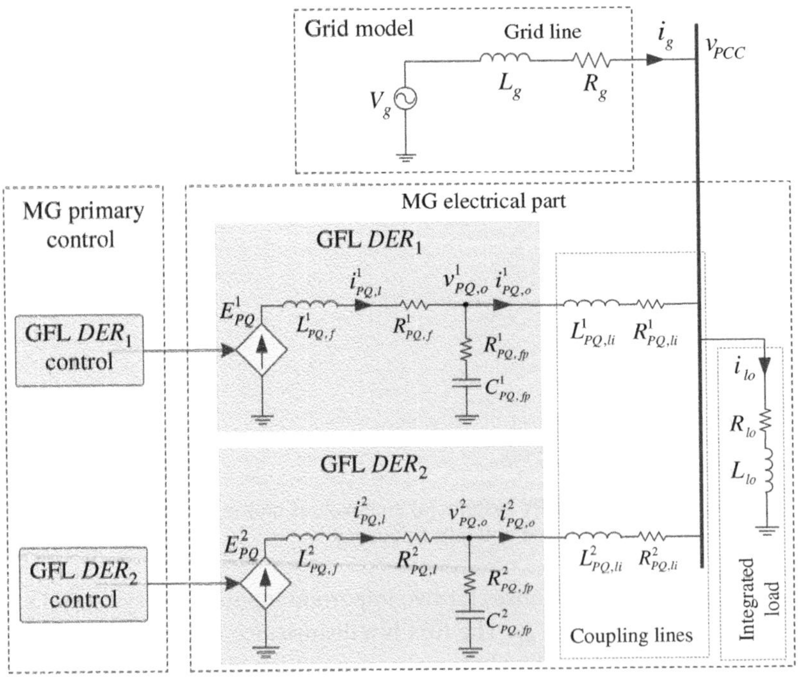

Figure 3.3 The studied grid-connected ac MG including two GFL DERs and an integrated constant RL load.

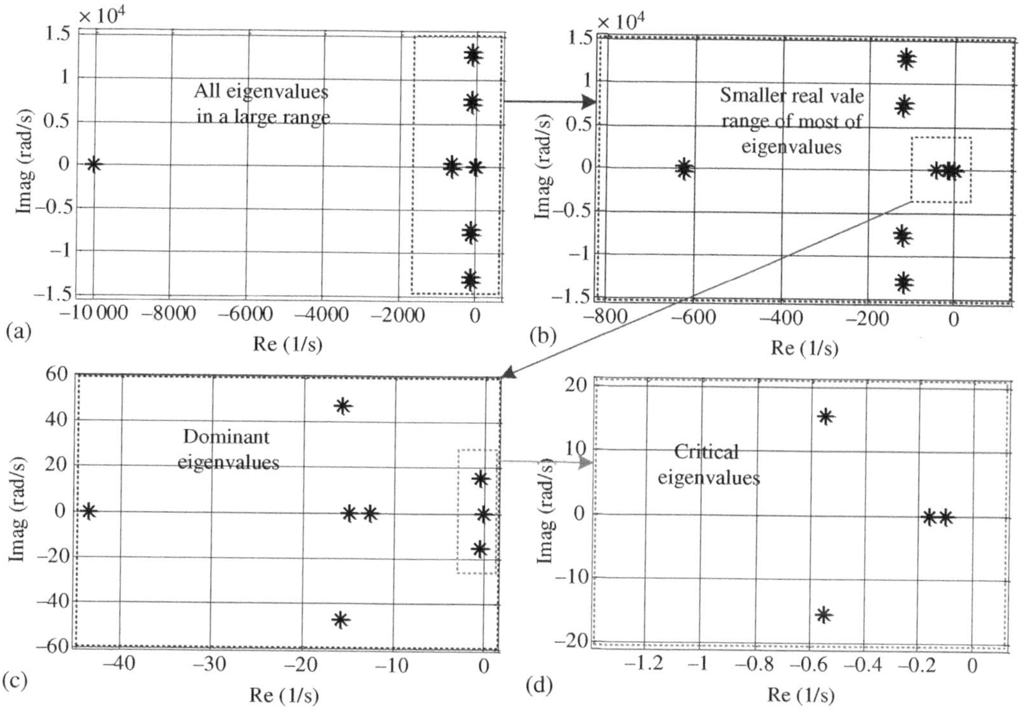

Figure 3.4 Eigenvalues of the studied grid-connected ac MG: (a) all 24 eigenvalues, (b) a smaller real value range of the most of eigenvalues, (c) dominant eigenvalues, (d) critical eigenvalues.

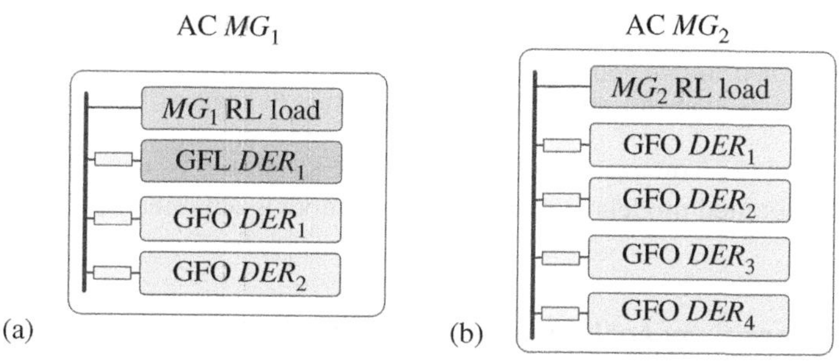

Figure 3.5 Case studies of islanded ac microgrids: (a) MG_1 with two GFO DERs and one GFL DERs, (b) MG_2 with four GFO DERs.

3.3.3 Islanded AC Microgrids: The Detailed Model

Figure 3.5 shows the configuration of two islanded ac MGs studied here. In Figure 3.5(a), the MG_1 with two GFO DERs and one GFL DER, and in Figure 3.5(b), the MG_2 with four GFO DERs are shown. A general topology of the studied ac MGs is also shown in Figure 3.6, where the power network comprises DER coupling lines and a PCC. Detailed small-signal models for these case studies including data of parameters and initial values can be found in [35]. In both cases, the primary controller is considered as the outermost control layer. The orders of the detailed model of MG_1 and MG_2 are 38 and 54, respectively. In this regard, the state variables include droop (3), voltage (2), and current (2) controllers, the output filter (4), and the coupling line (2) for GFO DERs, and include the current controller (2), the PLL (2), the output filter (4), and the coupling line (2)

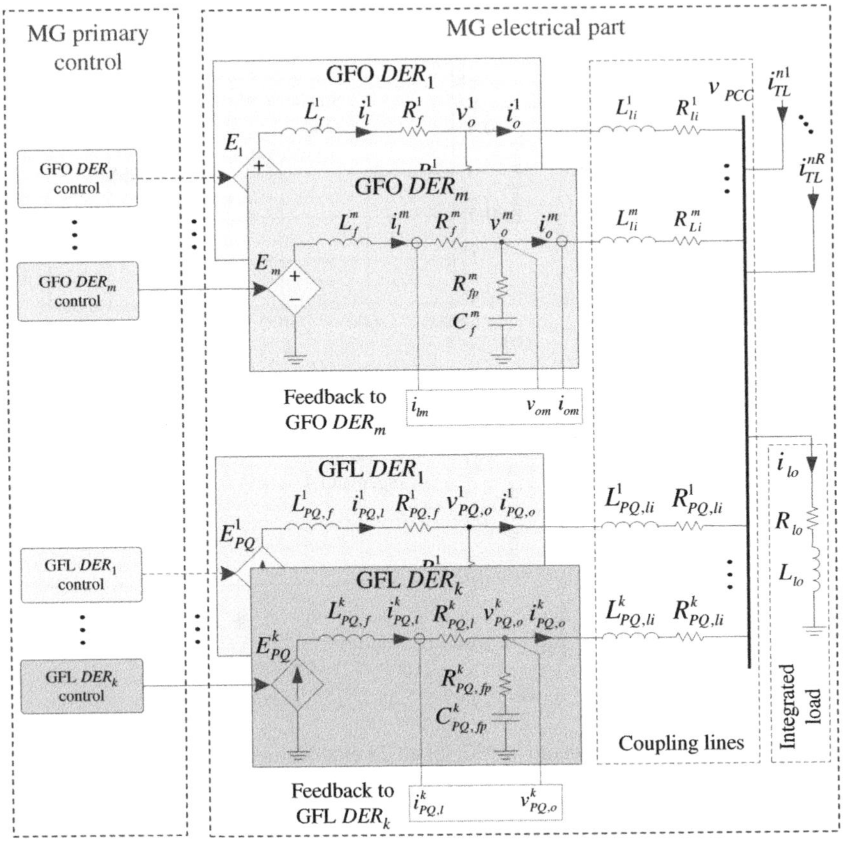

Figure 3.6 A general topology of the studied autonomous ac microgrids shown in Figure 3.5, including both GFO and GFL DERs, as well as a simplified network and an integrated constant RL load.

for GFL DERs, and include the RL load (2). The number of GFO/GFL DERs should be considered in each case study. In Chapter 2, the detailed models and the simplified models obtained there are compared using eigenvalues.

3.3.4 Islanded AC Microgrids: A Sensitivity Analysis-Based Simplified Model

Here, a simplified model of the studied islanded ac MGs is represented, which is a composition of model-based and module-based simplification methods. The methodology of the sensitivity analysis-based simplification method has been represented in Section 2.3.3.4.

3.3.4.1 Removing/Reconfiguration Process of Modules

Note that all of the insignificant modules for preserving the dominant low-frequency modes (DLFMs) cannot be removed using the regular perturbation method. It is due to the inevitable effects of such modules in the solving process of the set of governing MG equations. Thus, at least, their static model should be preserved, which can be found using the singular perturbation method. To this end, consider the SSR of a module with insignificant dynamics (X_{ID}) as

$$\varepsilon \dot{X}_{ID} = A_{ID}X_{ID} + B_{ID}U_{ID}, \tag{3.5a}$$

$$Y_{ID} = C_{ID}X_{ID} + D_{ID}U_{ID}, \tag{3.5b}$$

where U_{ID} and Y_{ID} are the input and output vectors and A_{ID}, B_{ID}, C_{ID}, and D_{ID} are the corresponding matrices with appropriate sizes. Since all of the dynamics are insignificant, if $\varepsilon \to 0$ and A_{ID} is nonsingular, one can consider $\dot{X}_{ID} \to 0$ and replace X_{ID} from (3.5a) to (3.5b) in order to find the static model as follows:

$$Y_{ID} = (D_{ID} - C_{ID}A_{ID}^{-1}B_{ID})U_{ID}. \tag{3.6}$$

The dynamics of the output filter and coupling line of GFL DERs as well as the integrated load are deleted from the simplified model using this method and their static models are used because they do not participate in DLFMs.

Reconfiguration of the dynamic and static models of the preserved modules is shown in Figure 3.7. The GFO DER_m including the power controller and the CRF transformations (DQ/dq and dq/DQ) with the coupling line and PCC voltage are restructured as shown in Figure 3.7(a). Since the dynamics of the current and voltage controllers are fast and can be removed, the input of the voltage controller v^m_{odq} will be equal to the output of the current controller v^m_{ref} (see Figure 2.25). The structure of the simplified model of GFL DER_k can be obtained by replacing the static models of the output filter and coupling line as shown in Figure 3.7(b).

In the representation of singular perturbation, the dynamics of compound modules such as GFO DER_m or GFL DER_k can be divided into fast dynamics (X_{FD}) and slow dynamics (X_{SD}), which their relationship is as

$$\varepsilon\dot{X}_{FD} = A_{11}X_{FD} + A_{12}X_{SD}, \tag{3.7a}$$

$$\dot{X}_{SD} = A_{21}X_{FD} + A_{22}X_{SD}, \tag{3.7b}$$

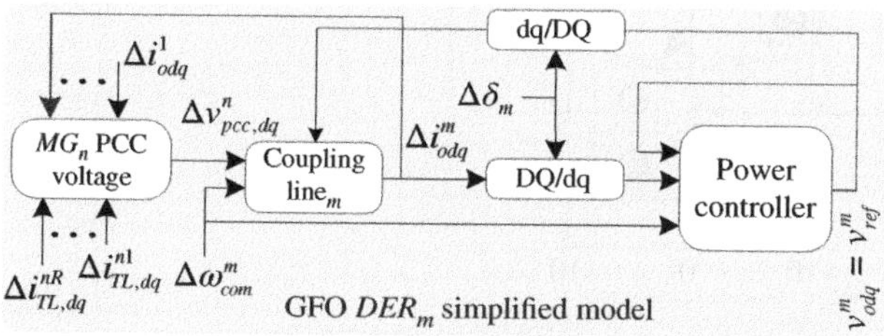

(a)

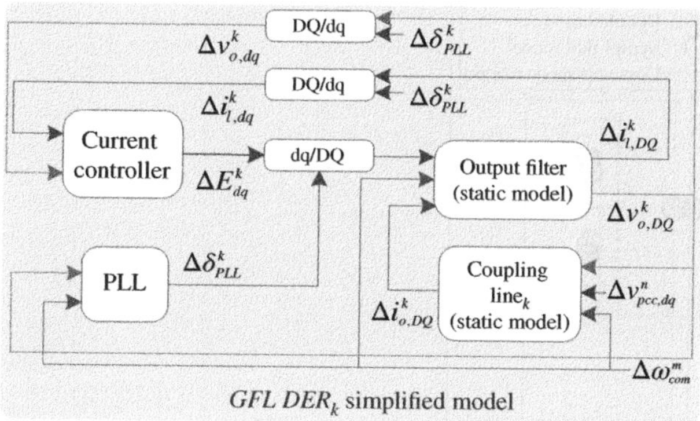

(b)

Figure 3.7 Reconfiguration of DER interconnections after removing insignificant modules for preserving the dominant low-frequency modes: (a) GFO DER_m simplified model in comparison with Figure 2.25, (b) GFL DER_k simplified model in comparison with Figure 2.28.

where A_{11}, A_{12}, A_{21}, and A_{22} are the submatrices of the compound module state matrix. If $\varepsilon \to 0$ and A_{11} is nonsingular, by replacing X_{FD} from (3.7a) to (3.7b), the dynamic modes of the simplified module model are approximated as follows:

$$\dot{X}_{SD} = (A_{22} - A_{21}A_{11}^{-1}A_{12})X_{SD}. \tag{3.8}$$

3.3.4.2 DLFMs Comparison of the Detailed and Simplified Models

In order to validate the simplified model for autonomous ac MGs, the DLFMs are compared to the DLFMs of the detailed model represented in Section 3.3.3. Figures 3.8(a) and 3.8(b) show a comparison of the MG_1 with two GFO DERs and one GFL DER and the MG_2 with four GFO DERs, respectively. The MG_1 and MG_2 are introduced in Section 3.3.3. The error can be calculated for each eigenvalue using the relative error percentage, which is 2.3 % in the worst case. In Figure 3.8(a), the dynamic mode in Group (1) is the inter-inverter mode [4] expressing the interaction between the two GFO DERs. Note that the CRF is based on the frequency for one of the DERs, e.g. the first droop-based DER, which is named the base DER here. The dynamic mode in Group (2) shows the interaction between the single GFL DER and the base DER. An important fact

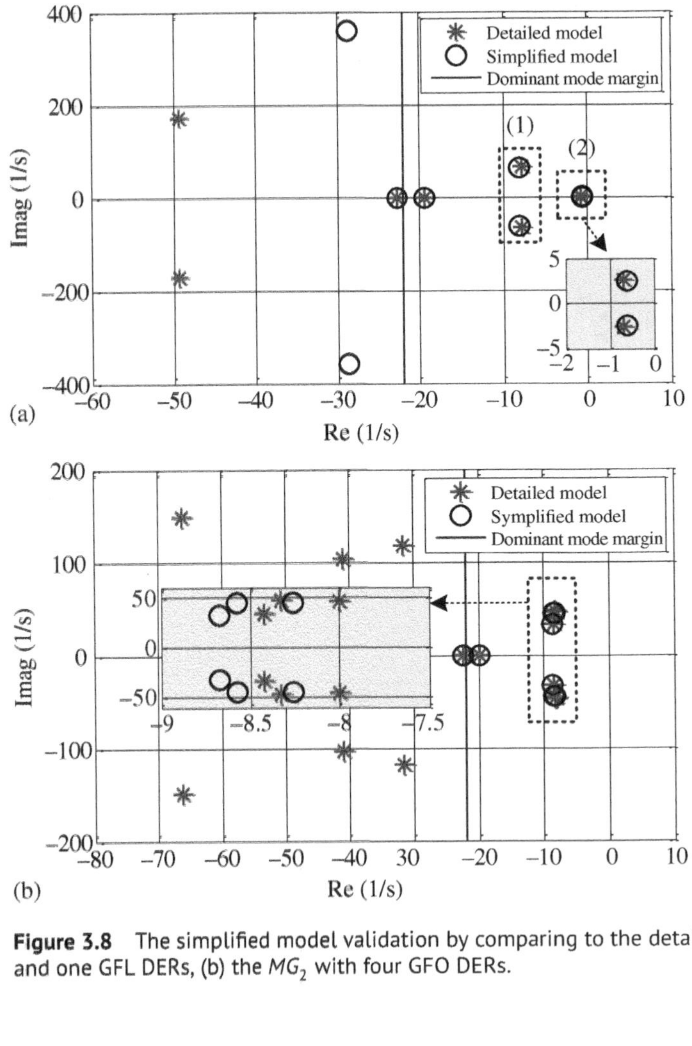

Figure 3.8 The simplified model validation by comparing to the detailed model: (a) the MG_1 with two GFO and one GFL DERs, (b) the MG_2 with four GFO DERs.

is that the DLFMs, which show the interaction among DERs, are complex conjugates (oscillatory). In Figure 3.8(b), there exist three oscillatory DLFMs showing the interaction between each GFO DER and the base DER.

The two case studies indicate an oscillatory DLFM emulating the interaction between each DER and the base DER within the MGs. It is noteworthy that the result is generalized for GFO DERs in the studied family of ac MGs with any number of DERs. Furthermore, the result is applicable to MGs with different power network structures [4]. Nevertheless, the oscillatory DLFMs expressing the interaction between GFL DERs and the base DER are limited to a relatively large subset of the studied MG family. Outside the specific parameter range, the oscillatory DLFMs will be transformed into two nonoscillatory modes. Here, the parameters of GFL DERs are considered to be within the range causing the oscillatory DLFMs. The parameter ranges can easily be obtained via sensitivity analysis.

3.3.4.3 The Oscillatory DLFM Comparison

A sensitivity analysis is provided in Figure 3.9 to compare the sensitivity-analysis-based simplified model with the existing simplified models, which are usually named by the model order of each inverter belonging to each DER. The simplified models include a fifth-order model [3], a fourth-order model [4], and a third-order model [7], whose preserved state variables are shown in Table 3.1. Moreover, the detailed model represented in Section 3.3.3 and [1] is considered to be the reference model. Here, the MG_1 with just two GFO DERs and without the GFL DER is considered. The oscillatory DLFM is of interest to be compared, and nondominant dynamic modes are not shown. For low values of the droop gain, m_p, the proposed model has a higher accuracy due to its more similar behavior to the detailed model behavior. Generally, the proposed model shows almost a similar behavior to the fifth-order model behavior to the m_p changes. It is more accurate than the fourth-order and third-order models. However, all the reduced-order models have a lower tendency to be unstable by increasing m_p than the detailed model.

Using the sensitivity analysis-based simplification method, the order of GFO and GFL DER models including the coupling line is reduced from 13 and 10 to 5 and 4, respectively. Moreover, the order of the case study MGs in Figures 3.8(a) and 3.8(b) is reduced from 38 and 54 to 14 and 20,

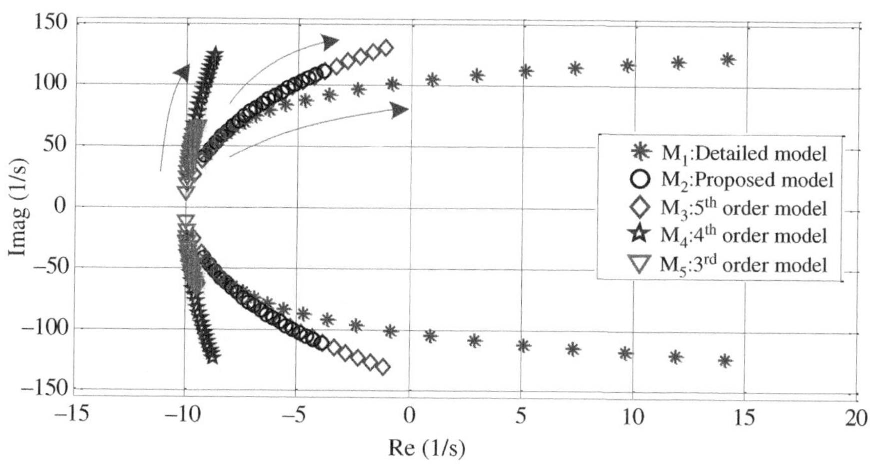

Figure 3.9 The oscillatory DLFM comparison among the sensitivity analysis-based simplified model, existing simplified models and the detailed model by the droop gain sensitivity analysis as $0.001\% < m_p < 0.02\%$.

Table 3.1 State variables of the compared models in Figure 3.9.

Model	State variables	References
M1	$\Delta\delta, \Delta P, \Delta Q, \Delta v_{odq}, \Delta i_{ldq}, \Delta i_{odq}$, CCIO[a], VCIO[b]	[1]
M2	GFO ER: $\Delta\delta, \Delta P, \Delta Q, \Delta i_{odq}$	Proposed method
	GFL DER: $\Delta\delta_{PLL}$, CCIO, PLLIO[c]	
M3	$\Delta\delta, \Delta P, \Delta Q, \Delta i_{odq}$	[3]
M4	$\Delta\delta, \Delta P, \Delta i_{odq}$	[4]
M5	$\Delta\delta, \Delta\omega, \Delta V_{od}$	[7]

a) current controller integrator outputs,
b) voltage controller integrator outputs,
c) PLL integrator output.

respectively. Although the low-frequency MG models are reduced well, the high-level dynamic studies, e.g. the frequency stability and control of IMGs justify the importance of finding a very low-order model.

3.3.5 Islanded AC Microgrids: Aggregated Single-Order Model

The main objective of developing a very low-order model, specifically a single-order model, for an MG, is its utilization within large-scale power grids, such as interconnected MGs. Here, a single-order model of ac MGs is represented for applications of the interconnected MGs, which is referred to in Section 10.3.

3.3.5.1 General Steps of Modeling

As mentioned in Section 3.3.4.1, the low-frequency dynamic behavior of autonomous ac MGs can be summarized in the oscillatory DLFMs, representing the interactions among DERs and the base DER. In fact, each DER excluding the base DER, whether GFO or GFL, is equivalent to an oscillatory DLFM. Aggregating any number of the oscillatory DLFMs to one oscillatory mode leads to a low-frequency second-order model for autonomous ac MGs, which in turn results in a single-order frequency model.

Figure 3.10 shows four steps for finding the proposed single-order model for autonomous ac MGs as follows:

- Step 1: Aggregate the dominant oscillatory modes obtained from the simplified low-frequency model presented in Section 3.3.4. Note that the aggregation method for simplifying detailed models is illustrated in Section 2.3.3.3.
- Step 2: Find the second-order model using the aggregated complex conjugate mode.
- Step 3: Calculate the virtual inertia and damping coefficient (J,D) by mapping the second-order model on the virtual swing equation.
- Step 4: Exclude the single-order frequency model for autonomous ac MGs from the second-order model.

The input and output of the transfer function, equivalent to the SSR of each DER, can be determined according to the simplified MG model using the sensitivity analysis method. Based on the

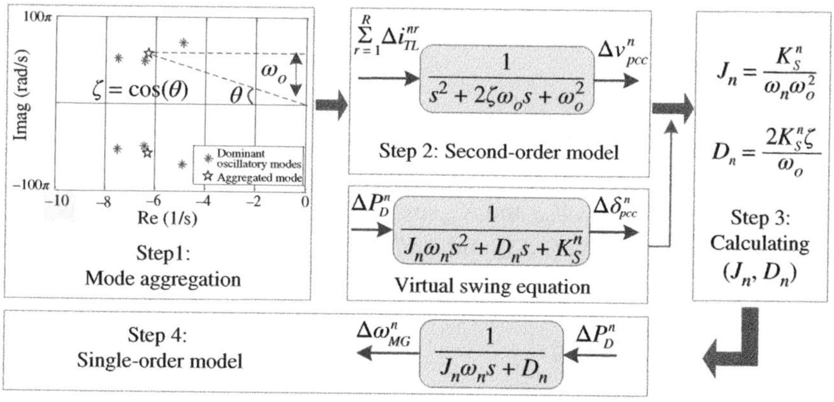

Figure 3.10 Four-step method for calculating the single-order frequency model of autonomous ac microgrids.

models presented in Figure 3.7 for GFO and GFL DERs, the simplified model of each DER is represented as

$$\Delta i_{odq}^{DER} = G_{com}^{DER}(s)\Delta\omega_{com} + G_{pcc}^{DER}(s)\Delta v_{pcc,dq}, \tag{3.9}$$

where $G_{com}^{DER}(s)$ and $G_{pcc}^{DER}(s)$ can be calculated employing the interconnection method. By applying KCL for the PCC in Figure 3.6, $\Delta v_{pcc,dq}$ is obtained as follows

$$\Delta v_{pcc,dq} = R_v \sum_{s=1}^{S} \Delta i_{odq}^{DER_s} + R_v \sum_{r=1}^{R} \Delta i_{TL,dq}^{nr} - R_v \Delta i_{Lo,dq}, \tag{3.10}$$

where R_v is the virtual resistor [1, 36], and $i_{TL,dq}^{nr}$ is the tie/interlinking line current between the studied MG called MG_n and the neighbor MG, i.e. MG_r. Substituting (3.9) and the static model of the integrated load in (3.10), all DER currents are aggregated at the PCC, and $\Delta v_{pcc,dq}$ can be indicated as

$$\Delta v_{pcc,dq} = G_{IL}^{pcc}(s) \sum_{r=1}^{R} \Delta i_{TL,dq}^{nr}, \tag{3.11}$$

where $G_{IL}^{pcc}(s)$ can also be calculated using the interconnection method [36]. Therefore, $\sum_{r=1}^{R} \Delta i_{TL,dq}^{nr}$ and $\Delta v_{pcc,dq}$ are the input and output of the second-order model of each MG, which is a model like (2.23).

3.3.5.2 Virtual Swing Equation-Based Single-Order Model

Here, a virtual swing equation is considered for fully inverter-based autonomous ac MGs as

$$J_n \omega \frac{d\Delta\omega_{MG}^n}{dt} = \Delta P_{MG}^n - D_n \Delta\omega_{MG}^n, \tag{3.12}$$

where J_n and D_n are the virtual moment of inertia and damping coefficient of MG_n. ΔP_{MG}^n is the net value of MG_n input power as

$$\Delta P_{MG}^n = \sum_{q=1}^{Q} T_{qn}(\Delta\delta_{pcc}^q - \Delta\delta_{pcc}^n), \tag{3.13}$$

where $\Delta\delta_{pcc}^{q}$ is the PCC voltage angle of the connected grid, e.g. an interconnected MG_q to the studied MG_n, and T_{qn} can be calculated as

$$T_{qn} = \frac{V_{pcc}^{q} V_{pcc}^{n}}{Z_{qn}^{2}} \left[R_{qn}\sin(\Delta\delta_{pcc}^{qn}) + X_{qn}\cos(\Delta\delta_{pcc}^{qn})\right], \tag{3.14}$$

where $\Delta\delta_{pcc}^{qn} = \delta_{pcc}^{q} - \delta_{pcc}^{n}$, $Z_{qn} = \sqrt{R_{qn}^2 + X_{qn}^2}$, R_{qn} and X_{qn} are the resistance and reactance of the tie-line between MG_q and MG_n. Substituting (3.13) and $\Delta\omega_{MG}^{n} = d\Delta\delta_{pcc}^{n}/dt$ in (3.12) and taking Laplace transform result in

$$G_{VSE}^{MG_n}(s) = \frac{\Delta\delta_{pcc}^{n}(s)}{\Delta P_{D}^{n}(s)} = \frac{1}{J_n\omega_n s^2 + D_n s + K_S^n}, \tag{3.15}$$

where VSE in the subscript indicates a virtual synchronous equation, and K_S^n and ΔP_D^n are the synchronizing power coefficient and disturbance power of MG_n as

$$K_S^n = \sum_{q=1}^{Q} T_{qn}, \tag{3.16a}$$

$$\Delta P_D^n = \sum_{q=1}^{Q} T_{qn}\Delta\delta_{pcc}^{q}. \tag{3.16b}$$

Note that the transfer functions (3.11) and (3.15) are equivalent. Therefore, each autonomous ac MG can be mapped on a virtual swing equation, and J_n and D_n can be calculated by comparing (3.11) and (3.15), which results in

$$J_n = \frac{K_S^n}{\omega_n\omega_o^2}, \tag{3.17a}$$

$$D_n = \frac{2K_S^n\zeta}{\omega_o}. \tag{3.17b}$$

Finally, considering $\Delta\omega_{MG}^{n} = d\Delta\delta_{pcc}^{n}/dt$ and (3.15), the single-order frequency model of autonomous ac MG_n is shown as

$$G_{freq}^{MG_n}(s) = \frac{\Delta\omega_{MG}^{n}(s)}{\Delta P_{D}^{n}(s)} = \frac{1}{J_n\omega_n s + D_n}. \tag{3.18}$$

Exercise 3.1 Obtain the single-order models of the studied ac MGs shown in Figure 3.5.

Solution

The single-order models of the studied MGs are obtained using the four-step method shown in Figure 3.10.

For the MG_1, $J_1 = 0.71$ (W·s²/rad) and $D_1 = 1.9$ (kW·s/rad), and for the MG_2, $J_2 = 1.45$ (W·s²/rad) and $D_2 = 5.6$ (kW·s/rad) are calculated using (3.17a) and (3.17b). Therefore, the single-order frequency models of the MG_1 and MG_2 for the nominal angular frequency as $\omega_n = 314$ (rad/s) can be given as follows using (3.18):

$$G_{freq}^{MG_1}(s) = \frac{\Delta\omega_{MG_1}(s)}{\Delta P_D 1(s)} = \frac{1}{222.9\,s + 1900}, \tag{3.19a}$$

$$G_{freq}^{MG_2}(s) = \frac{\Delta\omega_{MG_2}(s)}{\Delta P_D 2(s)} = \frac{1}{455.3\,s + 5600}. \tag{3.19b}$$

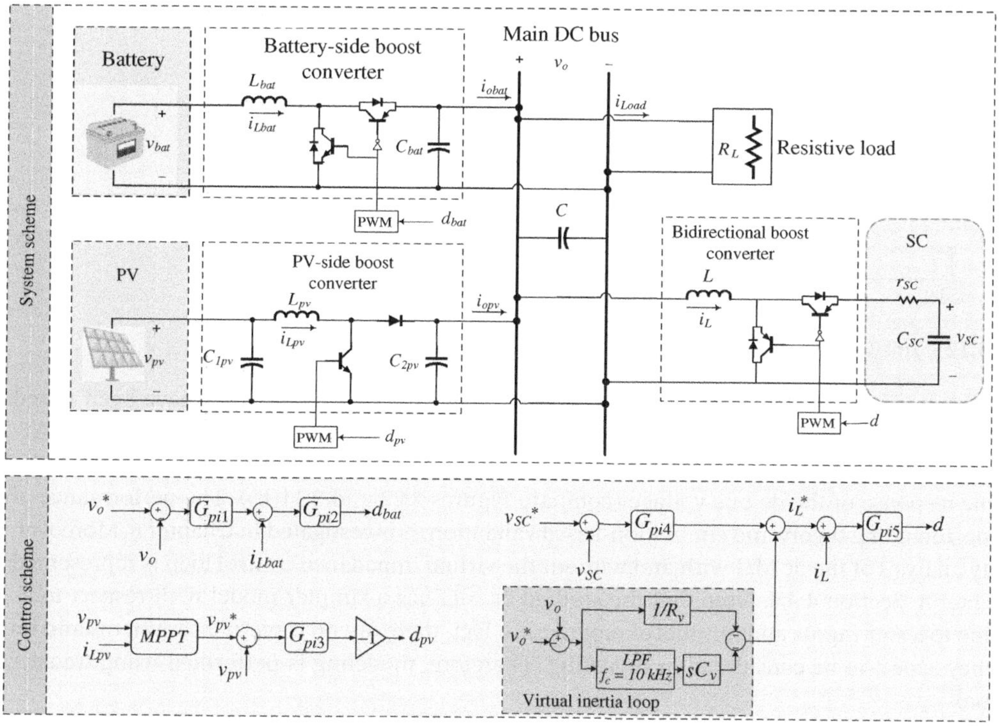

Figure 3.11 Structure of the studied dc MG: (a) power components and connections, and (b) control units of the PV, the battery, and the super-capacitor.

3.3.6 Islanded DC Microgrid

The studied dc MG is shown in Figure 3.11. In this structure, two types of energy sources, i.e. PV and battery ESS, are considered. The PV system is selected to provide the average load demand all the time and to store surplus power in the battery for backup. The battery is chosen such that it can regulate the oscillating output of PV and support the critical loads all the time regardless of renewable energy source (RES) conditions. The power flow between each source and the main grid is controlled via dc–dc converters. This improves the overall operation of the dc MG and cycle life and enhances the flexibility of the energy sources [37]. The dc MG shown in Figure 3.11 has a two-stage configuration. In the first level, namely the source stage, in order to absorb the maximum available power for PV, a unidirectional boost converter is employed. Meanwhile, the bidirectional boost converter is used for the battery to generate a constant output voltage. The second level is a common load bus, where the loads are connected to the main load bus in this stage. Different voltage levels can be used in dc MGs. The PV with the rated output voltage of 109.4 V and the battery with the rated output voltage of 100 V are selected in the first stage, and the output voltage, i.e. the main dc bus voltage is fixed at 200 V. A super-capacitor (SC) is considered as a secondary energy storage connected to the main dc bus to improve the performance and inertia of the overall system [38]. The SCs have been studied for start–stop energy system applications such as electrical vehicles (EV), heavy-duty trucks [39], and high pulse power applications [40]. It is well known that the SC-ESS rarely provides continuous load power. Indeed, it just cooperates during start-up and transient states from one power level to another. The power flow exchange between the SC and the dc MG is controlled through a bidirectional dc–dc converter.

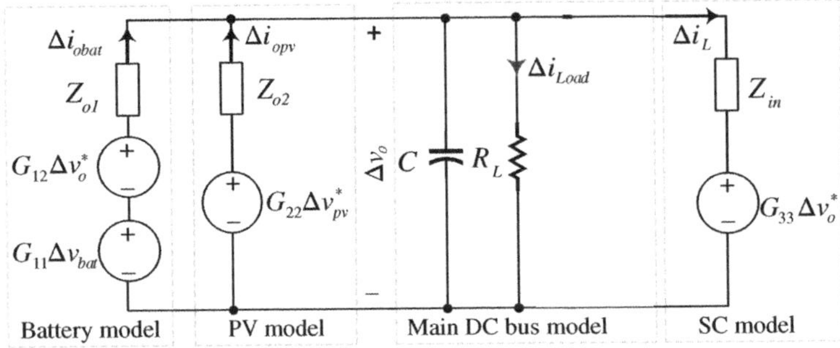

Figure 3.12 Small-signal model of dc microgrid in a simplified schematic.

In order to find a model for the studied dc MG, the SSRs of all its modules have been found in Chapter 2. Therefore, one can easily calculate the overall model using RCT functions in MATLAB. However, a virtual impedance loop is added to the basic control of the SC to improve the dynamic response of the dc bus voltage (compare Figures 2.22 and 3.11(b)). The performance of this loop, including theory and simulation-based validation, is investigated in Chapter 6. Moreover, stability analysis of the dc MG with and without the virtual impedance control loop is represented in Chapter 4, Section 4.4.4. Note that the studied dc MG has a simpler model with respect to ac MGs due to lesser inputs and outputs of modules. In fact, there are no three-phase systems and no frequency aspect to be considered in modeling. Therefore, modeling is performed using transfer functions.

The virtual impedance control loop can be modeled and an SSR can be given as follows:

$$\dot{x}_{SC,VI} = a_{SC,VI}x_{SC,VI} + B_{SC,VI}U_{SC,VI}, \tag{3.20a}$$

$$y_{SC,VI} = c_{SC,VI}x_{SC,VI} + D_{SC,VI}U_{SC,VI}, \tag{3.20b}$$

where $x_{SC,VI}$ is the output of the internal integrator of the low-pass filter (LPF), $U_{SC,VI} = [\Delta v_o\ \Delta v_o^*]^T$, $a_{SC,VI} = -\omega_c$, $B_{SC,VI} = C_v\omega_c^2[1\ -1]$, $c_{SC,VI} = 1$, and $D_{SC,VI} = (C_v\omega_c + 1/R_v)[-1\ 1]$.

Figure 3.12 shows a simplified schematic of the small-signal model of the dc MG, where the SSRs are transformed into the equivalent transfer functions. In this schematic, the models of power and control parts of each generation unit are merged. The battery model is given as follows:

$$\hat{v}_o(s) = G_{11}(s)\cdot\Delta v_{bat}(s) + G_{12}(s)\cdot\Delta v_o^*(s) - Z_{o1}(s)\cdot\Delta i_{obat}(s), \tag{3.21}$$

where

$$G_{11}(s) = \frac{D'_{bat} + i_{Lbat0}G_{pi2}(s)}{T_1(s)},\ G_{12}(s) = \frac{(v_{o0}D'_{bat} - i_{Lbat0}L_{bat}s)\,G_{pi1}(s)G_{pi2}(s)}{T_1(s)},$$

$$Z_{o1}(s) = \frac{L_{bat}s + v_{bat0}G_{pi2}(s)}{T_1(s)},$$

$$T_1(s) = {D'_{bat}}^2 + L_{bat}C_{bat}s^2 + G_{pi1}(s)G_{pi2}(s)\left(v_{o0}D'_{bat} - i_{Lbat0}L_{bat}s\right)$$
$$+ G_{pi2}(s)\left(v_{o0}C_{bat}s + D'_{bat}i_{Lbat0}\right)\cdot$$

and $D'_{bat} = 1 - D_{bat}$. The PV transfer function-based small-signal model is as follows:

$$\Delta v_o(s) = G_{22}(s)\cdot\Delta v_{pv}^*(s) - Z_{o2}(s)\cdot\Delta i_{opv}(s), \tag{3.22}$$

where

$$G_{22}(s) = i_{Lpv0}G_{pi3}(s) - T_2(s)v_{o0}G_{pi3}(s),\ Z_{o2}(s) = \frac{1}{C_{2pv}s + T_2(s)D'_{pv}},$$

$$T_2(s) = \frac{D'_{pv}\left(C_{2pv}s + 1/r_{pv}\right) + i_{Lpv0}G_{pi3}(s)}{1 + v_{o0}G_{pi3}(s) + L_{pv}s\left(C_{2pv}s + 1/r_{pv}\right)}, \tag{3.23}$$

and $D'_{pv} = 1 - D_{pv}$. The SC transfer function-based small-signal model can also be given as follows:

$$\Delta v_o(s) = G_{33}(s) \cdot \Delta v_o^*(s) + Z_{in}(s)\Delta i_L(s), \tag{3.24}$$

where

$$G_{33}(s) = \frac{v_{sc0}G_{pi5}(s)\left(C_v sG_{LPF}(s) + 1/R_v\right)}{T_3(s)},$$

$$Z_{in}(s) = \frac{Ls + D'_{sc}r_{sc} + v_{sc0}G_{pi5}(s) + \frac{D'_{sc}}{C_{sc}s}\left(D'_{sc} + v_{sc0}G_{pi4}(s)G_{pi5}(s)\right)}{T_3(s)},$$

$$T_3(s) = 1 + v_{sc0}G_{pi5}(s)\left(C_v sG_{LPF}(s) + 1/R_v\right),$$

and $D'_{sc} = 1 - D_{sc}$. Finally, one can find the closed-loop transfer function between the output voltage and the reference voltage as follows:

$$G_{cl,vv*}^{dcMG}(s) = \frac{\Delta v_o(s)}{\Delta v_o^*(s)} = -\frac{Z_{o1}{}^{-1}G_{12} + Z_{in}{}^{-1}G_{33}}{Cs + R_L{}^{-1} + Z_{o1}{}^{-1} + Z_{in}{}^{-1} + Z_{o2}{}^{-1}}. \tag{3.25}$$

In addition, the closed-loop transfer function-based small-signal model between the dc bus voltage and the load current is achieved as follows:

$$G_{cl,vi}^{dcMG}(s) = \frac{\Delta v_o(s)}{\Delta i_{Load}(s)} = -\frac{1}{Cs + R_L{}^{-1} + Z_{o1}{}^{-1} + Z_{in}{}^{-1} + Z_{o2}{}^{-1}}. \tag{3.26}$$

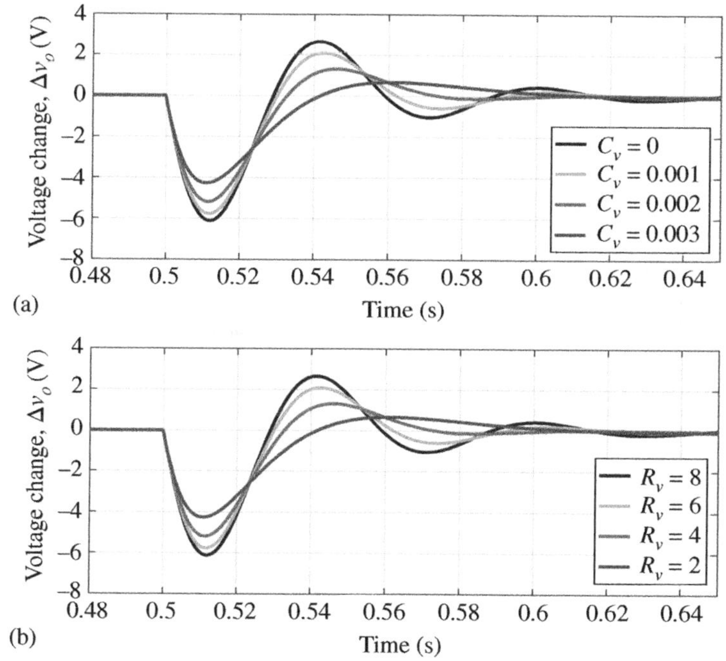

Figure 3.13 Step change in the load power of the studied dc MG: (a) various values of C_v, $R_v = 5$ and, (b) various values of R_v, $C_v = 0.002$.

Figure 3.13 shows the step responses of $G_{cl}^{dcMG}(s)$ for a 20% change in load power. From Figure 3.13(a), the dc bus voltage would smoothly change, and the rate of change of voltage (RoCoV) is lower when C_v increases. Thereby, the value of C_v denotes the synthetic inertia of the dc MG. However, by increasing C_v, the system damping decreases and is prone to cause the dc bus voltage oscillation. This problem can be solved by increasing the damping coefficient $(1/R_v)$ that effectively eliminates the low-frequency voltage oscillations, as shown in Figure 3.13(b). The proposed control parameters are determined to ensure small-signal stability due to small disturbances or load perturbations at steady-state operation.

3.4 Large-Signal Modeling of Microgrids

Figure 3.14 shows an ac MG including four GFO DERs, a power network with three lines, and a variety of load types comprising a constant resistive load, an active load, a motor load, and a CPL. All these modules are modeled in Sections 2.4.5, 2.4.6, 2.5.1.2. Nevertheless, here, their models before linearizing are used, i.e. nonlinear models, which are obtained using differential equations. However, a large number of modules are inherently linear, e.g. proportional-integral (PI) and droop controllers. Therefore, their SSRs can be used here. Tables 3.2–3.5 show complete information of the studied MG modules.

3.4.1 Model Validation

The studied MG is simulated in MATLAB/SimPowerSystems and utilizes differential and algebraic equations. Figure 3.15 shows the comparison results of these two different models. In

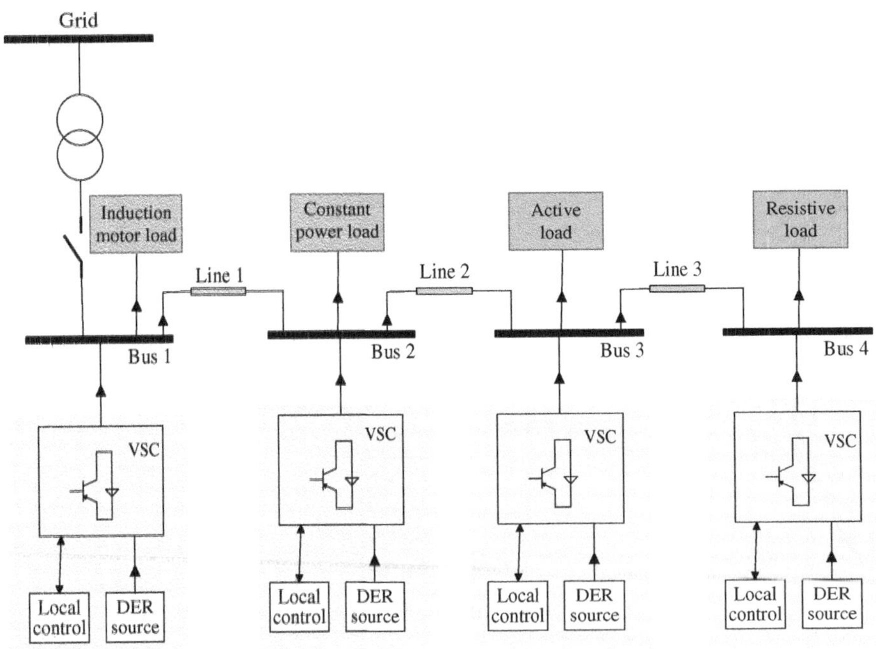

Figure 3.14 A comprehensive ac MG with four GFO DERs and various load types (VSC: voltage source converter).

Table 3.2 Parameters of the GFO DERs in the ac MG shown in Figure 3.14.

Parameter	Value	Parameter	Value	Parameter	Value
$P_{1,nom}$	10 (kW)	k_P1	0.628 rad/(kWs)	K_{pv}	0.05
$P_{2,nom}$	15 (kW)	k_P2	0.418 rad/(kWs)	K_{iv}	390
$P_{3,nom}$	20 (kW)	k_P3	0.314 rad/(kWs)	K_{pi}	10.5
$P_{4,nom}$	20 (kW)	k_P4	0.252 rad/(kWs)	K_{ii}	16 000
$Q_{1,nom}$	6 (kVAr)	k_Q1	1.66 V/(kVAr)	f_{sw}	8 (kHz)
$Q_{2,nom}$	9 (kVAr)	k_Q2	1.11 V/(kVAr)	f_{nom}	50 Hz
$Q_{3,nom}$	12 (kVAr)	k_Q3	0.833 V/(kVAr)	v_{nom}	230 (V_rms)
$Q_{4,nom}$	15 (kVAr)	k_Q4	0.666 V/(kVAr)	ω_c	31.4 (rad/s)
L_f	1.35 (mH)	C_f	50 μF	R_f	0.1 Ω

Table 3.3 Parameters of the motor load in the ac MG shown in Figure 3.14.

Parameter	Value	Parameter	Value	Parameter	Value
P_{nom}	10 (hp)	r_s	0.7834 (Ω)	L_{rr}	47.75 (mH)
V_{nom}	400 (V)	L_{ss}	127.1 (mH)	L_m	124.1 (mH)
f_{nom}	50 (Hz)	r_r	0.7402 (Ω)	P (pole)	4

Table 3.4 Parameters of the active load in the ac MG shown in Figure 3.14.

Parameter	Value	Parameter	Value	Parameter	Value
L_f	1.35 (mH)	K_{pv}	0.5	L_c	0.93 (mH)
C_f	8.8 (μF)	K_{iv}	150	P_{nom}	25 (kW)
R_f	0.1 (Ω)	K_{pi}	7	R_{lo}	40.83 (Ω)
f_{sw}	10 (kHz)	K_{ii}	25 000	v_{dc}	700 (V)
ω_c	31.4 (rad/s)	R_c	0.03 (Ω)		

Table 3.5 Parameters of the CPL and lines in the ac MG shown in Figure 3.14.

Parameter	Value	Parameter	Value
Z_{Li1}	0.23+j0.11 (Ω)	Z_{Li3}	0.30+j0.47 (Ω)
Z_{Li2}	0.35+j0.58 (Ω)	S_{CPL}	4.25 (kW)+j2.63 (kVAr)

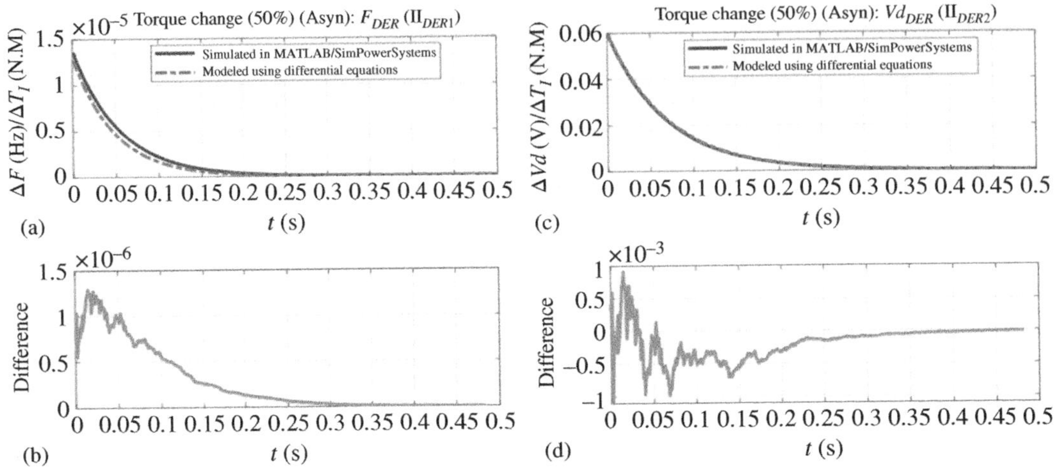

Figure 3.15 Model validation of the studied MG in Figure 3.14: (a) relative changes of the frequency of DER_3 due to 50% variation of the motor load torque, (b) the error between the nonlinear simulated system and the nonlinear model for the DER_3 frequency, (c) relative changes of the direct voltage component of DER_3 due to 50% variation of the motor load torque, (d) the error between the nonlinear simulated system and the nonlinear model for the DER_3 voltage.

Figure 3.15(a), relative changes in the frequency of DER_3 due to a 50% variation in the motor load torque are shown. It is obvious that the difference between these two models is low. Moreover, Figure 3.15(b) shows the error between the nonlinear simulated system and the nonlinear model for the DER_3 frequency changes shown in Figure 3.15(c), which demonstrates a very small error. Figure 3.15(c) shows relative changes in the direct component of the output DER_3 voltage due to a 50% variation in the motor load torque. Figure 3.15(d) also shows the corresponding error between the simulated system and the model outputs. Here, a small error can also be seen during the occurrence of the disturbance.

3.4.2 Time-Domain Simulations

Although time-domain simulations, real-time simulations, and experimental tests are used throughout the book for stability analysis and control design validation, here some time-domain simulation results of the studied MG shown in Figure 3.14 are represented in Figure 3.16. The initial conditions are zero without the frequency, which is started from 50 Hz. A disturbance occurs at $t = 0.45$ s and continues to $t = 0.65$ s as a 10% increase in the nominal voltage of the MG. The results include some waveforms of the both simulated MG in MATLAB/SimPowerSystems and the modeled MG using differential equations.

Figures 3.16(a) and 3.16(b) show the active power of DER_2 and the reactive power of DER_1 from the start of the simulation to $t = 1$ s. After a transient with 0.2 s settling time, the steady-state value of the active power is about 10 kW, and the reactive power is about 4.4 kVAr. The disturbance behavior can also be seen, where the reactive power shows larger oscillations. Figure 3.16(c) shows the frequency of DER_3. The transients are very small due to a considerable coupling between frequency and voltage in the studied MG. Figure 3.16(d) shows the electromagnetic torque of the asynchronous motor load. The oscillation pattern of the electromagnetic torque is negative to the frequency oscillation pattern. However, the amplitude of oscillation is considerable with respect to that of the frequency due to a direct relationship with squared voltage amplitude. The

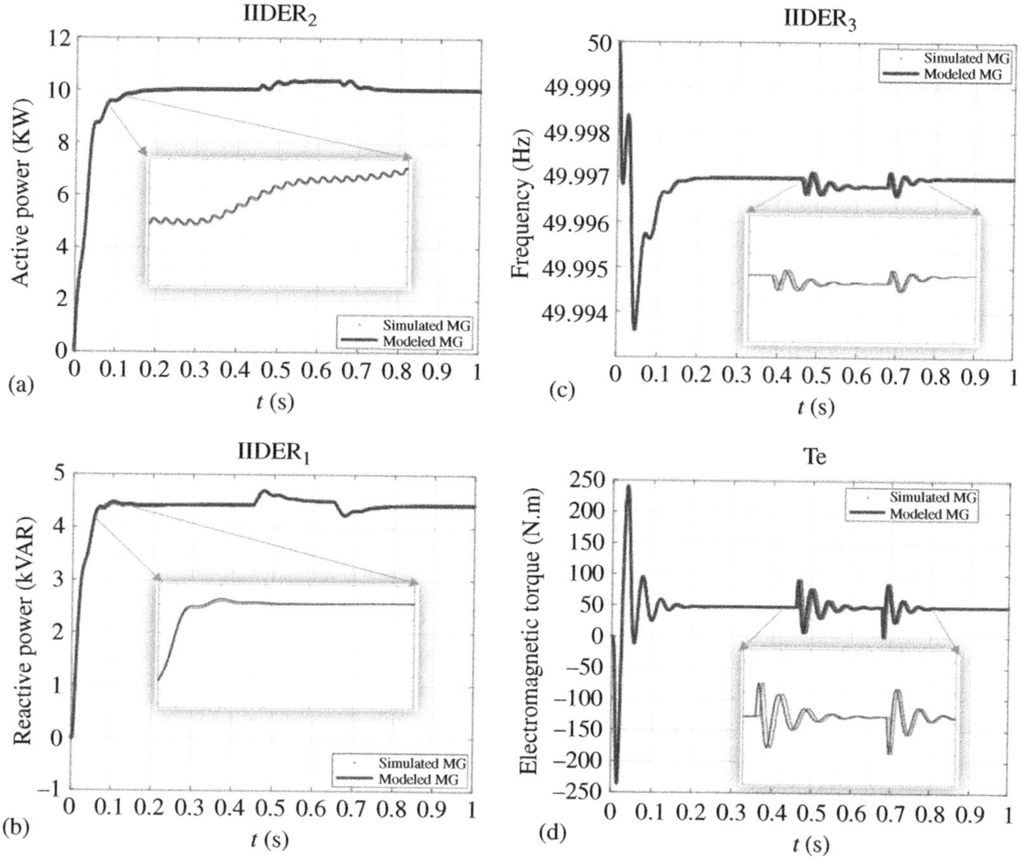

Figure 3.16 Time-domain simulation results of the studied ac MG shown in Figure 3.14 comparing the derived nonlinear model: (a) the active power of DER_2, (b) the reactive power of DER_1, (c) the frequency of DER_3, and (d) the electromagnetic torque of the asynchronous motor load.

results show a good match between the time-domain simulations and the obtained nonlinear model.

3.5 Summary

This chapter provides an overview of the dynamic modeling of different ac and dc MGs. The concept of a common reference frame is essential for small-signal modeling of ac MGs, which is explained in detail. In Chapter 2, separate small-signal models, represented as state-space representations, were calculated for the most important MG electrical and control subsystems. In this chapter, the submodel interconnections based on their real electrical, communication, and control connections in some case studies, using MATLAB codes, are presented. Although there are only a few case studies due to the limitations of the book pages, the general steps and basic codes are provided to enable researchers to easily generalize the modeling method for other case studies with different structures and control strategies.

Since the modeling method is based on separate module models and then interconnecting them in a generalized numerical method, its expandability is very high. Model validation is done using eigenvalue analysis and step response for some case studies. Prony analysis-based validation will be presented in Chapter 8 for some other cases, i.e. interconnected MGs. Note that more analytical results will be presented in Chapter 4 which focuses on the stability analysis of the studied MGs.

References

1 Nagaraju Pogaku, Milan Prodanovic, and Timothy C Green. Modeling, analysis and testing of autonomous operation of an inverter-based microgrid. *IEEE Trans. Power Electron.*, 22(2):613–625, 2007.

2 Juan C Vasquez, Josep M Guerrero, Mehdi Savaghebi, Joaquin Eloy-Garcia, and Remus Teodorescu. Modeling, analysis, and design of stationary-reference-frame droop-controlled parallel three-phase voltage source inverters. *IEEE Trans. Ind. Electron.*, 60(4):1271–1280, 2013.

3 Ling Luo and Sairaj V Dhople. Spatiotemporal model reduction of inverter-based islanded microgrids. *IEEE Trans. Energy Convers.*, 29(4):823–832, 2014.

4 Iraklis P Nikolakakos, H H Zeineldin, Mohamed Shawky El-Moursi, and James L Kirtley. Reduced-order model for inter-inverter oscillations in islanded droop-controlled microgrids. *IEEE Trans. Smart Grid*, 9(5):4953–4963, 2017.

5 Zhikang Shuai, Yelun Peng, Xuan Liu, Zuyi Li, Josep M Guerrero, and Z John Shen. Dynamic equivalent modeling for multi-microgrid based on structure preservation method. *IEEE Trans. Smart Grid*, 2018. doi: 10.1109/TSG.2018.2844107.

6 Md Rasheduzzaman, Jacob A Mueller, and Jonathan W Kimball. Reduced-order small-signal model of microgrid systems. *IEEE Trans. Sustainable Energy*, 6(4):1292–1305, 2015.

7 Petr Vorobev, Po-Hsu Huang, Mohamed Al Hosani, James L Kirtley, and Konstantin Turitsyn. High-fidelity model order reduction for microgrids stability assessment. *IEEE Trans. Power Syst.*, 33(1):874–887, 2017.

8 Zheng Yuan, Zhengchun Du, Chongtao Li, and Ting An. Dynamic equivalent model of VSC based on singular perturbation. *IET Gener. Transm. Distrib.*, 10(14):3413–3422, 2016.

9 Valerio Mariani, Francesco Vasca, Juan C Vásquez, and Josep M Guerrero. Model order reductions for stability analysis of islanded microgrids with droop control. *IEEE Trans. Ind. Electron.*, 62(7):4344–4354, 2014.

10 Xiaoqiang Guo, Zhigang Lu, Baocheng Wang, Xiaofeng Sun, Lei Wang, and Josep M Guerrero. Dynamic phasors-based modeling and stability analysis of droop-controlled inverters for microgrid applications. *IEEE Trans. Smart Grid*, 5(6):2980–2987, 2014.

11 Chengshan Wang, Hao Yu, Peng Li, Jianzhong Wu, and Chengdi Ding. Model order reduction for transient simulation of active distribution networks. *IET Gener. Transm. Distrib.*, 9(5):457–467, 2015.

12 Ernane Antonio Alves Coelho, Porfirio Cabaleiro Cortizo, and Pedro Francisco Donoso Garcia. Small-signal stability for parallel-connected inverters in stand-alone AC supply systems. *IEEE Trans. Ind. Appl.*, 38(2):533–542, 2002.

13 Iraklis P Nikolakakos, Hatem H Zeineldin, Mohamed Shawky El-Moursi, and Nikos D Hatziargyriou. Stability evaluation of interconnected multi-inverter microgrids through critical clusters. *IEEE Trans. Power Syst.*, 31(4):3060–3072, 2016.

14 Andre Pires Nobrega Tahim, Daniel J Pagano, Eduardo Lenz, and Vinicius Stramosk. Modeling and stability analysis of islanded DC microgrids under droop control. *IEEE Trans. Power Electron.*, 30(8):4597–4607, 2014.

15 Tomislav Dragičević, Xiaonan Lu, Juan C Vasquez, and Josep M Guerrero. DC microgrids—Part I: A review of control strategies and stabilization techniques. *IEEE Trans. Power Electron.*, 31(7):4876–4891, 2015.

16 Amr Ahmed A Radwan and Yasser Abdel-Rady I Mohamed. Linear active stabilization of converter-dominated DC microgrids. *IEEE Trans. Smart Grid*, 3(1):203–216, 2011.

17 Nasim Rashidirad, Mohsen Hamzeh, Keyhan Sheshyekani, and Ebrahim Afjei. A simplified equivalent model for the analysis of low-frequency stability of multi-bus DC microgrids. *IEEE Trans. Smart Grid*, 9(6):6170–6182, 2017.

18 Xiaoyong Chang, Yongli Li, Xuan Li, and Xiaolong Chen. An active damping method based on a supercapacitor energy storage system to overcome the destabilizing effect of instantaneous constant power loads in DC microgrids. *IEEE Trans. Energy Convers.*, 32(1):36–47, 2016.

19 Sajad Ghadiriyan and Mohsen Rahimi. Mathematical representation, stability analysis and performance improvement of DC microgrid system comprising hybrid wind/battery sources and CPLs. *IET Gener. Transm. Distrib.*, 13(10):1845–1855, 2019.

20 Allahyar Akhbari and Mohsen Rahimi. Control and stability analysis of DFIG wind system at the load following mode in a DC microgrid comprising wind and microturbine sources and constant power loads. *Int. J. Electr. Power Energy Syst.*, 117:105622, 2020.

21 Zeinab Karami, Qobad Shafiee, Yousef Khayat, Meysam Yaribeygi, Tomislav Dragičević, and Hassan Bevrani. Decentralized model predictive control of DC microgrids with constant power load. *IEEE J. Emerging Sel. Top. Power Electron.*, 9(1):451–460, 2019. doi: 10.1109/JESTPE.2019.2957231.

22 Mehran Jami, Qobad Shafiee, and Hassan Bevrani. Dynamic improvement of DC microgrids using a dual approach based on virtual inertia. *J. Modern Power Syst. Clean Energy*, 10(3):667–677, 2020.

23 Airan Frances, Rafael Asensi, Ó García, Roberto Prieto, and Javier Uceda. Modeling electronic power converters in smart DC microgrids—an overview. *IEEE Trans. Smart Grid*, 9(6):6274–6287, 2017.

24 Rui Wang, Qiuye Sun, Pengfei Tu, Jianfang Xiao, Yonghao Gui, and Peng Wang. Reduced-order aggregate model for large-scale converters with inhomogeneous initial conditions in DC microgrids. *IEEE Trans. Energy Convers.*, 36(3):2473–2484, 2021.

25 Qianwen Xu, Navid Vafamand, Linglin Chen, Tomislav Dragičević, Lihua Xie, and Frede Blaabjerg. Review on advanced control technologies for bidirectional DC/DC converters in DC microgrids. *IEEE J. Emerging Sel. Top. Power Electron.*, 9(2):1205–1221, 2020. doi: 10.1109/JESTPE.2020.2978064.

26 Srikanth Kotra and Mahesh K Mishra. Design and stability analysis of DC microgrid with hybrid energy storage system. *IEEE Trans. Sustainable Energy*, 10(3):1603–1612, 2019.

27 Mohd Alam, Kuldeep Kumar, Saket Verma, and Viresh Dutta. Renewable sources based DC microgrid using hydrogen energy storage: Modelling and experimental analysis. *Sustainable Energy Technol. Assess.*, 42:100840, 2020.

28 Syeda Shafia Zehra, Aqeel Ur Rahman, and Iftikhar Ahmad. Fuzzy-barrier sliding mode control of electric-hydrogen hybrid energy storage system in DC microgrid: Modelling, management and experimental investigation. *Energy*, 239:122260, 2022.

29 Vahidreza Nasirian, Seyedali Moayedi, Ali Davoudi, and Frank L Lewis. Distributed cooperative control of DC microgrids. *IEEE Trans. Power Electron.*, 30(4):2288–2303, 2014.

30 Lexuan Meng, Tomislav Dragicevic, Javier Roldan-Perez, Juan C Vasquez, and Josep M Guerrero. Modeling and sensitivity study of consensus algorithm-based distributed hierarchical control for DC microgrids. *IEEE Trans. Smart Grid*, 7(3):1504–1515, 2015.

31 Michele Tucci, Lexuan Meng, Josep M Guerrero, and Giancarlo Ferrari-Trecate. Stable current sharing and voltage balancing in DC microgrids: A consensus-based secondary control layer. *Automatica*, 95:1–13, 2018.

32 Mi Dong, Li Li, Yuwen Nie, Dongran Song, and Jian Yang. Stability analysis of a novel distributed secondary control considering communication delay in DC microgrids. *IEEE Trans. Smart Grid*, 10(6):6690–6700, 2019.

33 Xia Chen, Jianyu Zhou, Mengxuan Shi, Yin Chen, and Jinyu Wen. Distributed resilient control against denial of service attacks in DC microgrids with constant power load. *Renewable Sustainable Energy Rev.*, 153:111792, 2022.

34 Farhad Shahnia and Ali Arefi. Eigenanalysis-based small signal stability of the system of coupled sustainable microgrids. *Int. J. Electr. Power Energy Syst.*, 91:42–60, 2017.

35 Matlab codes for the microgrids dynamic modeling, stability and control book. www.wiley.com/go/shafiee/microgrids. Accessed: 2023-07-07.

36 Mobin Naderi, Yousef Khayat, Qobad Shafiee, Tomislav Dragicevic, Hassan Bevrani, and Frede Blaabjerg. Interconnected autonomous AC microgrids via back-to-back converters—Part I: Small-signal modeling. *IEEE Trans. Power Electron.*, 35(5):4728–4740, 2019.

37 Aitor Etxeberria, Ionel Vechiu, Haritza Camblong, and J-M Vinassa. Comparison of three topologies and controls of a hybrid energy storage system for microgrids. *Energy Convers. Manage.*, 54(1):113–121, 2012.

38 Ahmad Saudi Samosir and Abdul Halim Mohamed Yatim. Implementation of dynamic evolution control of bidirectional DC–DC converter for interfacing ultracapacitor energy storage to fuel-cell system. *IEEE Trans. Ind. Electron.*, 57(10):3468–3473, 2010.

39 Christopher R Lashway, Ahmed T Elsayed, and Osama A Mohammed. Hybrid energy storage management in ship power systems with multiple pulsed loads. *Electr. Power Syst. Res.*, 141:50–62, 2016.

40 Patricia H Smith, Raymond B Sepe Jr., Kyle G Waterman, and L Jeff Myron. Development and analysis of a lithium carbon monofluoride battery-lithium ion capacitor hybrid system for high pulse-power applications. *J. Power Sources*, 327:495–506, 2016.

4

Microgrids Stability

4.1 Introduction

Microgrids (MGs) are active distribution networks including both generation and consumption units. Generation units consist of different types of distributed energy resources (DERs) and energy storage systems (ESSs). DERs can be intermittent renewable converter-based or nonintermittent rotary generators. Consumption units can include a wide range of constant impedance loads, constant power loads, motor loads, or active loads. Some of them have complicated control structures or critical dynamic modes. On the other hand, MGs can be operated in grid-connected, islanded, or interconnected operating modes as well as transition modes, i.e. islanding and gridconnecting. Therefore, the stability of MGs including this wide range of components and operating modes is challenging and very important to have a secure and quality operation.

MG stability challenges have been taken into account in several research and technical reports. One can classify them according to MG operating modes. In grid-connected MGs, the most important challenges comprise grid synchronization, fault ride-through, and voltageride-through capabilities. Grid-forming (GFO) and grid-following (GFL) DER interconnections with stiff and weak grids are some other challenges of grid-connected DERs and MGs. Islanded MGs should cope with stability challenges like frequency stability, voltage stability, power sharing stability, and low inertia, as well as communication-related concerns such as delay-dependent stability and cyberattacks. Furthermore, smooth transitions between steady-state modes, i.e. islanding and grid-connecting, deal with some large-signal stability challenges.

Stiff grid-connected MGs including large-scale GFO DERs with respect to the grid may be involved in serious challenges due to interactions between voltage/frequency control loops and grid control loops. However, stiff grid-connected MGs, where DERs are dominantly based on the GFL control strategy, have fewer stability challenges. Nevertheless, weak grid-connected MGs including GFL DERs may experience challenges related to synchronization and voltage stability. On the other hand, islanded MGs are usually concerned with low inertia and low X/R ratio challenges. The low inertia of converter-based DERs results in very fast frequency responses, which are harmful during faults and related to protection devices. Low X/R ratios and low distance (impedance) of DER coupling lines lead to interactions between frequency and voltage-control loops and may cause power sharing stability challenges. Hence, selecting appropriate control strategies becomes challenging for such systems. In addition, some control loops are necessary to add to the control system, e.g. a virtual impedance loop.

MG stability challenges can also be classified based on the type of stability, e.g. large-signal stability and small-signal stability. Large-signal stability issues may be experienced due to large

Microgrids: Dynamic Modeling, Stability and Control, First Edition. Qobad Shafiee, Mobin Naderi, and Hassan Bevrani.

disturbances including faults, islanding, grid connecting, and large load connecting/disconnecting. Time-domain waveforms can generally show such issues. On the other hand, small-signal stability challenges have focused on control gains and system parameters. Although there exist some general rules for selecting control gains and considering system parameters in all MG types; the MG topology, type of DERs, and control strategy cause some differences resulting in special considerations for each MG. These considerations can independently be satisfied using a small-signal stability analysis of each MG.

This chapter includes a classification of MG stability and basic requirements for the MG stability analysis. Several small-signal stability analyses are provided using eigenvalue analysis (EA), participation matrix, and sensitivity analysis to study different types of MGs and discuss their parameters and control gains. MGs under study consist of the grid-connected and islanded ac MGs as well as the dc MG, which was already introduced in Chapter 3. In stability analyses, the most important parameters of the MGs are investigated, e.g. droop characteristic gains, phase-locked loop (PLL) gains, coupling line parameters, virtual impedance gains, and secondary control gains. A discussion on the weakness of ac MGs is also represented. The transient stability of ac MGs is followed by the power sharing stability challenge. Finally, the chapter ends with a stabilization case for a Synchronverter, which is a type of virtual synchronous machine.

4.2 Stability Definition and Classification

The power system stability is the ability of the system for an initial operating condition to maintain the balanced state or recover it after a disturbance such that all the variables of the system are limited to a certain value and practically the entire power system continues to work flawlessly [1]. This stability definition is generic and can be used generally for MGs. In [1], different types of power system stability are defined according to the variables undergoing change, the severity of the disturbance, and the duration of the instability. Considering the differences in the nature of MGs from traditional power systems, such as low inertia and strong dependence on voltage and frequency, the classification of MG stability is represented in [2], which is a little different from what has been proposed in [1]. Due to the strong dependence of variables in MGs, classification based on variables is not an easy task. For example, a voltage instability may immediately lead to a frequency instability or vice versa. Therefore, in [1], the stability classification is done based on the type of equipment involved in the instability.

Figure 4.1 shows the stability classification of islanded/interconnected MGs. Regarding the type and location of MG/intercommoned microgrid (IMG) components, stability is divided into control system stability and power supply/balance stability. According to the controlled device type, the first one can be classified into electric machine stability and power-electronics converter stability. According to control infrastructure, stability of physical parts, e.g. inner loops is addressed as communication-free stability, and stability of communication-based control loops is called communication-related stability. On the other hand, the second one is classified into voltage and frequency stabilities. MG voltage is stable if both dc link voltage of converters and ac voltage are stabilized using decentralized local controllers. Frequency stability is twofold including intra-MG and inter-MG balance stabilities. In terms of disturbance size, stability is generally small-signal or large-signal/transient, and in terms of disturbance duration, it is short-term or long-term. The main classification, which belongs to MG stability in islanded and grid-connected operating modes, has been presented in [2] and the complementary classification for Interconnected MGs is presented in [3].

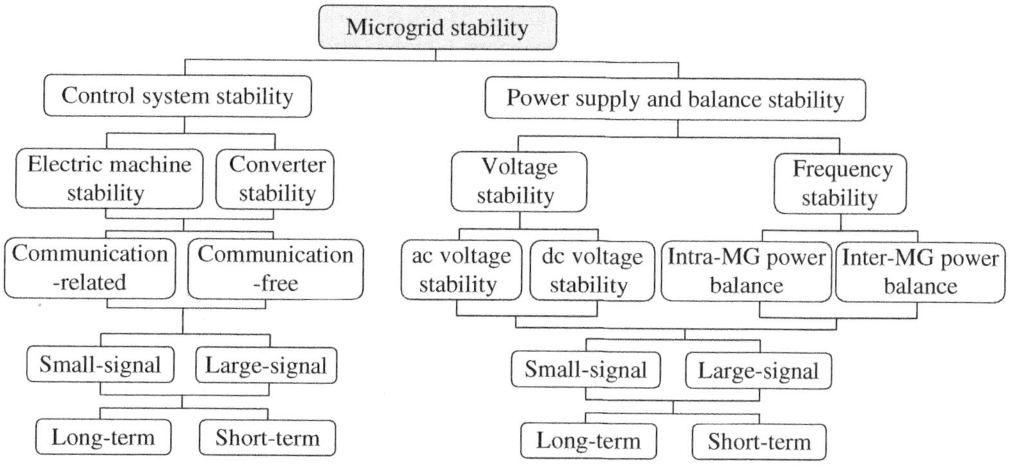

Figure 4.1 Stability classification for islanded/interconnected microgrids.

Power supply/balance stability is the ability of the MG to maintain power balance and share power among DERs such that the operation requirements are satisfied. This type of stability deals with issues such as the loss of a DER, exceeding the limits of DERs, and load interruptions without a fault occurrence. On the other hand, control system stability deals with issues such as the inappropriateness of the control strategies and the poor adjustment of the parameters.

Frequency stability is the ability of a power system to maintain the steady-state frequency of the system following the occurrence of a severe disturbance, which leads to a serious imbalance between generation and load. Frequency stability in islanded and interconnected MGs (IMGs) is very important due to the low inertia of the system. Because the low inertia of the microgrid compared to the large and interconnected power grids increases the frequency sensitivity of the system to the generation-load imbalances. In fact, small load-supply imbalances cause large frequency deviations. Therefore, one of the main goals of MG control in island/interconnected mode is to maintain the frequency of the system within the acceptable range.

Voltage stability is the power system's ability to maintain steady-state voltage in all buses within a specific range for normal operating conditions and after occurring a disturbance. When a disturbance, e.g., increase in load demand, or change in the system state, causes an increasing and uncontrollable drop in voltage, the system enters a voltage instability. The main reason for the instability is the inability of the power grid to provide the required reactive power and finally, the voltage drops occur in lines due to the large real/reactive power flows through the line impedances [4]. In a grid-connected MG, the voltage follows the main grid voltage. Since the rated power of the main grid is much larger than that of the MG, the main grid acts as an infinite bus for the MG. Therefore, in this mode, voltage stability is not challenging. However, in the islanded and interconnected modes, one of the main control objectives is to control, regulate, and restore the MG voltage.

Another type of voltage instability happens on the dc side of the inverter units, which depends on the ability of these units to maintain the dc link voltage. Depending on the type of DER, this voltage is supplied through a buck/boost converter from a dc source, e.g. a photovoltaic (PV), or from an ac to dc inverter connected to an ac source, e.g. a wind turbine. The ripples on the dc link voltage are strongly dependent on the instantaneous power passing through the link [2]. If there is a problem with the control of the converter/inverter or the power injected to the ac side exceeds its limits, the instability of the DC link voltage is probable. There is also a dc voltage stability challenge in the dc link of IMGs through back-to-back converters, which is studied in Chapter 9.

Small-signal stability shows the system's ability to maintain synchronism due to small disturbances. These disturbances generally occur due to small changes in load and generation. In such cases, the disturbances can be considered small enough to make it possible to linearize the system equations around the operating point in order to analyze the stability. In traditional power systems, small-signal stability challenges occur mainly due to the lack of damping of various types of oscillations, including local modes, inter-area modes, control modes, and torsional modes. However, in an autonomous MG, small-signal stability issues include low-frequency modes and medium/high-frequency modes. Low-frequency dominant modes are highly dependent on power sharing parameters and DER coupling lines. While medium- and high-frequency modes are affected by internal loops of voltage and current as well as load and filter dynamics [5, 6]. Small-signal stability analysis is usually done using EA.

Transient stability or large-signal stability shows the ability of the system to maintain synchronism due to the occurrence of a strong transient disturbance. In this case, the reaction of the system includes large changes in the rotor angle of the generators (voltage angle in power electronics-based MGs) and follows the nonlinear power-angle swing equation. The system stability depends on both the initial operating point and the disturbance intensity. After a large disturbance, the steady-state operating point is different from the operating point before the perturbation. The MGs may experience a large disturbance at the time of switching from grid-connected mode to islanded mode and vice versa, DER plug-and-play, large and sudden load changes, and frequent short-circuits. In such cases, linear controllers do not guarantee MG stability in different operating points, and basically instability is expected due to severe transients.

4.3 Basic Requirements

Several stability analysis methods are used in MG stability literature. EA is a fundamental tool of small-signal stability analysis, which can only determine stability/instability for a specific operating point. Participation factor (PF)/matrix provides the ability to determine the amount of being influenced for each eigenvalue and by each state variable, i.e. the corresponding module. Sensitivity analysis (SA) extends PF for a range of parameter values or operating points and results in loci of eigenvalues, which can visualize the impact of parameters or initial condition variations. Time-domain simulations (TDSs), real-time simulations (RTSs), and experimental results (ERs) are powerful methods to analyze large-signal stability using nonlinear models and real laboratory setups. Though the ERs have more accuracy, the TDS and RTS are more suitable for stability studies due to the possibility of instability. Lyapunov theorem (LT) is another tool with a powerful mathematical basis to assess large-signal stability. Nevertheless, it is practically limited to low-order models. Other MG stability analysis techniques include frequency response, critical clearing time (CCT), kernel reflection, phase portrait, formal analysis, Gershgorin theorem, and methods based on measured data [3].

4.3.1 Eigenvalue Analysis

For each type of state space representation (SSR), i.e. open-loop model in the presence of inputs or closed-loop free-motion model, one can analyze the small-signal stability of the system by calculating eigenvalues. Assuming A_{MG} as the state matrix of an MG, the eigenvalues can be calculated as follows:

$$|\lambda I - A_{MG}| = 0, \tag{4.1}$$

where λ shows the eigenvalues, I is the identity matrix with the same size as the A_{MG}, and $|.|$ indicates the matrix determinant operator. To achieve an asymptotic stability, all eigenvalues should have negative real values. The stability margin can also be increased by moving eigenvalues away from the imaginary axis to its left side. Two sets of eigenvalues, namely dominant and critical eigenvalues, have the most impact on the small-signal stability. The dominant eigenvalues can be defined as a group of eigenvalues that are nearer to the imaginary axes with respect to other eigenvalues that are considerably farther than the imaginary axis at least three times the dominant ones. The critical eigenvalues can also be defined as very near eigenvalues to the imaginary axis, which also have the potential to be unstable due to parameter fluctuations.

4.3.2 Participation Matrix

The participation matrix is used to find the contribution of dynamic modes (eigenvalues) in the state variables and vice versa, which can be calculated as

$$\boldsymbol{M}_{PF} = (\boldsymbol{\Phi}^{-1}) \bullet \boldsymbol{\Phi}, \tag{4.2}$$

where $\boldsymbol{\Phi}$ is the right eigenvector matrix of $\boldsymbol{A}_{MG}$ and $\bullet$ denotes the Hadamard product. $\boldsymbol{M}_{PF}(i,j)$ shows the involvement of the i'th state variable in the j'th eigenvector and also the involvement of the j'th eigenvector in the i'th state variable. Thus, in order to find each state variable, all participant eigenvalues should be taken into account. In fact, each participant in the i'th state variable is equivalent to a nonzero element of the i'th row in $\boldsymbol{M}_{PF}$, which is a positive number less than/equal to 1 if the matrix elements (participation factors) are normalized as follows [7]:

$$\boldsymbol{M}_{PF,n}(i,j) = |\boldsymbol{M}_{PF}(i,j)| \,/ \sum_{m} \boldsymbol{M}_{PF}(i,m), \tag{4.3}$$

where $\boldsymbol{M}_{PF,n}(i,j)$ is the normalized participation share of the j'th eigenvalue in the i'th state variable.

Exercise 4.1 Here, there is a simple example to understand the participation matrix and participation factor. Consider the fourth-order model of a droop-based DER presented in [8] and studied in Section 3.3.4.3, where the state matrix is obtained from the detailed model using singular perturbation theory. The preserved state variables are $\Delta\delta$, ΔP, Δi_{od}, and Δi_{oq}. The corresponding parameters are as $m_p = 0.0016$ W.s/rad, $n_q = 0.0653$ VAr/V, $\omega_c = 20$ rad/s, $R_{li} = 0.84\,\Omega$, $L_l i = 3.8$ mH, $v_{d0} = 316.5$ V, $v_{q0} = 0$ V, $i_{od0} = 1.02$ A, $i_{oq0} = -0.52$ A, and $\omega_0 = 313.4$ rad/s. Calculate the state matrix, eigenvalues, and participation matrix for the fourth-order model and discuss them.

Solution

For the given parameters and using Section 3.3.4.1, the state matrix for the forth-order model can be calculated as

$$A_{DER,4th-order} = \begin{bmatrix} 0 & -0.0016 & 0 & 0 \\ -5952.5 & -20 & 9495 & 902.1 \\ -7830.5 & 0 & -218.7 & 7990.4 \\ 82422 & 0 & -313.4 & -218.7 \end{bmatrix}. \tag{4.4}$$

The eigenvalues are obtained as $\lambda_{1,2} = -219.1 \pm j1581.3$ and $\lambda_{3,4} = -9.7 \pm j61.3$ using (4.1). Note that λ_1 and λ_2 are equivalent to a dynamic mode and λ_3 and λ_4 show the second dynamic mode. Negative real values of all eigenvalues demonstrate the small-signal stability of the model (4.4). It is obvious that λ_3 and λ_4 are the dominant eigenvalues. The normalized participation matrix, obtained using (4.3), is given below. As can be understood, $\Delta\delta$ and ΔP are the main participants in the dominant eigenvalues $\lambda_{3,4}$, and Δi_{od} and Δi_{oq} have low participation in them. The MATLAB code for this exercise can be found here.[1]

Eigenvalues

State variables	λ_1 ($-219.1 \pm j1581.3$)	λ_2	λ_3 ($-9.7 \pm j61.3$)	λ_4
$\Delta\delta$	0	0	0.498	0.498
ΔP	0	0	0.497	0.497
Δi_{od}	0.499	0.499	0.002	0.002
Δi_{oq}	0.499	0.499	0.003	0.003

Exercise 4.2 In Exercise 4.1, assume the $\omega - P$ droop gain, m_p, is changeable from 0.0003 W.s/rad to 0.0063 W.s/rad. Obtain the loci of eigenvalues for the forth-order model.

Solution

Considering the above-mentioned considerations about the *for* loop and the *color* code, and composing them with the main MATLAB code of the fourth-order model mentioned in Exercise 4.1, one can easily develop the MATLAB code for the sensitivity analysis of the fourth-order model.[1] The sensitivity analysis result is available within the MATLAB code, where two oscillatory non-dominant modes show low changes, and two oscillatory dominant modes show considerable changes with respect to the $\omega - P$ droop gain variations. The sensitivity analysis result for only the oscillatory dominant modes of the fourth-order model and some other models is shown in Fig. 3.9 of Chapter 3. Note that sensitivity analyses presented in the rest of this chapter are realized using the same approach.

4.3.3 Sensitivity Analysis

Although the participation matrix (factor) can show the impact of each state variable in each eigenvalue (dynamic mode), it can also be calculated for a single operating point and a set of parameters. Therefore, stability margins cannot be shown very well using the participation matrix. In order to improve analyzing stability for different operating points and parameter values, the participation matrix should be calculated several times and show the results for each calculation. It is not so easy to show a large number of participation factors for different conditions. Sensitivity analysis is an alternative to represent a visual output to show eigenvalues for different operating points or changing parameters.

Note that each small-signal model includes the data of the operating point and influencing parameters. In order to apply sensitivity analysis, a range of initial values of each state variable or a logical range of each parameter is considered. Using a MATLAB code, the range values are swept and the eigenvalues are shown on the same figure for each value. Therefore, one can understand the impact of various operating points and parameters on the eigenvalues, and also on the stability margins. There are two MATLAB programming considerations in order to have better results. A *for* loop should be used to change all values of the range and show results for each value. A *color* code is required within the *plot* code to have a better visual show of eigenvalue changes. In fact, using the *color* code, it is easier to know how the eigenvalues change by changing the range values.

1 Matlab codes for the microgrids dynamic modeling, stability and control book. www.wiley.com/go/shafiee/microgrids. Accessed: 2023-07-07.

4.4 Small-Signal Stability Analysis

The dynamic behavior of an MG is completely different from a synchronous generator (SG). A typical MG relies on control loops and algorithms with fast time responses. In this regard, the wide timescale related to the controls of MGs can result in cross-couplings with both the electromechanical dynamics of machines and the electromagnetic transients of the network, over a wide frequency range from few Hz to hundreds of Hz. The MG stability problems are mostly caused by incorrect settings of its various control loops, which can be characterized independently from the power grid. An MG must be designed to assure a stable operation; therefore, stability analysis of this system is a vital issue, which is usually reduced to the stability analysis of a grid-connected converter, and can be handled using different methods in *time* and *frequency* domains [9].

In the frequency domain, the *impedance method* is a simple, and common method to analyze the MG stability, which is based on the frequency characteristics of the system [10–13]. In this method, the frequency characteristics of the MG are investigated to satisfy a stability criterion. The system frequency characteristics can be determined using different methods [14–16], such as analytical, numerical, and simulation techniques [9, 17–19]. In addition to impedance methods, the frequency domain-based stability analysis class uses several powerful graphical tools such as the *closed-loop eigenvalues/poles graph* or *Root Locus (loci)*, *Bode*, and *Nyquist* diagrams. The graph of eigenvalues (or poles) provides a simple graphical tool, which is widely applied for stability and sensitivity analysis of the linearized MGs and is also used for the given examples in this chapter.

Due to the nonlinearity of MGs, sometimes it is difficult to analyze the system frequency characteristics for the stability analysis [20]. For example, the impedance method is only applicable to a linear model MGs. Unlike the frequency domain, in the time domain, it is usually desirable to analyze the system transient stability using the MGs nonlinear models. Several approaches are given for this purpose. One of them is based on the *Poincaré map* method, which is reduced to analysis of the system transition matrix or *sensitivity matrix* for perturbations along an operating trajectory [21–23].

The sensitivity analysis is an important part of the MG stability analysis. In the present chapter, this issue is considered in the study of system parameter/unit tolerance impacts on MG stability [24, 25]. Sensitivity analysis should be taken into account for a safer stability margin during the synthesis process. In the given examples, the importance of sensitivity analysis in the stability investigation of an MG is emphasized.

In this section, EA and participation matrix results are represented and discussed for case studies, while their modeling is represented in Chapter 3. Most important participant modules and state variables in dominant and critical dynamic modes are determined.

4.4.1 Grid-Connected PV

Figure 3.1 in Chapter 3 shows the grid-connected PV, where its small-signal model has been presented. Figure 3.2 also represents all 16 eigenvalues of the system. It is obvious that the studied grid-connected PV at the determined operating point is asymptotically stable. Figure 4.2 shows the results of the participation matrix analysis, where the horizontal axis indicates eigenvalues, and the vertical axis shows participation factors of each contributing state variable (module) in each dynamic mode, i.e. each couple of complex conjugate eigenvalues. The state variable(s) of each module is (are) shown in Figure 4.2, e.g. X_{lc} includes all four state variables of the output filter current and voltage. A nonimportant dynamic mode, i.e. $\lambda_{15,16} = -2.7 \times 10^7 \pm j313$, appears, which shows the impact of the virtual resistor in modeling the power network. This dynamic mode is not shown in Figure 4.2.

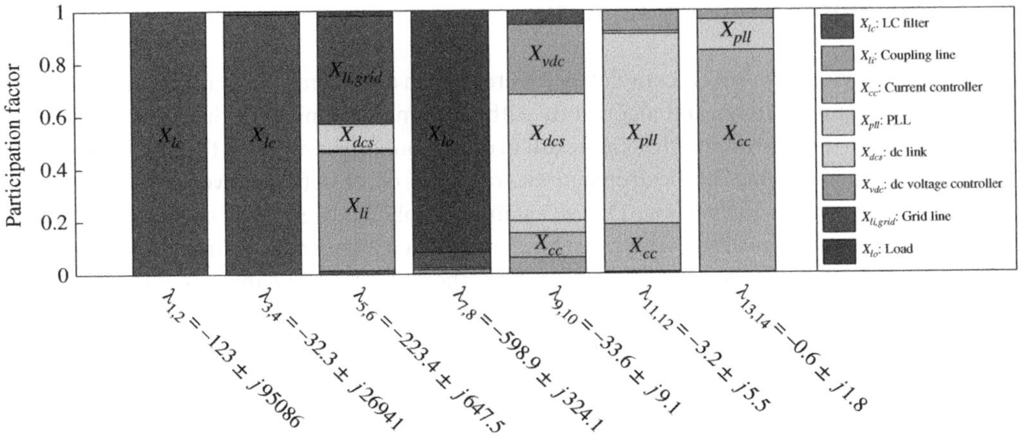

Figure 4.2 Participation matrix analysis results of the grid-connected PV system shown in Figure 3.1.

According to Figure 4.2, $\lambda_{1,2}$ and $\lambda_{3,4}$ are almost completely affected by X_{lc}, i.e. LC filter state variables. $\lambda_{1,2}$ are far from the imaginary axis and are not dominant at the studied operating point. $\lambda_{3,4}$ are nearer to the imaginary axis and their low damping has a longer impact on the dynamics than $\lambda_{1,2}$, which has a higher damping. $\lambda_{5,6}$ have three main participants, i.e. state variables of the PV coupling line (PF = 0.45), the grid line (PF = 0.41), and dc side elements (PF = 0.1). Its real value is −223.4, which shows that a portion of the effects of these modules is not considerable for the studied operating point in terms of stability. The main contributing state variables in $\lambda_{7,8}$ are X_{lo}, i.e. the direct and quadrature components of the load current. Regarding −598.9 real value for this mode, load dynamics of a constant resistive–inductive (RL) load do not contribute to small-signal stability challenges.

Regarding real values of λ_{9-14}, these eigenvalues are the dominant dynamic modes for the studied grid-connected PV. Modules on the dc side providing dc link voltage dynamics (PF= 0.48), dc voltage controller (PF= 0.26), and current controller (PF= 0.09) are the predominant modules contributing to $\lambda_{9,10}$. The interaction among voltage and current control loops as well as dc side modules can obviously be deduced in this mode. This mode is not critical in the studied operating point. However, the behavior of this mode should be investigated using sensitivity analysis of the corresponding parameters. It can also be observed that λ_{11-12} and λ_{13-14} are critical modes and are intensively affected by the current controller and the PLL. Therefore, the most important modules in the grid-connected PV are the current controller and the PLL in terms of small-signal stability employing the participation matrix. Therefore, their parameters should be tuned carefully considering stability margins.

Note that participation matrix analysis and corresponding results can be very useful for small-signal stability studies. Nevertheless, it has some disadvantages. It can only find the reciprocal effect of state variables and dynamic modes in a certain operating point. In fact, the results may have considerable change for different operating points, especially for systems with large values of nonlinearity. Some parameters of the system may change with time, due to aging, control goals, etc. The participation matrix cannot consider such parameter changes. Sensitivity analysis can investigate both changes in the operating point and parameters. Sensitivity analysis results can show stability margins, which is not possible using a participation matrix. In all sensitivity analysis results, arrows are used, which show the direction of parameter changes from lower values to higher values.

4.4.1.1 Sensitivity Analysis: LC Filter Parameters

Figure 4.3(a) shows the loci of the grid-connected PV eigenvalues for changing the filter inductance from 0.2 to 10.2 mH, where wide variations of a number of eigenvalues can be seen. Moreover, instability appears for the inductance values smaller than 1.6 mH. In Figure 4.3(b), changes in eigenvalues close to the imaginary axis are shown. Another mode with a large imaginary value gets closer to the imaginary axis and then becomes unstable by increasing the filter inductance. It is noteworthy that this mode has a large stability margin, which decreases very fast due to increasing the filter inductance. Therefore, there are both upper and lower limits for the filter inductance amount. Figure 4.3(c) zooms in the dominant dynamic modes with a small stability margin. Here two complex conjugate modes transform into four absolute real modes, where two of them become unstable. In order to find the upper limitation, this unstable mode and the unstable mode shown in Figure 4.3(b) are compared. The unstable mode shown in Figure 4.3(c) becomes unstable sooner for the marginal inductance value of 3.4 mH. Figure 4.3(d) shows the loci of the grid-connected PV dynamic modes for the filter capacitance variation from 30 to 120 μF, where instability appears for smaller values than 59 μF. Here, there is not a wide range of variations. Therefore, the sensitivity of dynamic modes to the filter capacitance changes is less. Furthermore, an upper band limitation does not exist for the capacitance values in terms of small-signal stability.

4.4.1.2 Sensitivity Analysis: Coupling/Grid Line Length

The eigenvalue loci of the studied grid-connected PV for changing coupling/grid line length are shown in Figure 4.4. As can be seen in Figure 4.4(a), where the PV coupling line length changes from 30 to 300 m, the dominant modes do not have considerable variations. Furthermore, the more sensitive modes are still far from the imaginary axis with an appropriate margin. Figure 4.4(b) shows the loci of eigenvalues due to changes in the main grid line length from 30 to 300 m.

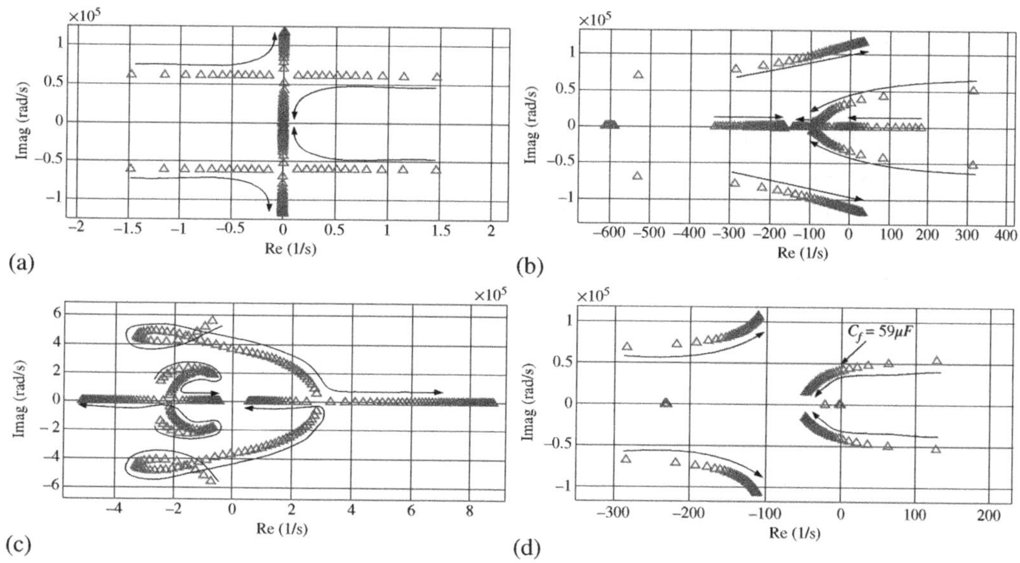

Figure 4.3 The loci of the grid-connected PV eigenvalues for changing the filter inductance from 0.2 to 10.2 mH: (a) wide variations of a number of eigenvalues, appearance of instability for values smaller than 1.6 mH, (b) changes of eigenvalues close to the imaginary axis, (c) emerging instability for values greater than 3.4 mH, and (d) the eigenvalue loci for changing the filter capacitance from 30 to 120 μF, where instability appears for smaller values than 59 μF.

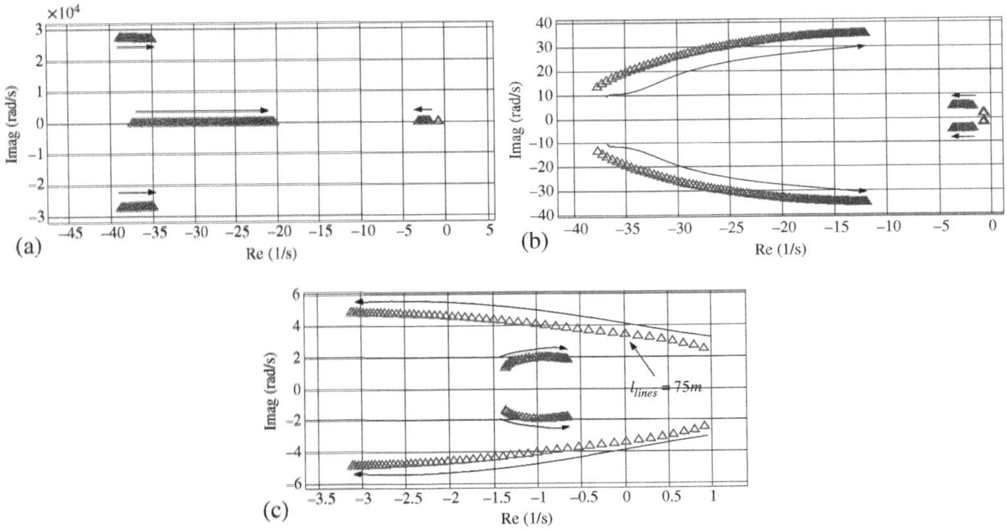

Figure 4.4 The loci of the grid-connected PV eigenvalues for: (a) changing the coupling line length from 30 to 300 m, (b) changing the main grid line length from 30 to 300 m, and (c) changing simultaneously both the coupling line length and the grid line length from 30 to 300 m.

Approximately, a similar behavior can be seen in this case. However, a smaller stability margin of the farther mode is seen. In each case, when a line length is being changed, the other line length is fixed at 280 m.

Figure 4.4(c) shows the dominant eigenvalue loci when both the coupling line and the grid line lengths are changed from 30 to 300 m. Here, instability appears for small values of the line lengths, i.e. when both lines are shorter than 75 m. Both dynamic modes shown in Figure 4.4(c) are intensively affected by the PLL state variables. In fact, interactions among PLL state variables and lines' state variables are a reason for instability. In Figures 4.4(a)–4.4(c), the proportional gain of the PLL controller equals 1. One can make the system stable by increasing the gain to 5 for the same changes in Figures 4.4(a)–4.4(c). The PLL has a considerable impact on the quadrature component of the point of common coupling (PCC) voltage during extracting the grid frequency. This action is very challenging when the grid is stiff, i.e. the PCC voltage can accept just a small tolerance. On the other hand, the DER control actions have stronger influences on the PCC voltage for smaller coupling lines. By decreasing both line lengths, such a situation is realized, and interactions of the grid and the DER, i.e. the grid-connected PV, become stronger on the PCC. The instability is probable in this situation. The next subsection represents more details of the sensitivity analysis of the PLL parameters and its impact on the small-signal stability.

4.4.1.3 Sensitivity Analysis: PLL Gains

Figure 4.5(a) shows the loci of the dominant dynamic modes due to changes in the integral gain of the PLL controller from 1 to 50. Although the system is stable for all values, the stability margin is very low for small values of the integral gain. Figure 4.5(b) shows the loci of the dominant dynamic modes when the proportional gain of the PLL controller varies from 1 to 20. For the base values of the studied system parameters, small-signal stability is preserved. However, the same scenario in Figure 4.5(c) shows that the system becomes unstable for small values of the proportional gain when the length of both coupling and grid lines are 60 m.

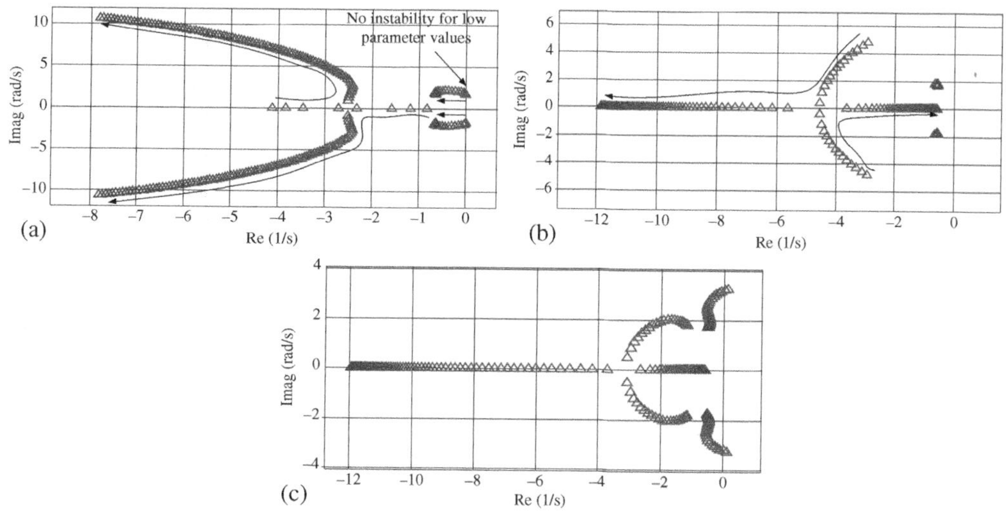

Figure 4.5 The loci of the grid-connected PV eigenvalues for: (a) changes in the integral gain of the PLL controller from 1 to 50, (b) changes in the proportional gain of the PLL controller from 1 to 20, (c) a similar scenario to (b) for the length of each coupling and grid lines as 60 m.

Generally, small-signal stability is less sensitive to the integral gain of proportional-integral (PI) controllers compared to the proportional gain. Another important point is about the upper limitation of the gains. Although the upper band is not limited based on the small-signal stability criterion even for larger values than those shown in Figure 4.5, transient stability will be endangered for the large gain values.

4.4.1.4 Sensitivity Analysis: Current Control Gains

In order to control direct and quadrature components of the output current of the grid-connected PV, two PI controllers are employed. Figure 4.6(a) shows the loci of the dominant eigenvalues for changes in the integral gain of the direct component current controller from 40 to 20000. The instability emerges for larger gains than 14430, which shows a wide stabilizing range of the gain. Nevertheless, the stability margin for this range is very small. Figure 4.6(b) shows the loci of the dominant eigenvalues when the integral gain of the quadrature component current controller changes from 40 to 430. The small-signal stability is lost for the gains larger than 220, which shows more sensitivity to the integral gain of the quadrature current component.

Figure 4.6(c) shows the loci of the eigenvalues for changes in the proportional gain of the direct/quadrature component current controller from 0.04 to 4. Here, a nondominant mode for the base values is unstable for very low values of the proportional gain as 0.58 and larger than that. It is obvious that the quadrature component current controller plays an important role in small-signal stability due to its interconnections with the PLL, which acts on the quadrature voltage component. Moreover, the proportional gain of this controller has a larger portion in stability challenges.

4.4.1.5 Sensitivity Analysis: DC Voltage Control gains

A PI controller is also used to regulate the dc link voltage. Figure 4.7(a) indicates the loci of the nearer eigenvalues to the imaginary axis, while the integral gain of the dc voltage controller varies from 500 to 2000. The dominant modes have negligible variations caused by the integral gain

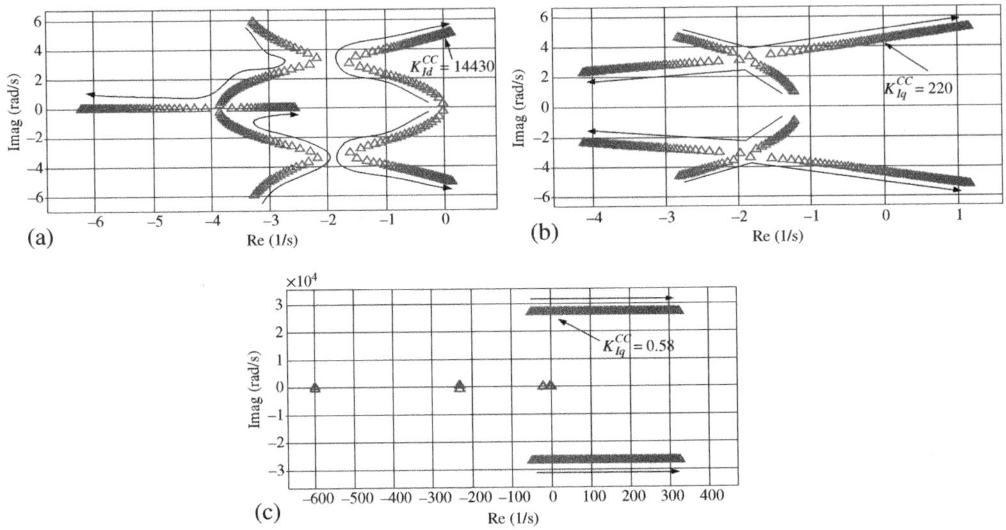

Figure 4.6 The loci of the grid-connected PV eigenvalues for: (a) changes in the integral gain of the direct component current controller from 40 to 20000, (b) changes in the integral gain of the quadrature component current controller from 40 to 430, and (c) changes of the proportional gain of the direct/quadrature component current controller from 0.04 to 4.

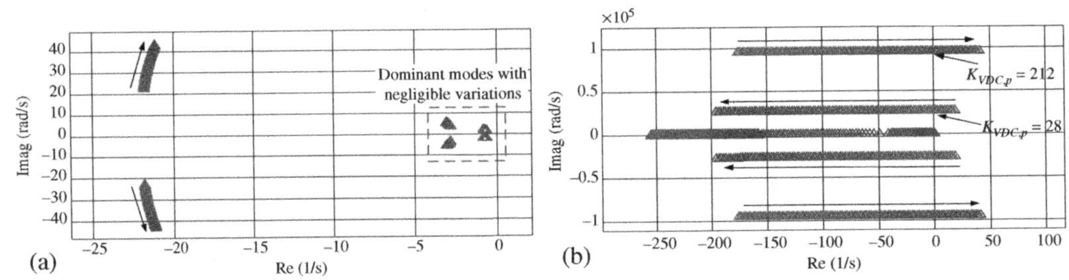

Figure 4.7 The loci of the grid-connected PV eigenvalues for: (a) changes in the integral gain of the dc voltage controller from 500 to 2000, (b) changes in the proportional gain of the dc voltage controller from 1 to 300.

changes. A dynamic mode around −22 real shows the largest variations, which is not remarkable in terms of stability margin.

In Figure 4.7(b), severe changes in a number of eigenvalues to the variations of the proportional gain of the dc voltage controller within 1 and 300 are shown. The stability is limited to the range of the proportional gain as $28 <= K_{VDC,p} <= 212$. Although the range is large, the sensitivity to the changes in this parameter is considerable. DC link voltage is hard to be controlled due to the disturbances from both dc and ac sides. Therefore, the dc voltage controller parameters play an important role in small-signal stability.

4.4.2 Grid-Connected AC Microgrids

As explained in Chapter 3, the studied grid-connected ac MG includes two GFL DERs with a constant dc link voltage. Each GFL DER includes 10 state variables. The constant RL impedance load and the grid line are modeled using four state variables in total. Hence, the small-signal model of the

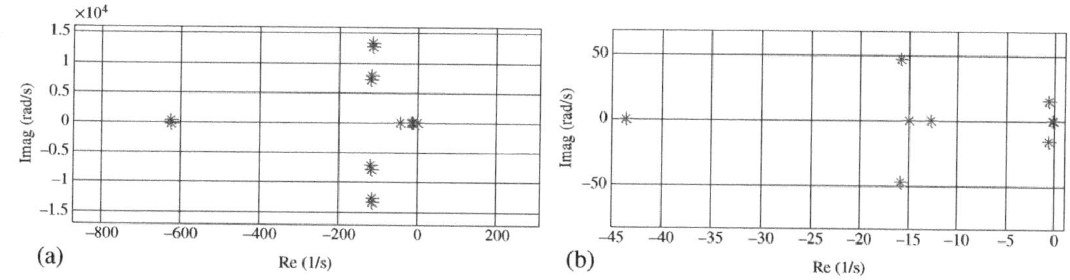

Figure 4.8 Eigenvalues of the grid-connected ac MGs including two DERs: (a) All analytical eigenvalues, (b) the dominant eigenvalues.

studied ac MG consists of 24 state variables, which result in 24 eigenvalues shown in Figure 4.8(a) excluding a complex conjugate mode affected by the virtual resistor. General analysis of this MG is similar to the grid-connected PV but with more eigenvalues. Two nonsignificant modes around −200 real in Figure 3.2(a) change to two oscillatory modes with low damping at a larger real value, i.e. around −160 real. Figure 4.8(b) shows the dominant eigenvalues, where there are five critical eigenvalues. Although sensitivity analysis of the studied MG shows some differences from the grid-connected PV sensitivity analysis, representing these results for parameter sensitivity analysis is not required.

4.4.2.1 Sensitivity Analysis: Grid Strength Study

Figure 4.9 shows the eigenvalue loci of the studied ac MG for changes in the proportional gain of the direct component current controller from 0.01 to 1, where the grid line length is 280 m for stars with solid lines and is 2.8 km for triangles with dashed lines. The instability happens for a large range of the proportional gain in both scenarios. Only very small gains can lead to a stable MG. The MG with a longer line of the upstream grid is a weaker grid because the PCC voltage may experience larger fluctuations. The weak MG with a grid line length of 2.8 km shows worse loci of the dominant modes with respect to the stiffer MG with a grid line length of 280 m. The oscillatory modes of the weak MG have a smaller stability margin and a lower damping.

The weak MG requires a GFO DER to control the voltage in the absence of an infinity bus or a stiff upstream grid. When there is no GFO DER, power injections of the GFL DERs and the corresponding controller gains and system parameters are disturbances and uncertainties for the MG voltage. In fact, both small-signal stability and transient stability are sensitive to these disturbances and uncertainties. On the other hand, it was explained in Section 4.4.1.2 that the small lengths of

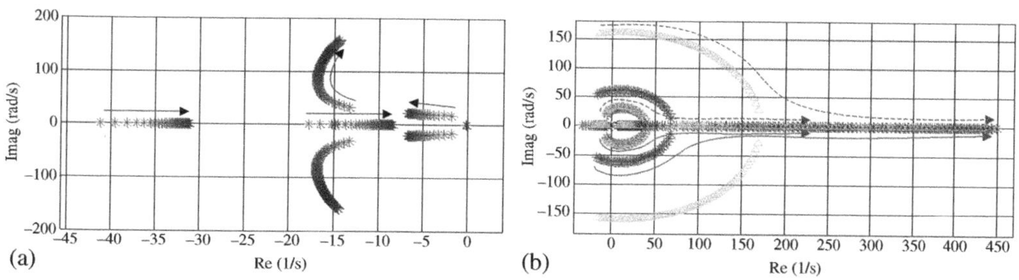

Figure 4.9 The eigenvalue loci of the studied ac MG for changes in the proportional gain of the direct component current controller from 0.01 to 1, where the grid line length is as 280 m for stars with solid lines and is as 2.8 km for triangles with dashed lines.

the GFL DER and the grid result in instability due to big interactions with PLL variables. In fact, PLLs cause stability challenges, especially for stiff upstream grids. Therefore, an ac MG formed by GFL DERs and connected to both weak and stiff grids can have small-signal stability challenges. Nevertheless, the challenges related to PLL in stiff grids are solved by improving the PLL performance [26, 27] or using free-PLL GFL control strategies according to the global positioning system (GPS) signals [28–30].

4.4.2.2 Sensitivity Analysis: Interaction of GFL DERs

In this scenario, the goal is to determine interactions among GFL DERs in the grid-connected mode. Although GFL DERs are operated as current sources which do not have interaction challenges of GFO DERs, their synchronization via PLLs causes some interactions with each other through the PCC voltage.

Figure 4.10 shows the eigenvalue loci of the studied ac MG for changes in the integral gain of the DER_1 direct component current controller shown by stars, and changes in the integral gain of the DER_2 direct component current controller shown by triangles, from 1 to 10. All other control gains for both DERs are fully the same. In Figure 4.10(a), the grid line length is 30 m, i.e. the grid is stiff. Variation patterns of the dynamic modes due to the same control gain of the DER_1 and the DER_2 are very different. In Figure 4.10(b), the grid line length is considered to be a larger amount 280 m, i.e. the grid is relatively weak. Here, variation patterns of the dynamic modes are more similar. However, there still exist some discrepancies. In Figure 4.10(c), an intensively weak grid is considered, where the grid line length is 2800 m. Here, the change patterns of the eigenvalues are very similar with minimum differences.

Comparing the results of Figures 4.10(a)–4.10(c), one can conclude that GFL DERs in stiff grids can show their maximum impact on the dynamics and stability of the grid-connected MG. In fact, interactions between the PLL and the current controller are maximized in small-signal studies

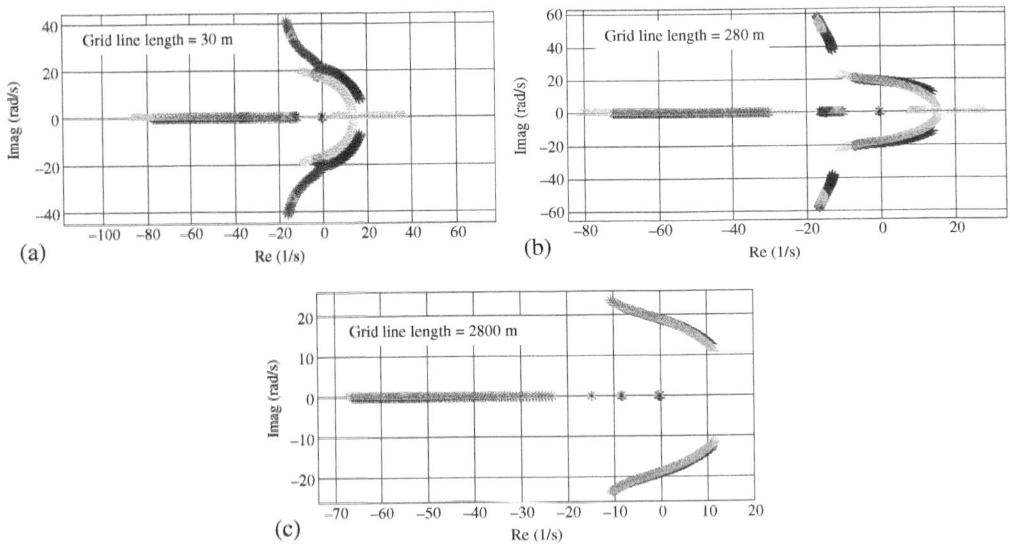

Figure 4.10 The eigenvalue loci of the studied ac MG for changes in the integral gain of the direct component current controller of the DER_1 shown by stars and of the DER_2 shown by triangles, from 1 to 10 when grid line length is as (a) 30 m (stiff grid), (b) 280 m (relatively weak grid), and (c) 2800 m (very weak grid).

when a stiff voltage is provided. PLLs try to make the quadrature component of the measured voltage zero in order to extract the frequency. This action has larger interactions when the voltage is forced to be another value by a stronger source, i.e. the main grid. Therefore, acceptable ranges of the parameters for an appropriate small-signal stability margin should be selected. Then, the control parameters can be optimized within the acceptable ranges. Moreover, required considerations should be provided for the power parameters.

4.4.3 Islanded AC Microgrids

Here, the ac MG shown in Figure 3.5(a) and introduced in Section 3.3.3 is assumed to include two GFO DERs and one GFL DER. All control layers are comprised of inner controllers, primary control, and secondary control layers shown in Figures 2.25 and 2.28. In some scenarios, comparisons between the control units with and without virtual impedance and control units with and without a secondary control layer are comprised. Sensitivity analysis of the most important parameters is accomplished including droop gains, virtual inductance/resistance, secondary control gains of GFO DERs, and GFL DER current controller and PLL gains.

4.4.3.1 Sensitivity Analysis of Droop Gains

According to the bandwidth of local controllers, in which a decrease is considered from the innermost layer to the outermost layer during controller design, the droop characteristics have the most dominant parameters in small-signal stability. Figure 4.11(a) shows the traces of dominant modes for variation of the $\omega - P$ droop gain such that maximum allowable frequency changes are considered as $0.2\% < \Delta\omega_{max}/\omega_{rated} < 2\%$, i.e. from 0.1 Hz ($\simeq$ 0.628 rad/s) to 1 Hz ($\simeq$ 6.28 rad/s) in $\Delta\omega_{max}/P_{rated}$ considering the rated frequency as 50 Hz. The secondary controller is not included in this study. The stars show the dominant modes when the virtual inductance exists and the triangles show the dominant modes when it is not connected. In both traces, solid arrows determine the direction of the gain increase. Generally, the value of the $\omega - P$ droop gain has a remarkable effect on the dominant modes of ac MGs. Within the interval, the gain increase leads to improving the stability margin and decreasing the oscillation damping. However, larger gains than the interval (not shown) cause the oscillatory mode to come back to the lower real values and may lead to instability [5]. The most important role of the virtual inductance is to increase the damping, which can be understood by comparing stars and triangles after changing the two real dynamic modes to the one oscillatory mode.

Figure 4.11(b) shows the trace of dominant modes for variation of the $v - Q$ droop gain such that maximum allowable voltage changes are assumed as $1\% < \Delta v_{max}/V_{rated} < 10\%$, i.e. from 3.26

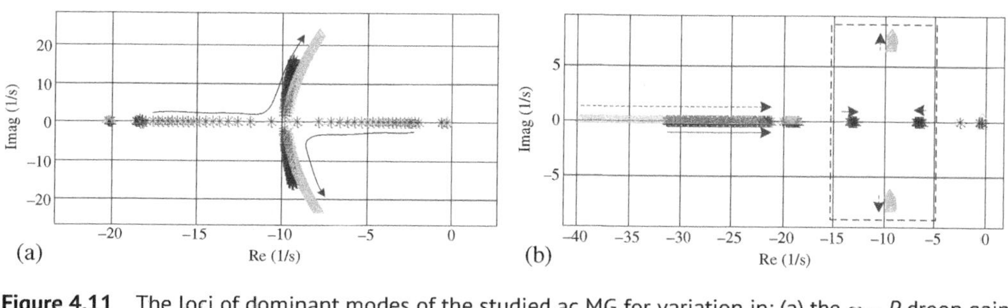

Figure 4.11 The loci of dominant modes of the studied ac MG for variation in: (a) the $\omega - P$ droop gain through maximum allowable frequency changes as $0.2\% < \Delta\omega_{max}/\omega_{rated} < 2\%$, (b) the $v - Q$ droop gain through maximum allowable voltage changes as $1\% < \Delta v_{max}/V_{rated} < 10\%$.

to 32.60 V considering the rated peak voltage as $400\sqrt{2/3}$. Solid arrows show the trace with gain increase direction when the virtual inductance exists and the dashed arrows show it without considering the virtual inductance. Changing the $v - Q$ droop gain does not have a considerable impact on the dominant modes with and without virtual inductance. Moreover, the comparable dominant modes in the box show better damping when the virtual inductance is used as Figure 4.11(a) also shows. This feature of the virtual impedance in the small-signal stability of ac MGs is studied in more detail in the next subsection.

4.4.3.2 Sensitivity Analysis of Virtual Impedance

Virtual impedance is added to the common primary control layer to basically improve power sharing. In fact, the parallel operation of GFO DERs based on $\omega - P$ and $v - Q$ droop characteristics intensively relies on the characteristics of the output impedance of DER coupling lines. These droop mechanisms are applicable for high X/R ratios of the network lines such as transmission lines. Nevertheless, there may exist a range of low X/R ratios to the unit ratio in distribution lines, where ac MGs are operated. Virtual impedance/inductance can provide an appropriate X/R ratio for the common droop characteristics through control loops [31]. For a detailed explanation about the impact of DER coupling line X/R ratios and virtual impedance on the droop control performance and in turn, the MG stability, see Sections 5.5.1 and 5.5.2 of Chapter 5. Although the simplest structures use fixed virtual inductance as shown in Figure 2.25 and fixed virtual impedance as shown in Figure 4.12 with variables related to Figure 2.25, its structure and parameters can be systematically designed. In order to improve the performance of the basic virtual impedance loop, several control methods have been employed. Generally, adaptive virtual impedance design is performed to consider output current amplitude and phase to improve power sharing, which is a sense of the output impedance characteristic of coupling lines [31–34]. Iterative methods are also used to have a variable virtual impedance with [35] and without communication [36]. The optimal design of the virtual impedance has been studied considering several operating constraints, including steady-state voltage limits, power decoupling, damping of dominant modes, and reactive power sharing [37]. Robust synthesis can be used to see the unknown values of the line impedances as uncertainties, which has not been studied yet. Virtual impedance is also used for supplementary objectives such as harmonic power sharing, voltage unbalance mitigation, voltage regulation [38], and smooth transition [39]. Several aspects of virtual impedance strategy and other power sharing methods based on impedance reshaping are reviewed in [38, 40].

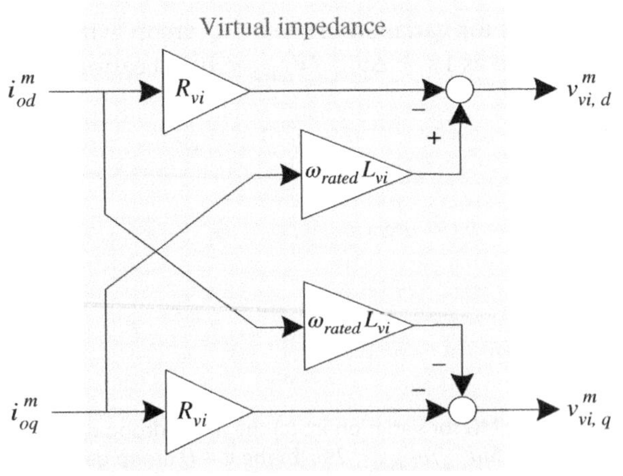

Figure 4.12 Basic structure of the virtual impedance loops including both resistance and inductance (see Figure 2.25 for more details of the outer loop connections).

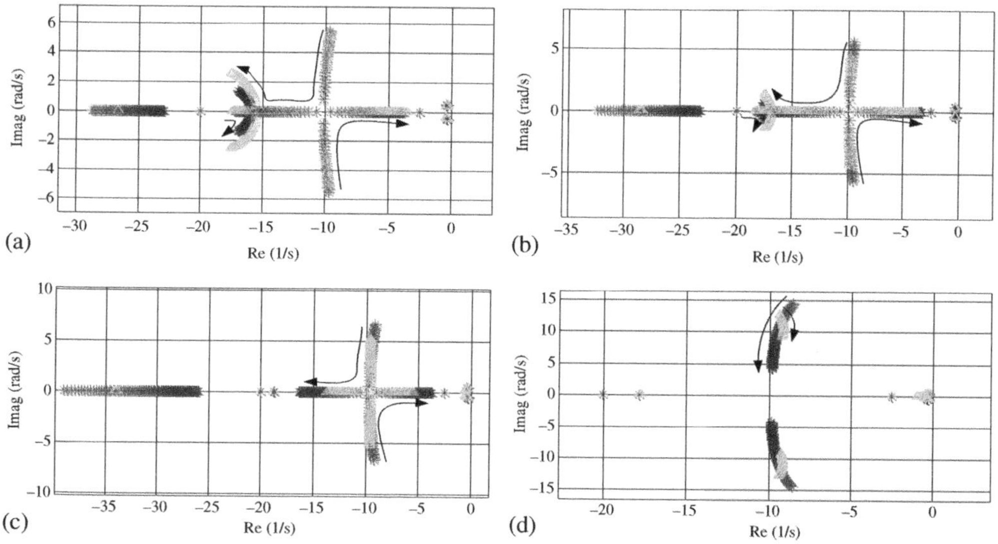

Figure 4.13 The loci of dominant modes of the studied ac MG for virtual inductance variation as $0 < L_{vi} < 12$ mH for DER_1 (stars) and DER_2 (triangles) in: (a) scenario 1, (b) scenario 2, (c) scenario 3, and (d) scenario 4, described in Table 4.1.

Table 4.1 Information of the scenarios considered for stability analysis of virtual inductance changing in Figure 4.13.

Scenario	Z_{DER_1}/Z_{DER_2}	S_{DER_1}/S_{DER}	$L_{li,1}$ (mH)	$R_{li,1}$ (Ω)	$P_{rated,1}$ (kW)	$Q_{rated,1}$ (kVAr)
1	2	2	3.8	0.8	40	20
2	0.5	2	1.9	0.4	40	20
3	2	0.5	3.8	0.8	20	10
4	2	0.5	3.8	0.8	10	5

The loci of dominant modes of the studied ac MG for virtual inductance variation as $0 < L_{vi} <$ 12 mH are shown in Figure 4.13 for different scenarios, which are described in Table 4.1. The scenarios provide different impedance ratios combined with different rated power ratios of the GFO DER_1 and the GFO DER_2. Because the impact of virtual impedance is related to these ratios directly for DER coupling impedances and indirectly through droop gains for rated powers. On the other hand, the virtual inductance value is changed one time just for GFO DER_1 shown by stars and another time just for GFO DER_2 shown by triangles.

Regarding all scenarios, a constant feature can be deduced, which is an increase in oscillation damping of the oscillatory dominant modes. In the three scenarios shown in Figures 4.13(a)–4.13(c), increasing the virtual inductance changes the complex conjugate oscillatory mode to two real modes, where one of them will approximately be a nondominant mode and another one approaches the imaginary axis. Therefore, an increase of the virtual inductance can be divided into two intervals, which the first is useful to decrease damping and the last is harmful to small-signal stability of the ac MG. Comparing Figures 4.13(c) and 4.13(d), one can conclude that the sensitivity of the trace of eigenvalues and consequently stability of ac MGs to the virtual inductance magnitude is decreasing with reducing the rated power of DERs. This can be

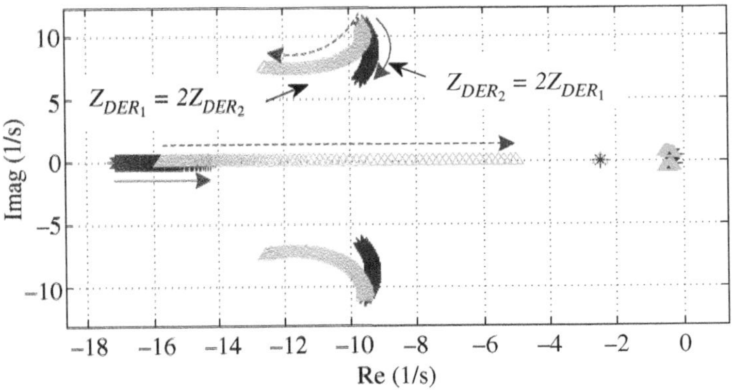

Figure 4.14 The loci of dominant modes of the studied ac MG for virtual resistance variation as $0 < R_{vi} < 1\Omega$ when $S_{DER_1} = 2S_{DER_2}$ in both situations and $Z_{DER_2} = 2Z_{DER_1}$ for stars and $Z_{DER_1} = 2Z_{DER_2}$ for triangles.

understood by analyzing the relationships between droop gains and the virtual inductance loops in Figure 2.25. According to the rate of changes of stars and triangles, it is obvious that changing the virtual inductance for both DERs does not result in the same rate of changes, i.e. one of them is larger. Nevertheless, different ratios represented in Table 4.1 cannot help to infer a conclusion except by showing a nonlinearity among the real impedances, virtual impedances, and the rated powers of DERs.

Figure 4.14 shows the loci of dominant eigenvalues of the studied ac MG for the virtual resistance variation $0 < R_{vi} < 1\Omega$ in the structure shown in Figure 4.12. The virtual inductance is 2 mH throughout the changes. Here, $Z_{DER_2} = 2Z_{DER_1}$ for stars and $Z_{DER_1} = 2Z_{DER_2}$ for triangles as well as $S_{DER_1} = 2S_{DER_2}$ in both situations. Dashed arrows show stars are changing and solid arrows show triangles are changing when the virtual resistance increases. In both cases, an oscillatory mode and a real mode have considerable variations due to the parameter changes. In fact, the virtual resistance increase leads to improving oscillation damping via the oscillatory mode and reducing the stability margin by the real mode.

Note that the behavior of dynamics is better when $Z_{DER_2} = 2Z_{DER_1}$, especially in terms of the stability margin caused by the real mode. In fact, the real mode of the situation $Z_{DER_2} = 2Z_{DER_1}$ has a lower tendency to instability than the equivalent mode of the situation $Z_{DER_1} = 2Z_{DER_2}$. It is due to the compatibility of the conditions $Z_{DER_2} = 2Z_{DER_1}$ and $S_{DER_1} = 2S_{DER_2}$. In other words, the DER_1 has a larger output power/current than the DER_2 and thereby its coupling line output impedance should be lesser than that of the DER_2 in order to prevent circulating currents and to effectively share powers.

4.4.3.3 Stability Analysis of Secondary Control

In this section, two challenges are studied. First, the impact of the most important control and power grid parameters on small-signal stability is analyzed. These parameters are the $\omega - P$ droop characteristic gain and the coupling line length of the GFO DERs. Then, sensitivity analysis of the secondary control gain is done, i.e. the secondary integrator parameter in both frequency and voltage loops.

The traces of participating dominant modes of the studied ac MG by varying the $\omega - P$ droop gain are shown in Figure 4.15(a). The stars are considered for existing the secondary controller according to Figure 2.25 and the triangles show the dynamic modes of the studied MG without the

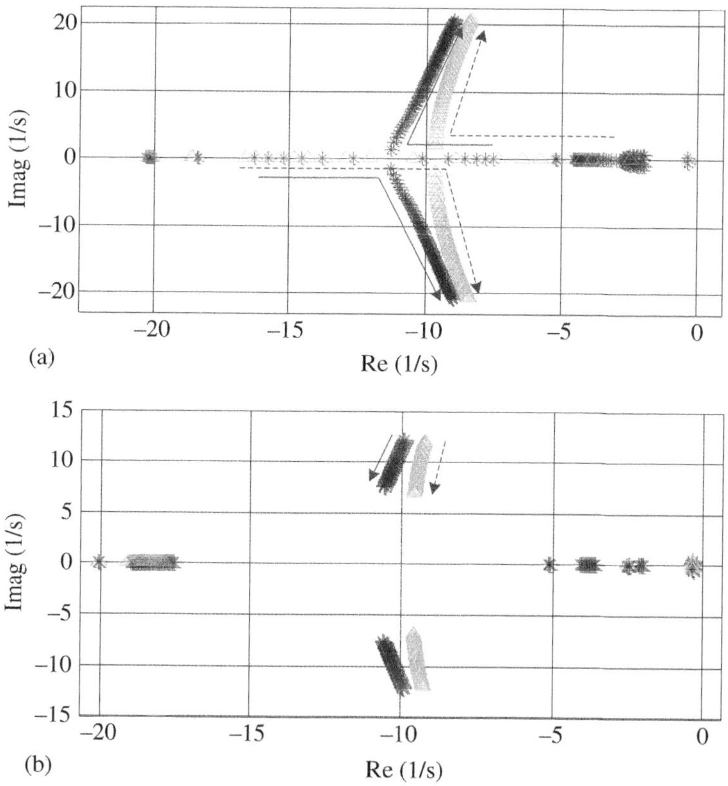

Figure 4.15 The loci of dominant modes of the studied ac MG with (stars) and without (triangles) the secondary controller for (a) the $\omega - P$ droop gain through maximum allowable frequency changes as $0.2\% < \Delta\omega_{max}/\omega_{rated} < 2\%$, and (b) $0.5 < Z_{DER_2}/Z_{DER_1} < 2$.

secondary controller. The gain varies based on maximum allowable frequency changes as $0.2\% < \Delta\omega_{max}/\omega_{rated} < 2\%$, i.e. from 0.1 to 1 Hz. The behavior of the oscillatory dynamic modes is similar, i.e. they are real modes, which will be oscillatory by increasing the gain. However, the closed-loop secondary control out of the primary control leads to improving the stability margin. Some other affected dynamic modes by the gain changes, which are placed between −5 real and −1 real, are dynamic modes added due to the state variables of the secondary controller. They cannot be seen in the scenario without the secondary controller, i.e. shown by the triangles. These modes are more sensitive to the secondary control gain. Hence, the rule of decremental bandwidth in designing the cascaded control loops of GFO DERs causes placing these modes nearer to the imaginary axis as the dominant critical modes.

The loci of affected dominant modes by varying the ratio of the GFO DER lengths are shown in Figure 4.15(b). The stars are still considered to show the existence of the secondary controller, while the triangles show the dynamic modes without the secondary controller. The length/impedance ratio of the coupling lines is assumed as $0.5 < Z_{DER_2}/Z_{DER_1} < 2$. In fact, the arrows state a smaller coupling line for the DER_2 with respect to the DER_1 as 0.5 at first and its gradual increase to twice of the length of DER_1 at last. The secondary control also leads to better stability margins for affected modes by the corresponding changes. Nevertheless, the stability of the secondary-related critical dynamic modes needs to be investigated by sensitivity analysis of corresponding control gains.

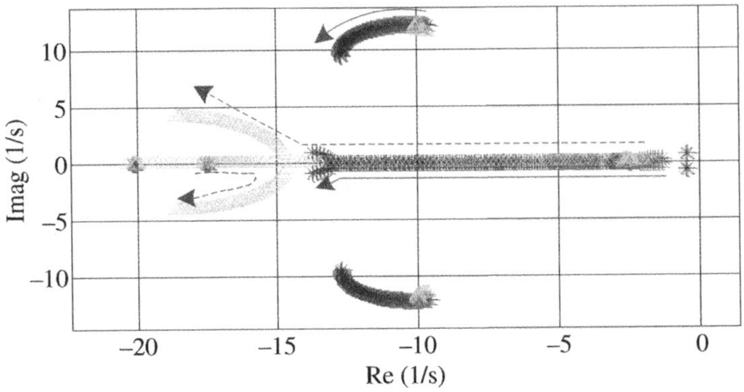

Figure 4.16 The loci of dominant modes of the studied ac MG for variations of DER_1 secondary control gain (stars) and variation of DER_2 secondary control gain (triangles) as $1 < \tau^{-1} < 10$.

Figure 4.16 shows the loci of critical modes affected by the secondary control gain (τ^{-1}) variation of DER_1 shown by stars and that of DER_2 indicated by triangles. By increasing the gain, the corresponding critical modes will be far from the imaginary axis. In other words, these critical modes transform into nondominant modes. The noncritical oscillatory mode around −10 real is also improved by only increasing the DER_1 secondary control gain in terms of damping and stability margin. Therefore, increasing τ^{-1}, i.e. decreasing the time constant of the secondary control loop, totally results in a faster response and a larger stability margin related to the secondary control-affected dynamic modes. However, an oscillatory dynamic mode can be seen in Figure 4.16 and Figs. 4.11–4.16, where it is located between −1 real and 0, and is the most dominant mode of the studied ac MG. Note that this mode is not sensitive to none of the above investigated parameters such as the secondary and primary control gains.

4.4.3.4 Sensitivity Analysis of GFL DER Parameters

The most dominant mode located very close to the imaginary axis is strongly participated by the GFL DER state variables. The loci of the most dominant modes are shown in Figure 4.17(a) for the integral gain changes of the GFL DER current controller as $100 < K_{ii} < 1000$. The larger gains lead to improving overall stability margin due to moving the most dominant modes to the left side of the *s* plane. It can also be deduced that lower gains than 100 may cause instability. In addition to the most dominant oscillatory mode, another dominant mode almost located at −2 real is improved in terms of stability, but its changes are not considerable. Figure 4.17(b) shows the trace of this

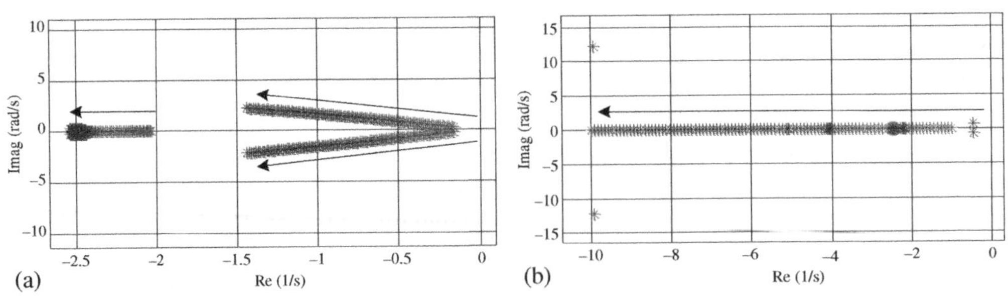

Figure 4.17 The loci of the most dominant modes of the studied ac MG for (a) the changes in the integral gain of the GFL DER current controller as $100 < K_{ii} < 1000$, (b) the changes in the integral gain of the GFL DER PLL as $20 < K_{ii} < 200$.

real dominant mode for the changes of the integral gain of the GFL DER PLL as $20 < K_{ii} < 200$. Increasing the integral gain of the PLL leads to a remarkable improvement in the behavior of this dynamic mode. Therefore, using these sensitivity analyses, one can find suboptimal ranges of the control parameters, which can be used in advanced control methods to find an optimal set of control parameters.

4.4.3.5 Weakness of AC Microgrids

Here MG weakness is studied as a main reason for instability. Generally, three main features can be listed for weak grids. The first is the high line impedance, which causes large voltage drops within the grid and typically far from generation units. The second is the high R/X ratio of grid lines. Power sharing among generation units is usually based on the conventional frequency-active power droop characteristics, which is acceptable for lines with low R/X ratios [34, 41]. Therefore, power sharing may not be desired in MGs with high R/X ratios only using the conventional droop. Although some methods like virtual impedance [42, 43] and generalized/modified droop controls [40, 44, 45] are represented to solve power sharing challenges in MGs, they are related to MG configuration and grid data. The third feature of weak grids is low inertia, which causes large frequency changes, oscillations, and rate of change of frequency. Such large frequency disturbances lead to unfavorable power fluctuations as well as the misoperation of protection devices.

Voltage Weakness The studied ac MG as a weak grid is similar to the MG shown in Figure 3.5(a), including two GFO and one GFL DERs, which has a small difference in the network as shown in Figure 4.18. The data of line and load are also included in Figure 4.18. Such a configuration is a usual case and is not a worst case from a weakness viewpoint to have a fair study. The complementary information is represented for MG_1 of the studied system in Section 3.3.3. Figure 4.19(a) shows the load bus voltage V_{B5} versus changes of the GFO_1 bus voltage from 0.5 to 1.5 pu for a steady-state model of the MG and considering ideal voltage and current sources for GFO and GFL DERs without control units. Obviously, small lengths of Z_2 result in considerable variations of V_{B5} due to the high resistance of the low-voltage lines.

In the second scenario, the DERs are not assumed ideal, i.e. they have the rated powers as 10 kW (GFO_1), 20 kW (GFO_2), and 5 kW (GFL). The current limiting feature of DERs is also activated as a more real case to show the unfavorable impact of a high R/X ratio. Figure 4.19(b) shows the GFO_1 voltage, GFO_1 and GFO_2 currents, and their limitations, where the MG load increases at B_5 by 40% of the rated MG power. Thus, the overall MG load increases to 90% of the rated power. Although the virtual impedance loops are used for GFO units, due to the high R/X ratio, smaller rating of GFO_1, and smaller distance of the disturbance to GFO_1, power sharing fails and GFO_1 output current reaches its limitation, while the GFO_2 current is much smaller than its limitation.

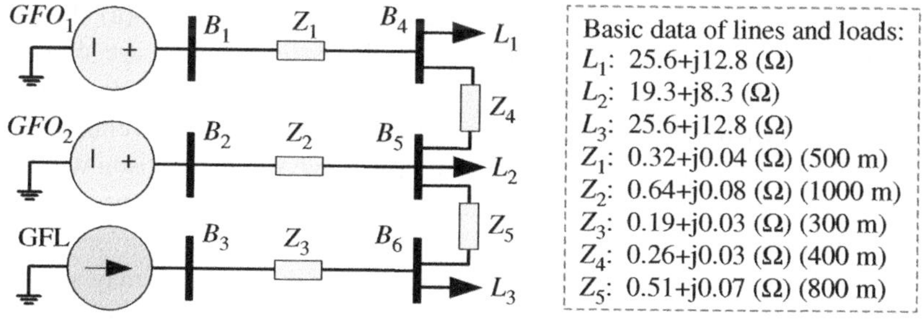

Figure 4.18 A typical MG with grid-forming and grid-following generation units in order to study weak voltage/frequency.

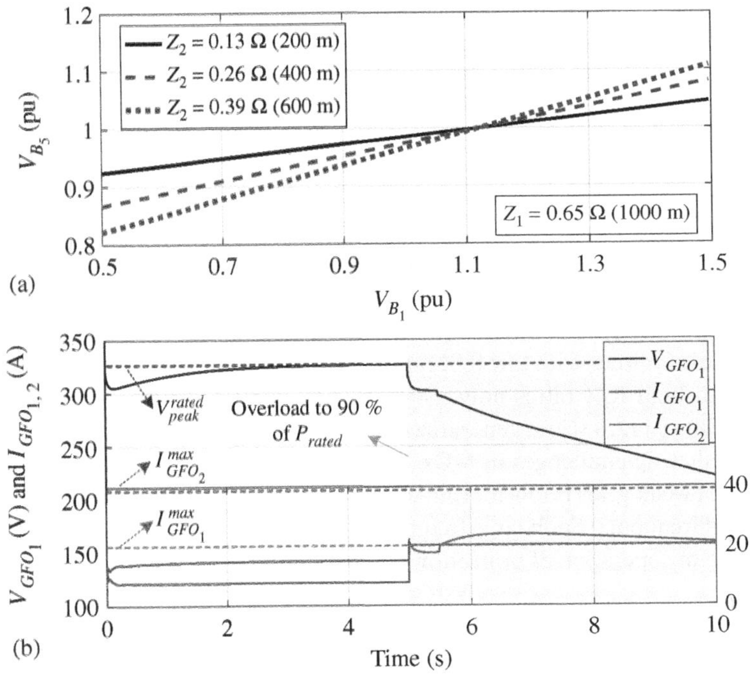

Figure 4.19 Voltage weakness in the studied MG shown in Figure 4.18: (a) steady-state load bus voltage (V_{B5}) variation due to the GFO_1 voltage (V_{B1}) changes and (b) transient instability of GFO_1 voltage due to the load increase at B_4 to 90 % of the MG rated power, lack of power sharing, and finally current limitation of GFO_1 as the nearest generation unit to the disturbance.

Frequency Weakness To study frequency weakness, the delay of the controllers and power parts of a fully converter-based MG can be considered as virtual inertia, and one can estimate it using the virtual swing equation [46] as follows:

$$J = (1/\omega_{mg}\dot{\omega}_{mg})(P_{sourse} - P_{load}), \tag{4.5}$$

where J is the virtual moment of inertia, P_{source} is the sum of the active power of DERs, P_{load} is the sum of active load powers, and ω_{MG} is the MG's angular frequency. Such a data-driven estimation is done in the sample MG using ω_{MG} measurement employing a PLL [46], as well as voltage and current measurement, and applying corresponding calculation relationships [47] to achieve P_{source} and P_{load} online. Due to large instantaneous changes of ω_{MG} per sample time during disturbances and appearing $\dot{\omega}_{MG}$ at the denominator of (4.5), an averaging process is added. Figure 4.20(a) shows three common power disturbances at $t = 5$ s. Inertia responses to these disturbances are shown using the MG frequency in Figure 4.20(b), which are kept on to the first overshoot/undershoot. Figure 4.20(c) shows the estimated virtual inertias and their average values during the overall inertia responses. The amounts of 0.29, 2.85, and 10.46 kg/ m^2 are very low even in comparison with a low-power wind turbine, e.g. 8.5 kW wind generator with 2 m diameter and 20 kg weight, which has a virtual inertia of 10 kg/m^2 [48].

Based on typical studies on MGs, they can be considered as very weak grids and may suffer from large voltage/frequency variations. In the synchronization process of two weak MGs, where the voltage magnitude, phase angle, and frequency differences should be controlled to be zero, such variations make the process harder and more challenging. Hence, a robust

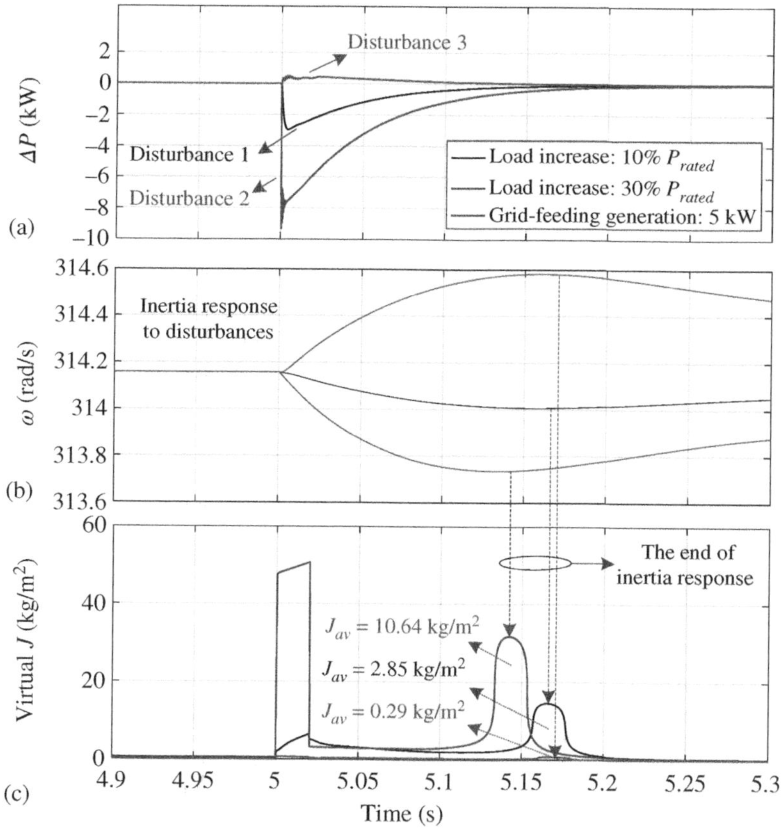

Figure 4.20 Inertia estimation of the studied MG shown in Figure 4.18: (a) active power disturbances, (b) inertia responses to the disturbances on the frequency, and (c) estimation of the virtual moment of inertia based on disturbance analysis.

controller is required to deal with the aforementioned uncertainties and disturbances during the synchronization procedure.

4.4.3.6 Relative Stability Improvement Using Grid-Supporting Control Strategy

Eigenvalue analysis and sensitivity analysis were used in Sections 4.4.3.1–4.4.3.5 to show stabilizing ranges of MG parameters and to improve the stability margin in the frequency domain. Regarding thresholds and constraints for frequency and voltage in power systems, relative stability or stability margin can be improved. In IEEE Standard 1547 [49], the continuous operation capability of DERs is allowed for voltage and frequency ranges as $0.88(p.u.) \leq V \leq 1.1(p.u.)$ and $58.8(Hz) \leq f \leq 61.2(Hz)$ for a rated frequency of 60 Hz, respectively. The equivalent allowed frequency range for a rated frequency of 50 Hz can be deduced as $49(Hz) \leq f \leq 51(Hz)$. In fact, the continuous operation is standardized for the frequency of DERs as $0.98(p.u.) \leq f \leq 1.02(p.u.)$ for both operational frequencies, i.e. 50 and 60 Hz. Therefore, the relative stability of islanded MGs can be improved by operating DERs with smaller frequency and voltage variations from the nominal values, i.e. farther values from the margins.

GFO DERs are necessary in isolated MGs to control voltage and frequency. Unintermittent and reliable primary sources such as ESSs are usually operated using the GFO control strategy for secure voltage and frequency control. However, GFL DERs are operated to maximize the output power of renewable energies, which are intermittent energy sources. GFL DERs do not contribute to voltage and frequency control and aim to track references of active and reactive powers. At least, a GFO DER with appropriate nominal power is required to support maximum allowable voltage and frequency changes due to sudden load/generation variations. Other DERs can be operated as GFO or GFL strategy depending on the MG topology and requirements. In case of possibility, some GFL DERs can be rescheduled as grid-supporting (GSU) DERs when the controlled power of the existing GFO DERs is not enough. Thus, the central control of MG distinguishes to add controlled power by rescheduling GFL DERs as GSU DERs. As shown in Figure 2.26, the GSU control strategy provides a term of difference between the nominal frequency/voltage and the real frequency/voltage multiplied by droop gains as follows:

$$\Delta P_{gsu} = k_{\omega}(\omega_{rated} - \omega_{gsu}), \tag{4.6}$$

$$\Delta Q_{gsu} = k_{v}(v_{rated} - v_{gsu}). \tag{4.7}$$

Regarding (2.94) from Chapter 2, this droop mechanism presents an operating point between GFO and GFL operating points, in which tracking of both sets of active and reactive powers, and frequency and voltage is considered using a trade-off mechanism tuned by the droop gains k_{ω} and k_v. Larger droop gains result in more precise frequency and voltage control as well as larger errors in active and reactive power reference tracking.

The ac MG studied in Sections 4.4.3.1–4.4.3.5 including two GFO and one GFL DERs is considered, but the GFL DER is replaced with a similar GSU DER with the possibility to increase the droop gains. The GFO DERs are operated in the isolated MG to support the load through the MG network at $t = 0$ s. A large load increase happens at $t = 1$ s. The GSU DER is synchronized with the MG at $t = 2$ s, where $P_{ref} = 2$ kW and the $P - \omega$ droop gain is 382 s.W/rad. The gain increases to 1910 s.W/rad at $t = 3$ s and then to 3438 s.W/rad at $t = 4$ s, respectively. The droop gains are equivalent to 10%, 50%, and 90% of the nominal active power of the GSU DER, respectively. The GSU reactive power is scheduled to be zero throughout the simulation. Figure 4.21(a) shows the active powers of GFO and GSU DERs as well as scheduled and final references of the GSU DER known as P_{GFL}^{ref} and P_{GSU}^{ref}. Note that the scheduled reference is referred to as a GFL DER. This means the GSU DER without droop gains can be considered as a GFL DER, and thereby these simulations also show a comparison between GFL and GSU DERs. Operation of the GSU DER and the droop gain increase lead to an increase of the final reference of the GSU active power in three steps and subsequently increase of the real GSU power.

Figure 4.21(b) shows the frequency of DERs, where the GFO DER frequencies are internally calculated in their primary control loops, while the GSU frequency is measured through its PLL. The last has greater overshoot/undershoot affected by the PLL dynamics. As a result, the frequency stability margin increases due to operating the GSU DER, and its k_{ω} gain increases. Therefore, the isolated MG frequency can be more tolerable against disturbances when a part of GFL DERs switches their operation to GSU DER operation. Figure 4.21(c) shows the voltage magnitude of GFO DERs and the PCC. Although GSU DER operation at $t = 2$ s results in a small voltage increase, the droop gain increase has no considerable effect. In a normal operation of common $\omega - P$ and $v - Q$ droop characteristics of GFO DERs in low-voltage isolated ac MGs, active power changes cannot have a remarkable effect on the MG voltage.

In a similar simulation, the autonomous MG is operated, its load increases and the GSU DER is synchronized with the MG with a zero active power and zero $P - \omega$ droop gain of the GSU DER. However, the scheduled reactive power reference increases as $Q_{GFL}^{ref} = 1$ kVAr at $t = 2$ s. The $Q - v$ droop gain is as 9.2 VAr/v at this time. The gain increases to 46 VAr/v at $t = 3$ s and

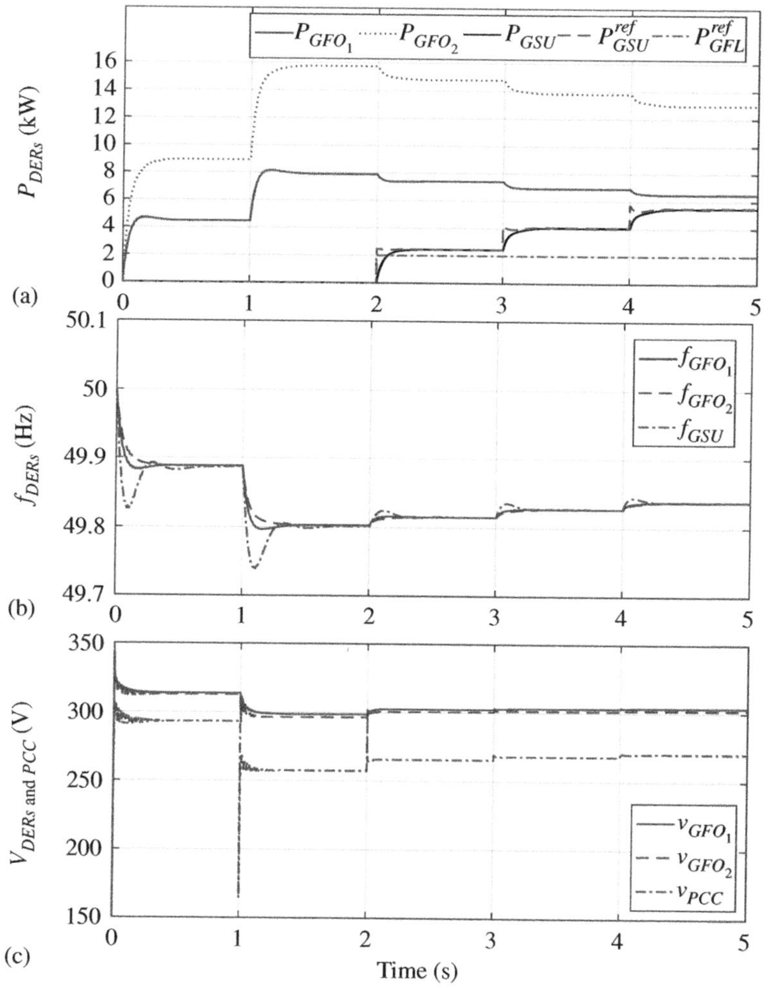

Figure 4.21 GSU DER support in the MG stability margin by increasing the $P-\omega$ droop gain: (a) active powers of GFO and GSU DERs, (b) DER frequencies, and (c) DER and PCC voltages.

then to 82.7 VAr/v at $t = 4$ s, respectively. The corresponding droop gains are equivalent to 10%, 50%, and 90% of the nominal reactive power of the GSU DER, respectively. In Figure 4.22(a), the reactive powers of GFO and GSU DERs as well as scheduled and final references of the GSU DER, namely, Q_{GFL}^{ref} and Q_{GSU}^{ref}. Both operation of the GSU DER and the $Q-v$ droop gain increase lead to an increase of the Q_{GSU}^{ref} in three steps and consequently increase the reactive GSU DER power.

Figure 4.22(b) shows the DER frequencies for the reactive power generation of the GSU DER, where the gain increase has a negative impact on the frequency stability margin. However, the GSU $Q-v$ droop gain variation cannot cause considerable changes in the MG frequency. On the other hand, Figure 4.22(c) shows the PCC and DER voltages. The voltages are lower than the standard upper limit due to the load increase. The GSU operation and the $Q-v$ gain increase result in an improvement in the voltages. In fact, injecting reactive power controlled by a term of voltage feedback leads to a voltage increase. Note that injecting reactive power without any voltage feedback

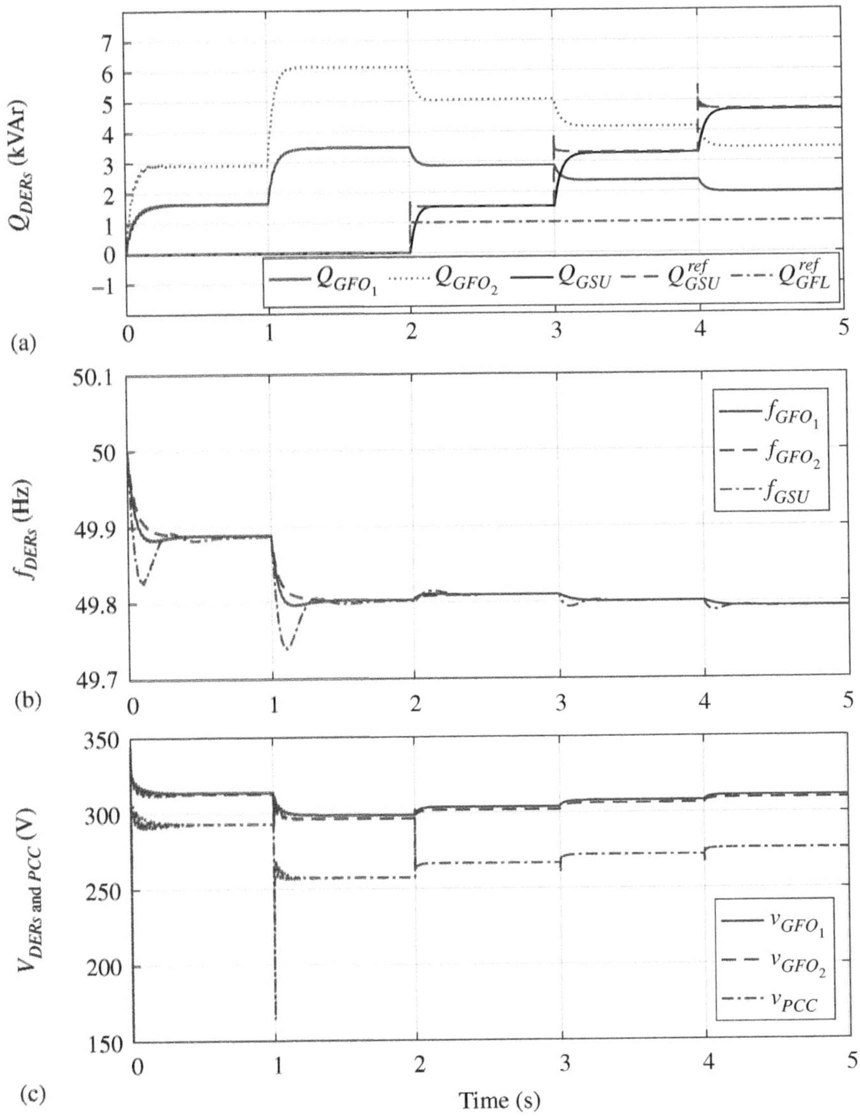

Figure 4.22 GSU DER support in improvement of the studied MG stability margin by increasing the $Q-v$ droop gain: (a) reactive powers of GFO and GSU DERs, (b) DER frequencies, and (c) DER and PCC voltages.

through GFL DERs also increases the MG voltage, and larger reactive power values lead to larger voltage values. Nevertheless, the voltage increase is not controlled and may lead to overvoltage. In a more real MG, the operator cannot select an appropriate reactive power reference for GFL DERs due to changing situations of the MG configuration and consumption.

In another simulation, both active and reactive power injections are considered which are shown in Figures 4.21(a) and 4.22(a) due to the GSU operation by activating both control loops and increase of both $P-\omega$ and $Q-v$ droop gains are considered. Figure 4.23(a) compares the GSU frequency of all three situations, i.e. increasing only the $P-\omega$ droop gain, increasing only the $Q-v$ droop gain, and increasing both the droop gains. Increasing only the $P-\omega$ droop gain

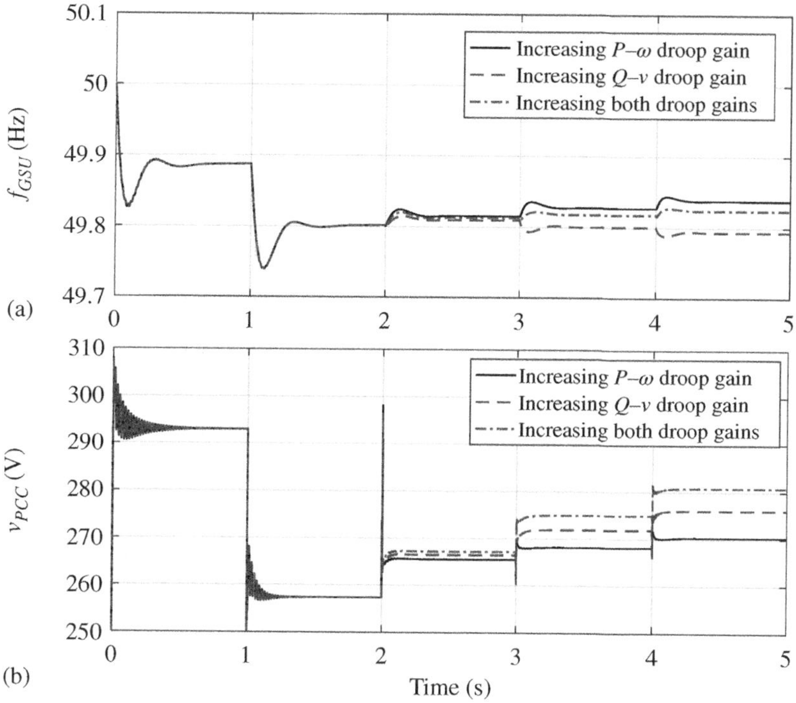

Figure 4.23 GSU DER support in improvement of the studied MG stability margin by increasing both $P-\omega$ the $Q-v$ droop gains: (a) comparison of DER frequencies with previous two scenarios and (b) comparison of PCC voltages with previous two scenarios.

results in a greater frequency stability margin. Figure 4.23(b) compares the PCC voltage of the three situations. Here, increasing both droop gains leads to a greater voltage stability margin.

4.4.4 Islanded DC Microgrids

The studied dc MG is shown in Figure 3.11 and has been introduced and modeled in Section 3.3.6. Here, the small-signal stability of the dc MG loaded by a resistive load is studied via sensitivity analysis of the eigenvalues using the transfer function derived in (3.25) and (3.25). Therefore, for the virtual impedance-oriented control strategy of the super-capacitor, there are two important parameters to be analyzed, namely, the virtual capacitance (C_v) and the virtual resistance (R_v).

The system parameters are given in Tables 4.1 and 4.2. In Figure 4.24(a), the eigenvalues of the entire system are plotted for two different values of R_v when C_v increases in an acceptable range. By increasing C_v, the damping ratio of the complex conjugate eigenvalues (λ_3 and λ_4) continually decreases, thus the system is more prone to low-frequency oscillations. Meanwhile, the damping effect increases with decreasing R_v as can be seen for $R_v = 5$ with respect to $R_v = 10$.

In order to improve the transient response, the required virtual capacitance can be calculated based on the transient response and the system stability. The transient state during the first moment after $t = 0^+$ depends on the overall capacitance connected to the main dc bus and virtual parameters (Table 4.3). Accordingly, $Z_{o1} = 1/(sC_{bat})$, $Z_{o2} = 1/(sC_{2pv})$, and $Z_{in} = 1/(sC_v + 1/R_v)$. From (3.25), the small signal of the main dc bus voltage during the first moment after $t = 0^+$ is obtained as

$$\Delta v_o(s) = -\frac{1}{C_T s + 1/R_T}\Delta i_{Load}(s), \tag{4.7}$$

Table 4.2 Power component parameters of the studied dc MG.

Module	Parameter	Value (Unit)	Parameter/ nominal value	Value (Unit)
PV DER	L_{pv}	5 (mH)	V_{C1pv}	109.4 (V)
	C_{1pv}	0.5 (μF)	C_{2pv}	0.5 (μF)
Battery ESS	L_{bat}	4 (mH)	V_{bat}	100 (V)
	C_{bat}	0.1 (mF)	V_{Cbat}	200 (V)
Super-capacitor ESS	L_{sc}	10 (mH)	V_{sc}	400 (V)
	C_{sc}	2 (mF)	r_{sc}	0.01 (Ω)
Load	R_L	133 Ω		
Main dc bus	C	0.4 (mF)	$V\_\{o, n\}$	200 (V)

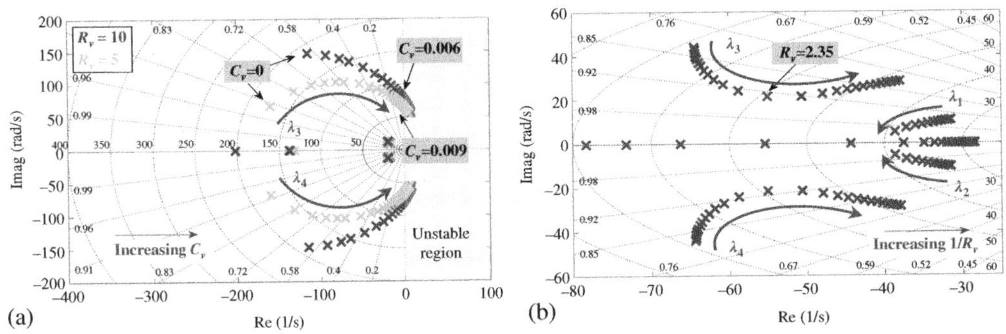

Figure 4.24 Trace of the dominant eigenvalues of the studied dc MG for: (a) varying C_v while $R_v = 10$ and $R_v = 5$, (b) varying R_v while $C_v = 0.002$.

Table 4.3 Control parameters of the studied dc MG.

Module	Parameter	Value	Parameter	Value
Battery ESS	K_{p1}	0.01	K_{i1}	100
	K_{p2}	0.3	K_{i2}	100
PV DER	K_{p3}	0.01	K_{i3}	2
Super-capacitor ESS	K_{p4}	0.15	K_{i4}	2
	K_{p5}	0.1	K_{i5}	100

where $C_T = C + C_{bat} + C_{2pv} + C_v$ and $1/R_T = 1/R_v + 1/R_L$. Thus, the maximum rate of change of voltage can be expressed as follows:

$$|RoCoV_{max}| = \left.\frac{|\Delta v_{o,max}|}{\Delta t}\right|_{\Delta t \to 0} = \frac{|\Delta i_{Load,max}|}{C_{T,max}}. \quad (4.8)$$

It is assumed that the value of $RoCoV_{max}$ does not exceed 6000 v/s. The dc output current through the converters varies from 0 to their nominal value (15 A), i.e. $\Delta i_{load,max} = 15$ A. Hence,

$C_{T,min} = \Delta i_{Load,max}/RoCoV_{max} = 0.0025$. To guarantee the transient performance, $C_v > 0.0015$ must be satisfied. The selected C_v should ensure the stable operation of the whole system, i.e. all the dominant eigenvalues are located on the left side of the s domain. Therefore, the acceptable range is $0.0015 < C_v < 0.009$ (for $R_v = 5$).

Figure 4.24(b) shows the trajectory of dominant eigenvalues when $1/R_v$ increases from 0.4 to 0.6. By increasing $1/R_v$, the damping factor of the studied dc MG is improved. As it is observed, there is also an inflection point for changing the virtual resistance in the eigenvalue locus. Therefore, this inflection point will be the optimal one for R_v, i.e. $R_v = 2.35$ in the case study.

4.5 Transient Stability

The transient stability of the power system can be analyzed in four general methods, including graphical, direct, automatic learning, and time-domain methods [50]. Graphical methods consist of equal-area criterion and phase portrait mostly used in one-machine systems [51]. The direct method employs Lyapunov functions, which is more applicable for low/reduced-order systems [52, 53]. Automatic learning methods benefit from intelligent algorithms such as genetic algorithm and artificial neural networks to assess transient stability [54]. Usual time-domain assessment criteria are the CCT and the power transfer limit (PTL) [55–57]. In general, time-domain methods have been mostly used for MG transient stability analysis as common methods [57, 58]. However, Lyapunov function [53], bifurcation theory [59], and unsymmetrical fault analysis [60] have been utilized rarely. Another common transient stability assessment method is sensitivity analysis, which is realized by changing the equilibrium point based on a linearized state space model for systems with low nonlinearity degrees and the nonlinear model for even highly nonlinear systems [50].

Here, two case studies of transient stability challenges are represented. First is the power sharing problem in islanded ac MGs. In this case, we aim to show instability due to an inappropriate rate of coupling line impedances of GFO DERs with respect to the rate of their rated powers as well as to demonstrate an effective role of virtual impedance in power sharing improvement. Second is the stabilization of a type of virtual synchronous generator (VSG) controller, namely, synchronverter using Lyapunov function and adaptive backstepping method, which is also a transient stability analysis based on Lyapunov theorem.

4.5.1 Power Sharing Stability in AC Microgrids

In this section, the isolated ac MG is the studied MG shown in Figure 4.25, including two GFO DERs and one GFL DER, while the GFL DER is not operated to focus on power sharing of the GFO DERs. The coupling line impedances of GFO DER_1 and GFO DER_2 are $Z_{li}^{GFO_1} = 0.09 + j0.12\Omega$ and $Z_{li}^{GFO_2} = 0.18 + j0.24\Omega$, respectively. The rated power of GFO DER_1 is half of the rated power of GFO DER_2, i.e. $S_{r1}/S_{r2} = 0.5$. Please see Sections 5.5.1 and 5.5.2 of Chapter 5 for a complete discussion on the impact of network line types and virtual impedance on the droop control performance and power sharing. According to the MG configuration shown in Figure 4.25, the ratio of DER coupling line impedances should be $Z_{li}^{GFO_1}/Z_{li}^{GFO_2} = 2$ in order to have appropriate power sharing between the GFO DERs. However, in the studied case, the worst situation of power sharing is considered, where $Z_{li}^{GFO_1}/Z_{li}^{GFO_2} = 0.5$, i.e. both impedances and rated powers have the same ratios. In addition, the coupling lines have low impedances, which causes larger disturbances during synchronization. Four scenarios are assumed to show the effect of virtual impedance on power sharing stability,

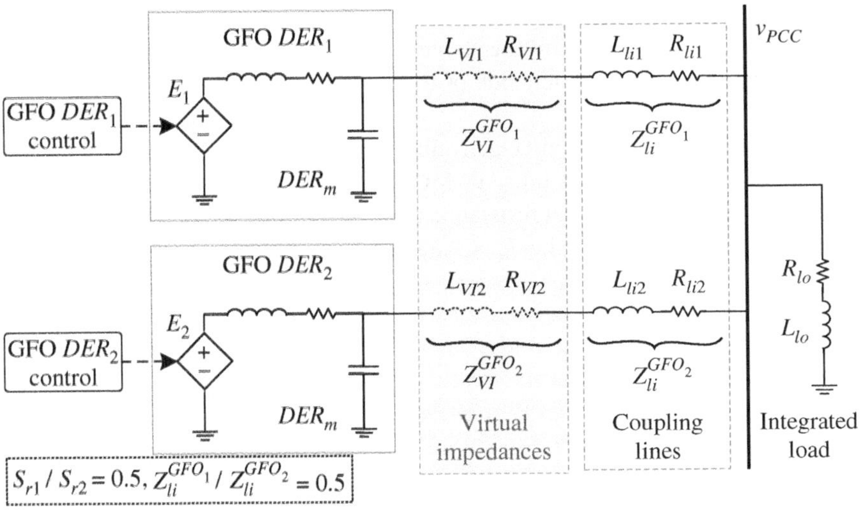

Figure 4.25 A schematic of the studied ac MG shown in Figure 3.5(a), including the two GFO DERs without the GFL DER, and focusing on the virtual impedance effect.

Table 4.4 The considered scenarios for power sharing stability study using a virtual impedance in Figure 4.26.

Scen- ario	R_{VI1} (Ω)	L_{VI1} (mH)	R_{VI2} (Ω)	L_{VI2} (mH)	$Z_{VI}^{GFO_1}/Z_{VI}^{GFO_2}$	$Z_{li}^{GFO_1}/Z_{li}^{GFO_2}$	S_{r1}/S_{r2}
1	0	0	0	0	—	0.5	0.5
2	0.5	2	0.5	2	1	0.5	0.5
3	1	4	0.5	2	2	0.5	0.5
4	1.5	6	0.5	2	3	0.5	0.5

which are introduced in Table 4.4. Two indices are considered for active and reactive power sharing accuracy as follows:

$$\Delta P_{ps} = (P_{r2}/P_{r1})P_{GFO_1} - P_{GFO_2}, \tag{4.9a}$$

$$\Delta Q_{ps} = (Q_{r2}/Q_{r1})Q_{GFO_1} - Q_{GFO_2}, \tag{4.9b}$$

where P_{r1} and P_{r2} are the rated active powers, and Q_{r1}, and Q_{r2} are the rated reactive powers of DERs. P_{GFO_1} and P_{GFO_2} are the measured active powers, and Q_{GFO_1} and Q_{GFO_2} are the measured reactive powers of the DERs. The indices ΔP_{ps} and ΔQ_{ps} are calculated in W and VAr, respectively, such that index values close to zero demonstrate better power sharing.

Figures 4.26(a) and 4.26(b) indicate the active and reactive power sharing indices for the four scenarios introduced in Table 4.4, respectively. The isolated MG is operated with the two GFO DERs at $t = 0$ s and a load increase is experienced at $t = 1$ s. In Scenario 1, the MG cannot be stable without virtual impedance, and thus power sharing instability appears. The ΔP_{ps} is reduced in steady state, but the ΔQ_{ps} remains still large, about −20 kVAr in steadystate, which shows a very weak reactive power sharing. In Scenario 2, the same virtual impedance is added to both DERs. Therefore, power sharing stability is achieved. The active power sharing index is zero in the steady

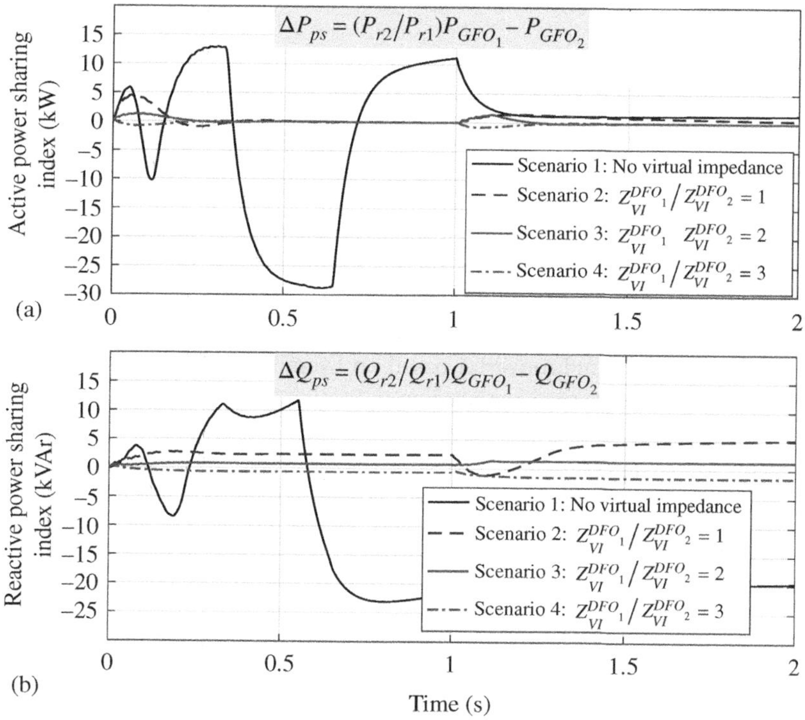

Figure 4.26 Power sharing indices for transient stability analysis of the two-DER isolated MG: (a) active power sharing index, and (b) reactive power sharing index.

state of the primary MG operation and is reduced to around zero after the load change. However, the reactive power sharing index has a steady-state error, and it increases after the load disturbance. Note that the ratio of total line impedances, i.e. sum of real and virtual impedances, does not change in Scenario 1 and can be shown as follows:

$$\left| \frac{Z_{li}^{GFO_1} + Z_{VI}^{GFO_1}}{Z_{li}^{GFO_2} + Z_{VI}^{GFO_2}} \right| = 0.5.$$

Therefore, power sharing is not improved. In Scenario 2, the ratio of the virtual impedances is selected as $Z_{VI}^{GFO_1}/Z_{VI}^{GFO_2} = 2$, which is the reverse of the ratio of the coupling lines. The total line impedance ratio is obtained as 1.6, which is an appropriate impedance ratio for $S_{r1}/S_{r2} = 0.5$. The power sharing indices also show good results for Scenario 3. The active power is shared with a zero index, i.e. maximum accuracy. The reactive power sharing is performed better than Scenario 2 with a lower index value.

Another virtual impedance ratio as $Z_{VI}^{GFO_1}/Z_{VI}^{GFO_2} = 3$ is tested in Scenario 4. This ratio leads to a total impedance ratio of 2.3. According to Figure 4.26, the accuracy of power sharing in Scenarios 3 and 4 is approximately the same. The positive values in Scenario 3 indicate that the output power of GFO DER_1 is larger than its droop-based-determined reference. However, the negative values in Scenario 4 show that the output power of GFO DER_2 is larger than its droop-based-determined reference. Furthermore, by comparing the total impedance ratios in Scenarios 3 and 4, i.e. 1.6 and 2.3, one can find the optimal value of the virtual impedance to achieve the total impedance ratio as 2, which provides the best power sharing for the MG under study.

4.5.2 Synchronverter Stabilization

Section 2.6 represented the nonlinear modeling of the synchronverter, which is a type of inertia emulation control strategy like VSG. Equations (2.114)(a)–(2.114)(d) show the nonlinear behavior of the synchronverter topology, which may cause instability in the transients, e.g. islanding and large load changes. Since the synchronverter model is extensively nonlinear, its stabilization using linear controllers/compensators is not possible easily. Therefore, a nonlinear controller design is required to stabilize the synchronverter. The additional term d is considered as an uncertain disturbance accumulating all the possible uncertainties for the synchronverter control strategy. If the disturbance is estimated in a finite time, the stable operation of the controller can be preserved by applying an appropriate supplementary control signal. To this end, an adaptive backstepping method is used, which can stabilize the synchronverter based on Lyapunov theorem. In order to follow Section 4.5.2.1, see (2.114) and its relevant explanation in Section 2.6.

4.5.2.1 Adaptive Backstepping Stabilizing Method

Since the control input u exists in (2.114)(c) and (2.114)(d), X_3 is defined as $X_3 = [x_3\ x_4]^T$. Therefore, the dynamics of X_3 can be represented as

$$\dot{X}_3 = f(X) + g(X)u + D, \tag{4.10}$$

where $D = [0\ d]^T$, and

$$f(X) = [a_3 x_3\ a_3 x_4 + a_6 \dot{I} x_3 \cos x_1 - a_6 I x_3 \sin x_1]^T, \tag{4.11a}$$

$$g(X) = [a_4\ a_5 I \cos x_1]^T. \tag{4.11b}$$

According to the adaptive backstepping method, in the first step, the Lyapunov function v_1 is selected as a positive semi-definite function as follows:

$$v_1 = \frac{z_1^2}{2}, \tag{4.12}$$

where $z_1 = x_1$. Note that 0 is selected as a reference for $x_1 = \Delta\delta_s$. In order to stabilize z_1, it should be proved that $\dot{v}_1 = z_1 \dot{z}_1 = z_1 x_2$ is negative definite. In this way, z_2 is selected as

$$z_2 = x_2 - \alpha_1. \tag{4.13}$$

By substituting $x_2 = z_2 + \alpha_1$ in $\dot{v}_1$ and considering $\alpha_1 = -k_1 z_1, k_1 > 0$, the derivative of v_1 is obtained as

$$\dot{v}_1 = -k_1 z_1^2 + z_1 z_2, \tag{4.14}$$

where $z_1 z_2$ cannot be proved to be negative definite.

In the second step, a new Lyapunov function is chosen such that it is positive semi-definite:

$$v_2 = v_1 + \frac{z_2^2}{2}. \tag{4.15}$$

The derivative of v_2 is as

$$\dot{v}_2 = \dot{v}_1 + z_2 \dot{z}_2, \tag{4.16}$$

where $\dot{z}_2$ is required, which can be found from (4.13) as

$$\dot{z}_2 = (a_1 + k_1)(z_2 + \alpha_1) + a_2 (x_4), \tag{4.17}$$

where x_4 needs to be substituted by z_4 as

$$z_4 = xz_4 - \alpha_2. \tag{4.18}$$

Substituting (4.18) in (4.17) and the changed form of (4.17) in (4.16), the derivative of v_2 is calculated as

$$\dot{v}_2 = -k_1 z_1^2 + z_1 z_2 + z_2 \left((a_1 + k_1) \left(z_2 + \alpha_1 \right) + a_2 \left(z_4 + \alpha_2 \right) \right). \tag{4.19}$$

In order to stabilize both z_1 and z_2, the supplementary design variable α_2 should be selected as follows:

$$\alpha_2 = \frac{1}{a_2} \left(-z_1 - (a_1 + k_1) z_2 - (a_1 + k_1) \alpha_1 - k_2 z_2 \right), k_2 > 0. \tag{4.20}$$

Therefore, (4.19) is obtained as

$$\dot{v}_2 = -k_1 z_1^2 - k_2 z_2^2 + a_2 z_2 z_4, \tag{4.21}$$

where, $a_2 z_2 z_4$ cannot be proved to be negative definite.

By considering $\hat{D}$ as the estimation of D, the error signal $\tilde{D}$ is defined as follows:

$$\tilde{D} = \hat{D} - D. \tag{4.22}$$

Then, the final step of the adaptive backstepping design process is started by defining the final Lyapunov function as

$$v_3 = v_2 + \frac{1}{2} Z_3^T P Z_3 + \frac{1}{2} \tilde{D}^T \Gamma^{-1} \tilde{D}, \tag{4.23}$$

where $P \in \Re^{2\times 2}$ is a positive definite constant matrix, $\Gamma \in \Re^{2\times 2}$ is an arbitrary design parameter matrix, and

$$Z_3 = X_3 - \Lambda, \tag{4.24}$$

where $\Lambda = [0 \ \alpha_2]^T$. Hence, the derivative of (4.23) can be calculated as

$$\dot{v} = -k_1 z_1^2 - k_2 z_2^2 + a_2 z_2 z_4 + \dot{Z}_3^T P Z_3 + \dot{\hat{D}}^T \Gamma^{-1} \tilde{D}. \tag{4.25}$$

Substituting $X_3 = Z_3 + \Lambda$ in (4.25) and then (4.10), the derivative of v_3 is found as

$$\dot{v} = -k_1 z_1^2 - k_2 z_2^2 + a_2 z_2 z_4 + \left(f(X) + g(X) u + D - \dot{\Lambda} \right)^T P Z_3 + \dot{\hat{D}}^T \Gamma^{-1} \tilde{D}. \tag{4.26}$$

In order to stabilize the set of $\{z_1, z_2, Z_3\}$ by making $\dot{v}$ negative definite, the control input shown in Figure 4.27 should be selected as

$$u = \left(g^T(X) g(X) \right)^{-1} g^T(X) \left[M - f(X) - \hat{D} + \dot{\Lambda} \right], \tag{4.27}$$

where

$$M = \left(Z_3^T P \right)^+ \left(-k_3 Z_3^T Z_3 + a_2 z_2 z_4 \right), \tag{4.28a}$$

$$\left(Z_3^T P \right)^+ = \left[\left(Z_3^T P \right)^T Z_3^T P \right]^{-1} \left(Z_3^T P \right)^T. \tag{4.28b}$$

$\left(Z_3^T P \right)^+$ is nonsingular at any point except origin, $\left(g^T(X) g(X) \right) = a_4^2 + a_5^2 I^2 \cos^2 x_1$ is always nonzero, and $k_3 > 0$. By substituting (4.27) in (4.26), one can find the modified form of $\dot{v}_3$ as

$$\dot{v}_3 = -k_1 z_1^2 - k_2 z_2^2 - k_3 Z_3^T Z_3 - Z_3^T P \tilde{D} + \dot{\hat{D}}^T \Gamma^{-1} \tilde{D}. \tag{4.29}$$

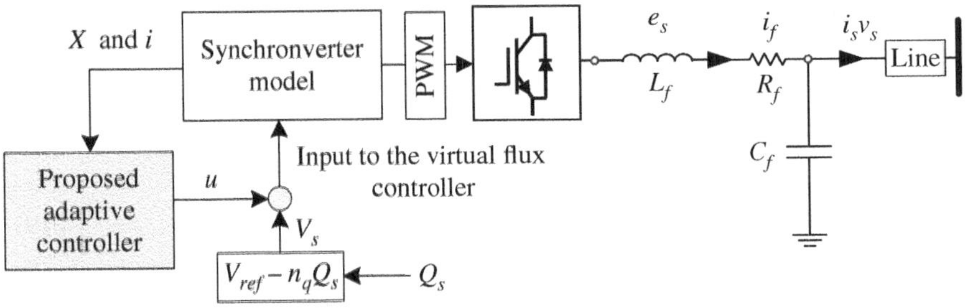

Figure 4.27 Schematic of a synchronverter controller stabilized by an adaptive backstepping controller (see Figure 2.29 for the synchronverter model).

Finally, the adaptation law can be obtained as

$$\dot{\hat{D}} = \Gamma P Z_3, \tag{4.30}$$

and the final form of $\dot{v}_3$ by substituting (4.30) in (4.29) is given as follows:

$$\dot{v}_3 = -k_1 z_1^2 - k_2 z_2^2 - Z_3^T P Z_3. \tag{4.31}$$

Note that (4.31) is negative definite for all positive definite amounts of k_1, k_2, and P, which are the design parameters. Moreover, Γ can be selected freely to achieve a better estimation of d. Generally, z variables are the error values of the x variables. Therefore, zero values for the z variables mean the x variables track the references properly, which leads to the stabilization of the synchronverter system.

4.5.2.2 Simulation Results

The synchronverter parameters are listed in Table 4.5. The adaptive controller parameters are obtained by a simple trial and error method, which can also be seen in Table 4.5. In order to analyze the stabilizing performance of the adaptive controller, a disturbance is considered as

Table 4.5 Parameters of the studied synchronverter and the stabilizing adaptive backstepping control.

Parameter	Symbol (unit)	Value
Parameters/variable of the studied synchronverter		
Virtual moment of inertia	J (kg.m^2)	5.68
Damping coefficient	m_p (W.s/rad)	42
Flux control gain	k_a	0.05
Flux control time constant	τ_a (s)	0.01
Output inverter current amplitude	I (A)	*
Design parameters of the adaptive backstepping control		
First design parameter	k_1	4
second design parameter	k_2	1
Third design parameter	P	$200I_{2\times2}$
Adaptation law matrix	Γ	$8I_{2\times2}$

* $1 + (1 - e^{-5(t-10)})u(t-10)$

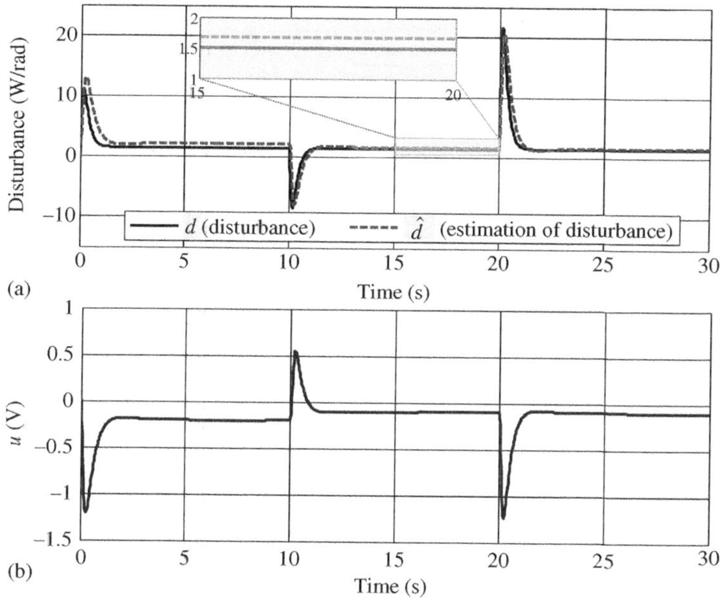

Figure 4.28 The stabilizing adaptive controller features during a permanent disturbance: (a) the disturbance and its estimation obtained by the adaptation law, (b) the control effort of the stabilizing controller.

$20e^{-5t}u(t) - 20e^{-5(t-10)}u(t-10) + 40e^{-5(t-20)}u(t-20) + 1.5$, which is shown in Figure 4.28(a). The disturbance estimation calculated via (4.30) is also shown, which is obtained with a small error of less than 5%. This disturbance has a permanent nonzero term that emulates a real situation, in which the disturbance takes long for a remarkable time without protection devices trip. Figure 4.28(b) indicates the control effort of the adaptive controller, which shows a constant amount to attenuate the permanent disturbance effects.

The state variables of the studied synchronverter with and without the stabilizing adaptive controller are shown in Figure 4.29. $\Delta\delta_s$ shows a slow instability for the synchronverter without the stabilizing controller. However, the stabilization is realized by the adaptive backstepping design. Nevertheless, all four variables have experienced a steady-state error, which is caused by the permanent disturbance. Another improvement of the stabilization process is lesser transient states due to the disturbance changes.

4.6 Summary

This chapter covers the basic requirements for small-signal stability analysis of MGs, including eigenvalue analysis, participation matrix, and sensitivity analysis. The eigenvalue analysis of each case study modeled in Chapter 3 is presented, and stability analysis is improved using sensitivity analysis for the most effective parameters and gains on the stability margins. In ac MGs, the most challenging unit is the PLL, where sensitivity analysis for its parameters and the parameters of related modules to the PLL, such as the coupling line, shows probable instability for some very

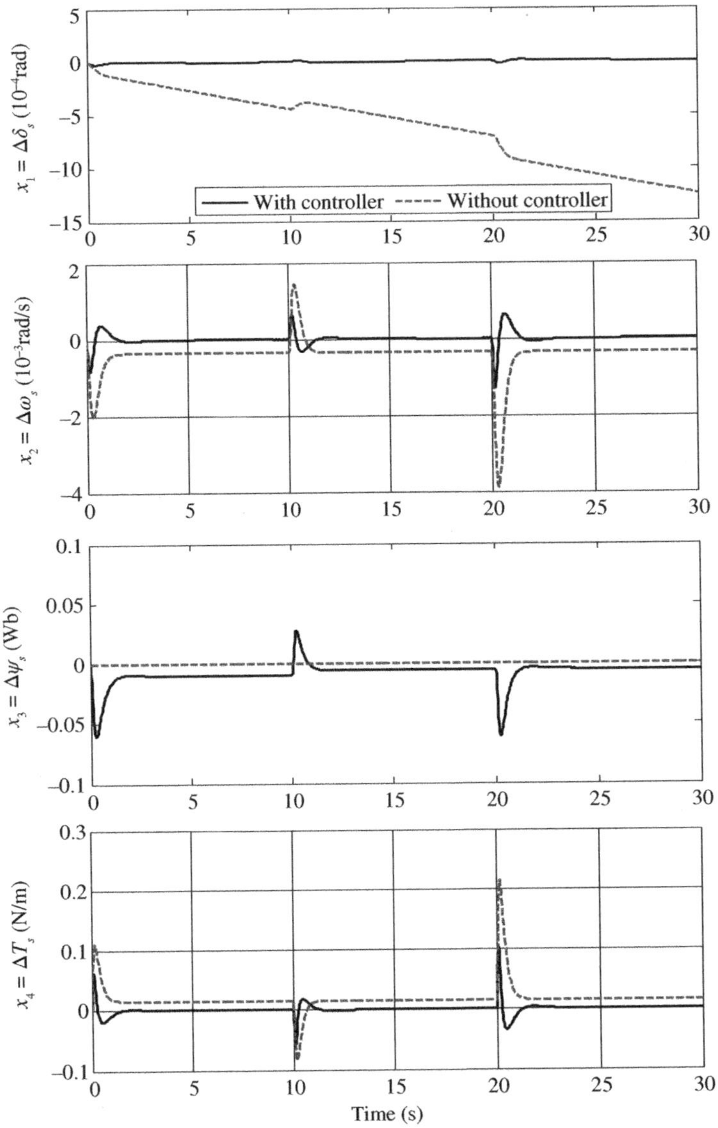

Figure 4.29 State variables of the studied synchronverter with and without the stabilizing controller under the mentioned disturbance.

high or very low values. Additionally, the control parameters of the current controller and the DC link voltage controller have a considerable impact on the stability of grid-connected MGs and GFL DERs, especially in weak grids. On the other hand, islanded MGs and GFO DERs are more sensitive to higher control-level parameters, such as droop control and secondary control gains.

Using time-domain simulation, it is shown that the relative stability of islanded ac MGs can be improved using different GFL control strategies, e.g. VSG control strategy, which is vital for weak MGs with a sensitive frequency and vulnerable voltage to sudden changes. Another stability and dynamic response improvement is presented for the dc MG case study using other virtual dynamic methods, such as virtual capacitor and virtual resistor. The role of virtual impedance in enhancing the transient stability of the ac MG is investigated using power sharing stability analysis in the time domain. Finally, a case study of nonlinear Lyapunov-based stabilization is presented for synchronverters (another form of VSGs) to show the effectiveness of adaptive control methods in stabilizing nonlinear control systems.

References

1 Prabha Kundur, John Paserba, Venkat Ajjarapu, Göran Andersson, Anjan Bose, Claudio Canizares, Nikos Hatziargyriou, David Hill, Alex Stankovic, Carson Taylor, et al. Definition and classification of power system stability IEEE/CIGRE joint task force on stability terms and definitions. *IEEE Trans. Power Syst.*, 19(3):1387–1401, 2004.

2 Mostafa Farrokhabadi, Claudio A Cañizares, John W Simpson-Porco, Ehsan Nasr, Lingling Fan, Patricio A Mendoza-Araya, Reinaldo Tonkoski, Ujjwol Tamrakar, Nikos Hatziargyriou, Dimitris Lagos, et al. Microgrid stability definitions, analysis, and examples. *IEEE Trans. Power Syst.*, 35(1):13–29, 2019.

3 Mobin Naderi, Yousef Khayat, Qobad Shafiee, Frede Blaabjerg, and Hassan Bevrani. Dynamic modeling, stability analysis and control of interconnected microgrids: A review. *Appl. Energy*, 334:120647, 2023.

4 Prabha Kundur, Neal J Balu, and Mark G Lauby. *Power system stability and control*. McGraw-Hill, New York, 1994.

5 Nagaraju Pogaku, Milan Prodanovic, and Timothy C Green. Modeling, analysis and testing of autonomous operation of an inverter-based microgrid. *IEEE Trans. Power Electron.*, 22(2):613–625, 2007.

6 Zhikang Shuai, Yelun Peng, Xuan Liu, Zuyi Li, Josep M Guerrero, and Z John Shen. Dynamic equivalent modeling for multi-microgrid based on structure preservation method. *IEEE Trans. Smart Grid*, 2018. doi: 10.1109/TSG.2018.2844107

7 Farhad Shahnia and Ali Arefi. Eigenanalysis-based small signal stability of the system of coupled sustainable microgrids. *Int. J. Electr. Power Energy Syst.*, 91:42–60, 2017.

8 Iraklis P Nikolakakos, H H Zeineldin, Mohamed Shawky El-Moursi, and James L Kirtley. Reduced-order model for inter-inverter oscillations in islanded droop-controlled microgrids. *IEEE Trans. Smart Grid*, 9(5):4953–4963, 2017.

9 H Bevrani, T Kato, T Ise, and K Inoue. *Grid connected converters: Modeling, stability and control*. Elsevier Science, 2022. ISBN 9780323999021.

10 Yakov A Familiant, Jing Huang, Keith A Corzine, and Mohamed Belkhayat. New techniques for measuring impedance characteristics of three-phase AC power systems. *IEEE Trans. Power Electron.*, 24(7):1802–1810, 2009.

11 Jian Sun. Impedance-based stability criterion for grid-connected inverters. *IEEE Trans. Power Electron.*, 26(11):3075–3078, 2011.

12 Mauricio Cespedes and Jian Sun. Impedance modeling and analysis of grid-connected voltage-source converters. *IEEE Trans. Power Electron.*, 29(3):1254–1261, 2013.

13 Atle Rygg, Marta Molinas, Chen Zhang, and Xu Cai. A modified sequence-domain impedance definition and its equivalence to the dq-domain impedance definition for the stability analysis of AC power electronic systems. *IEEE J. Emerging Sel. Top. Power Electron.*, 4(4):1383–1396, 2016.

14 Richard Tymerski. Frequency analysis of time-interval-modulated switched networks. *IEEE Trans. Power Electron.*, 6(2):287–295, 1991.

15 Richard Tymerski. Application of the time-varying transfer function for exact small-signal analysis. *IEEE Trans. Power Electron.*, 9 (2):196–205, 1994.

16 Billy K H Wong and H Shu-Hung Chung. A systematic graphing technique for small-signal low-frequency characterization of PWM DC/DC converters. *IEEE Trans. Ind. Electron.*, 47(1):45–54, 2000.

17 Chung-Wen Ho, Albert Ruehli, and Pierce Brennan. The modified nodal approach to network analysis. *IEEE Trans. Circuits Syst.*, 22(6):504–509, 1975.

18 Leon O Chua. Computer aided analysis of electronic circuits. *Algorithms and computational techniques*, 1975.

19 Jiri Vlach and Kishore Singhal. *Computer methods for circuit analysis and design*. Springer Science & Business Media, 1983.

20 Hassan Bevrani, Bruno François, and Toshifumi Ise. *Microgrid dynamics and control*. John Wiley & Sons, Hoboken, NJ, 2017.

21 Octavian Dranga, Balázs Buti, and István Nagy. Stability analysis of a feedback-controlled resonant DC-DC converter. *IEEE Trans. Ind. Electron.*, 50(1):141–152, 2003.

22 Y Kuroe. Computer method to analyze stability and bifurcation phenomena. *Nonlinear Phenomena in Power Electronics-Attractors, Bifurcations, Chaos, and Nonlinear Control*, 2001.

23 Suzan Eren, Majid Pahlevaninezhad, Alireza Bakhshai, and Praveen K Jain. Composite nonlinear feedback control and stability analysis of a grid-connected voltage source inverter with LCL filter. *IEEE Trans. Ind. Electron.*, 60(11):5059–5074, 2012.

24 Jia Liu, Yushi Miura, Hassan Bevrani, and Toshifumi Ise. A unified modeling method of virtual synchronous generator for multi-operation-mode analyses. *IEEE J. Emerging Sel. Top. Power Electron.*, 9(2):2394–2409, 2020. doi: 10.1109/JESTPE.2020.2970025.

25 Jia Liu, Hassan Bevrani, and Toshifumi Ise. Design-oriented Q-V response modeling approach for grid-forming distributed generators considering different operation modes. *IEEE J. Emerging Sel. Top. Power Electron.*, 10(1):387–401, 2021.

26 Saeed Golestan, Josep M Guerrero, and Juan C Vasquez. Three-phase PLLs: A review of recent advances. *IEEE Trans. Power Electron.*, 32(3):1894–1907, 2016. doi: 10.1109/TPEL.2016.2565642.

27 Saeed Golestan, Jose Matas, Abdullah M Abusorrah, and Josep M Guerrero. More-stable EPLL. *IEEE Trans. Power Electron.*, 37(1):1003–1011, 2021.

28 Sushil S Thale and Vivek Agarwal. Controller area network assisted grid synchronization of a microgrid with renewable energy sources and storage. *IEEE Trans. Smart Grid*, 7(3):1442–1452, 2015.

29 Soo Hyoung Lee, DongHee Choi, and Jung-Wook Park. Inertia-free stand-alone microgrid—Part I: Analysis on synchronized GPS time-based control and operation. *IEEE Trans. Ind. Appl.*, 54(5):4048–4059, 2018.

30 Dayan B Rathnayake, Milad Akrami, Chitaranjan Phurailatpam, Si Phu Me, Sajjad Hadavi, Gamini Jayasinghe, Sasan Zabihi, and Behrooz Bahrani. Grid forming inverter modeling, control, and applications. *IEEE Access*, 9:114781–114807, 2021.

31 Moudud Ahmed, Lasantha Meegahapola, Arash Vahidnia, and Manoj Datta. Adaptive virtual impedance controller for parallel and radial microgrids with varying X/R ratios. *IEEE Trans. Sustainable Energy*, 13(2):830–843, 2021. doi: 10.1109/TSTE.2021.3133413.

32 Hisham Mahmood, Dennis Michaelson, and Jin Jiang. Accurate reactive power sharing in an islanded microgrid using adaptive virtual impedances. *IEEE Trans. Power Electron.*, 30(3):1605–1617, 2014. doi: 10.1109/TPEL.2014.2314721.

33 Xiaodong Liang, Chowdhury Andalib-Bin-Karim, Weixing Li, Massimo Mitolo, and Md Nasmus Sakib Khan Shabbir. Adaptive virtual impedance-based reactive power sharing in virtual synchronous generator controlled microgrids. *IEEE Trans. Ind. Appl.*, 57(1):46–60, 2020. doi: 10.1109/TIA.2020.3039223

34 A S Vijay, N Parth, Suryanarayana Doolla, and Mukul C Chandorkar. An adaptive virtual impedance control for improving power sharing among inverters in islanded AC microgrids. *IEEE Trans. Smart Grid*, 12(4):2991–3003, 2021. doi: 10.1109/TSG.2021.3062391

35 Fei Deng, Xiangke Li, Xiaobin Zhang, and Paolo Mattavelli. An iterative virtual impedance regulation strategy in islanded microgrids for enhanced balanced, unbalanced and harmonic current sharing. *IEEE Trans. Sustainable Energy*, 13(1):514–526, 2021. doi: 10.1109/TSTE.2021.3118555

36 Fei Deng, Wenli Yao, Xiaobin Zhang, and Paolo Mattavelli. A decentralized current sharing strategy for islanded resistive microgrids based on iterative virtual impedance regulation. *IEEE Trans. Ind. Inf.*, 18(6):3958–3969, 2021. doi: 10.1109/TII.2021.3110951

37 Xiangyu Wu, Chen Shen, and Reza Iravani. Feasible range and optimal value of the virtual impedance for droop-based control of microgrids. *IEEE Trans. Smart Grid*, 8(3):1242–1251, 2016. doi: 10.1109/TSG.2016. 2519454

38 Fei Deng, Wenli Yao, Xiaobin Zhang, Yi Tang, and Paolo Mattavelli. Review of impedance-reshaping-based power sharing strategies in islanded AC microgrids. *IEEE Trans. Smart Grid*, 14(3):1692–1707, 2022. doi: 10.1109/TSG.2022.3208752

39 Mohsen Eskandari and Li Li. Microgrid operation improvement by adaptive virtual impedance. *IET Renew. Power Gener.*, 13(2):296–307, 2019.

40 Hua Han, Xiaochao Hou, Jian Yang, Jifa Wu, Mei Su, and Josep M Guerrero. Review of power sharing control strategies for islanding operation of AC microgrids. *IEEE Trans. Smart Grid*, 7(1):200–215, 2015. doi: 10.1109/TSG.2015.2434849

41 Mohsen Eskandari, Li Li, and Mohammad H Moradi. Improving power sharing in islanded networked microgrids using fuzzy-based consensus control. *Sustainable Energy Grids Netw.*, 16:259–269, 2018.

42 Jinwei He, Yun Wei Li, Josep M Guerrero, Frede Blaabjerg, and Juan C Vasquez. An islanding microgrid power sharing approach using enhanced virtual impedance control scheme. *IEEE Trans. Power Electron.*, 28(11):5272–5282, 2013. doi: 10.1109/TPEL.2013.2243757

43 A S Vijay, Suryanarayana Doolla, and Mukul C Chandorkar. Varying negative sequence virtual impedance adaptively for enhanced unbalanced power sharing among DGS in islanded AC microgrids. *IEEE Trans. Energy Convers.*, 36(4):3271–3281, 2021.

44 Kim Diefrei Remboski Felisberto, Paulo Thiago de Godoy, Diogo Marujo, Adriano Batista de Almeida, and Rodrigo de Barros Iscuissati. Trends in microgrid droop control and the power sharing problem. *J. Control Autom. Electr. Syst.*, 33(3):719–732, 2022.

45 Yang Han, Hong Li, Pan Shen, Ernane Antônio Alves Coelho, and Josep M Guerrero. Review of active and reactive power sharing strategies in hierarchical controlled microgrids. *IEEE Trans. Power Electron.*, 32(3):2427–2451, 2016. doi: 10.1109/TSG.2015.2434849

46 Mobin Naderi, Qobad Shafiee, Hassan Bevrani, and Frede Blaabjerg. Low-frequency small-signal modeling of interconnected AC microgrids. *IEEE Trans. Power Syst.*, 36(4):2786–2797, 2020.

47 Mobin Naderi, Yousef Khayat, Qobad Shafiee, Tomislav Dragicevic, Hassan Bevrani, and Frede Blaabjerg. Interconnected autonomous AC microgrids via back-to-back converters—Part I: Small-signal modeling. *IEEE Trans. Power Electron.*, 35(5):4728–4740, 2019.

48 USDOE. Small wind electric systems: A guide produced for the Tennessee Valley Authority. *National Renewable Energy Laboratory (NREL)*, pages 1–27, DOE/GO-102004-1897, 2009.

49 IEEE Standards Coordinating Committee 21. IEEE standard for interconnection and interoperability of distributed energy resources with associated electric power systems interfaces. *IEEE Std 1547*, 2018.

50 Mania Pavella, Damien Ernst, and Daniel Ruiz-Vega. *Transient stability of power systems: A unified approach to assessment and control*. Springer Science & Business Media, 2012.

51 Atif Zaman Khan and Farrukh Shahzad. A PC based software package for the equal area criterion of power system transient stability. *IEEE Trans. Power Syst.*, 13(1):21–26, 1998.

52 Thanh Long Vu and Konstantin Turitsyn. Lyapunov functions family approach to transient stability assessment. *IEEE Trans. Power Syst.*, 31 (2):1269–1277, 2016.

53 Yun Zhang, Le Xie, and Qifeng Ding. Interactive control of coupled microgrids for guaranteed system-wide small signal stability. *IEEE Trans. Smart Grid*, 7(2):1088–1096, 2016.

54 J Q James, David J Hill, Albert Y S Lam, Jiatao Gu, and Victor O K Li. Intelligent time-adaptive transient stability assessment system. *IEEE Trans. Power Syst.*, 33(1):1049–1058, 2018.

55 L A S Pilotto, M Szechtman, and A E Hammad. Transient AC voltage related phenomena for HVDC schemes connected to weak AC systems. *IEEE Trans. Power Delivery*, 7(3):1396–1404, 1992.

56 Raouia Aouini, Bogdan Marinescu, Khadija Ben Kilani, and Mohamed Elleuch. Stability improvement of the interconnection of weak AC zones by synchronverter-based hvdc link. *Elect. Power Syst. Res.*, 142:112–124, 2017.

57 A H Kasem Alaboudy, Hatem H Zeineldin, and James Kirtley. Microgrid stability characterization subsequent to fault-triggered islanding incidents. *IEEE Trans. Power Delivery*, 27(2):658–669, 2012.

58 Jaber Alipoor, Yushi Miura, and Toshifumi Ise. Stability assessment and optimization methods for microgrid with multiple VSG units. *IEEE Trans. Smart Grid*, 9(2):1462–1471, 2018.

59 Zhikang Shuai, Yang Hu, Yelun Peng, Chunming Tu, and Z John Shen. Dynamic stability analysis of synchronverter-dominated microgrid based on bifurcation theory. *IEEE Trans. Ind. Electron.*, 64(9):7467–7477, 2017.

60 Ting-Chia Ou. A novel unsymmetrical faults analysis for microgrid distribution systems. *Int. J. Electr. Power Energy Syst.*, 43(1):1017–1024, 2012.

5

Microgrid Control: Concepts and Fundamentals

5.1 Introduction

Microgrid (MG) is a vital component of the energy transition from fossil fuel-based energy systems to renewable energy and electric transportation systems. However, due to the intermittent, stochastic, and distributed nature of the distributed energy resources (DERs), this small-scale dynamic power system requires advanced and updated control algorithms to ensure the stability and cost-effectiveness operation in both islanded and grid-connected modes. The control system must regulate the system outputs, e.g. frequency and voltage, distribute the load among MG units, and optimize operating costs while ensuring smooth transitions between operating modes.

This chapter provides an overview of the main control challenges and solutions for MGs, covering all control levels and strategies, with a focus on simple and linear control solutions that are more accessible to power grids and power electronics communities. The chapter also presents different control strategies for MGs, including hierarchical control and different levels of the control hierarchy. Since power converters play an important role in controlling DERs, control schemes for these systems are presented simply as the base for primary control of MGs. The chapter includes examples and MATLAB simulations to support the presented control methodologies. The MATLAB/Simulink source files are also provided to allow readers to replicate the examples and develop and validate their own control algorithms.

Overall, this chapter provides a comprehensive introduction to the fundamental control techniques used in MGs and lays the groundwork for Chapter 6, which focuses on the latest advances and developments in MG control.

5.2 Fundamentals and Requirements

This section covers the fundamental concepts needed to understand the control systems of MGs. It begins by introducing the control systems followed by the key control objectives and challenges faced by MGs and an overview of the main control architectures used in these systems.

5.2.1 Introduction to Control Systems

Control systems are essential for managing the behavior of complex and dynamic systems like MGs. A control system continuously measures the output(s) or states of a system and adjusts the input(s)

Microgrids: Dynamic Modeling, Stability and Control, First Edition. Qobad Shafiee, Mobin Naderi, and Hassan Bevrani.

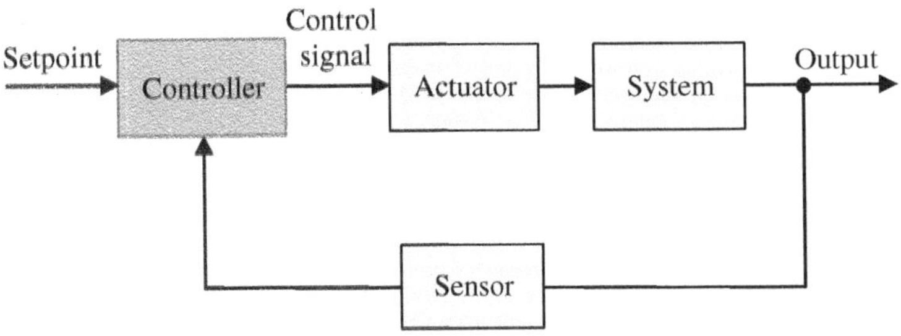

Figure 5.1 A general scheme for closed-loop control systems.

to achieve the desired response. It is made up of four main components: the system being controlled, sensors, which measure the output(s)/states of the system, a controller that processes the measurements and generates a control signal(s), and an actuator(s) that applies the control signal to the system. The block diagram of a closed-loop control system is shown in Figure 5.1, where the setpoint is the desired value of the output of the system being controlled. The controller processes the measurement (or estimations) from the sensor(s) and generates a control signal that is applied by the actuator(s) to the system. The sensor measures the output of the system and provides feedback to the controller to adjust the control signal as necessary. It should be noted that in cases where the states/outputs are not measurable or measurements are not available, estimated data may be used [1].

Cascade control is a type of classical control system that uses multiple controllers in a series to achieve more precise control. In a cascade control system, there are two or more controllers in a nested loop structure. The output of the first controller is used as the setpoint for the second controller, and so on. Each controller adjusts the input to the next controller to achieve the desired response. Cascade control is commonly used in MG control because it provides more precise control of the system. By using multiple controllers in a nested loop structure, it is possible to utilize linear and simple control techniques achieving faster response times and better disturbance rejection. MG control systems commonly use cascade control and linear control methods (e.g. proportional or proportional–Integral controllers) to regulate voltage, frequency, and power flow within the MG. It is worth noting that the cascade control loops can be replaced with advanced control methods where several control objectives can be included in a single control loop. This will be discussed in more detail in Chapter 6.

5.2.2 Control Objectives and Challenges

MGs present a number of significant challenges that must be adequately addressed in the design of their control system to ensure a stable, reliable, secure, and cost-effective operation. Some of these challenges stem from conventional power system assumptions, while others are the result of stability issues posed by this new concept [2]. The following are the most important challenges in the MG control area:

- *Low inertia*: In contrast to large power systems with a large number of synchronous generators and thus large inertia, MGs with a significant share of power converter-based DERs present low inertia characteristics. Because of their low inertia, these systems may experience severe voltage and frequency fluctuations in the aftermath of a disturbance. Therefore, conventional control systems will be ineffective, and proper control techniques must be employed.

- *Dynamic modeling*: MGs, like any other systems, need to be modeled to study their behavior over different time scales and for various control objectives. In most cases, modeling is necessary before implementing designed control systems to demonstrate unknown challenges and verify control performance. While simple and practical methods can be used to design linear control techniques like PI controllers, accurate system modeling is crucial for appropriate control design, particularly for advanced control techniques. Conventional power system modeling assumptions, such as three-phase balanced conditions and inductive networks, are no longer valid for MGs. To better study the behavior and stability of these systems and design suitable control approaches, conventional modeling needs to be updated to account for new types of sources (such as renewable energy sources [RESs] and energy storage systems [ESSs]), loads (including constant power, constant impedance, motor, and active loads), power converters, and power networks (low X/R characteristics). For more details, refer to Chapter 3.
- *Stability*: MG stability challenges can be classified according to MG operating modes. Grid-synchronization and fault/voltage ride-through capabilities are the most significant stability challenges in grid-connected mode. In islanded mode, MGs must deal with issues such as frequency stability, voltage stability, and power sharing stability, as well as some recent communication-related issues such as delay-dependent stability and cyber attacks. Interaction of the control systems of MG units may also lead to local oscillations and accordingly system instability. Additionally, smooth transitions between modes of operation (i.e. islanding and grid-connected modes) as well as MG interconnection address some large-signal stability issues. As a result, the control system must be designed in such a way that it guarantees system stability under all possible conditions. Small-signal and transient stability analysis, as described in Chapter 4, can be used to make sure that designed controllers ensure system stability.
- *Uncertainty*: Unlike in the bulk power system, the uncertainty associated with demand and generation is significant in MGs. Because of the intermittent nature of RESs and fluctuating load profiles, an MG's power can become unbalanced. If the DERs of an MG do not properly coordinate, some may experience power deficiency while others have surplus energy, resulting in frequency and voltage instability. To ensure a reliable and cost-effective operation with stochastic generation and demand, appropriate control strategies must be used.
- *Cyber security*: Whatever electrical configuration is chosen for an MG, it should be tightly coupled with a cyber network to meet communication requirements and take advantage of various control strategies. The topology of a cyber network is not always the same as that of a physical power network. It should, however, be flexible enough to accommodate the scalability of MGs with plug-and-play (PnP) capability. An MG is essentially a cyber-physical system in which the communication network allows data exchange among DERs for proper monitoring and control. The majority of the literature assumes that data is transferred to and from the control loops or between DERs via an ideal communication network. Cyber vulnerabilities (e.g. data dropouts, time delays, transmission intervals, and failures) that may occur in a real communication network, on the other hand, can have severe economic and physical consequences for system operations. Moreover, cyber-attacks can have far more harmful and devastating effects on the operation and stability of power electronics-based MGs, especially in the islanded operation. State estimation, voltage and frequency control, and protection systems are of targets for cyber attacks [3]. To deal with the aforementioned cyber-security challenges, the control system must be well-equipped with appropriate techniques.

To meet the aforementioned challenges, the control system must ensure the MG's reliable operation. The following are the most desirable features of the MG control system, as well as the main functions that can be requested:

- *Output control*: The DERs output variables, namely currents and voltages, must be controlled locally in order to track the provided references in a fast response while damping oscillations appropriately.
- *Frequency/voltage regulation*: The system's frequency and voltage must be properly regulated close to their nominal values, both in islanded and grid-connected modes.
- *Power quality*: Voltage/frequency stabilization is affected by nonlinear loads and unbalanced grid conditions. Due to high-frequency switching operations, power converters are major sources of harmonics that pollute the power grid. However, if properly controlled, they can be used to address power quality issues such as voltage/current harmonics and unbalances.
- *Power balance*: The MG's DERs must be able to deal with sudden active power imbalances while keeping frequency and voltage deviations within acceptable ranges. The load power should be properly shared among the DERs based on their power rates.
- *Economic dispatch*: An appropriate dispatch of MG's DERs while considering reliability can significantly reduce the operating costs and accordingly increase the profit.
- *Power flow management*: Power flows between the MG and the main grid as well as possibly with other MGs should be managed/optimized with respect to grid requirements and constraints, in order to reduce energy costs, avoid undesired power loss, and supply the grid with the power needed.
- *Smooth transitions between operation modes*: An MG should be capable of operating in islanded and grid-connected modes and handling the transitions between the two. For each mode of operation, appropriate control strategies should be developed, ensuring smooth transitions. High-speed islanding detection and accurate synchronization algorithms (for reconnection to the main grid or other MGs) are essential.
- *Demand-side management*: In order to accommodate MG's needs, demand-side management or demand response mechanisms should be utilized to allow certain variations in the demand of a portion of the loads.

Given the various time constants involved, including fast dynamics in output controls and slower dynamics in economic dispatch, the deployment of a hierarchical control structure is quite appealing. The complexity and sophistication of the solutions for the MG's control requirements will be highly influenced by whether it is designed to operate in an islanded or grid-connected mode. While the interaction with the main grid is essential in the grid-connected mode of operation, reliability issues are of main interest in the islanded mode of operation. This chapter provides various control strategies and techniques to address the aforementioned challenges and objectives.

5.2.3 Control Architectures

MGs, like most large systems, can employ a variety of control architectures, including *centralized*, *decentralized*, and *distributed* to address their control challenges and objectives. Figure 5.2 presents general schemes for different control architectures that can be applied to MGs. The control system, as depicted in Figure 5.2, provides control actions u_i based on system information and feedback measurements from subsystems y_i.

In the centralized structure, a single entity is responsible for the system's control and management. The control scheme gathers data from various remote locations ($y_1, y_2, \dots, y$) in order to decide on all control actions ($u_1, u_2, \dots, u_i$) in a single place (see Figure 5.2(a)). For example, in an MG control system, the central controller assigns the setpoints to the loads and DERs in order to attain the MG's goals. The benefit of this approach is that it enables global decision-making

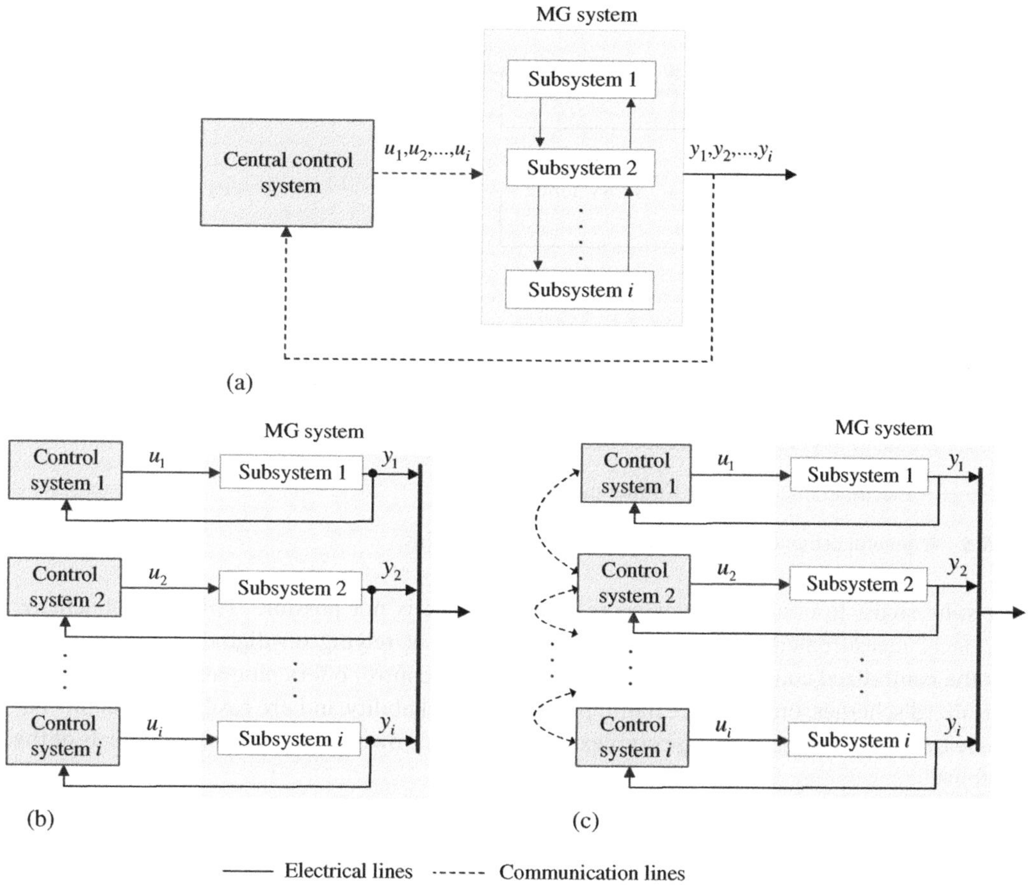

Figure 5.2 Control architectures for microgrids: (a) centralized control scheme, (b) decentralized control scheme, and (c) distributed control scheme.

and offers a global optimal solution since it has full knowledge of the system and full control over all of the units [4]. Nevertheless, this architecture suffers from several drawbacks: (i) it is computationally expensive because a significant amount of data from various subsystems needs to be exchanged and analyzed, (ii) because the entire system is controlled by a single central controller, it is unreliable and has a single point of failure, and (iii) it does not account for plug and play functionality, which means that the controller must be updated when a new subsystem enters or leaves the system.

Due to the complexity and the aforementioned drawbacks of central control for large systems, a control system distributed among several agents appears to be a promising solution (e.g. MGs with a number of DERs or interconnected MGs). A more useful alternative solution is the decentralized architecture, in which each subsystem is managed by a local control agent (Figure 5.2(b)). An agent is only aware of and makes decisions regarding its own subsystem while working toward its own benefit; however, local controllers do not take into account the behavior of the entire network, which could lead to unexpected results. Because the various subsystems typically have different objectives, decisions are made locally. The subsystems need to be decoupled in order to implement the decentralized control scheme, which has an impact on the entire system's performance. Since no information is exchanged between the local controllers, a cooperative strategy

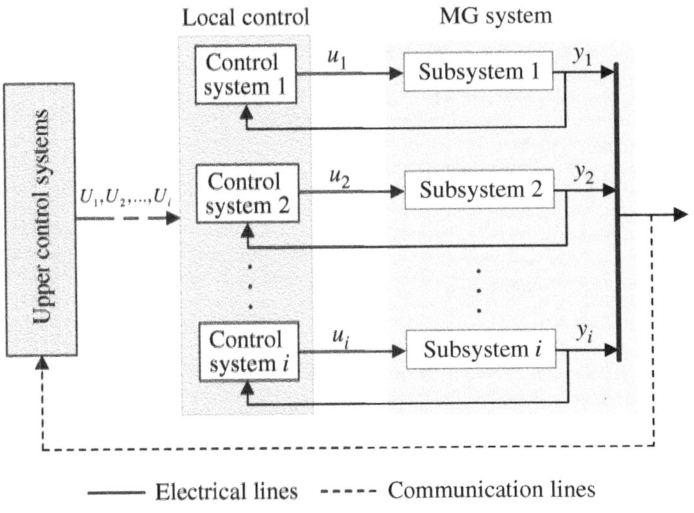

Figure 5.3 A general scheme for hierarchical control of microgrids.

among subsystems to obtain a better overall performance is not possible. This control scheme utilizes local measurements and control actions instead of relying on digital communication. Unlike the centralized control architecture, decentralized control offers plug-and-play capability. Decentralized schemes, on the other hand, provide high reliability and are easily expandable, but they have the disadvantage of being complex due to multiple ownership and different goals of the subsystems.

The distributed control architecture is a compromise between the two mentioned approaches. It employs local controllers to manage each subsystem, with the subsystems sharing information via communication links, as illustrated in Figure 5.2(c). The nature and amount of information exchanged can vary greatly between implementations, for example, knowledge about the dynamic behavior of other subsystems, decisions made on other different subsystems, or a global criterion to control the entire system. Then, cooperation and negotiation among control agents based on an agreement (e.g. consensus algorithm) are possible in order to improve overall system performance. The performance of the control architecture is affected by the number of nodes, the number of messages exchanged, the structure of the system, and the accuracy and optimality of the solution in all of the above-mentioned architectures.

Through the use of a hierarchical control architecture, a compromise between fully centralized and fully decentralized control architectures is facilitated in MGs. Each of the three architectures can be used in the hierarchical control scheme, depending on the level of hierarchy and the control objectives. While the first level of the hierarchy employs a decentralized control structure, the higher levels make use of digital communication, and thus centralized or distributed solutions may be employed. A general scheme of hierarchical control structure for MGs is depicted in Figure 5.3.

Primary (local), secondary (supplementary), central, and global controls are the four levels that make up hierarchical control in MGs. While the primary controls are known as decentralized controllers, the central, and global controllers are operating as centralized controllers, whereas secondary control can be implemented in either of the three architectures (i.e. centralized, decentralized, and distributed). More details on the hierarchical control of MGs are given in Section 5.4.

5.3 Control Strategies for Power Converters

5.3.1 Introduction

MGs consist of different types of DERs because of many benefits, most of which are interfaced and controlled by power converters. The power converters can provide more flexible operation compared to the direct connection of conventional generators, e.g. synchronous and induction generators to the grid. They are effectively used to convert the primary energy source to electric energy and transfer it to the MG bus in order to be used by the local loads in islanded mode or to be injected into the main grid or other MGs. Figure 5.4 represents the general structure of a power converter-based DER inside an MG. As shown in Figure 5.4, the power converter-based DERs are normally equipped with a back-to-back structure including a typical source-side power converter (dc–dc or ac–dc depending on the type of source) to control the primary source and link it to a dc bus, and a (dc–dc or dc–ac) power converter to provide required power output for the MG. For example, PV and wind turbines are interfaced with dc–dc and ac–dc source side converters, respectively, in order to extract the maximum power in the given environmental conditions. For dc systems, e.g. dc MGs, the grid-side converter is replaced by a dc–dc converter to provide dc output. The primary source is responsible for providing the power demand in order to keep a constant dc voltage bus. The grid-side power converter is fed by a dc link and interfaced with a passive (LC or LCL) filter to suppress switching harmonics, while its switches are controlled by appropriate control loops for a desirable system operation. A separate control system must be also designed for the primary source in order to control its output. The control systems can be simply designed to achieve all the system objectives. Here, our focus is on control strategies for the grid-side power converters. Interested readers can find details of modeling and control of the source-side power converter in [5].

Depending on their operation and control objectives in an MG, power converters can be classified into *grid-forming* and *grid-following* power converters. Grid-forming (GFO) converters are controlled to either form an autonomous grid (e.g. MGs in island mode), providing frequency and voltage control or shape the output voltage delivering other needed ancillary services. The GFO power converters are able to shape dynamics in the MG, for example, they can provide virtual inertia and damping to frequency changes, a similar behavior to synchronous generators, avoiding the problems associated with low inertia. This is simply possible since the GFO converter creates its own internal voltage reference angle based on the output power of the converter.

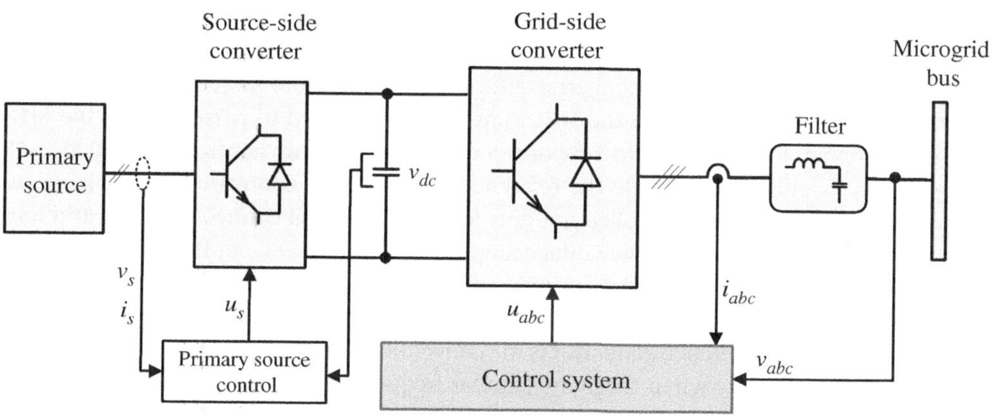

Figure 5.4 General structure of a power converter-based DER in an MG.

Grid-following (GFL) converters are designed to inject a certain amount of power into the MG bus. Therefore, they are connected to the MG bus as current injectors, while they follow the voltage and frequency of the coupling point. These converters should be perfectly synchronized with the ac voltage at the connection point using a synchronization algorithm, e.g. a phase-locked loop (PLL), in order to deliver the accurate active and reactive power. Generally, GFO converters can be represented as voltage sources with a low-output impedance, while GFL converters are assumed as current sources with a shunt high impedance [6].

GFL converters cannot operate independently and require the support of other components to function in MG operations. They would need GFO converters in an islanded MG operation or the main grid in a grid-connected MG operation to form the required voltage for synchronization. On the contrary, the GFO converters can operate autonomously. Therefore, each MG may consist of a number of GFL converters, while at least one GFO converter is required in order to play the role of voltage/frequency control and stabilization. In an MG, the ac voltage generated by the GFO power converter will be used as a reference for the rest of connected GFL converters. In grid-connected mode, the utility grid provides the necessary voltage and frequency control, whereas, in grid-isolated mode, at least one GFO is required to play the role of voltage/frequency control and stabilization. The GFL converters can be used in both islanded and grid-connected operation modes of MGs for the sake of supplying active and reactive powers.

To sum up, the GFO converters can support the frequency and shape the output voltage, while the GFL converters are only able to supply the maximum power into the MG bus like a current source. It should be remarked that since both GFL and GFO converters can be controlled in the available power range to support the grid voltage and supply the scheduled power, they are sometimes referred to as grid-supporting voltage/current source converters in some research studies [6]. In such studies, grid-supporting converters are identified as the third control strategy for power converters. The GFL converters are also defined as grid-feeding converters in the literature.

5.3.2 Grid-Following Power Converters

A simple and general structure of the control system for GFL power converters is represented in Figure 5.5. As shown in this figure, the GFL power converter contains two main subsystems, a synchronization algorithm (e.g. PLL) and a current controller. The other two blocks, i.e. power control and *P*–*Q* reference generator, provide the reference for the current controller. The current controller regulates the current injected into the grid, according to the reference provided by the power control loop i^*. The synchronization algorithm estimates the instantaneous frequency/angle of the measured MG bus voltage and provides the angular frequency and angle of the current reference signal. As mentioned, in the GFL converters, the system voltage and frequency are regulated by the connected external grid. In case the GFL converter is supposed to participate in the MG support (in some research known as grid-supporting converter), another control loop called *P*–*Q* reference generator is utilized to provide the reference for the power control/calculator loop. In such a case, the GFL converter is controlled not only to supply the load connected to the grid but also to contribute to the regulation of the voltage amplitude and frequency of the MG and utility grid in grid-connected mode. For example, the grid codes for wind turbines are defined so that they should be designed to provide a given amount of power to the utility grid for regulatory purposes [6]. An example of the *P*–*Q* reference generator is the conventional droop control that provides the active/reactive power reference when receiving angular frequency ω and voltage amplitude E of the point of common coupling (PCC) via the synchronization algorithm. Figure 5.6 represents a typical structure of control loops for a three-phase GFL converter.

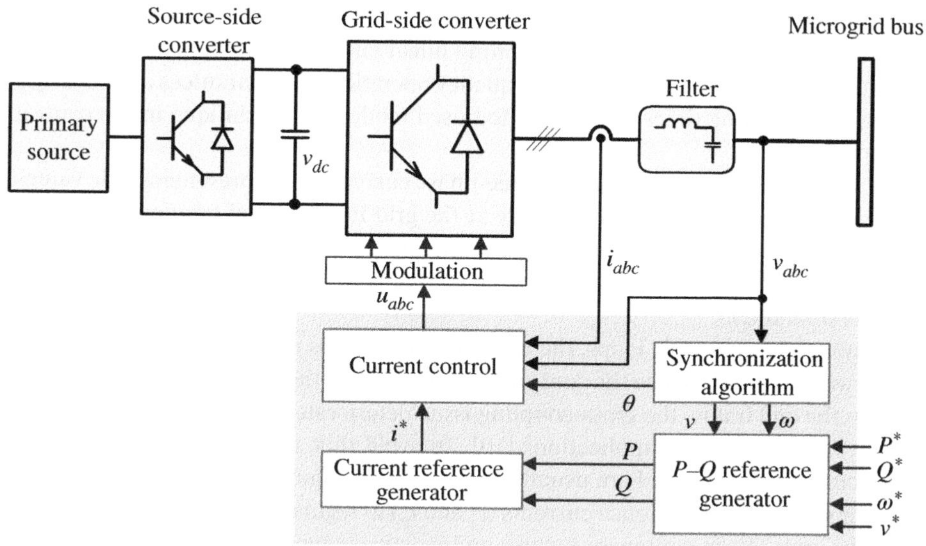

Figure 5.5 General scheme for control system of a grid-following power converter.

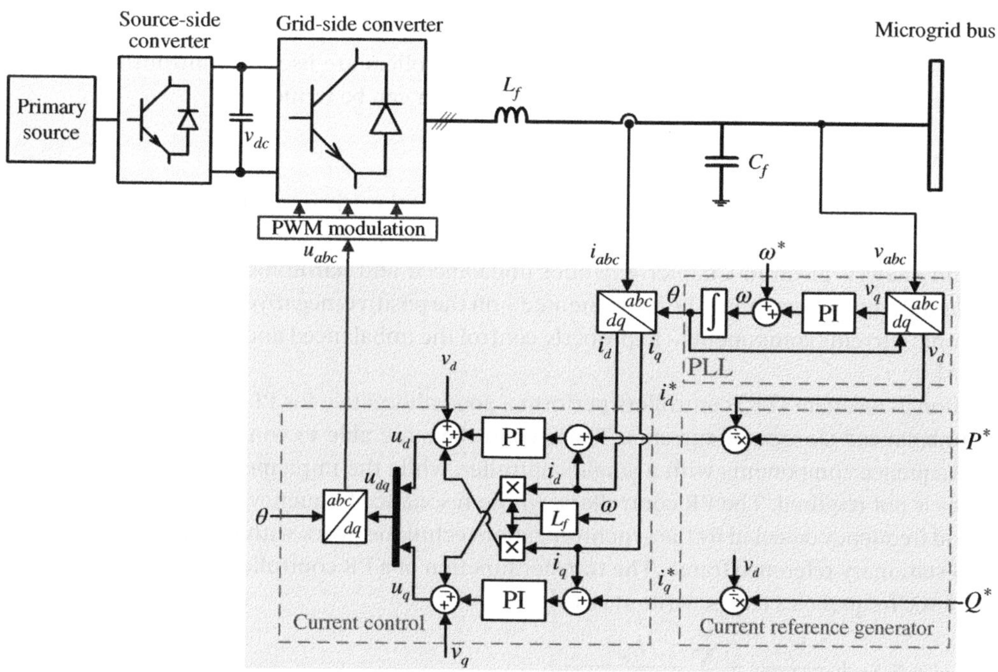

Figure 5.6 Typical control structure for a three-phase grid-following power converter.

5.3.2.1 Current Control

Different control techniques, from linear to nonlinear control (e.g. hysteresis, sliding-mode [7], deadbeat control, or predictive control [8]), have been proposed for the current control of power converters [9]. Since the model of power converters is usually available, model-based methods such as model predictive control and deadbeat control have been widely employed to achieve the control objectives. However, regular PI controllers implemented on the *dq* synchronous reference frame or resonant controllers implemented on the $\alpha\beta$ stationary reference frame, in combination with PWM, are the most commonly used methods for implementing linear current controllers in three-phase systems [6]. Since the PWM module allows fixed frequency operation and guarantees a decoupling between switching and sampling times, it is very often used modulation technique in the control of power converters.

Using the Park transformation, the sinusoidal three-phase currents are represented as dc values in an orthogonal *dq0* frame, rotating synchronously at the grid fundamental frequency detected by the synchronization algorithm. In this reference frame, instead of three control loops for the *abc* frame, two independent control loops are implemented to regulate the direct and quadrature current components (i_d and i_q).

When a control system has multiple loops, the various control loops may upset each other, thus the use of a feed-forward loop can isolate the control loops from each other. In the current control of power converters in the *dq0* frame, the cross-coupling issue deteriorates the system performance, particularly in low PWM pulse-ratio applications [10]. To avoid this, the decoupling techniques based on inductor current feed-forward are usually employed (see Figure 5.6). In GFL converters, the power controller provides the reference currents (i_d^* and i_q^*) to regulate the delivered active and reactive power to the grid. The instantaneous active and reactive power components, used by the power control loop, in this reference frame are calculated by

$$p = \frac{3}{2}(v_d i_d + v_q i_q); \quad q = \frac{3}{2}(v_d i_q - v_q i_d). \tag{5.1}$$

In this standard structure, proportional-integral (PI) controllers are used to control the d and q current components. The transfer function of a PI controller can be written as

$$G_{PI}(s) = k_P + \frac{k_I}{s}, \tag{5.2}$$

where k_P is the proportional gain, and k_I is the integral gain at the grid frequency. These parameters can be simply designed to provide an appropriate performance for the current controller using the models and analysis given in Chapter 4. Under unbalanced and harmonic conditions, however, independent PI controllers should be implemented – on the positive/negative sequence current and the harmonic currents components – to properly control the unbalanced and harmonic currents to the grid [6].

Proportional–resonant (PR) controllers perform a good alternative for PI controllers when controlling unbalanced sinusoidal currents. These controllers are able to control both positive and negative sequence components with a single controller, while the implementation of decoupling techniques is not required. The PR controllers, whose resonance frequency is tuned at the fundamental grid frequency detected by the synchronization technique, work with ac variables expressed on the $\alpha\beta$ stationary reference frame. The transfer function of a PR controller for the fundamental and harmonic frequencies can be written as [6]

$$G_{PR}(s) = k_P + \frac{k_R s}{s^2 + \omega_0^2} + \sum_{h=2}^{n} \frac{k_{ih} s}{s^2 + (h\omega_0)^2}, \tag{5.3}$$

where k_P is the proportional gain, k_R is the resonant gain at the grid frequency, k_{ih} is the resonant gain at the h-harmonic to be controlled, and ω_0 is the fundamental frequency detected by the synchronization algorithm. To compensate for the harmonics of the injected current, additional PR harmonic compensators can be implemented in parallel at the desired harmonic frequencies. The instantaneous active and reactive power components in the $\alpha\beta$ frame are expressed by

$$p = \frac{3}{2}(v_\alpha i_\alpha + v_\beta i_\beta); \quad q = \frac{3}{2}(v_\beta i_\alpha - v_\alpha i_\beta). \tag{5.4}$$

5.3.2.2 Synchronization Algorithm

In order to deliver accurate active and reactive power to the grid, a precise synchronization algorithm is needed for GFL converters to estimate the grid voltage parameters, i.e. voltage amplitude, frequency, and phase angle. The synchronization algorithm is generally implemented based on two different structures depending on the feedback parameter: phase-locked loop (PLL) synchronization structures, and frequency-locked loop (FLL) synchronization structures. The former adjusts the phase difference, while the latter adjusts the frequency difference. The FLL provides an improved performance under abnormal grid conditions compared to the PLL because it is less sensitive to phase angle jumps of the grid voltage, which may occur during transient grid faults [11].

The PLL converts the three-phase voltage waveforms from the *abc* reference frame into the rotating *dq0* reference frame, using the Park transformation. The angular position of this *dq0* reference frame is controlled through a feedback control loop, e.g. a PI, which drives the quadratic component of voltage v_q to zero, and accordingly provides the required voltage parameters for the other loops. There have been many studies on PLL [12] and FLL [13], most of which aim at improving the structure of the loops to improve the synchronization performance. The implementation of a typical PLL is shown in Figure 5.6.

5.3.3 Grid-Forming Power Converters

The GFO converters, known as voltage source converters, are represented as controllable voltage sources with low-output impedance, much like the grid-tied synchronous generators (SGs). However, they can respond much faster to any contingencies comparing to the SGs. GFO converters are designed to provide functionalities that are traditionally offered by SGs, enabling their smooth operation in low-inertia grids such as MGs. One notable example is the virtual synchronous generator (VSG), a well-known concept that employs GFO converters to provide virtual inertia and damping for low-inertia grids. As voltage sources with low-output impedance, they need an extremely accurate synchronization system to operate with other GFO converters. Power sharing among these converters is a significant challenge, which is a function of their output impedances. Since a constant dc link is necessary for the GFO converters, in industrial applications, these power converters are fed by stable dc voltage sources driven by batteries, fuel cells, or other primary sources like hydro turbines, gas-microturbines, where the availability of the primary energy is guaranteed.

A variety of control strategies have been developed for GFO converters over the last few years [14]. Figure 5.7 shows a general scheme for the control of GFO power converters. As shown in this figure, the control system of GFOs contains three main subsystems: a voltage control unit, a current control unit, and a dynamic characteristics unit, which provides the main reference signals for the voltage controller. The inputs to the control system are either the amplitude E^* and the angular frequency ω^* of the voltage to be formed by the power converter at the point of common coupling (PCC) or the active power P^* and active power Q^* to be injected into the grid for the sake of shaping it and providing required services (e.g. frequency or reactive power support). The voltage controller controls the grid voltage to follow its reference, while the current controller regulates the current supplied by the converter. It is noteworthy that the internal current control loop is avoided for GFO converters in some research; however, the cascaded inner control loops guarantee the converter from over-current in case of a fault [15]. The dynamic characteristics unit contains a control mechanism to emulate required dynamics. This control loop provides the elements (i.e. E and ω) of the voltage reference v^* for the voltage controller. It could be a droop control mechanism, a VSG, a virtual oscillator controller, or any other desirable dynamics. More details on the dynamic characteristics will be provided in Section 5.5.

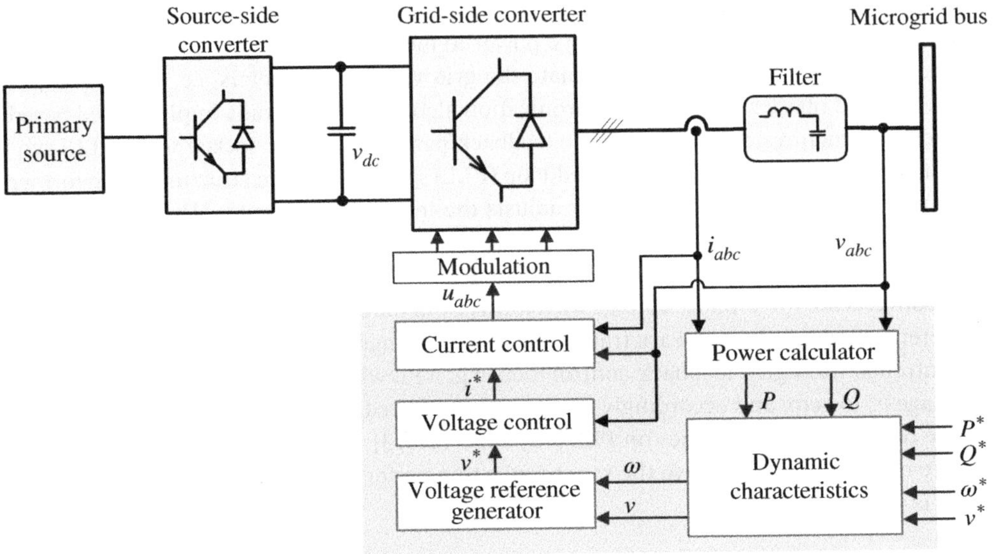

Figure 5.7 General scheme for control system of a grid-forming power converter.

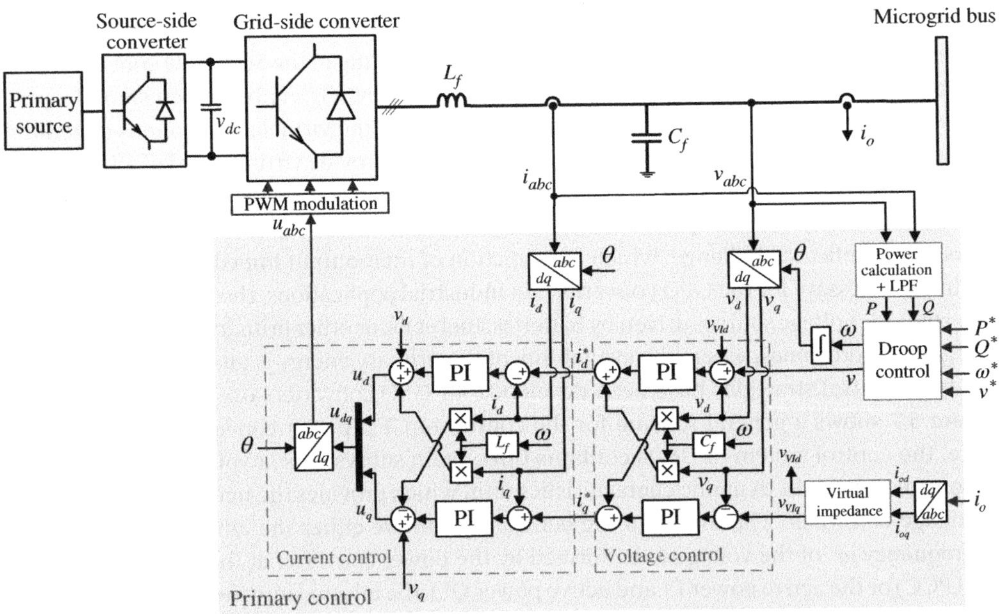

Figure 5.8 Typical control structure for a three-phase grid-forming power converter.

Figure 5.8 represents a typical structure of control loops for GFO converters. As this figure shows, following the reference provided by the droop control, the voltage control loop which is in charge of regulating the output voltage sets the current reference i^* for the current control loop. The inner current loop regulates the current supplied by the converter, tracking the reference current provided by the outer voltage loop. The output of the current control loop is transformed to the *abc* reference frame and then given to the modulation block, which is a pulse width modulation (PWM).

A similar linear control structure as presented for the current controller in Section 5.3.2.1 can be implemented for the voltage control loop. The transformation between the *abc* and *dq0* reference frames is performed based on the voltage angle θ produced by the droop mechanism. The virtual impedance loop can be also classified as a part of the dynamic characteristics unit shown in Figure 5.7, which helps improve the power sharing and damping characteristics of the droop control mechanism. This control loop translates the grid current i_o to voltage values, v_{VId} and v_{VIq}, directly subtracted from the direct and quadratic voltage references of the voltage controller. Readers can find more details about virtual impedance in Section 5.5.

It should be noted that both the *dq0* synchronous and the $\alpha\beta$ stationary reference frames can be used to control the GFO power converters. Unlike the GFL converters, the GFO converters do not require a PLL in the stand-alone mode, and the needed angle/frequency of the system can be extracted from the output of the dynamic characteristics unit. Nevertheless, it needs an accurate synchronization algorithm to be connected to other power converters or to the grid. It should be also pointed out that the voltage control loop of the GFO power converters will be disabled when the MG is connected to the main grid.

Chapter 2 provides a thorough analysis of GFL and GFO converters, including a detailed modeling approach that enables a deeper understanding of their behavior and performance in diverse operating conditions.

5.4 Hierarchical Control

5.4.1 The Control Hierarchy

From a control system design viewpoint, an MG is a complex system consisting of a variety of linear/nonlinear subsystems with strong cross-coupling between them. Different control loops must be implemented in order to ensure the performance and stability of MGs. Several feedback variables are used in these control loops, including current, voltage, frequency/angle, and active/reactive power. The control loops can be implemented in decentralized, distributed, or centralized frameworks depending on the performance required, as discussed in Section 5.2.3.

Taking the idea from the conventional power systems, the control loops are implemented using a hierarchy and the aforementioned control strategies. The hierarchical control of MGs involves dividing the control and management tasks into different levels or stages, with each level responsible for specific functions. The hierarchical control structure is designed to ensure that the MG operates efficiently, securely, and reliably. As shown in Figure 5.9, the hierarchical control for MGs consists of four levels [5]: primary (local), secondary (supplementary), central, and global controls. In some studies, this hierarchical architecture has been introduced in three levels, with central and global control being referred to as tertiary control [16]. The MGs' hierarchical control structure is in charge of addressing various control objectives described in Section 5.2.2, such as proper load sharing and DERs coordination, voltage/frequency regulation, MG synchronization with the main grid and other MGs, operating cost optimization, and power flow control between the MG, neighbor MGs, and the main grid.

In contrast to primary control which operates without communication, secondary, central, and global controls may need communication channels. While the primary control benefits from a decentralized structure, the central and global controllers are centralized. Depending on the performance required, secondary control can be implemented using centralized, distributed, or decentralized strategies.

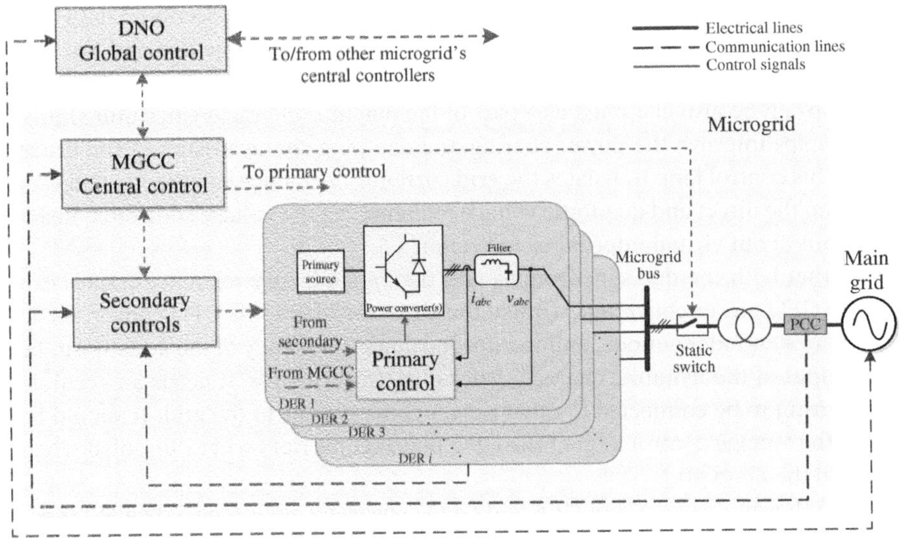

Figure 5.9 Hierarchical control of microgrids.

This section gives an overview of the hierarchy control levels for MGs, along with information on their functions and timescales. The following sections provide detailed information and examples about each control layer.

5.4.2 Control Layers

Primary Control: Primary control is the first control layer in the hierarchical control platform. Due to the decentralized structure of this control layer, digital communication is not required for its operation; instead, it relies on locally measured signals. Primary control responds to system dynamics and ensures that the system's voltage and frequency follow their nominal values. Output (current and voltage) control, power sharing between DERs, reducing circulating currents, and seamless transition between operation modes are other control objectives that lie in this control level.

Primary control employs different control loops depending on the operation mode (islanded or grid-connected) and type of power converters (i.e. GFO or GFL). As described in Section 5.3, primary control for DERs controlled by GFO converters typically consists of inner current/voltage control loops and dynamic characteristics loops, such as droop control mechanism and virtual impedance loop. However, the primary control for GFL converters includes a current control loop, a power control loop, and a synchronization algorithm.

Output control and system stabilization are generally achieved by the inner control loops, while simultaneous frequency/voltage regulation and power sharing are achieved by the droop control mechanism. Voltage and current loops can also be embedded in a single controller, which can dramatically improve the dynamic performance of the system [17]. The virtual impedance loop, as an optional loop, improves the power quality and power sharing accuracy of MGs with mismatched inductive/resistive impedance [18]. Droop control is inspired by conventional power systems and mimics the steady-state characteristics of a synchronous generator [6]. Recently, some other dynamic characteristics and virtual inertia emulation mechanisms such as virtual synchronous generator (VSG) [19], virtual oscillator control (VOC) [20] and synchronverter [21] have been introduced for implementation at the primary level.

Secondary Control: The secondary control layer is responsible for ensuring power quality and mitigating steady-state voltage and frequency deviations by determining the set points for the primary control [2, 22, 23]. It can, however, be designed to improve reactive power sharing accuracy, suppress circulating currents, and eliminate harmonics and unbalances [2]. Secondary control is typically implemented using low-bandwidth peer-to-peer communication; it operates at a slower rate to decouple from primary control and reduce communication bandwidth requirements. Depending on the way digital communication is utilized, three secondary control strategies namely centralized, distributed, and decentralized, have been introduced in the literature [24].

Central Control: The central control covers MG power/energy control and all possible emergency control schemes and special protection plans to maintain the MG stability and availability in the face of contingencies [5]. Emergency control measures in an MG may include automatic load shedding, islanding detection, connection/disconnection to/from the main grid, and system reconfiguration. For example, in the case of a power outage, the MG may automatically switch to island mode and provide power to critical loads. It also operates as an MG energy management system (EMS) and monitors/manages the MG's primary and secondary controllers. This control level is implemented in the microgrid central controller (MGCC).

Global Control: The global control facilitates the long-term, economically optimal operation by controlling the power flow between the MG, other interconnecting MGs, and the main grid. This level is operating outside the MGs in the distribution network and is responsible for scheduling and planning. The global control center connects MGCCs to the main grid's distribution network and manages power flow control and market operations.

The interconnection of different control layers as well as the control layers with the MGs components and the main grid is illustrated in Figure 5.9. Moreover, Figure 5.10 summarizes the functions, control strategies, and time scales of the MGs' control hierarchy.

MGs, unlike conventional power systems, employ power-electronic interfaces. Although these interfaces are generally fast enough to provide a quick control response to a disturbance (such as load/generation changes or contingencies), DER-activated power has several limitations [25]. Due to the low inertial feature of MGs and the capacity limitations of DERs, a change in load has a significant impact on system operation. When a disturbance occurs in MGs, both the time scale and the amount of activated power must be considered. The dynamic response and time scale imposed by the hierarchical control loops of the MGs are depicted in Figure 5.10.

5.5 Primary Control

Primary control, also known as local control or internal control, is the first layer of the MGs' control hierarchy. This control layer is typically implemented in a decentralized manner based on local measurements with no need for digital communication. This control loop appears in different forms depending on the type of DERs, e.g. synchronous generators, or converter-based generators, the type of power converters in the converter-based generators (i.e. GFO or GFL), and the mode of operation (i.e. islanded or grid-connected). As mentioned in Section 5.4, output control (current and voltage variables), frequency and voltage stability, power sharing, and P/Q control (in grid-connected mode) are the main control functions that lie in this control layer. Over the past decade, virtual inertia emulation (e.g. using VSG) has been introduced as a new function for primary control of power converter-based systems to improve the dynamic performance of MGs in islanded mode and to provide ancillary services in grid-connected mode [19, 26].

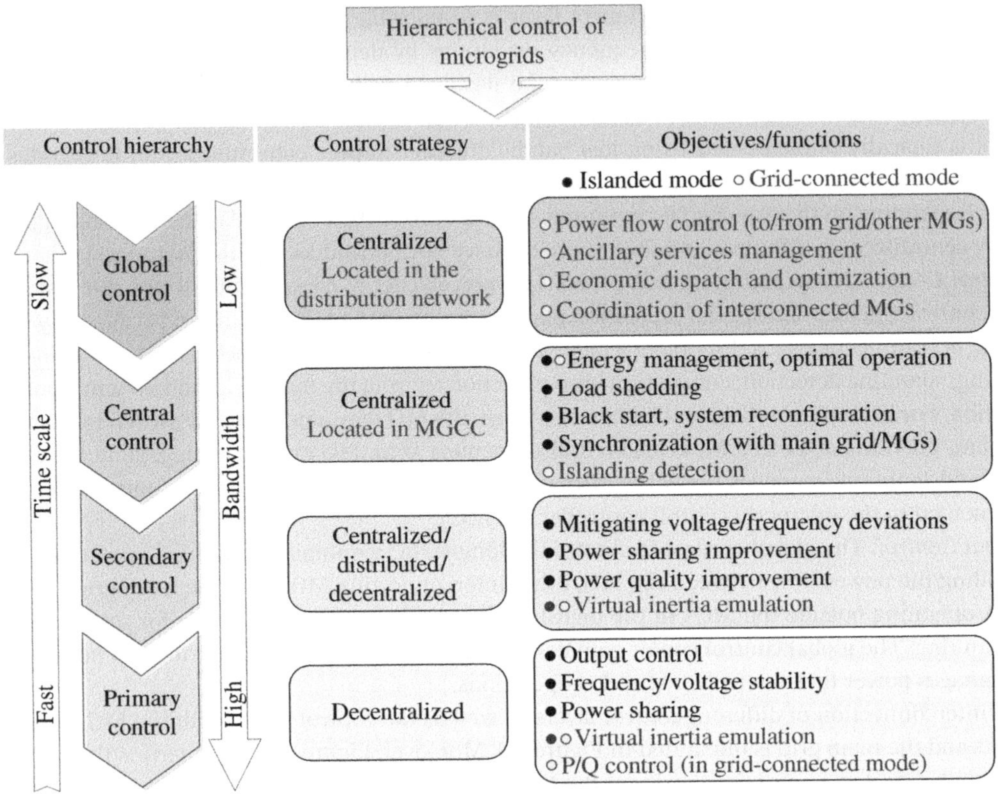

Figure 5.10 Functions, strategies, and time scales of the microgrids' control hierarchy.

In synchronous generators, output control and power sharing are performed in the primary control layer by the voltage regulator, governor, and the inertia of the machine [5]. In the primary control of power electronics-based generators, similar functions are carried out using inner current and voltage control loops, and virtual droop control. A converter-based MG, as mentioned in Section 5.3, consists of a number of DERs interfaced with GFO and GFL power converters, depending on their control objectives (see Figure 5.11). At least one GFO power converter is required to form the MG in the stand-alone condition.

A typical primary control scheme for ac MGs is shown in Figure 5.11. The MG is composed of two types of DERs, some of which are controlled by GFO power converters to form the MG and control the system's frequency and voltage, and the others controlled by GFL converters to provide maximum power from the primary sources (e.g. PVs or wind turbines). The GFO converters are equipped with all the basic primary control loops, i.e. current/voltage loops, droop control, and virtual impedance. The obtained active/reactive powers from the DER's outputs are given to the droop mechanism after passing through a low-pass filter (LPF). The droop control mechanism together with the virtual impedance loop provides the voltage reference for the inner loops. The cascaded voltage and current loops then produce the proper reference for the PWM block.

The control loops are implemented in the *dq* frame, as shown in Figure 5.11, however, one can use other reference frames to apply the control units. Park transformation has been utilized to transform the sinusoidal three-phase output variables (i_{abc}, v_{abc}, and i_o) to dc values in an orthogonal *dq* frame, as well as to convert the control signal u_{dq} to *abc* frame to be given to the PWM modulation. Regular PI controllers are implemented on the current and voltage loops as discussed

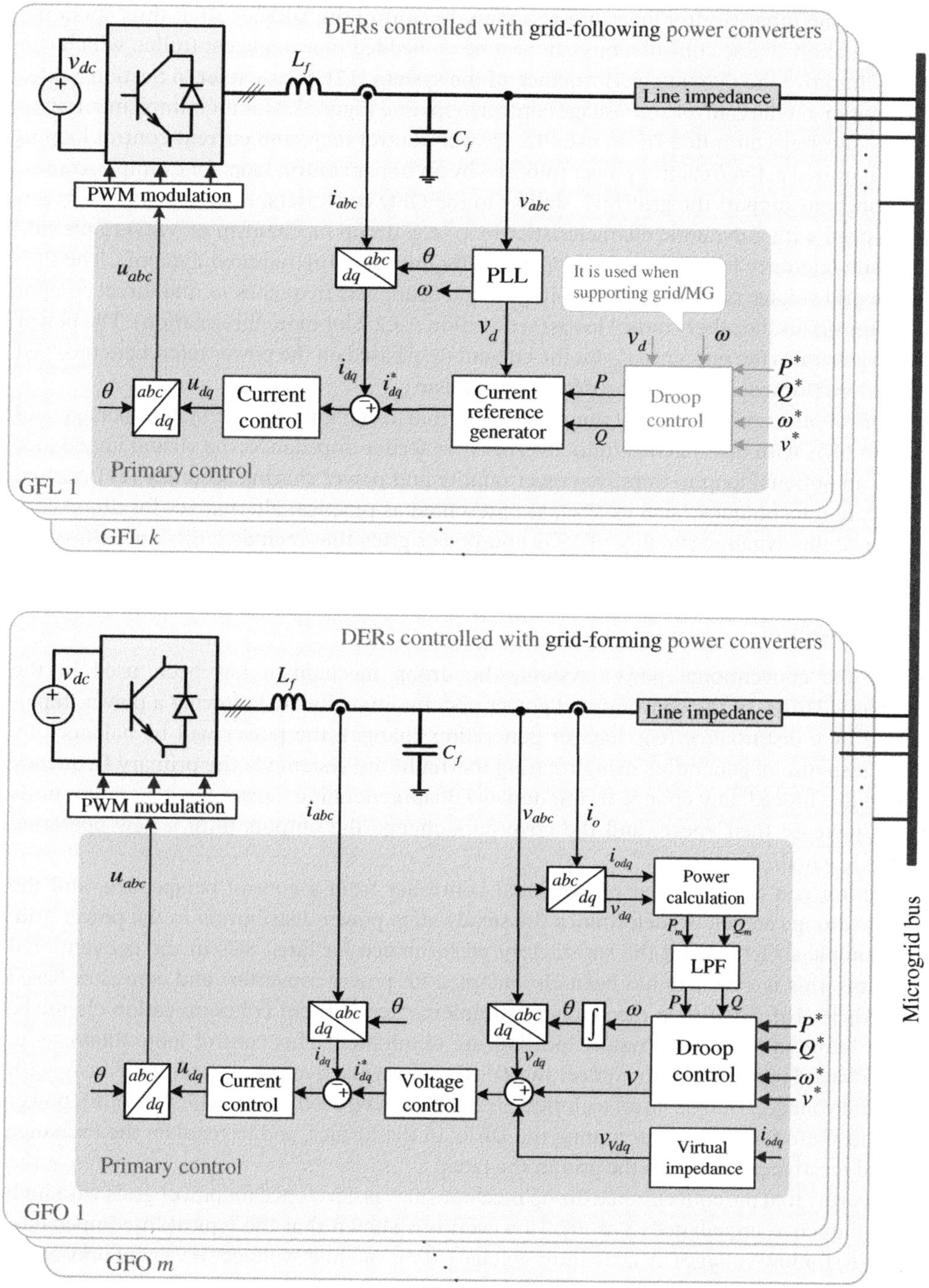

Figure 5.11 Typical primary control scheme of MGs, including DERs controlled by both GFO and GFL power converters.

in Section 5.3. The inner control loops are necessary to control the DER's output thus operating in normal operation. These control loops can also be embedded in a single controller, which can significantly improve the dynamic performance of the system [17]. Please refer to Section 5.3 for more information on the control and voltage control loops and Figure 5.8 for their implementation.

The GFL converter-controlled DERs use PLL, power control loop, and current control loop to provide the required active/reactive power imposed by an upper control loop, for example, central control. In order to support the grid/MG, similar to the GFO converters, the GFL converters can be also equipped with a dynamic characteristics loop (e.g. droop mechanism or VSG) to provide an appropriate reference for the current loop, virtually imposing the required dynamic. The PLL provides the grid voltage parameters (i.e. voltage angle θ, angular frequency ω, and direct component of voltage v_d) for the other control loops (see Section 5.3.2.2 for more information). The power control loop generates the reference i^*_{dq} for the current loop based on the power references received from other loops (for instance from the droop mechanism).

Droop control offers power sharing among DERs while also regulating the system voltage and frequency. In MGs with mismatched inductive/resistive feeder impedance, the virtual impedance loop acts as an optional loop to improve power quality and power sharing accuracy [27]. Virtual inertia emulation techniques, such as VSG, are presented as practical alternatives for droop control to enhance the dynamic stability of MGs and power grids that include DERs and MGs. The following subsections provide more details about these control loops.

5.5.1 Droop Control

Inspired by the conventional power system, the droop mechanism has been used in the converter-based DERs. In the conventional power system, immediately following a power imbalance caused by a disturbance (e.g. load or generation change), the power will be balanced by the natural response of generators using rotating inertia in the system via the primary frequency control loop [5]. Indeed, any change in the network load/generation causes the generating units to decrease/increase their speed, and the governors change the outputs until a new operating frequency is reached.

Droop control is a decentralized proportional controller from a control perspective, and the control gain (droop coefficient) determines the steady-state power distribution in the power grid. Due to its simple structure and the satisfactory performance for large SGs in the conventional power systems, this mechanism has been also adapted for power converters and converter-based DERs in MGs [6, 16]. The droop mechanism is implemented without communication channels, thereby the limits imposed by physical location are eliminated. This control loop allows us to create a feedback loop in order to control the DER's active/reactive power simultaneously with frequency and voltage. Droop control techniques [28, 29] can be used in both GFO and GFL power converters, to share the load power among the DERs in the former, and to regulate the exchange of active and reactive powers with the grid in the latter.

It is noteworthy that the conventional droop mechanism was developed for power grids with high inertia and inductive impedance ($X \gg R$). However, it is shown that the typical line impedance ratio (i.e. X/R) for low-voltage (LV), medium-voltage (MV), and high-voltage (HV) networks are in the range of 0.13, 1.18, and 3.18, respectively [6]. This clearly shows that the X/R ratio of the line impedances in the HV networks (e.g. in conventional power grids) is much higher than the one in LV networks. In MGs which usually operate at LV and MV networks, with low inertia DERs, the line impedance is not dominantly inductive (i.e. $X \ll R$ or $X \cong R$). Therefore, the conventional droop mechanism does not work properly for such a system and different formulations and strategies must be used for MGs.

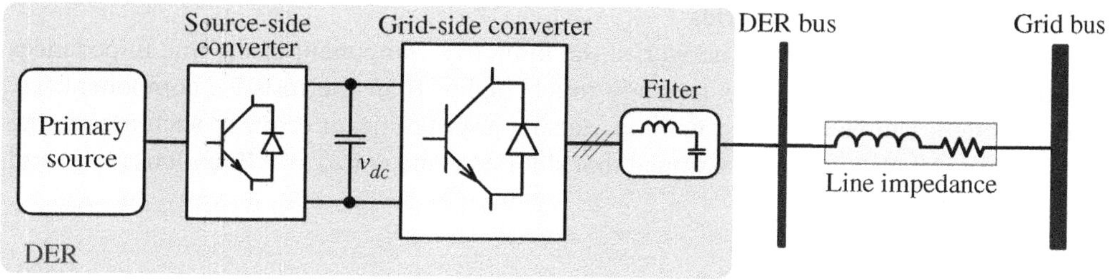

(a)

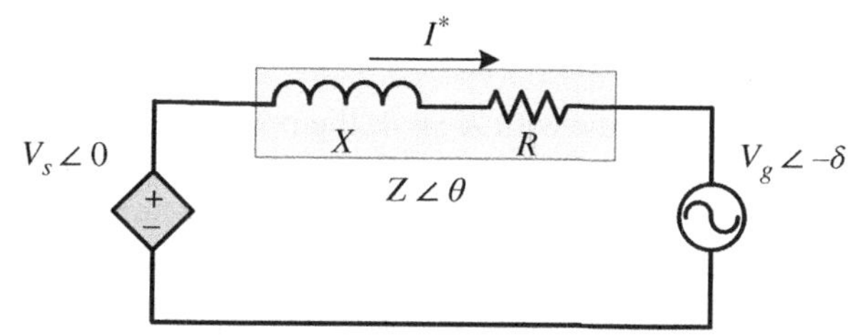

(b)

Figure 5.12 General representation of an inverter-based DER connected to a grid: (a) electrical scheme and (b) equivalent model.

Figure 5.12 represents the power flow of an inverter-based DER connected to a grid/MG through line impedance. The DER contains three basic elements: a primary power source, a source-side converter to provide the DC interface, and a grid-side converter (inverter), connected to the rest of MG or a distribution network through a line impedance $Z\angle\theta$. A simplified equivalent model of the scheme considering the power converter-based DER as an ideal voltage source is shown in Figure 5.12, where V_s and V_g represent the DER voltage (as source voltage) and grid bus voltage, respectively. Using the information provided in Figure 5.12(a), the complex power delivered to the grid bus can be calculated as

$$P + jQ = V_s I^* = V_s \left(\frac{V_s - V_g e^{j\delta}}{Ze^{-j\theta}} \right) = \frac{V_s^2}{Z} e^{j\theta} - \frac{V_s V_g}{Z} e^{j(\theta+\delta)}, \tag{5.5}$$

from which the active and reactive powers can be obtained as

$$P = \frac{V_s^2}{Z} \cos\theta - \frac{V_s V_g}{Z} \cos(\theta + \delta), \tag{5.6a}$$

$$Q = \frac{V_s^2}{Z} \sin\theta - \frac{V_s V_g}{Z} \sin(\theta + \delta), \tag{5.6b}$$

where V_s, V_g, θ, and δ are the voltage of the DER, voltage of the grid bus, line impedance angle, and bus angle, respectively. Considering the line impedance $Ze^{j\theta} = R + jX$, and replacing it in (5.5), the general description of active and reactive powers can be written as

$$P = \frac{V_s}{R^2 + X^2}[R(V_s - V_g \cos\delta) + XV_g \sin\delta], \tag{5.7a}$$

$$Q = \frac{V_s}{R^2 + X^2}[-RV_g \sin\delta + X(V_s - V_g \cos\delta)]. \tag{5.7b}$$

From (5.7a) and (5.7b), one can see the influence of the line impedance on the dependency of active and reactive powers on the DER output voltage and power angle δ.

5.5.1.1 Droop Control for Inductive Grids

As previously noted, in HV and MV networks, the inductive component of the line impedances (between the source and the grid) is typically much higher than the resistive component, i.e. $X \gg R$. Therefore, R can be neglected without causing any significant error. In such a case, the power angle δ is small, thus it can be assumed that $\sin(\delta) \approx \delta$ and $\cos(\delta) \approx 1$. Equations (5.7a) and (5.7b) then become

$$P \approx \frac{V_s}{X}(V_g \sin\delta) \;\Rightarrow\; \delta \approx \frac{XP}{V_s V_g}, \tag{5.8a}$$

$$Q \approx \frac{V_s}{X}(V_s - V_g \cos\delta) \;\Rightarrow\; V_s - V_g \approx \frac{XQ}{V_s}. \tag{5.8b}$$

According to (5.8a) and (5.8b), for systems with $X \gg R$, the power angle depends predominantly on active power P, while the voltage difference depends on reactive power Q. In other words, the angle or frequency can be controlled by regulating P, whereas Q controls the inverter voltage, independently. These conclusions form the basis for the well-known frequency-active power ($f - P$) and voltage-reactive power ($v - Q$) droop control mechanisms [30]

$$f - f^* = -k_p(P - P^*), \tag{5.9a}$$

$$v - v^* = -k_q(Q - Q^*), \tag{5.9b}$$

where f^* and v^* are rated system frequency and voltage, respectively, P^* and Q^* are the set points for active and reactive of the inverter, and k_p and k_q are droop coefficients. In other words, $f - f^*$ and $v - v^*$ represent the system frequency and voltage deviations, respectively, from their rated values, and $P - P^*$ and $Q - P^*$ are the variations in the active and reactive powers delivered by the power converter to compensate for such deviations [6]. Thus, the droop control mechanism provides a fast control action to maintain an instantaneous balance between the system's production and consumption [5].

The relationships (5.9a) and (5.9b) can be graphically represented by the droop characteristics shown in Figure 5.13, where the slope of the frequency and voltage droop characteristic is set by the k_p and k_q parameters, respectively.

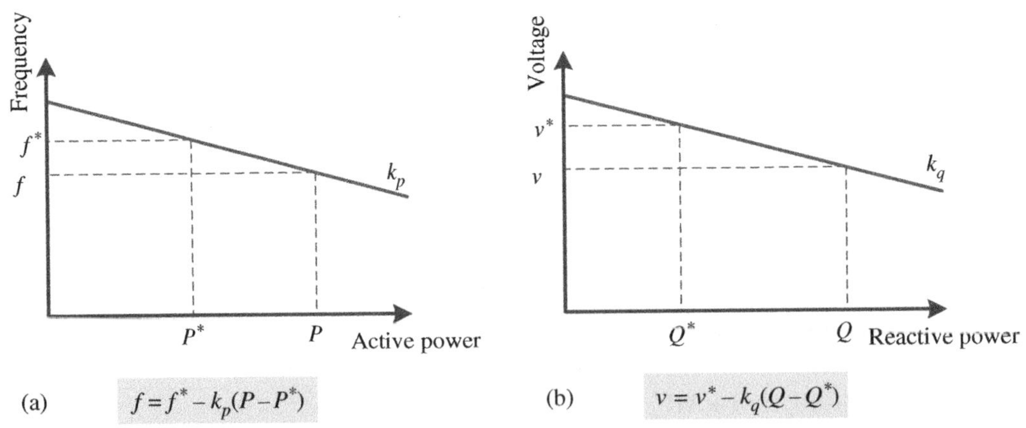

Figure 5.13 Droop characteristics for MGs with dominant inductive behavior ($X \gg R$): (a) frequency-active power droop mechanism and (b) voltage-reactive power droop mechanism.

Each MG's DER controlled by GFO or GFL power converters can adjust its active and reactive power reference based on the droop characteristics shown in Figure 5.13 to participate in the system frequency and voltage regulation, or to support the grid (see Figure 5.11).

5.5.1.2 Droop Control for Resistive Grids

The grid impedance in LV networks is mainly resistive, and thus, in contrary to the previous case, the inductive part can be neglected. Ultimately, given the assumption that the power angle δ has a small value, one can rewrite Eqs. (5.7a) and (5.7b) as

$$P \approx \frac{V_s}{R}(V_s - V_g \cos\delta) \;\; \Rightarrow \;\; V_s - V_g \approx \frac{RP}{V_s}, \tag{5.10a}$$

$$Q \approx -\frac{V_s}{R}(V_g \sin\delta) \;\; \Rightarrow \;\; \delta \approx -\frac{RQ}{V_s V_g}. \tag{5.10b}$$

Equations (5.10a) and (5.10b) show that adjusting active power influences the voltage amplitude, while adjusting reactive power influences the frequency. One can conclude that in contrary to the inductive power grids, the conventional $f - P$ and $v - Q$ droop control mechanisms do not work properly for resistive grids such as LV MGs. Using (5.10a) and (5.10b), the following droop control expressions can be written as

$$f - f^* = k_q(Q - Q^*), \tag{5.11a}$$

$$v - v^* = -k_p(P - P^*). \tag{5.11b}$$

Figure 5.14 represents the $v - P$ and $f - Q$ droop control characteristics for resistive networks.

5.5.1.3 Droop Control for Resistive–Inductive Grids

In a general case, where the grid networks have neither predominant inductive nor resistive behavior, the combined effect of the resistive and inductive line impedance components should be considered in the droop control equations [6, 30]. For this purpose, a rotational transformation matrix T is used to transform the active and reactive powers, P and Q, into the modified rotational power components, P' and Q', as follows:

$$\begin{bmatrix} P' \\ Q' \end{bmatrix} = [T]\begin{bmatrix} P \\ Q \end{bmatrix} = \begin{bmatrix} \sin\phi & -\cos\phi \\ \cos\phi & \sin\phi \end{bmatrix}\begin{bmatrix} P \\ Q \end{bmatrix}, \tag{5.12}$$

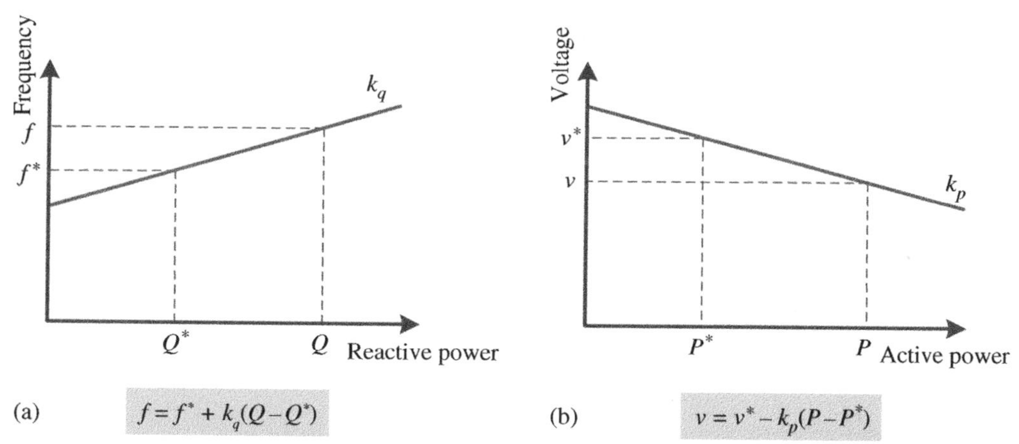

Figure 5.14 Droop characteristics for resistive MGs with dominant resistive behavior ($R \gg X$): (a) frequency-reactive power droop mechanism and (b) voltage-active power droop mechanism.

where $\phi = \pi/2 - \theta$ is the ratio angle of matrix T, and θ is the power angle of line impedance. Thus, one can write:

$$P' = \frac{X}{Z}P - \frac{R}{Z}Q, \tag{5.13a}$$

$$Q' = \frac{R}{Z}P + \frac{X}{Z}Q. \tag{5.13b}$$

Given that δ takes a small value, the application of the above-mentioned transformation to (5.7a) and (5.7b) results in

$$P' \approx \frac{V_s}{Z}(V_g \sin\delta) \quad \Rightarrow \quad \delta \approx -\frac{ZP'}{V_s V_g}, \tag{5.14a}$$

$$Q' \approx \frac{V_s}{Z}(V_s - V_g \cos\delta) \quad \Rightarrow \quad V_s - V_g \approx \frac{ZQ'}{V_s}. \tag{5.14b}$$

Equations (5.14a) and (5.14b) indicate that the power angle δ and voltage difference $V_s - V_g$ can be controlled by regulating the rotating active and reactive powers, respectively. Therefore, in a general case, the droop control equations can be expressed as

$$f - f^* = -k_p(P' - P'^*) = -k_p\frac{X}{Z}(P - P^*) + k_p\frac{R}{Z}(Q - Q^*), \tag{5.15a}$$

$$v - v^* = -k_q(Q' - Q'^*) = -k_q\frac{R}{Z}(P - P^*) + -k_q\frac{X}{Z}(Q - Q^*). \tag{5.15b}$$

From (5.15a) and (5.15b), one can conclude that frequency and voltage variations can be compensated by changing the k_p and k_q parameters. It is obvious that these general equations are also valid for resistive and inductive grids. A good example of implementing the general droop control can be found in [31].

The general droop control in (5.15a) and (5.15b) and the other two above-mentioned droop mechanisms are obtained based on the measured active and reactive powers of the system. However, these droop mechanisms require LPFs to provide an appropriate measured power implementing the virtual control loops. Using such filters for power calculations, on the other hand, slows down the system dynamic. To avoid this slow dynamic, the droop control can be similarly derived for active and reactive currents [30]. This approach has benefits when DER voltage differs significantly from its rated value (e.g. when a short-circuit occurs). Implementing droop control through active and reactive currents would avoid infinitely high currents during the fault. Section 5.5.1.4 discusses various issues and challenges associated with the conventional droop control.

5.5.1.4 Discussion on the Conventional Droop Control

Droop control, as a decentralized approach, can be deployed on DERs controlled by both GFO and GFL power converters (see Figure 5.11). For one or more DERs contributing to the MG's voltage and frequency stabilization, the conventional $f - P$ and $v - Q$ droop controls, which are valid for inductive networks, are used for the purpose of accurate active and reactive power sharing among the DERs. Thus, the well-known droop equations (5.9a) and (5.9b) are implemented on each GFO converter as follows:

$$\omega = \omega^* - k_p(P - P^*), \tag{5.16a}$$

$$v = v^* - k_q(Q - Q^*), \tag{5.16b}$$

where ω^* and v^* are the nominal angular frequency and nominal voltage amplitude of the MG, respectively, P and Q are the actual loading, P^* and Q^* are the active and reactive power rating of GFO converters, respectively, and k_p and k_q are the droop coefficients. The average active and reactive powers (P and Q) are obtained after passing the instantaneous measured powers through an LPF as

$$P = G_{LPF}(s)p, \tag{5.17a}$$

$$Q = G_{LPF}(s)q, \tag{5.17b}$$

where $G_{LPF}(s) = \omega_c(s + \omega_c)^{-1}$ is the LPF with cutoff frequency ω_c. The instantaneous active and reactive powers of each converter in the dq frame are calculated as

$$p = \frac{3}{2}(v_d i_{od} + v_q i_{oq}), \tag{5.18a}$$

$$q = \frac{3}{2}(v_q i_{od} - v_d i_{oq}), \tag{5.18b}$$

where v_d, v_q, i_{od}, and i_{oq} are the output voltage and current of DERs in the $dq0$ frame, respectively.

It is very important to remark that in stand-alone conditions, the power references (P^* and Q^*) of the droop control mechanism should be considered as zero. It is also worth noting that in conventional droop representations, angular frequency ω is commonly used instead of frequency f. A general scheme for implementing droop control in primary control of MGs is shown in Figure 5.11. Figure 5.15 represents the detailed implementation of the conventional droop mechanism in (5.16) for a GFO-controlled DER in an MG.

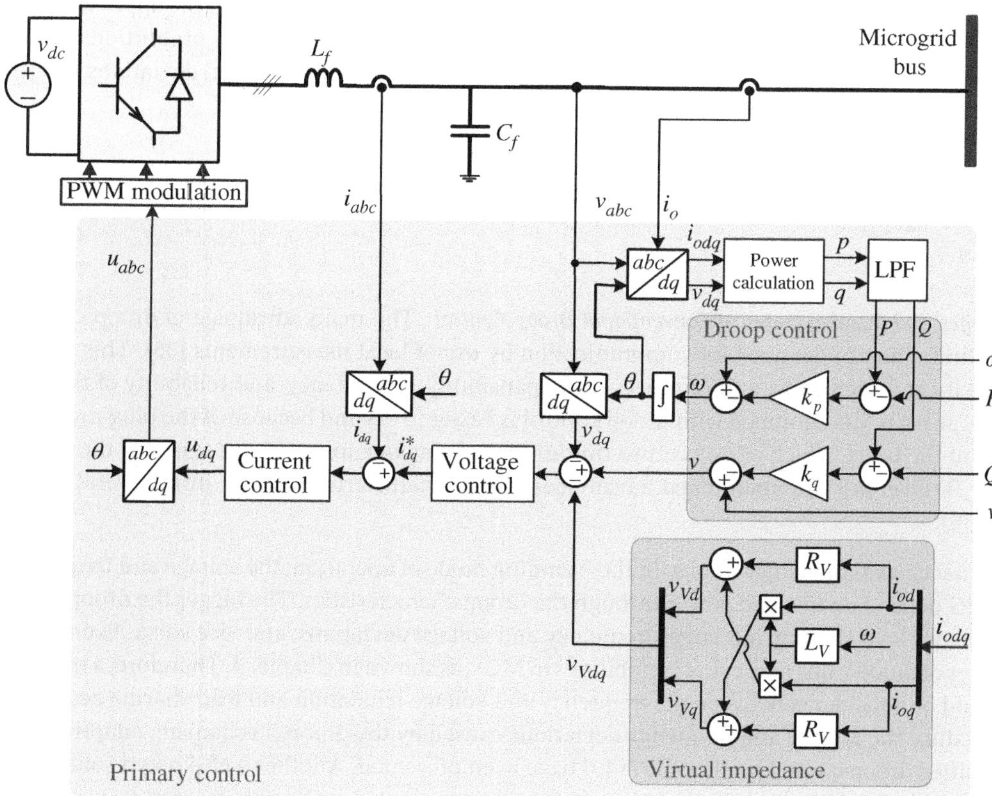

Figure 5.15 A GFO power converter equipped with primary control, including detailed implementation of conventional droop mechanism and virtual impedance.

The conventional droop control in (5.16) is commonly used for DERs controlled by GFO converters since measuring instantaneous real power is more feasible than obtaining an accurate measurement of instantaneous frequency. Additionally, a frequency reference is required in the common control strategy of GFO converters, which is provided through the conventional droop mechanism. On the other hand, in a GFL converter control system, there is a measured frequency from the point of coupling, and power references are needed. Therefore, the inverse droop is a suitable choice to match the requirements and the available data/measurements in such applications. In GFL power converters (see Figure 5.11) and some schemes of VSGs, the droop mechanism is implemented inversely, with P being a function of ω and Q being a function of v, as follows:

$$P = P^* - k_f(\omega - \omega^*), \tag{5.19a}$$

$$Q = Q^* - k_v(v - v^*), \tag{5.19b}$$

where k_ω and k_v are the frequency and voltage droop coefficients, respectively.

As shown in Chapter 4, the choice of droop coefficients affects the system stability. It was proven that the value of the droop gains has a remarkable effect on the dominant modes of ac MGs (e.g. see Figure 3.9 and Section 4.4.3.1); within the acceptable range, the gain increase improves the stability margin, while larger gains may lead the system to instability. On the other hand, large droop gains result in more frequency and voltage deviations. Therefore, the droop parameters must be appropriately designed considering the small signal stability analysis [32, 33] (please refer to Chapter 4 for more details). For the conventional droop mechanism, one practical and simple approach is to choose the droop parameters in a way that the DER's active (reactive) power is proportional to its power rate P_{rated} (Q_{rated}), considering the maximum allowed frequency (voltage) deviations $\Delta\omega_{max}$ (Δv_{max}) [34]:

$$k_p = \frac{\Delta\omega_{max}}{P_{rated}}, \tag{5.20a}$$

$$k_q = \frac{\Delta v_{max}}{Q_{rated}}. \tag{5.20b}$$

Advantages and Disadvantages of Conventional Droop Control The main advantage of droop control is the elimination of the need for communication by use of local measurements [35]. This avoids complexity and high costs, and improves the expansibility, redundancy, and reliability of the system [34, 36]. An MG equipped with droop control is easier to expand because of the plug-and-play feature of the units, which allows connecting/disconnecting one unit without stopping the whole system. Besides the aforementioned advantages, droop characteristics suffer from several drawbacks [36]:

1) *Frequency and voltage deviations*: In the islanding mode of operation, the voltage and frequency of MG depend on the load power through the droop characteristics. The bigger the droop gains the better load sharing, but larger frequency and voltage deviations, and vice versa. Even some values of droop gains may cause instabilities in MGs, as shown in Chapter 4. Therefore, a tradeoff should be considered between the frequency and voltage regulation and load sharing accuracy. To reduce the voltage and frequency deviations caused by the droop mechanism, adaptive and modified droop control methods [37–40] have been presented. Another well-known solution is to utilize secondary control loop [24], which will be presented in detail in Section 5.6.
2) *Slow dynamics and poor transient performance*: Power droop control methods have slow dynamics. This is because of the procedure of power calculation, i.e. the usage of LPFs for

active/reactive power measurements. Droop control based on active and reactive currents [30] and $V - I$ droop characteristics [41] are some solutions to avoid this issue.

3) *No harmonic load sharing*: The original droop method focuses on fundamental power sharing while it does not take harmonic sharing into account in the case of nonlinear loads. This would lead to harmonic circulating currents and poor power quality. Moreover, the calculation and smoothing of active and reactive power take some delays, leading to a slower dynamic response, as stated above. Therefore, to count for nonlinear loads, the conventional droop method should be modified in order to share the harmonic currents [42].
4) *Inherent conflict of voltage regulation and reactive power sharing*: Opposed to frequency, voltage is not a global variable in MGs. Therefore, in the conventional $\omega - P$ and $v - Q$ droop control, there exists a conflict between voltage regulation and reactive power sharing. By regulating voltage into the nominal value, the reactive power cannot be shared properly among the converters, and this may result in circulating reactive current [22, 43, 44]. A similar problem may occur in MGs with highly resistive lines, for the circulation of active current controlled through the voltage.
5) *Poor performance with output impedance uncertainties*: As mentioned, the power sharing performance in MGs depends on the line impedances and the utilized droop control method. Considering a suitable droop mechanism, when the line impedances between the converters and PCC are different and in some cases unknown, it may result in a large circulating current and poor power sharing among converters [43, 44]. Virtual structure-based methods, e.g. virtual impedance control loops have been proposed to cope with this issue. Section 5.5.2 introduces the concept and structure of virtual impedance for MGs.

5.5.1.5 Droop Control for DC Grids

Similar to the ac MGs, hierarchical control structure can be implemented for dc MGs. The primary control in dc MGs is responsible for improving voltage stability, current sharing, and preventing circulating current among converters. In addition to inner control loops, droop control is the main control loop in the primary layer of dc MGs. The voltage–power or voltage–current droop scheme, similar to the frequency-power droop control of a synchronous generator, can be employed in the primary layer of dc grids. Load changes in the system cause dc link voltage variation, thus the droop controller can change the output power of the converter by varying the voltage.

The basic idea behind droop control is to allow the output voltage of each converter in the MG to vary proportionally to the output current, in order to share the load among the sources and maintain system stability. In droop control, each converter is operated with a droop characteristic that determines the amount of voltage deviation from the nominal voltage as a function of the output current. The droop characteristic is expressed mathematically as

$$v_{ref} = v^* - R_d i_o, \tag{5.21}$$

where v^* is the nominal voltage, i_o is the output current of the converter, v_{ref} is the provided voltage reference by droop control for the lower internal loops, and R_d is the droop resistance (gain or slope) of the converter. The load current/power is shared between the converters on the basis of slope of their droop characteristics.

The droop resistance determines the slope of the voltage–current characteristic and is typically set by the system operator based on the desired load-sharing ratio. A smaller droop resistance results in less voltage deviation for a given current deviation, which means the converter will carry a larger share of the load. Conversely, a larger droop resistance results in more voltage deviation for a given current deviation, which means the converter will carry a smaller share of the load. The droop gain can be determined according to the rated current (i_{rated}) or rated power of the converter.

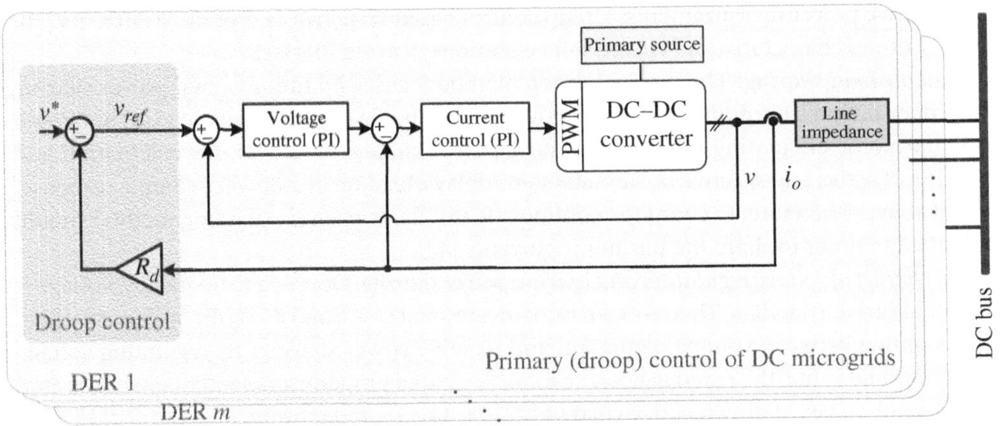

Figure 5.16 Implementation of droop control in the primary layer of a dc microgrid.

Considering Δv_{max} as the maximum allowable voltage deviation of the MG, one can calculate the droop slope (R_d) as

$$R_d = \frac{\Delta v_{max}}{i_{rated}}. \tag{5.22}$$

To implement droop control in a dc MG, each converter in the system is equipped with a local controller that measures the output current and provides the voltage reference (v_{ref}) for the voltage control loop. Figure 5.16 shows the implementation of droop control in the primary layer of dc MGs.

It should be noted that a droop controller can be implemented in an inverse form as well. In the inverse droop method, the output voltage of the converter is measured and the droop mechanism provides a current reference for the current controller. The primary control of the inverse droop method only contains a current controller and is faster than the droop method in (5.21). In this droop method, the droop gains must be carefully chosen since they can affect the stability margin of the current controller [45].

5.5.2 Virtual Impedance

An initiative solution to avoid the active and reactive power coupling and the strong dependence of the conventional droop control performance on the line impedance value is to use large inductors in order to connect the power converter to the ac bus, thereby making the line impedance predominantly inductive. However, this is not an efficient solution because besides increasing the size and the costs, the overall efficiency is reduced. A more effective solution, known as *virtual impedance*, has been proposed in the literature which updates the primary control loop by virtually emulating the effect of an inductive component (instead of a real component) in the line impedance [6, 46]. As shown in Figure 5.11, the virtual impedance updates the power converter output voltage reference given to the inner control loops, by subtracting the virtual voltage drop across the virtual impedance from the voltage reference originally provided by the droop control mechanism in (5.16) as

$$v_{ref} = v - Z_V i_o, \tag{5.23}$$

where Z_V is the dynamic of the virtual impedance loop, v is the voltage reference generated by the droop control loop, and i_o is the output current of the power converter passed through the line impedance. This virtual loop can be implemented in either of the reference frames; in Figure 5.11, it is implemented in a dq frame and its output is $v_{Vdq} = Z_V i_{odq}$.

The virtual impedance can be achieved by drooping the output voltage reference proportionally to the derivative of the output current with respect to the time, so that Z_V is purely inductive, i.e. $Z_V(s) = sL_V$. However, the derivation can simply amplify the high-frequency noises, which may lead the system to instability during transients. To overcome this issue and to improve the performance of the virtual impedance, different dynamics can be utilized. Indeed, virtual impedance is a virtual dynamic loop that can shape the voltage reference for the inner loops improving the power sharing among the DERs in an MG. For example, one can use an LPF instead of a pure derivative term of the output current, i.e. $Z_V(s) = L_V \frac{s}{s+\omega_c}$ [47]. It is worth mentioning that the value of the virtual impedance, i.e. L_V should be larger than the actual line impedance; otherwise, it will not have a predominant effect in the power flow equations [6]. If the virtual impedance Z_V is properly adjusted, it regulates the power sharing among power converters and prevents overcurrents when the DER is initially connected to the MG.

As presented in Chapter 4, the virtual impedance affects the damping and stability margin of the system, thus its parameters can be simply designed using the provided analysis. A basic virtual impedance loop consists of a virtual resistance and an inductance. Figure 5.15 shows an example of the basic virtual impedance implementation.

In order to improve the performance of the basic virtual impedance loop, several control methods have been employed. Generally, adaptive virtual impedance design is done to consider output current amplitude and phase, which is a sense of the output impedance characteristic of coupling lines [48–51]. Iterative methods are also used to have a variable virtual impedance with [52], and without communication [53]. The optimal design of the virtual impedance has been studied considering several operating constraints, including steady-state voltage limits, power decoupling, damping of dominant modes, and reactive power sharing [54]. Robust synthesis can be used to see the unknown values of the line impedances as uncertainties.

Virtual impedance is also used for supplementary objectives such as harmonic power sharing [55], voltage unbalance mitigation, voltage regulation [56], soft starting of the DERs [36], and smooth transition [57]. Several aspects of virtual impedance strategy and other power sharing methods based on impedance reshaping are reviewed in [56, 58].

5.5.3 A Simulation Study for Primary Control of AC Microgrids

Here, a simple example to understand the concept of droop control and virtual impedance is provided.

5.5.3.1 Case Study

Consider an islanded ac MG consisting of two GFO DERs connected to a local load, Z_l, with the parameters listed in Table 5.1. The performance of the conventional droop control and the virtual impedance loops in controlling the system's voltage and frequency, the active and reactive powers of the DERs, and the PCC voltage under variation of the local load as an unexpected load, Z_u, and for different types of line impedances, i.e. inductive ($X/R \gg 1$), resistive ($X/R \ll 1$), and resistive–inductive ($X/R = 1$) are investigated in the following.

5.5.3.2 Simulation Results

The control system presented in Figure 5.15 is implemented on the DERs inverters of the test system in the MATLAB environment. As mentioned, the conventional droop control offers accurate power sharing in systems with inductive lines ($X/R \gg 1$), while its performance is affected by increasing the resistive characteristics of the network (i.e. in the presence

Table 5.1 The information of the studied system.

Parameter	DER_1	DER_2	Z_l	Z_u
Active power (KW)	1	2	1.99	0.85
Reactive power (KVAR)	0.5	1	0.99	0.71

Table 5.2 The droop and VI characteristics of the DER units.

Unit	m_p	n_q	L_{VI} (H)	R_{VI} (Ω)
DER_1	0.0016	0.0653	0.05	0.2
DER_2	0.0008	0.0327	0.05	0.2

of resistive ($X/R \ll 1$) and resistive–inductive ($X/R = 1$) lines). The power sharing accuracy of resistive and resistive–inductive lines can therefore be increased by using virtual impedance. Table 5.2 lists the parameters of the droop control and the virtual impedance of the DERs.

Figures 5.17–5.21 show the system outputs under different scenarios for different types of line impedances. The conventional droop mechanism is performed from the beginning, while the virtual impedance loop is activated at $t = 2.5$ s followed by a load disturbance at $t = 5$ s. Figures 5.17 and 5.18 depict the active and reactive powers of DER_1 and DER_2, respectively, which show the related capacity of the DERs for the applied scenarios. The ratio of the active and reactive powers

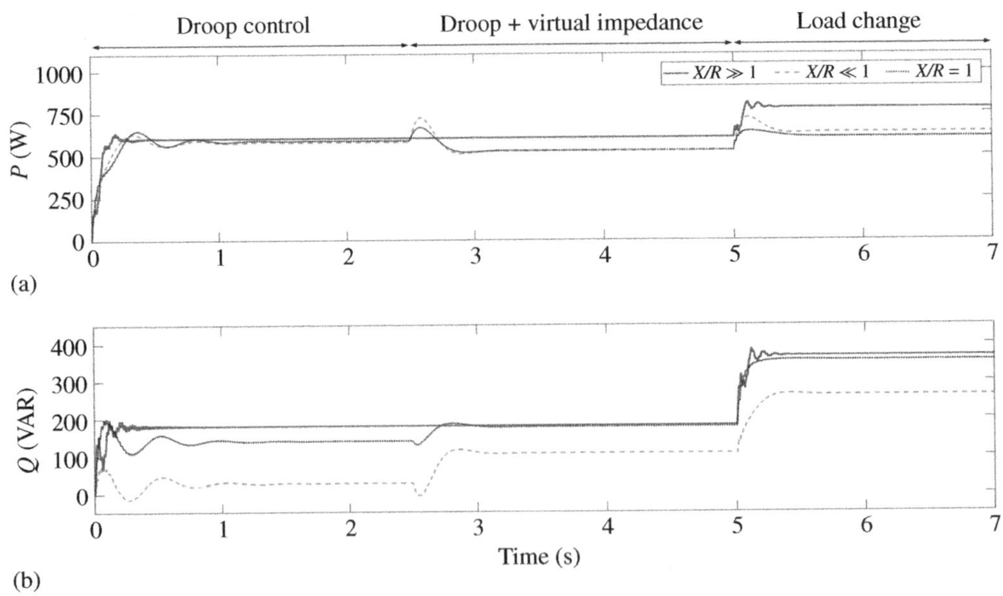

Figure 5.17 Performance of droop control and virtual impedance in the presence of different line impedances: (a) active and (b) reactive powers of DER_1.

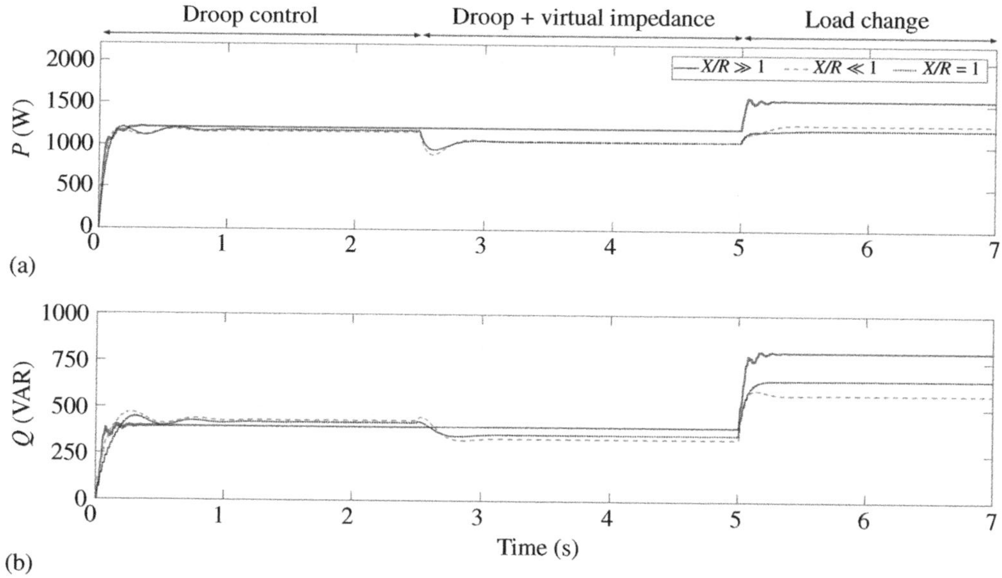

Figure 5.18 Performance of droop control and virtual impedance in the presence of different line impedances: (a) active and (b) reactive powers of DER_2.

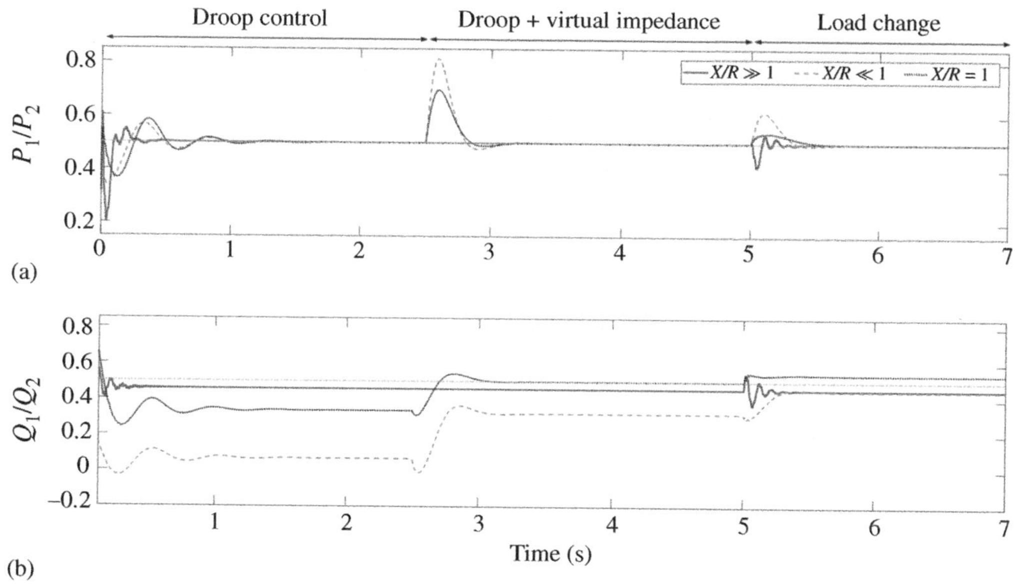

Figure 5.19 Performance of droop control and virtual impedance in the presence of different line impedances: (a) active and (b) reactive powers capacity ratio of the DERs.

of the DERs is shown in Figure 5.19. This figure shows that accurate power sharing is achieved for inductive lines, unlike the cases of resistive–inductive and resistive lines. Power sharing for these cases approaches the nominal ratio after applying the virtual impedance.

The system's frequency and voltage are subjected to steady-state errors at the primary control layer, as shown in Figures 5.20 and 5.21. Figure. 5.21 reports that systems with inductive lines experience the least voltage variation (especially in PCC) in facing load disturbances compared to the

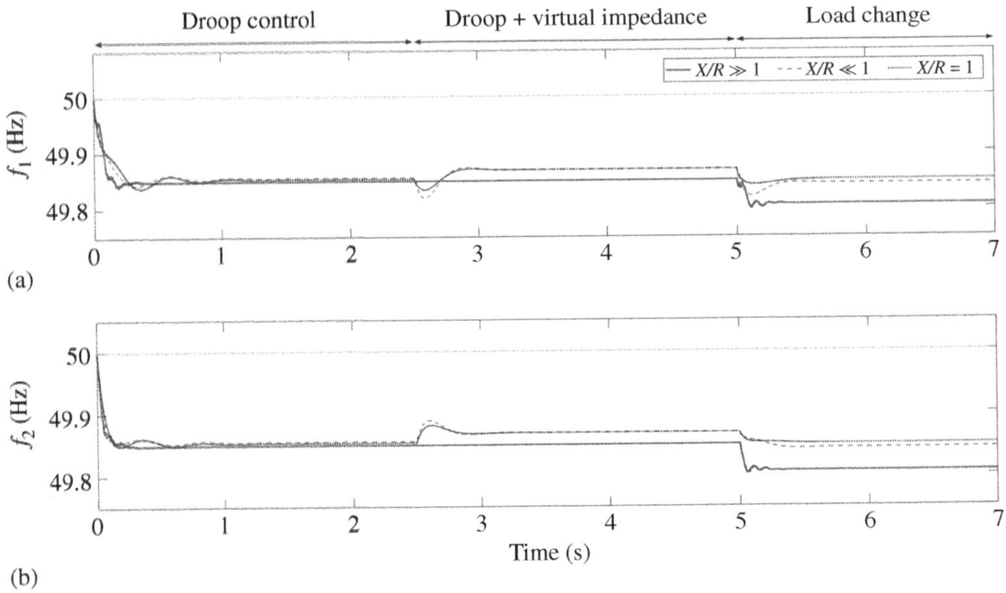

Figure 5.20 Performance of droop control and virtual impedance in the presence of different line impedances: (a) DER_1 frequency, (b) DER_2 frequency.

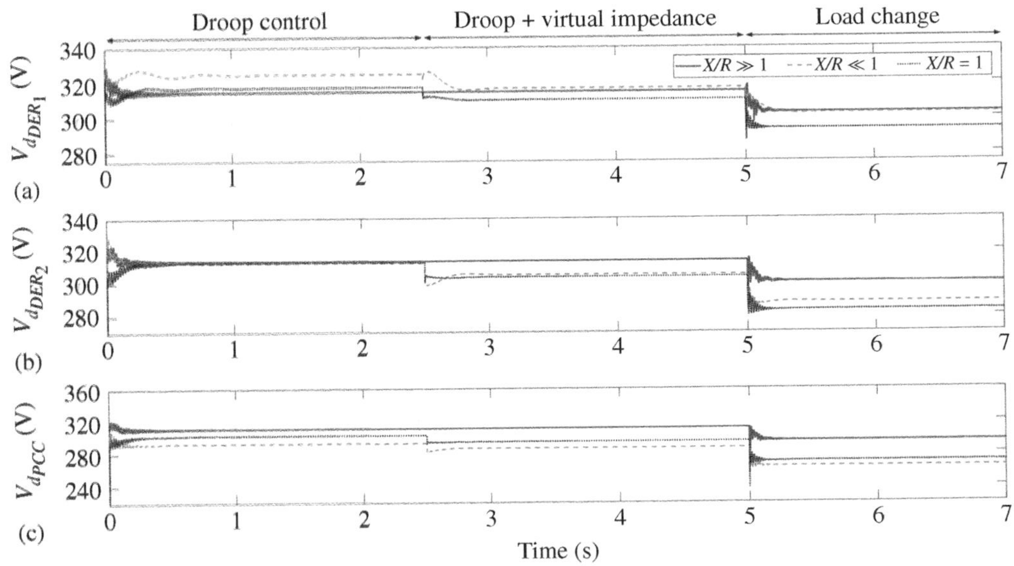

Figure 5.21 Performance of droop control and virtual impedance in the presence of different line impedances: (a) direct component of DER_1 voltage, (b) direct component of DER_2 voltage, and (c) direct component of voltage at the point of common coupling.

other two cases. As these results demonstrate, the virtual impedance cannot help with the voltage and frequency deviations caused by the droop mechanism. Secondary control is often implemented on the top of the primary control to restore the system's frequency and voltage to their nominal values (see the provided example in Section 5.6.5).

5.6 Secondary Control

In an MG, secondary control is responsible for maintaining power quality and enhancing system performance by regulating system frequency and voltage, especially during sudden changes in load or generation. This includes improving the stability of the MG and ensuring that power is delivered reliably and efficiently to connected loads.

In islanded mode, the MG operates as a stand-alone system, disconnected from the larger power grid. In this mode, secondary control complements the task of primary control to improve the power quality inside MGs and to enhance the system performance by removing the steady-state errors of the frequency and voltage. This control layer is basically introduced to eliminate the system's frequency/voltage deviations [2, 23]. Secondary control can also be used to improve (reactive) power sharing accuracy, suppress circulating currents, and harmonic elimination [2], there are fundamental conflicts between accurate reactive power sharing and voltage regulation [59].

In grid-connected mode, an MG is connected to the larger power grid and has the ability to both import and export power. The primary objective in this mode of operation is to coordinate the power flow between the MG and the main grid, with the MG following the voltage and frequency of the main grid. To achieve this, GFO and GFL power converters may be utilized. Some GFO control strategies incorporate feedback of voltage and frequency in their control to provide ancillary services for the grid, similar to certain types of VSGs.

This section will provide an overview of secondary control functions and architectures in the islanded mode of operation, along with several helpful examples.

5.6.1 Secondary Control Functions and Strategies

5.6.1.1 Secondary Control Functions

The droop control mechanism in (5.16) can be applied to inverter-interfaced DERs in both islanded and grid-connected modes. As previously noted, when DERs contribute to the system's frequency and voltage regulation in islanded mode, the power references (i.e. P^* and Q^*) of the droop mechanism can be set to zero. Thus, the conventional droop equations in (5.16) can be expressed as follows:

$$\omega_i = \omega^* - k_{p_i} P_i, \tag{5.24a}$$

$$v_i = v^* - k_{q_i} Q_i. \tag{5.24b}$$

The main function of secondary control is to eliminate the steady-state voltage and frequency deviation caused by droop control. The main role of secondary control is to eliminate the deviations of ω_i and v_i while simultaneously maintaining the stability of power sharing, voltage, and frequency of the MG. If the set of DERs in the MG is labeled $\mathcal{N} = \{1, \dots, N\}$, then this may be expressed as

$$\lim_{t \to t_f} \omega_i(t) = \omega^*, \tag{5.25a}$$

$$\lim_{t \to t_f} v_i(t) \approx v^*, \tag{5.25b}$$

$$\lim_{t \to \infty} (k_{p_i} P_i(t) - k_{p_j} P_j(t)) = 0, \tag{5.25c}$$

$$\lim_{t \to \infty} (k_{q_i} Q_i(t) - k_{q_j} Q_j(t)) = 0, \tag{5.25d}$$

for all $i,j = 1, \ldots, N$, where (5.25a)–(5.25d) represent frequency restoration and voltage regulation in finite time t_f and power sharing in steady state, respectively. In order to share reactive power accurately, the voltage values of DERs must not be perfectly regulated to their reference values, hence the ≈ symbol is used in (5.25b) [59].

The secondary control addresses the above limits by adding correction terms to the droop controller (5.24) as

$$\omega_i = \omega^* - k_{p_i} P_i + \delta u_{\omega,i}, \tag{5.26a}$$

$$v_i = v^* - k_{q_i} Q_i + \delta u_{v,i}, \tag{5.26b}$$

where $\delta u_{\omega,i}$ and $\delta u_{v,i}$ are the control signals provided by the secondary control layer of the DER_i. The secondary control layer is composed of two main control loops: the frequency control loop and the voltage control loop. These loops work together to regulate the frequency and voltage of the MG. Since voltage is a local variable opposed to frequency, there is always a conflict between the system's voltage regulation and reactive power sharing among the DERs. Therefore, two separate control loops are often implemented in the secondary layer to guarantee appropriate voltage regulation and reactive power sharing [22]. In such a case, the correction term $\delta u_{v,i}$ is produced by both control loops.

It is worth noting that the secondary control loops mentioned can be implemented using a single advanced control method that incorporates the control objectives into a cost function. This approach allows for a unified control strategy that optimizes both frequency and voltage regulation, while also taking into account other objectives [60].

Figure 5.22 depicts the secondary control behavior versus droop control response in regulating the system's voltage and frequency. As shown in this figure, by increasing the load power, the voltage and frequency drop due to the droop mechanism characteristic. Then, the secondary control shifts up/down the droop characteristics using the produced correction terms (δu_ω and δu_v) to remove the steady-state deviations, thus restoring the frequency and voltage to their nominal values.

In addition, power quality improvement is another function introduced under the name of secondary control [61–68]. For this function, individual control loops must be implemented in the

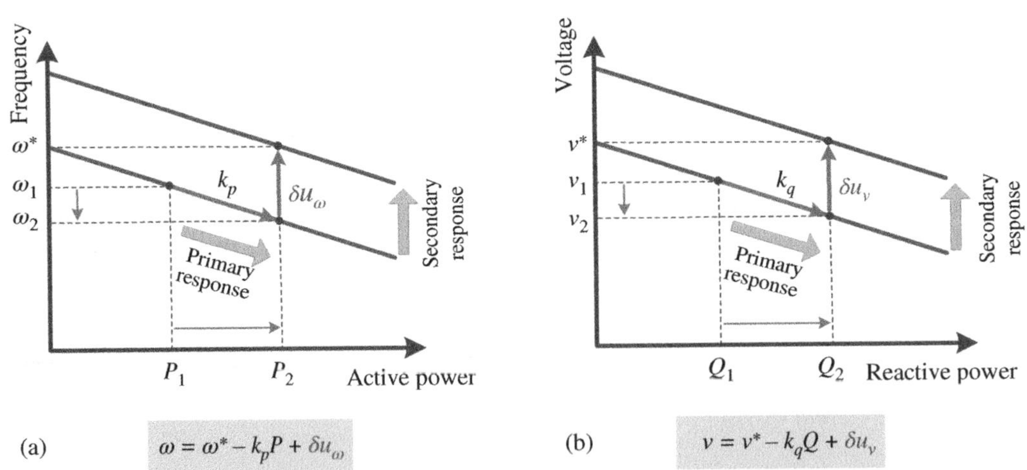

Figure 5.22 The behavior of secondary control versus droop control in regulating (a) frequency and (b) voltage.

secondary layer. Each control loop then provides a setpoint for the primary control and the droop mechanism similar to (5.26).

Sharing of data – such as frequency, voltage and active/reactive power – between DERs can be used to design the secondary control signals and to achieve the control objectives above. Depending on the utilized secondary control strategy (see Sections 5.6.2–5.6.4), these data are available either locally for each DER or via communication links. For each control objective, the related measured variable (e.g. voltage) or its averaged value [59] is compared with its nominal value and then passed through a controller (usually PI) to provide a control signal, i.e. the correction term [69, 70]. However, the simple classic *PI* controller can be replaced with other advanced control techniques [24, 71] to improve the secondary control performance.

5.6.1.2 Secondary Control Strategies

Secondary control can be implemented using all three control strategies mentioned in Section 5.2.3. The majority of existing secondary control strategies are based on communication infrastructures to provide a desirable performance. The communication-based solutions are classified into centralized and distributed secondary controllers. All frameworks based on these two control strategies are vulnerable to the failure of communication links. The conventional secondary control solutions are structured centrally, e.g. those used for frequency and voltage restoration [72], reactive power sharing [73], and voltage unbalance/harmonic compensation [74]. This structure, however, is highly constrained by complex point-to-point communication links between all DERs, no scalability, single point of failure, and low reliability. The centralized controller typically needs global knowledge about the system parameters and the load. Thus, it is unable to meet the PnP operational requirement of the system. It is usually costly both in communication and computation when the number of sources increases. Most importantly, the central controller exposes a single point-of-failure, i.e. any failure in the controller renders the entire system inoperable.

Unlike the centralized secondary control (CSC) structure, the distributed one provides more scalability, and higher reliability, while using a spare communication network [59]. Decentralized secondary control (DESC) is implemented locally, either using individual local states or estimating neighbor variables. In this framework, a secondary control can be designed without remote measurements and communication networks. Although this control strategy requires no communication infrastructure, they often have complex calculations. Figure 5.23 illustrates the three secondary control architectures, namely centralized secondary control (CSC), distributed secondary control (DISC), and decentralized secondary control (DESC), which will be introduced in the following sections.

5.6.2 Centralized Secondary Control

In the CSC structure, a central controller coordinates the DERs and provides a reference (δu_i) for the primary loop in order to achieve the control objectives (y_i), as shown in Figure 5.23(a). Active/reactive power management, voltage/frequency regulation, and power quality improvement are the main features of the CSC. In this structure, all required data, i.e. the DERs frequency/voltage, are generally transmitted through a high data rate communication network. Therefore, any deficiency in communication or failure degrades the MG efficiency.

The main challenges in the CSC architecture have been identified to be harmonic cancellation, unbalanced current reduction, and other power quality and management issues to enhance the system's performance.

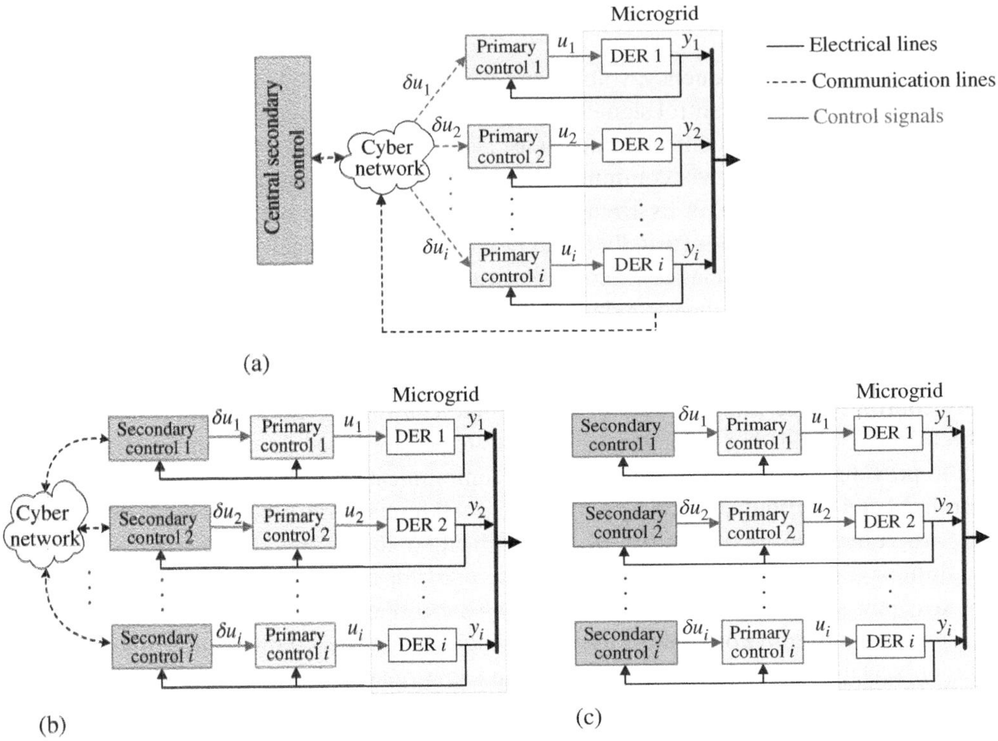

Figure 5.23 Three main secondary control architectures: (a) CSC, (b) DISC, and (c) DESC.

5.6.3 Distributed Secondary Control

As units in an MG are heterogeneous and spatially distributed, distributed control and multiagent system (MAS) control are promising approaches to enhance MG stability and performance while addressing reliability and enhancing the scalability of MGs. In distributed control of MASs, such as MGs, a number of DERs (as agents) work together to cooperatively control the MG and fulfill a set of objectives. In this case, the behavior of the MG depends on the agent (i.e. DER) dynamics and the topology of the employed communication infrastructure. In the following, DISC policies are classified into averaging, distributed consensus, and event-triggered methods. To better understand these distributed control policies, some basic information about communication networks is provided later.

5.6.3.1 Communication Network as a Graph

The communication network of an MG can be regarded as a directed graph with the DERs and the communication links playing the roles of its nodes and edges, respectively. Consider the graph $\mathscr{G} \subseteq (\mathscr{N}, \mathscr{E}, \mathscr{A})$, where, $\mathscr{N} = \{1, \dots, n\}$, $\mathscr{E} \subseteq \mathscr{N} \times \mathscr{N}$, and $\mathscr{A} = [a_{ij}] \in \mathbb{R}^{n \times n}$ are its node set, edge set, and adjacency matrix, respectively. If node i obtains data from node j, then node j is an in-neighbor of node i, node i is an out-neighbor of node j, $(j, i) \in \mathscr{E}$, and $a_{ij} = 1$; otherwise, nodes j and i are not neighbors, $(j, i) \notin \mathscr{E}$, and $a_{ij} = 0$. Let $N_i = \{j \mid (j, i) \in \mathscr{E}\}$, $N_i^o = \{j \mid (i, j) \in \mathscr{E}\}$, $d_i = \sum_{j \in N_i} a_{ij}$, and $d_i^o = \sum_{j \in N_i^o} a_{ji}$ be the in-neighbor set, out-neighbor set, in-degree, and out-degree of node i, respectively. Laplacian matrix of $\mathscr{G}$ is $\mathscr{L} = \mathscr{D} - \mathscr{A}$, where $\mathscr{D} = diag\{d_i\}$ (proper diagonal matrix). A path from node j to node i is a sequence of pairs, belonging to $\mathscr{E}$, expressed as $\{(j, n_1), \ldots, (i, n_m)\}$. A graph has a spanning tree, if there exists a node (called root node) from which there exists a path to any other node within the graph [75].

5.6.3.2 Average-Based DISC

In the averaging-based structure, each DER measures the required variables, e.g. frequency/voltage amplitudes, and communicates them to all other adjacent DERs. Let x_i denote a variable of interest (frequency, voltage, active power, etc.), with nominal value x^*. By averaging the values received from other adjacent DERs, a control signal can be built as

$$\delta u_i = K_i(s)\left(x^* - \frac{1}{N}\sum_{i\in\mathcal{N}} x_i\right), \tag{5.27}$$

to form the secondary control signals (i.e. the correction terms for the droop mechanism) in Eq. (5.26) [22, 60, 69, 76–78]. The controller $K_i(s)$ can be designed in different ways (e.g. PI) and tuned for dynamic performance.

5.6.3.3 Consensus-Based DISC

For MASs, the structure of communication infrastructure shows the direction and information of the agents as DERs in an MG. Algorithms leading to agreement among all DERs are called consensus, agreement, or distributed averaging algorithms [75]. The robustness of this technique is proved in several research works even in the presence of communication delays [72, 76, 79, 80]. In order to restore the frequency/voltage in the MG applying the consensus-based techniques, the following distributed protocol is used [59, 60, 81–83]

$$\delta u_i = K_i(s) \sum_{j\in\mathcal{N}, j\neq i} \alpha_{ij}(x_j - x_i) \tag{5.28}$$

to produce the control signals in (5.26). Fundamental performance limitations of such controllers have been examined in [84–86]. If the controller $K_i(s)$ in (5.28) contains integral control, then in steady-state it will hold that $x_i = x_j$ for all $i, j \in \mathcal{N}$. As a concrete example, this could correspond to active power sharing if $x_i = m_i P_i$, which would meet the control objective in (5.25c). In (5.28), the coefficients α_{ij} are nonnegative and are the elements of the so-called adjacency matrix of the graph describing the communication network. DER_i receives information from DER_j if $\alpha_{ij} \neq 0$; see [75] for more details on the consensus and cooperation in networked multi-agent systems. The summary of methods employing consensus protocols in the SC design procedure can be found in [87].

For a MAS such as an MG with many DERs, the conventional consensus is often not sufficient, as the communication network topology may be very dynamic. Recently, pinning-consensus (or leader-following consensus) problems [61, 88–90], and multiple objective problems for leader-following systems (or containment control) [91] have been presented. The term "pinning control" in distributed cooperative control means merely a fraction of nodes in the studied network, which are pinned by the designer to the objective trajectory and the rest of the nodes communicate with each other to achieve the expected networked tracking. To achieve the cooperative objectives, the leader-following consensus can be expressed as

$$e_i = \sum_{j\in N} \alpha_{ij}(x_j(t) - x_i(t)) - g_i(x_i(t) - x^*). \tag{5.29}$$

Then, the secondary correction term can be calculated as

$$\delta u_i = -cK_i e_i, \tag{5.30}$$

where $i = 1, 2, \dots, N$, x^* is the desired value for the consensus state i, $c \in \Re$ is the coupling gain and K_i is the feedback control vector. The pinning control gain $g_i \geq 0$ shows there is no control over the DER_i.

Unlike the leader-following consensus control, where there exists only one leader in the system, containment control protocol is presented in the presence of multiple leaders and followers in an MAS. The role of containment control can be highlighted where the control objective is to bring all the followers into a convex hull spanned by the leaders [91]. There are many open research challenges in this direction, such as choosing the (optimal) pinned DERs, convergence rate based on the MG scale, the optimal number of follower DERs, and challenges of communication network switching for interconnected MGs [92].

5.6.3.4 Event-Triggered DISC

Over the past years, event-triggered control (ETC) has been increasingly employed at the secondary control level of MGs, as it reduces information exchange among DERs while retaining stability [93–104]. Practically, instead of continuous data exchange among DERs, the required data can be shared when a criteria is satisfied or an event is triggered. Then, a sampled-data control method is performed, and data is exchanged by a designed mechanism on ETC [105, 106]. As illustrated in Figure 5.24, time-triggered, event-triggered, and self-triggered sampling methods are three presented approaches to realize the DISC.

In time-triggered secondary control, the control is driven by a clock [107]. This periodic paradigm can be seen as an open-loop sampling (see Figure 5.24(a)). In the event-triggered secondary control, a signal is sent to the controller if an event has occurred, rather than a continuous signal transmission. This can be considered as an introducing feedback in the sampling process (see Figure 5.24(b)). It requires the permanent monitoring of the state(s) to determine current performance. Finally, in the self-triggered control, the current state is employed not only to compute the control signal (the input to the system) but also to calculate the next time for recomputation of control law. Though this mechanism is still closed-loop based on the current performance, permanent monitoring of the state(s) is no longer required (see Figure 5.24(c)). A common structure of the ETC can be computed based on the errors between the current instant and the last event of the state variable. The errors are based on the observed and measured values of the state variable [108]. In the ETC designs, the following practical issues should be considered.

Event Mechanism The design of a mechanism for event detection is a key challenge for computing event time instants, which directly depends on reducing the recomputation of control law and communication of neighbors. Basically, the structure of the event mechanism should be physically explainable and easy to implement from a practical perspective. As an open research

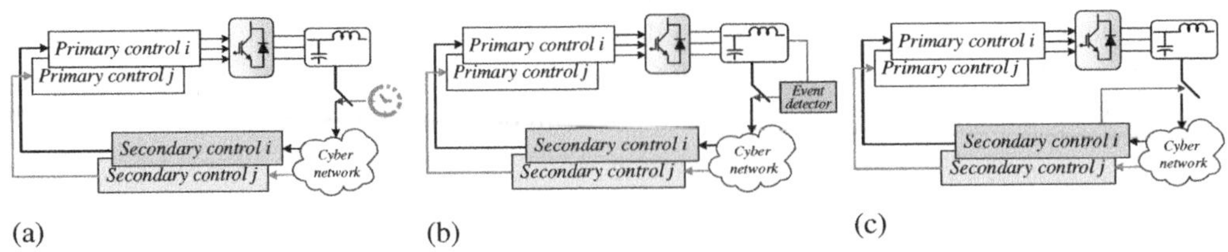

Figure 5.24 Distributed secondary control strategies: (a) time-triggered DISC, (b) event-triggered DISC, and (c) self-triggered DISC.

problem, in the MG applications, estimation and observation approaches can be utilized to reduce the employed communication network by designing an event mechanism for the secondary control [109, 110].

Consensus ETC If ETC is combined with consensus, the measurements of neighboring DERs are available only at event time instants. Therefore, an important concern for consensus-ETC designs is how to efficiently employ such information to recompute control law under the designed ETCs. Besides, it is a great challenge to design both the controller in (5.28) and a threshold-based mechanism for event detection in a unified framework, while the stability of the system must be maintained without the Zeno phenomenon.

Zeno Phenomenon When the ETC generates an infinite number of sampling instants (events) during a finite time, the Zeno phenomenon occurs [111] and makes the solution inapplicable for real-world systems. Because of limited communication and computational capabilities, no control system can be implemented on a digital platform, if this behavior exists. This concern should be considered in the control design procedure. To cope with this issue, switching approaches are addressed which by finding a waiting time that guarantees a lower bound for the interevent times in the event mechanism [112].

Clock Drift A key challenge toward DER synchronization, connecting an MG to the utility grid, simultaneous frequency/voltage regulation, and accurate power sharing in an MG is the implementation of the designed controller through appropriate hardware. Since each converter in the converter-based MGs operates based on the clock of its processor, a prevalent phenomenon in digital control implementation is clock drift [6, 113, 114], which may degrade the performance of the system (for example, see [115] for recognition of clock synchronization challenges). Because of the popularity of crystal oscillator-based clock processors, with nominal frequency f_c, a small uncertainty in clock drift makes the clock cycle of the processor as $\Delta t_c = 1/f_c(1 + \mu)$, where μ is the relative drift of the crystal-oscillator of processor clock [116]. This is an important challenge not only for the ETC but also for other secondary control strategies.

Several works investigate clock drift's influence on secondary control policies. The clock drift phenomenon in the MG was modeled and investigated first in [117]. Then, in [118] and [119] a comprehensive analysis of the performance of secondary control policies with respect to clock drifts is presented. Considering this phenomenon in different secondary controllers and different conditions such as PnP is an open area for research.

5.6.4 Decentralized Secondary Control

In DESC approaches, communication infrastructures are not used for frequency/voltage restoration; each DER restores its frequency/voltage amplitudes to nominal values separately. Although communication is not used in the secondary control operation, the communication infrastructures are required for transferring data from the MGCC and upper layers. For instance, a global communication infrastructure is required for coordination of the DERs during black start and PnP processes, real-time monitoring or functionalities of the MGCC, and the higher control loops. Design procedures employed in this structure are divided into the three main categories based on washout filter, local variable, and estimation methods.

5.6.4.1 Washout Filter-Based DESC

This approach is implemented on the droop control layer, as follows [120]:

$$\omega = \omega^* - m\left(\frac{s}{s+k_p}\right)\left(\frac{\omega_c}{s+\omega_c}\right)p, \tag{5.31a}$$

$$v = v^* - n\left(\frac{s}{s+k_q}\right)\left(\frac{\omega_c}{s+\omega_c}\right)q. \tag{5.31b}$$

If the MG is stable, then by applying the final value theorem to the equations (5.31), one can show that ω and v converge to their reference values.

A simplified DESC based on a generalized band-pass washout filter is introduced as [121]:

$$\omega = \omega^* - \frac{m}{k_{p\omega}+1}\cdot\underbrace{\underbrace{\frac{\omega_c}{s+\omega_c}}_{\text{low-pass}}\cdot\underbrace{\frac{s}{s+\dfrac{k_{i\omega}}{k_{p\omega}+1}}}_{\text{high-pass}}}_{\text{band-pass filter}}\cdot p, \tag{5.32a}$$

$$v = v^* - \frac{n}{k_{pv}+1}\cdot\underbrace{\underbrace{\frac{\omega_c}{s+\omega_c}}_{\text{low-pass}}\cdot\underbrace{\frac{s}{s+\dfrac{k_{iv}}{k_{pv}+1}}}_{\text{high-pass}}}_{\text{band-pass filter}}\cdot q. \tag{5.32b}$$

By adopting appropriate values for the high-pass filter as $k_{p\omega} = k_{pv} = 0$, $k_{i\omega} = k_p$, and $k_{iv} = k_q$, the washout filter-based secondary control (5.31) can be implemented as well. Indeed, the generalized band-pass washout filter for the secondary control is realized by cascading a low-pass filter and a high-pass filter. Its frequency characteristics affect the secondary control transient response. Some recent studies have presented an analogy between angle droop and frequency droop with new perspectives on virtual impedance using a washout filter [121].

Second-order washout filters can be also implemented on $\omega - Q$ and $v - P$ droop control strategy [122]:

$$\omega = \omega^* + n_q\frac{s}{A_\omega s^2 + B_\omega s + C_\omega}\left(\frac{\omega_c}{s+\omega_c}\cdot q\right), \tag{5.33a}$$

$$v = v^* - m_p\frac{s}{A_v s^2 + B_v s + C_v}\left(\frac{\omega_c}{s+\omega_c}\cdot p\right). \tag{5.33b}$$

Similar to [123], the effect of a fully decentralized second-order washout filter-based can be investigated for the future research by providing a comprehensive performance and robustness analysis as well as optimal tuning of the coefficients of the filter.

5.6.4.2 Local Variable-Based DESC

In this DESC strategy, local signals are used to design a secondary control without any communication network according to a time-dependent protocol [124] or optimal control-based DESC [125].

5.6.4.3 Estimation-Based DESC

Estimation-based methods serve as alternatives for communication infrastructures to inform the dynamics of the rest of MG to the local secondary control units. Although the estimation/

observation methods depend on the modeling of the system, they have drawn significant attention in MG applications because of their communication-free feature. In general, the estimation-based DESC replaces the measurements in (5.28) with estimates as

$$\delta u_i = K_i(s) \sum_{j \in \mathcal{N}} \alpha_{ij}(\hat{x}_j - x_i), \tag{5.34}$$

where $\hat{x}_j(t)$ is the estimated value of variable x_j to design the DESC.

Recently, fully DESC schemes have been presented for autonomous MGs utilizing the state estimation methods [126–129], and state variable estimation [129–133], as well as for hybrid AC/DC MGs [126]. The promise of observer/estimator-based controllers to achieve stability margins and reliability of the system under diverse uncertainty, will introduce new secondary control architectures for MGs.

To sum up, a simple overview of the secondary control implementation for master-slave, averaging, consensus and estimation-based control architectures is demonstrated in Figure 5.25.

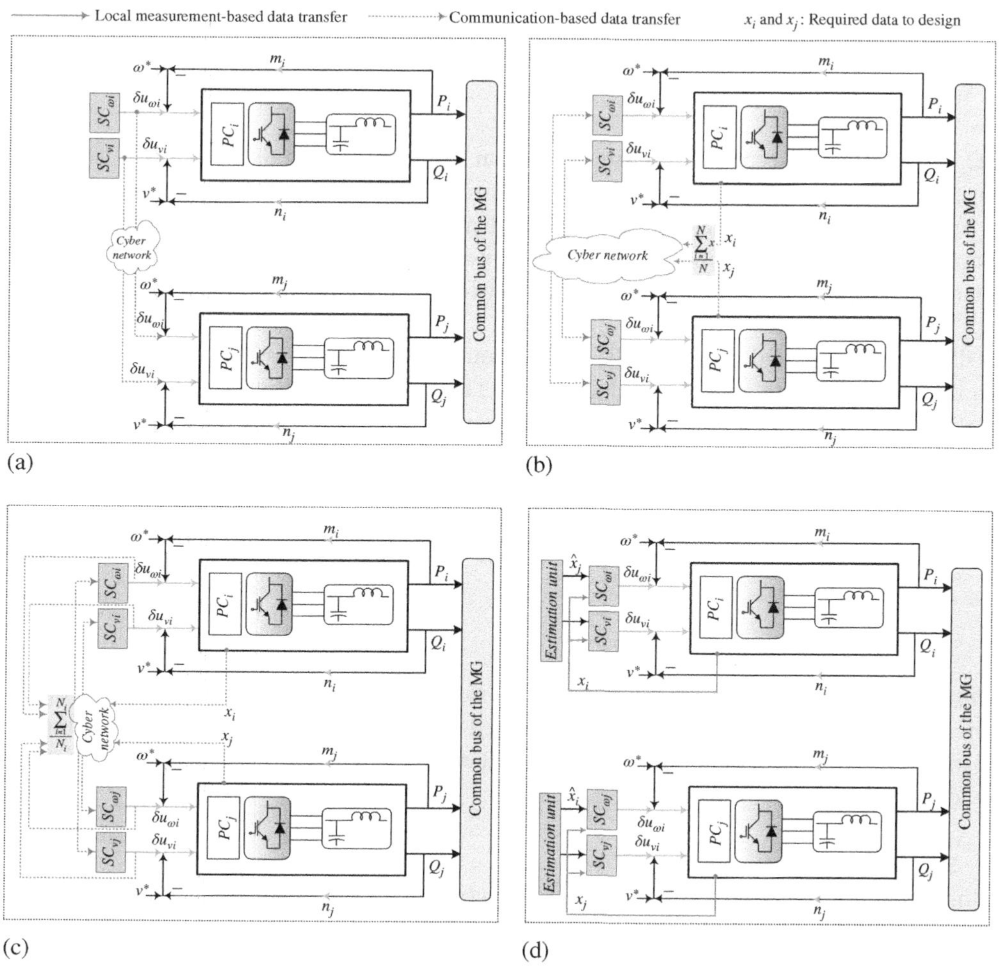

Figure 5.25 Secondary control structures: (a) centralized master-slave method, (b) distributed averaging method, (c) distributed consensus method, and (d) decentralized-estimation-based method.

Based on the presented architectures, the controller $K_i(s)$ can be designed considering the stability and performance issues presented in Chapter 4. The controller can also be replaced by advanced controllers as will be presented in the next chapter.

5.6.5 A Simulation Study for Secondary Control of AC Microgrids

A simulation study is performed to investigate the behavior of the secondary control loop on the system performance.

5.6.5.1 Case Study and Controller Implementation

The consensus-based DISC framework presented in [59] is chosen for the simulation study:

$$\omega_i = \omega^* - k_{p_i} P_i + \Omega_i , \tag{5.35a}$$

$$k_{\omega_i} \frac{d\Omega_i}{dt} = -(\omega_i - \omega^*) - \sum_{j=1}^{n} a_{ij} \left(\Omega_i - \Omega_j\right) , \tag{5.35b}$$

$$v_i = v^* - k_{q_i} Q_i + e_i , \tag{5.36a}$$

$$k_{v_i} \frac{de_i}{dt} = -\beta_i (v_i - v^*) - \sum_{j=1}^{n} b_{ij} \left(\frac{Q_i}{Q_i^*} - \frac{Q_j}{Q_j^*} \right) , \tag{5.36b}$$

where Ω_i and e_i are the secondary control correction terms ($\delta u_{\omega,i}$ and $\delta u_{v,i}$ in (5.26)) added to the conventional droop controls, Q_i^* is the ith DERs reactive power rating, β_i is a positive gain, and k_{ω_i} and k_{v_i} are frequency and voltage integral gains. The $n \times n$ matrix B with elements $b_{ij} > 0$ is the adjacency matrix of a communication network between the DERs. The secondary controller (5.36b) achieves a tunable compromise between voltage regulation and reactive power sharing.

The DISC is implemented on an ac MG test system composed of four GFO DERs connected through line impedances to support two local loads. The loads are present locally at DERs 1 and 4, and DERs 1 and 4 are rated for twice as much power as DERs 1 and 3 (see Table 5.3). The test system is equipped with the communication network with the graph adjacency matrices $A = [a_{ij}]$ in (5.35b) and $B = [b_{ij}]$ in (5.36b) being

$$A = \begin{pmatrix} 0 & 1 & 0 & 1 \\ 1 & 0 & 1 & 0 \\ 0 & 1 & 0 & 1 \\ 1 & 0 & 1 & 0 \end{pmatrix} , \quad B = b \cdot \begin{pmatrix} 0 & 1 & 0 & 1 \\ 1 & 0 & 1 & 0 \\ 0 & 1 & 0 & 1 \\ 1 & 0 & 1 & 0 \end{pmatrix} , \tag{5.37}$$

where $b \geq 0$ varies depending on the study under consideration (refer to Section 5.6.3.1). All other parameters are reported in Table 5.3. The distributed controllers (5.35) and (5.36) were implemented in Simulink®. Details of the secondary controller and primary control loop implementation are provided in the simulation file available in [134].

5.6.5.2 Simulation Results

Some simulation results are provided in the following (see Figures 5.26 and 5.27) to study the performance of the DISC in the regulation of the system's voltage and frequency and sharing the DERs' active/reactive powers under load disturbances and communication link failures.

Figure 5.26 shows the controller performance comparing droop control under load disturbances. In this case, only primary droop controllers (5.35a) and (5.36a) are running up to $t = 7$ s, at which

Table 5.3 Electrical and control parameters.

Parameter	Symbol	Value
	Electrical setup	
Nominal frequency	$\omega^*/2\pi$	50 Hz
DC voltage	V_{dc}	650 V
Nominal voltages	E^*	325.3 V (230 V rms)
Filter capacitance	C	25 μF
Filter inductance	L_f	1.8 mH
Output impedance	L_0	1.8 mH
Line impedance (1,2)	Z_{12}	$R = 0.8\,\Omega$, $L = 3.6$ mH
Line impedance (2,3)	Z_{23}	$R = 0.4\,\Omega$, $L = 1.8$ mH
Line impedance (3,4)	Z_{34}	$R = 0.7\,\Omega$, $L = 1.9$ mH

Parameter	Symbol	DERs 1 and 4	DERs 2 and 3
	Control parameters		
Rated active power	P_i^*	1400 W	700 W
Rated reactive power	Q_i^*	800 VAr	400 VAr
$P - \omega$ Droop coeff.	k_{p_i}	$2.5 \times 10^{-3}\ \frac{\text{rad}}{\text{W}\cdot\text{s}}$	$5 \times 10^{-3}\ \frac{\text{rad}}{\text{W}\cdot\text{s}}$
$Q - v$ Droop coeff.	k_{q_i}	$1.5 \times 10^{-3}\ \frac{\text{V}}{\text{VAr}}$	$3 \times 10^{-3}\ \frac{\text{V}}{\text{VAr}}$
Int. frequency gain	k_i	1.7 s	1.7 s
Int. voltage gain	κ_i	1 s	1 s

time the secondary controllers (5.35b) and (5.36b) are activated. The local load at DER 4 is detached at $t = 22$ s, then reattached at $t = 36$ s. As shown in Figure 5.26, the frequency deviation experienced under primary droop control is quickly eliminated by the secondary controller (5.35a) and (5.35b), and frequency regulation is maintained throughout load changes with minimal transients. Active power is accurately shared among the heterogeneous DERs throughout the entire runtime.

The inherent conflict between the voltage and reactive power is shown in the plots related to these two variables in Figure 5.26. The controller maintains the voltage amplitudes of DERs within the acceptable range while sharing the reactive power among the DERs properly. This is achieved by well tuning the controller parameters (β and b) in (5.36b). One can change the controller parameters to control the voltage and reactive power in a different way. The voltage and frequency correction terms (Ω_i and e_i) causing the aforementioned behaviors are presented at the bottom of Figure 5.26. These terms are simply added to the droop controls to shift up/down the droop characteristics.

Figure 5.27 shows the controller performance under communication link failure. The communication link between DERs 3 and 4 fails at $t = 2$ s. At $t = 7$ s the local load at DER 4 is detached, and is reattached at $t = 18$ s. The control parameters are the same as in the first study. As the results in Figure 5.27 show, the controllers (5.35a) and (5.36b) maintain the high performance shown in the first study despite the absence of the communication link between DERs 3 and 4.

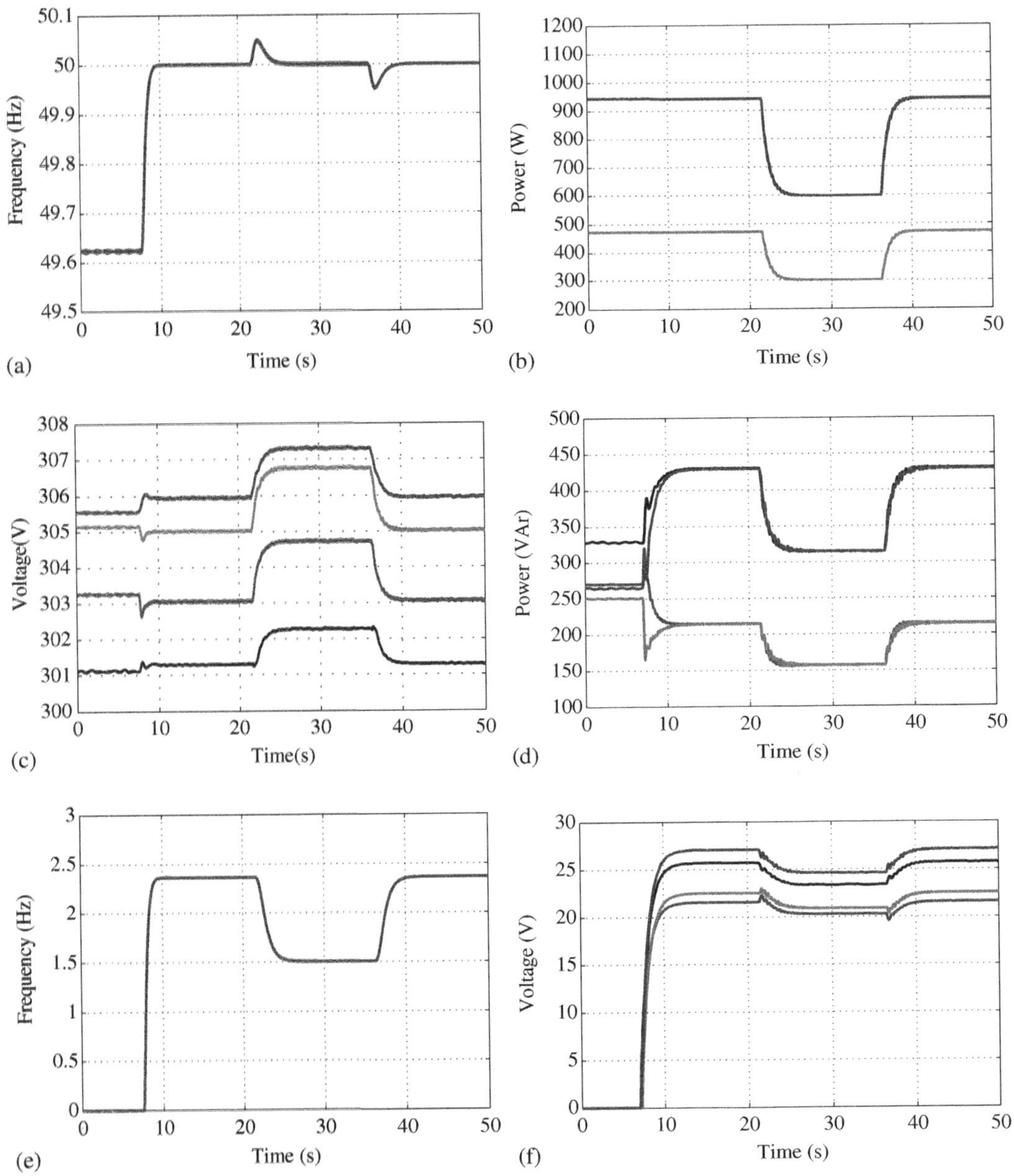

Figure 5.26 Performance of the distributed secondary controller in regulating frequency and voltage and sharing the active/reactive powers of DERs, with control parameters $\beta_1 = \beta_3 = \beta_4 = 0$, $\beta_2 = 4$, and $b = 100$ V: (a) frequency, (b) active powers, (c) voltage magnitudes, (d) reactive powers, (e) the DISC outputs (Ω), and (f) the DISC outputs (e).

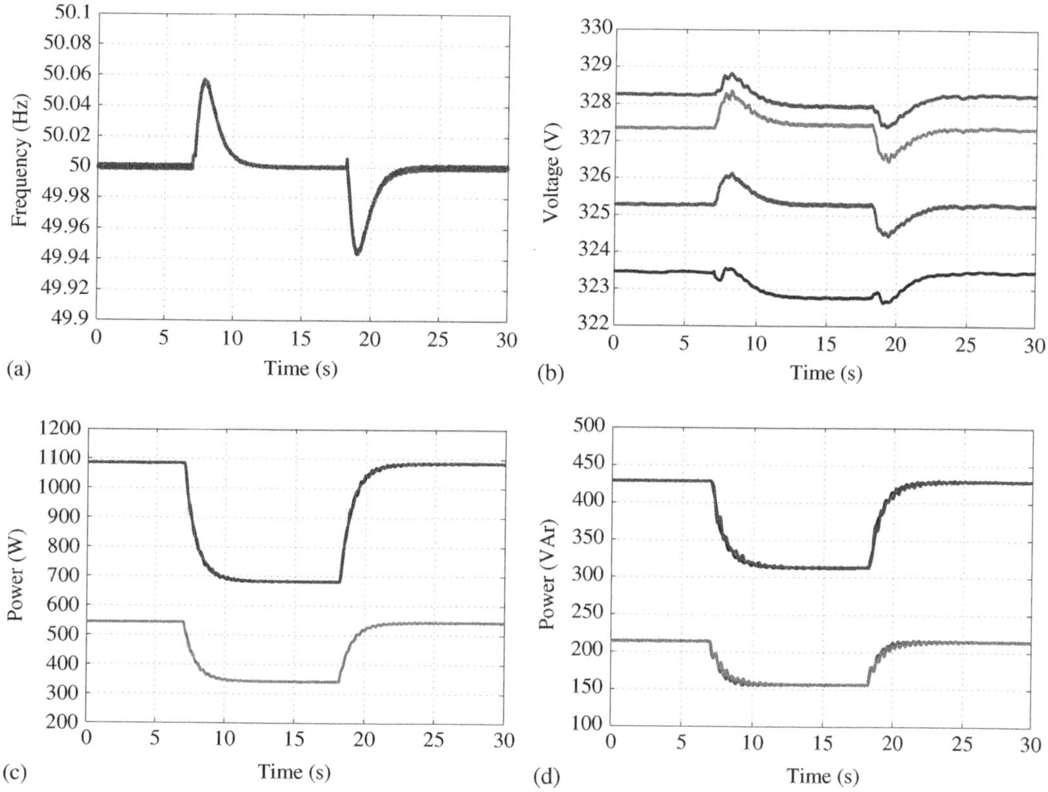

Figure 5.27 The controller performance under communication link failure: (a) frequency, (b) voltage magnitudes, (c) active powers, and (d) reactive powers.

5.7 Central Control

Following the hierarchical control of large power systems, higher control layers of MGs are classified into two separate layers, each with predefined tasks and objectives. Central control and global control, known as tertiary control layers in some research studies, are used to achieve long-term, high-level objectives defined by MGs and the main grid.

Central control, refers to the MGCC, which is responsible for ensuring the reliable, secure, and cost-effective operation of MGs in both grid-connected and islanded modes of operation. The MGCC coordinates the operation of DERs and loads within the MG and is responsible for ensuring their reliable and cost-effective operation, as well as for managing their power exchanges with the main grid. The lower secondary and primary controllers manage the DERs within the MG in coordination with the MGCC [5]. The MGCC manages the long-term power and energy flow between the MG sources and loads. It assigns active and reactive power references, as well as other appropriate control signals, to the DERs and controllable loads.

The MGCC also incorporates the optimal power flow control strategy, which ensures the reliable and efficient long-term operation of the MG. The optimal power flow control aims to minimize the MG's costs (including operational, environmental, and unreliability costs) by optimally dispatching DERs, reactive power resources, demand response resources, and even droop coefficients [135]. This control strategy can also be implemented in a distributed way to include other functionalities, such as electricity price forecasting, state estimation, or storage state-of-charge estimation. Once the optimal power flow is solved, the setpoints are sent to the secondary and primary control layers to be executed locally.

The central control also facilitates the transition between islanded and grid-connected modes of operation in coordination with the secondary control. In the grid-connected mode, the power flow between the MG and the main grid can be managed by the central control by providing appropriate voltage and frequency references for the secondary control layer.

The MGCC plays a crucial role in ensuring the smooth operation of MGs during major disturbances, such as the transition between grid-connected and islanded operation modes. MGs should take advantage of operating in islanding mode during power outages, blackouts, or emergency conditions in the main grid. In emergency conditions, an immediate change in the output power control of the MG is required, as it changes from power control strategy to frequency/voltage control in islanding mode. The islanding plan can be considered the most important emergency control scheme in MGs. To ensure system survival following islanding, controllable DERs, local loads, load shedding schemes, and special protection plans must be exploited in a cooperative manner [5].

Load shedding is an effective emergency control scheme in MGs that should be implemented following a significant drop in frequency and/or voltage. When an MG lacks generation and storage power to restore its frequency and/or voltage to an acceptable range following highly disruptive disturbances, load shedding must be applied. This process must be fast enough to prevent system collapse. Once a shortage is detected following a severe disturbance, the load shedding scheme starts operating and sheds the load that cannot be served. The load shedding strategy should be designed to respond quickly, shed the minimum necessary load, and minimize outages for electricity users.

5.8 Global Control

Global control is the highest level of control used to coordinate the long-term operation of an MG, including economic issues and electricity markets, with the main grid and other MGs. At the global control level, the MG can be controlled to interact with the distribution grid as a dispatchable and constant impedance load. This level of control is responsible for optimizing the MG's operation and defining its interaction with the distribution network and neighboring MGs by controlling the active and reactive power references of DERs through the MGCC and the lower control layers.

The MGCC and global control work together to manage the exchange of power between the MG and other MGs, as well as the main grid, and to provide ancillary services to the main grid. Global control also supervises the MGs market activities, including buying and selling active and reactive power to the grid, as well as possible network congestion, not only within the MG itself but also by transferring energy to nearby feeders of the distribution network and other MGs.

5.9 Summary

This chapter provides an overview of control challenges, strategies, and existing methodologies in MGs. The control strategies for power converters as main components for MG control are introduced in detail. Hierarchical control systems used in MGs are classified, and the control functions and features at different levels are described. The MG's hierarchical control structure consists of four control layers: primary, secondary, central, and global controls. The primary control manages the internal voltage/frequency and current control loops for each DER, while the secondary control is responsible for voltage and frequency regulation, power sharing, and power quality improvement. The central control deals with energy management, optimal control of MGs in islanded mode, emergency control, and overall protection schemes. Lastly, the global control supervises power flow between the given MG, other interconnecting MGs, and the main grid for economic optimal operation.

The primary and secondary controls are described in detail, with a focus on droop control and virtual inertia for primary control and various functions and strategies for secondary control. Simulation examples support the presented topics.

References

1 Katsuhiko Ogata. *Modern control engineering*, volume 5. Prentice Hall, Upper Saddle River, NJ, 2010.

2 Daniel E Olivares, Ali Mehrizi-Sani, Amir H Etemadi, Claudio A Cañizares, Reza Iravani, Mehrdad Kazerani, Amir H Hajimiragha, Oriol Gomis-Bellmunt, Maryam Saeedifard, Rodrigo Palma-Behnke, Guillermo A Jiménez-Estévez, and Nikos D Hatziargyriou. Trends in microgrid control. *IEEE Trans. Smart Grid*, 5(4):1905–1919, 2014. doi: 10.1109/TSG.2013.2295514.

3 Farzam Nejabatkhah, Yun Wei Li, Hao Liang, and Rouzbeh Reza Ahrabi. Cyber-security of smart microgrids: A survey. *Energies*, 14(1):27, 2021. ISSN 1996-1073. doi: 10.3390/en14010027. URL https://www.mdpi.com/1996-1073/14/1/27.

4 Carlos Bordons, Félix Garcia-Torres, and Miguel A Ridao. *Model predictive control of microgrids*, volume 358. Springer, 2020.

5 Hassan Bevrani, Bruno François, and Toshifumi Ise. *Microgrid dynamics and control*. John Wiley & Sons, Hoboken, NJ, 2017.

6 Joan Rocabert, Alvaro Luna, Frede Blaabjerg, and Pedro Rodriguez. Control of power converters in AC microgrids. *IEEE Trans. Power Electron.*, 27(11):4734–4749, 2012.

7 V I Utkin. Sliding mode control design principles and applications to electric drives. *IEEE Trans. Ind. Electron.*, 40(1):23–36, 1993. doi: 10.1109/41.184818.

8 Samir Kouro, Patricio Cortes, RenÉ Vargas, Ulrich Ammann, and JosÉ Rodriguez. Model predictive control—a simple and powerful method to control power converters. *IEEE Trans. Ind. Electron.*, 56(6):1826–1838, 2009. doi: 10.1109/TIE.2008.2008349.

9 M P Kazmierkowski and L Malesani. Current control techniques for three-phase voltage-source PWM converters: A survey. *IEEE Trans. Ind. Electron.*, 45(5):691–703, 1998. doi: 10.1109/41.720325.

10 Sizhan Zhou, Jinjun Liu, Linyuan Zhou, and Yan Zhang. DQ current control of voltage source converters with a decoupling method based on preprocessed reference current feed-forward. *IEEE Trans. Power Electron.*, 32(11):8904–8921, 2017. doi: 10.1109/TPEL.2017.2651139.

11 Chen Zhang, Sjur Føyen, Jon Are Suul, and Marta Molinas. Modeling and analysis of SOGI-PLL/FLL-based synchronization units: Stability impacts of different frequency-feedback paths. *IEEE Trans. Energy Convers.*, 36(3):2047–2058, 2021. doi: 10.1109/TEC.2020.3041797.

12 Saeed Golestan, Josep M Guerrero, and Juan C Vasquez. Three-phase PLLs: A review of recent advances. *IEEE Trans. Power Electron.*, 32(3):1894–1907, 2017. doi: 10.1109/TPEL.2016.2565642.

13 Hao Yi, Xiongfei Wang, Frede Blaabjerg, and Fang Zhuo. Impedance analysis of SOGI-FLL-based grid synchronization. *IEEE Trans. Power Electron.*, 32(10):7409–7413, 2017. doi: 10.1109/TPEL.2017.2673866.

14 Yashen Lin, Joseph H Eto, Brian B Johnson, Jack D Flicker, Robert H Lasseter, Hugo N Villegas Pico, Gab-Su Seo, Brian J Pierre, and Abraham Ellis. Research roadmap on grid-forming inverters. Technical report, National Renewable Energy Lab. (NREL), Golden, CO (United States), 2020.

15 H Bevrani, T Kato, T Ise, and K Inoue. *Grid connected converters: Modeling, stability and control.* Elsevier Science, 2022. ISBN 9780323999021.

16 Josep M Guerrero, Juan C Vasquez, José Matas, Luis García De Vicuña, and Miguel Castilla. Hierarchical control of droop-controlled AC and DC microgrids—a general approach toward standardization. *IEEE Trans. Ind. Electron.*, 58(1):158–172, 2011.

17 Tomislav Dragičević. Model predictive control of power converters for robust and fast operation of AC microgrids. *IEEE Trans. Power Electron.*, 33(7):6304–6317, 2018.

18 Xiongfei Wang, Yun Wei Li, Frede Blaabjerg, and Poh Chiang Loh. Virtual-impedance-based control for voltage-source and current-source converters. *IEEE Trans. Power Electron.*, 30(12):7019–7037, 2015.

19 Hassan Bevrani, Toshifumi Ise, and Yushi Miura. Virtual synchronous generators: A survey and new perspectives. *Int. J. Electr. Power Energy Syst.*, 54:244–254, 2014. ISSN 0142-0615. doi: 10.1016/j.ijepes.2013.07.009. URL https://www.sciencedirect.com/science/article/pii/S0142061513003062.

20 Brian B Johnson, Sairaj V Dhople, Abdullah O Hamadeh, and Philip T Krein. Synchronization of parallel single-phase inverters with virtual oscillator control. *IEEE Trans. Power Electron.*, 29 (11):6124–6138, 2014. doi: 10.1109/TPEL.2013.2296292.

21 Qing-Chang Zhong and George Weiss. Synchronverters: Inverters that mimic synchronous generators. *IEEE Trans. Ind. Electron.*, 58(4):1259–1267, 2011. doi: 10.1109/TIE.2010.2048839.

22 Qobad Shafiee, Josep M Guerrero, and Juan C Vasquez. Distributed secondary control for islanded microgrids—a novel approach. *IEEE Trans. Power Electron.*, 29(2):1018–1031, 2014.

23 Mehrdad Yazdanian and Ali Mehrizi-Sani. Distributed control techniques in microgrids. *IEEE Trans. Smart Grid*, 5(6):2901–2909, 2014.

24 Yousef Khayat, Qobad Shafiee, Rasool Heydari, Mobin Naderi, Tomislav Dragičević, John W Simpson-Porco, Florian Dörfler, Mohammad Fathi, Frede Blaabjerg, Josep M Guerrero, et al. On the secondary control architectures of AC microgrids: An overview. *IEEE Trans. Power Electron.*, 35(6):6482–6500, 2019. doi: 10.1109/TPEL.2019.2951694.

25 Hassan Bevrani. *Robust power system frequency control.* Springer, 2014.

26 Jia Liu, Yushi Miura, and Toshifumi Ise. Comparison of dynamic characteristics between virtual synchronous generator and droop control in inverter-based distributed generators. *IEEE Trans. Power Electron.*, 31(5):3600–3611, 2016. doi: 10.1109/TPEL.2015.2465852.

27 Chengshan Wang, Hao Yu, Peng Li, Jianzhong Wu, and Chengdi Ding. Model order reduction for transient simulation of active distribution networks. *IET Gener. Transm. Distrib.*, 9(5):457–467, 2015.

28 Jiefeng Hu, Yinghao Shan, Ka Wai Cheng, and Syed Islam. Overview of power converter control in microgrids—challenges, advances, and future trends. *IEEE Trans. Power Electron.*, 37(8):9907–9922, 2022. doi: 10.1109/TPEL.2022.3159828.

29 Xin Meng, Jinjun Liu, and Zeng Liu. A generalized droop control for grid-supporting inverter based on comparison between traditional droop control and virtual synchronous generator control. *IEEE Trans. Power Electron.*, 34(6):5416–5438, 2019. doi: 10.1109/TPEL.2018.2868722.

30 Karel De Brabandere, Bruno Bolsens, Jeroen Van den Keybus, Achim Woyte, Johan Driesen, and Ronnie Belmans. A voltage and frequency droop control method for parallel inverters. *IEEE Trans. Power Electron.*, 22(4):1107–1115, 2007. doi: 10.1109/TPEL.2007.900456.

31 Hassan Bevrani and Shoresh Shokoohi. An intelligent droop control for simultaneous voltage and frequency regulation in islanded microgrids. *IEEE Trans. Smart Grid*, 4(3):1505–1513, 2013.

32 Ernane Antonio Alves Coelho, Porfirio Cabaleiro Cortizo, and Pedro Francisco Donoso Garcia. Small-signal stability for parallel-connected inverters in stand-alone AC supply systems. *IEEE Trans. Ind. Appl.*, 38(2):533–542, 2002.

33 Nagaraju Pogaku, Milan Prodanovic, and Timothy C Green. Modeling, analysis and testing of autonomous operation of an inverter-based microgrid. *IEEE Trans. Power Electron.*, 22(2):613–625, 2007.

34 M C Chandorkar, D M Divan, and R Adapa. Control of parallel connected inverters in standalone AC supply systems. *IEEE Trans. Ind. Appl.*, 29(1):136–143, 1993. doi: 10.1109/28.195899.

35 A Mehrizi-Sani. Chapter 2 - Distributed control techniques in microgrids. In Magdi S Mahmoud, editor, *Microgrid*, pages 43–62. Butterworth-Heinemann, 2017. ISBN 978-0-08-101753-1. doi: 10.1016/B978-0-08-101753-1.00002-4. URL https://www.sciencedirect.com/science/article/pii/B9780081017531000024.

36 Yang Han, Hong Li, Pan Shen, Ernane Antônio Alves Coelho, and Josep M Guerrero. Review of active and reactive power sharing strategies in hierarchical controlled microgrids. *IEEE Trans. Power Electron.*, 32(3):2427–2451, 2017. doi: 10.1109/TPEL.2016.2569597.

37 Yasser Abdel-Rady Ibrahim Mohamed and Ehab F El-Saadany. Adaptive decentralized droop controller to preserve power sharing stability of paralleled inverters in distributed generation microgrids. *IEEE Trans. Power Electron.*, 23(6):2806–2816, 2008. doi: 10.1109/TPEL.2008.2005100.

38 Qing-Chang Zhong. Robust droop controller for accurate proportional load sharing among inverters operated in parallel. *IEEE Trans. Ind. Electron.*, 60(4):1281–1290, 2013. doi: 10.1109/TIE.2011.2146221.

39 Mohammad B Delghavi and Amirnaser Yazdani. An adaptive feedforward compensation for stability enhancement in droop-controlled inverter-based microgrids. *IEEE Trans. Power Delivery*, 26 (3):1764–1773, 2011. doi: 10.1109/TPWRD.2011.2119497.

40 Jaehong Kim, Josep M Guerrero, Pedro Rodriguez, Remus Teodorescu, and Kwanghee Nam. Mode adaptive droop control with virtual output impedances for an inverter-based flexible AC microgrid. *IEEE Trans. Power Electron.*, 26(3):689–701, 2011. doi: 10.1109/TPEL.2010.2091685.

41 Mohammad S Golsorkhi and Dylan D C Lu. A control method for inverter-based islanded microgrids based on V-I droop characteristics. *IEEE Trans. Power Delivery*, 30(3):1196–1204, 2015. doi: 10.1109/TPWRD.2014.2357471.

42 A Tuladhar, Hua Jin, T Unger, and K Mauch. Control of parallel inverters in distributed AC power systems with consideration of line impedance effect. *IEEE Trans. Ind. Appl.*, 36(1):131–138, 2000. doi: 10.1109/28.821807.

43 Juan C Vasquez, Josep M Guerrero, Alvaro Luna, Pedro Rodriguez, and Remus Teodorescu. Adaptive droop control applied to voltage-source inverters operating in grid-connected and islanded modes. *IEEE Trans. Ind. Electron.*, 56(10):4088–4096, 2009. doi: 10.1109/TIE.2009.2027921.

44 Yun Wei Li and Ching-Nan Kao. An accurate power control strategy for power-electronics-interfaced distributed generation units operating in a low-voltage multibus microgrid. *IEEE Trans. Power Electron.*, 24(12):2977–2988, 2009. doi: 10.1109/TPEL.2009.2022828.

45 Magdi S Mahmoud. *Microgrid: Advanced control methods and renewable energy system integration*. Elsevier, 2016.

46 Xiaoxiao Yu, Ashwin M Khambadkone, Huanhuan Wang, and Siew Tuck Sing Terence. Control of parallel-connected power converters for low-voltage microgrid—Part I: A hybrid control architecture. *IEEE Trans. Power Electron.*, 25(12):2962–2970, 2010. doi: 10.1109/TPEL.2010.2087393.

47 J M Guerrero, Luis Garcia de Vicuna, J Matas, M Castilla, and J Miret. Output impedance design of parallel-connected UPS inverters with wireless load-sharing control. *IEEE Trans. Ind. Electron.*, 52(4):1126–1135, 2005. doi: 10.1109/TIE.2005.851634.

48 Hisham Mahmood, Dennis Michaelson, and Jin Jiang. Accurate reactive power sharing in an islanded microgrid using adaptive virtual impedances. *IEEE Trans. Power Electron.*, 30(3):1605–1617, 2015. doi: 10.1109/TPEL.2014.2314721.

49 Xiaodong Liang, Chowdhury Andalib-Bin-Karim, Weixing Li, Massimo Mitolo, and Md Nasmus Sakib Khan Shabbir. Adaptive virtual impedance-based reactive power sharing in virtual synchronous generator controlled microgrids. *IEEE Trans. Ind. Appl.*, 57(1):46–60, 2021. doi: 10.1109/TIA.2020.3039223.

50 A S Vijay, N Parth, Suryanarayana Doolla, and Mukul C Chandorkar. An adaptive virtual impedance control for improving power sharing among inverters in islanded AC microgrids. *IEEE Trans. Smart Grid*, 12(4):2991–3003, 2021. doi: 10.1109/TSG.2021.3062391.

51 Moudud Ahmed, Lasantha Meegahapola, Arash Vahidnia, and Manoj Datta. Adaptive virtual impedance controller for parallel and radial microgrids with varying X/R ratios. *IEEE Trans. Sustainable Energy*, 13(2):830–843, 2022. doi: 10.1109/TSTE.2021.3133413.

52 Fei Deng, Xiangke Li, Xiaobin Zhang, and Paolo Mattavelli. An iterative virtual impedance regulation strategy in islanded microgrids for enhanced balanced, unbalanced and harmonic current sharing. *IEEE Trans. Sustainable Energy*, 13(1):514–526, 2022. doi: 10.1109/TSTE.2021.3118555.

53 Fei Deng, Wenli Yao, Xiaobin Zhang, and Paolo Mattavelli. A decentralized current sharing strategy for islanded resistive microgrids based on iterative virtual impedance regulation. *IEEE Trans. Ind. Inf.*, 18(6):3958–3969, 2022. doi: 10.1109/TII.2021.3110951.

54 Xiangyu Wu, Chen Shen, and Reza Iravani. Feasible range and optimal value of the virtual impedance for droop-based control of microgrids. *IEEE Trans. Smart Grid*, 8(3):1242–1251, 2017. doi: 10.1109/TSG.2016.2519454.

55 Jinwei He, Yun Wei Li, Josep M Guerrero, Frede Blaabjerg, and Juan C Vasquez. An islanding microgrid power sharing approach using enhanced virtual impedance control scheme. *IEEE Trans. Power Electron.*, 28(11):5272–5282, 2013. doi: 10.1109/TPEL.2013.2243757.

56 Fei Deng, Wenli Yao, Xiaobin Zhang, Yi Tang, and Paolo Mattavelli. Review of impedance-reshaping-based power sharing strategies in islanded AC microgrids. *IEEE Trans. Smart Grid*, 14(3):1692–1707, 2022. doi: 10.1109/TSG.2022.3208752.

57 Mohsen Eskandari, Li Li, Mohammad Hassan Moradi, Pierluigi Siano, and Frede Blaabjerg. Active power sharing and frequency restoration in an autonomous networked microgrid. *IEEE Trans. Power Syst.*, 34(6):4706–4717, 2019. doi: 10.1109/TPWRS.2019.2923797.

58 Hua Han, Xiaochao Hou, Jian Yang, Jifa Wu, Mei Su, and Josep M Guerrero. Review of power sharing control strategies for islanding operation of AC microgrids. *IEEE Trans. Smart Grid*, 7(1):200–215, 2015. doi: 10.1109/TSG.2015.2434849.

59 John W Simpson-Porco, Qobad Shafiee, Florian Dörfler, Juan C Vasquez, Josep M Guerrero, and Francesco Bullo. Secondary frequency and voltage control of islanded microgrids via distributed averaging. *IEEE Trans. Ind. Electron.*, 62(11):7025–7038, 2015.

60 Florian Dörfler, John W Simpson-Porco, and Francesco Bullo. Breaking the hierarchy: Distributed control and economic optimality in microgrids. *IEEE Trans. Control Netw. Syst.*, 3(3):241–253, 2016.

61 S Manaffam, M K Talebi, A K Jain, and A Behal. Synchronization in networks of identical systems via pinning: Application to distributed secondary control of microgrids. *IEEE Trans. Control Syst. Technol.*, 25(6):2227–2234, 2017.

62 Fen Tang, Josep M Guerrero, Juan C Vasquez, Dan Wu, and Lexuan Meng. Distributed active synchronization strategy for microgrid seamless reconnection to the grid under unbalance and harmonic distortion. *IEEE Trans. Smart Grid*, 6(6):2757–2769, 2015.

63 Kuntal Mandal and Soumitro Banerjee. Synchronization phenomena in interconnected power electronic systems. *IEEE Trans. Circuits Syst. Express Briefs*, 63(2):221–225, 2016.

64 Shankar Abhinav, Ioannis D Schizas, Frank Ferrese, and Ali Davoudi. Optimization-based AC microgrid synchronization. *IEEE Trans. Ind. Inf.*, 13(5):2339–2349, 2017.

65 Changhee Cho, Jin-Hong Jeon, Jong-Yul Kim, Soonman Kwon, Kyongyop Park, and Sungshin Kim. Active synchronizing control of a microgrid. *IEEE Trans. Power Electron.*, 26(12):3707–3719, 2011.

66 Kuntal Mandal and Soumitro Banerjee. Synchronization phenomena in microgrids with capacitive coupling. *IEEE J. Emerging Sel. Top. Circuits Syst.*, 5(3):364–371, 2015.

67 Seyed Abbas Taher, Mahdi Zolfaghari, Changhee Cho, Mehrdad Abedi, and Mohammad Shahidehpour. A new approach for soft synchronization of microgrid using robust control theory. *IEEE Trans. Power Delivery*, 32(3):1370–1381, 2017. doi: 10.1109/TPWRD.2016.2596106.

68 Shankar Abhinav, Hamidreza Modares, Frank L Lewis, Frank Ferrese, and Ali Davoudi. Synchrony in networked microgrids under attacks. *IEEE Trans. Smart Grid*, 9(6):6731–6741, 2017.

69 Qobad Shafiee, Čedomir Stefanović, Tomislav Dragičević, Petar Popovski, Juan C Vasquez, and Josep M Guerrero. Robust networked control scheme for distributed secondary control of islanded microgrids. *IEEE Trans. Ind. Electron.*, 61(10):5363–5374, 2014.

70 Qobad Shafiee, Vahidreza Nasirian, Juan C Vasquez, Josep M Guerrero, and Ali Davoudi. A multi-functional fully distributed control framework for AC microgrids. *IEEE Trans. Smart Grid*, 9(4):3247–3258, 2018.

71 Guannan Lou, Wei Gu, Yinliang Xu, Ming Cheng, and Wei Liu. Distributed MPC-based secondary voltage control scheme for autonomous droop-controlled microgrids. *IEEE Trans. Sustainable Energy*, 8(2):792–804, 2017. doi: 10.1109/TSTE.2016.2620283.

72 Shichao Liu, Xiaoyu Wang, and Peter Xiaoping Liu. Impact of communication delays on secondary frequency control in an islanded microgrid. *IEEE Trans. Ind. Electron.*, 62(4):2021–2031, 2015.

73 Alexander Micallef, Maurice Apap, Cyril Spiteri-Staines, Josep M Guerrero, and Juan C Vasquez. Reactive power sharing and voltage harmonic distortion compensation of droop controlled single phase islanded micro-grids. *IEEE Trans. Smart Grid*, 5(3):1149–1158, 2014.

74 Mehdi Savaghebi, Alireza Jalilian, Juan C Vasquez, and Josep M Guerrero. Secondary control for voltage quality enhancement in microgrids. *IEEE Trans. Smart Grid*, 3(4):1893–1902, 2012.

75 Reza Olfati-Saber, J Alex Fax, and Richard M Murray. Consensus and cooperation in networked multi-agent systems. *Proc. IEEE*, 95(1):215–233, 2007.

76 Johannes Schiffer, Florian Dörfler, and Emilia Fridman. Robustness of distributed averaging control in power systems: Time delays & dynamic communication topology. *Automatica*, 80:261–271, 2017.

77 Stefano Riverso, Michele Tucci, Juan C Vasquez, Josep M Guerrero, and Giancarlo Ferrari-Trecate. Stabilizing plug-and-play regulators and secondary coordinated control for AC islanded microgrids with bus-connected topology. *Appl. Energy*, 210:914–924, 2018.

78 F Dörfler and S Grammatico. Gather-and-broadcast frequency control in power systems. *Automatica*, 79:296–305, 2017.

79 Johannes Schiffer, Emilia Fridman, Romeo Ortega, and Jörg Raisch. Stability of a class of delayed port-Hamiltonian systems with application to microgrids with distributed rotational and electronic generation. *Automatica*, 74:71–79, 2016.

80 Ernane Antônio Alves Coelho, Dan Wu, Josep M Guerrero, Juan C Vasquez, Tomislav Dragicevic, Cedomir Stefanovic, and Petar Popovski. Small-signal analysis of the microgrid secondary control considering a communication time delay. *IEEE Trans. Ind. Electron.*, 63(10):6257–6269, 2016.

81 John W Simpson-Porco, Florian Dörfler, and Francesco Bullo. Synchronization and power sharing for droop-controlled inverters in islanded microgrids. *Automatica*, 49(9):2603–2611, 2013.

82 Martin Andreasson, D V Dimarogonas, K H Johansson, and Henrik Sandberg. Distributed vs. centralized power systems frequency control under unknown load changes. In *European Control Conference*, pages 3524–3529, Zürich, Switzerland, July 2013.

83 S T Cady, A D Domínguez-García, and C N Hadjicostis. A distributed generation control architecture for islanded AC microgrids. *IEEE Trans. Control Syst. Technol.*, 23(5):1717–1735, 2015.

84 E Tegling, M Andreasson, J W Simpson-Porco, and H Sandberg. Improving performance of droop-controlled microgrids through distributed PI-control. In *American Control Conference*, pages 2321–2327, Boston, MA, USA, July 2016. doi: 10.1109/ACC.2016.7525264.

85 M Andreasson, E Tegling, H Sandberg, and K H Johansson. Coherence in synchronizing power networks with distributed integral control. In *2017 IEEE 56th Annual Conference on Decision and Control (CDC)*, pages 6327–6333, Melbourne, Australia, December 2017. doi: 10.1109/CDC.2017.8264613.

86 H Flamme, E Tegling, and H Sandberg. Performance limitations of distributed integral control in power networks under noisy measurements. In *2018 Annual American Control Conference (ACC)*, pages 5380–5386, Milwaukee, WI, USA, June 2018. doi: 10.23919/ACC.2018.8431122.

87 Yousef Khayat, Qobad Shafiee, Rasool Heydari, Mobin Naderi, Tomislav Dragičević, John W Simpson-Porco, Florian Dörfler, Mohammad Fathi, Frede Blaabjerg, Josep M Guerrero, and Hassan Bevrani. On the secondary control architectures of AC microgrids: An overview. *IEEE Trans. Power Electron.*, 35(6):6482–6500, 2020. doi: 10.1109/TPEL.2019.2951694.

88 Wei Liu, Wei Gu, Qiang Huang, Liang Chen, and Xiaodong Yuan. Pinning group consensus-based distributed coordination control for active distribution systems. *IEEE Access*, 6:2330–2339, 2018.

89 Nelson Fabian Avila and Chia-Chi Chu. Distributed pinning droop control in isolated AC microgrids. *IEEE Trans. Ind. Appl.*, 53(4):3237–3249, 2017.

90 Saeed Manaffam, Morteza Talebi, Amit K Jain, and Aman Behal. Intelligent pinning based cooperative secondary control of distributed generators for microgrid in islanding operation mode. *IEEE Trans. Power Syst.*, 33(2):1364–1373, 2018.

91 Renke Han, Lexuan Meng, Giancarlo Ferrari-Trecate, Ernane Antonio Alves Coelho, Juan C Vasquez, and Josep M Guerrero. Containment and consensus-based distributed coordination control to achieve bounded voltage and precise reactive power sharing in islanded AC microgrids. *IEEE Trans. Ind. Appl.*, 53(6):5187–5199, 2017.

92 Mobin Naderi, Yousef Khayat, Qobad Shafiee, Frede Blaabjerg, and Hassan Bevrani. Dynamic modeling, stability analysis and control of interconnected microgrids: A review. *Appl. Energy*, 334:120647., 2023.

93 Shengxuan Weng, Dong Yue, Chunxia Dou, Jing Shi, and Chongxin Huang. Distributed event-triggered cooperative control for frequency and voltage stability and power sharing in isolated inverter-based microgrid. *IEEE Trans. Cybern.*, 49(4):1427–1439, 2018.

94 Jing Shi, Dong Yue, and Shengxuan Weng. Distributed event-triggered mechanism for secondary voltage control with microgrids. *Trans. Inst. Meas. Control*, 41(6), 2018. doi: 10.1177/0142331218770715.

95 Xiaohua Ge and Qing-Long Han. Distributed formation control of networked multi-agent systems using a dynamic event-triggered communication mechanism. *IEEE Trans. Ind. Electron.*, 64(10):8118–8127, 2017.

96 Yuan Fan, Guoqiang Hu, and Magnus Egerstedt. Distributed reactive power sharing control for microgrids with event-triggered communication. *IEEE Trans. Control Syst. Technol.*, 25(1):118–128, 2017.

97 Magdi S Mahmoud and Mohamed Saif Ur Rahman. Event triggered of microgrid control with communication and control optimization. *J. Franklin Inst.*, 353(16):4114–4132, 2016.

98 Chun-xia Dou, Bin Liu, and Josep M Guerrero. Event-triggered hybrid control based on multi-agent system for microgrids. *IET Gener. Transm. Distrib.*, 8(12):1987–1997, 2014.

99 Babak Abdolmaleki, Alireza Seifi, Mohammad Mehdi Arefi, and Qobad Shafiee. Event-triggered voltage control of inverter-based microgrids. In *2018 9th Annual Power Electronics, Drives Systems and Technologies Conference (PEDSTC)*, pages 522–528. IEEE, 2018.

100 C X Dou, Bin Liu, and Josep M Guerrero. MAS based event-triggered hybrid control for smart microgrids. In *IECON 2013 - 39th Annual Conference of the IEEE Industrial Electronics Society*, pages 1712–1717. IEEE, 2013.

101 Yuan Fan, Mingwei Sheng, Chuanbao Dong, and Yang Zhang. Sampling-based event-triggered control for distributed generators. In *2017 29th Chinese Control and Decision Conference (CCDC)*, pages 5556–5560. IEEE, 2017.

102 Yuan Fan, Chengxiao Zhang, and Cheng Song. Sampling-based self-triggered coordination control for multi-agent systems with application to distributed generators. *Int. J. Syst. Sci.*, 49(15):3048–3062, 2018.

103 Meng Chen, Xiangning Xiao, and Josep M Guerrero. Secondary restoration control of islanded microgrids with a decentralized event-triggered strategy. *IEEE Trans. Ind. Inf.*, 14(9):3870–3880, 2018.

104 Meng Chen and Xiangning Xiao. Secondary voltage control in islanded microgrids using event-triggered control. *IET Gener. Transm. Distrib.*, 12(8):1872–1878, 2018.

105 W P M H Heemels, Karl Henrik Johansson, and Paulo Tabuada. An introduction to event-triggered and self-triggered control. In *2012 IEEE 51st IEEE Conference on Decision and Control (CDC)*, pages 3270–3285. IEEE, 2012.

106 W P M H Heemels, Romain Postoyan, M C F Donkers, Andrew R Teel, Adolfo Anta, Paulo Tabuada, and Dragan Nešić. Periodic event-triggered control. *Event-based control and signal processing*, pages 105–119, 2015.

107 W P M Heemels Heemels, M C F Donkers, and Andrew R Teel. Periodic event-triggered control for linear systems. *IEEE Trans. Autom. Control*, 58(4):847–861, 2013.

108 Hao Zhang, Gang Feng, Huaicheng Yan, and Qijun Chen. Observer-based output feedback event-triggered control for consensus of multi-agent systems. *IEEE Trans. Ind. Electron.*, 61(9):4885–4894, 2014.

109 Antoine Girard. Dynamic triggering mechanisms for event-triggered control. *IEEE Trans. Autom. Control*, 60(7):1992–1997, 2015.

110 Dong Yue, Engang Tian, and Qing-Long Han. A delay system method for designing event-triggered controllers of networked control systems. *IEEE Trans. Autom. Control*, 58(2):475–481, 2013.

111 Xian-Ming Zhang, Qing-Long Han, and Bao-Lin Zhang. An overview and deep investigation on sampled-data-based event-triggered control and filtering for networked systems. *IEEE Trans. Ind. Inf.*, 13(1):4–16, 2017.

112 Anton Selivanov and Emilia Fridman. Event-triggered H_∞ control: A switching approach. *IEEE Trans. Autom. Control*, 61(10):3221–3226, 2016.

113 Jiho Han and Deog-Kyoon Jeong. A practical implementation of IEEE 1588-2008 transparent clock for distributed measurement and control systems. *IEEE Trans. Instrum. Meas.*, 59(2):433–439, 2010.

114 John Eidson and Kang Lee. IEEE 1588 standard for a precision clock synchronization protocol for networked measurement and control systems. In *2nd ISA/IEEE Sensors for Industry Conference*, pages 98–105. IEEE, July 2002.

115 Luca Schenato and Federico Fiorentin. Average timesynch: A consensus-based protocol for clock synchronization in wireless sensor networks. *Automatica*, 47(9):1878–1886, 2011.

116 Chao Lu and Alvin Warren Czanderna. *Applications of piezoelectric quartz crystal microbalances*, volume 7. Elsevier, 2012.

117 Johannes Schiffer, Romeo Ortega, Christian A Hans, and Jörg Raisch. Droop-controlled inverter-based microgrids are robust to clock drifts. In *Proceedings of the American Control Conference*, pages 2341–2346. IEEE, 2015.

118 Pau Martí, Manel Velasco, Enric Xavier Martín, Luis García de Vicuña, Jaume Miret, and Miguel Castilla. Performance evaluation of secondary control policies with respect to digital communications properties in inverter-based islanded microgrids. *IEEE Trans. Smart Grid*, 9(3):2192–2202, 2018.

119 Pau Marti, Javier Torres-Martinez, Carlos Xavier Rosero, Manel Velasco, Jaume Miret, and Miguel Castilla. Analysis of the effect of clock drifts on frequency regulation and power sharing in inverter-based islanded microgrids. *IEEE Trans. Power Electron.*, 33(12):10363–10379, 2018.

120 Yao Sun, Xiaochao Hou, Jian Yang, Hua Han, Mei Su, and Josep M Guerrero. New perspectives on droop control in AC microgrid. *IEEE Trans. Ind. Electron.*, 64(7):5741–5745, 2017.

121 Yang Han, Hong Li, Lin Xu, Xin Zhao, and Josep M Guerrero. Analysis of washout filter-based power sharing strategy—an equivalent secondary controller for islanded microgrid without LBC lines. *IEEE Trans. Smart Grid*, 9(5):4061–4076, 2018.

122 Jinghang Lu, Mehdi Savaghebi, and Josep M Guerrero. Second order washout filter based power sharing strategy for uninterruptible power supply. In *IECON 2017 - 43rd Annual Conference of the IEEE Industrial Electronics Society*, pages 7854–7859. IEEE, 2017.

123 Erik Weitenberg, Yan Jiang, Changhong Zhao, Enrique Mallada, Claudio De Persis, and Florian Dorfler. Robust decentralized secondary frequency control in power systems: Merits and trade-offs. *IEEE Trans. Autom. Control*, 64(10):3967–3982, 2018.

124 Juan M Rey, Pau Martí, Manel Velasco, Jaume Miret, and Miguel Castilla. Secondary switched control with no communications for islanded microgrids. *IEEE Trans. Ind. Electron.*, 64(11):8534–8545, 2017.

125 Yousef Khayat, Mobin Naderi, Qobad Shafiee, Yazdan Batmani, Mohammad Fathi, Josep M Guerrero, and Hassan Bevrani. Decentralized optimal frequency control in autonomous microgrids. *IEEE Trans. Power Syst.*, 34(3):2345–2353, 2018. doi: 10.1109/TPWRS.2018.2889671.

126 Nan Xia, Hoay Beng Gooi, Shuaixun Chen, and Wuhua Hu. Decentralized state estimation for hybrid AC/DC microgrids. *IEEE Syst. J.*, 12(1):434–443, 2018.

127 Md Masud Rana and Li Li. An overview of distributed microgrid state estimation and control for smart grids. *Sensors*, 15(2):4302–4325, 2015.

128 Arvin Morattab, Ouassima Akhrif, and Maarouf Saad. Decentralised coordinated secondary voltage control of multi-area power grids using model predictive control. *IET Gener. Transm. Distrib.*, 11(18):4546–4555, 2017.

129 Guannan Lou, Wei Gu, Liufang Wang, Bin Xu, Ming Wu, and Wanxing Sheng. Decentralised secondary voltage and frequency control scheme for islanded microgrid based on adaptive state estimator. *IET Gener. Transm. Distrib.*, 11(15):3683–3693, 2017.

130 Kaveh Dehghanpour, Zhaoyu Wang, Jianhui Wang, Yuxuan Yuan, and Fankun Bu. A survey on state estimation techniques and challenges in smart distribution systems. *IEEE Trans. Smart Grid*, 10(2):2312–2322, 2018.

131 Daniel A Haughton and Gerald Thomas Heydt. A linear state estimation formulation for smart distribution systems. *IEEE Trans. Power Syst.*, 28(2):1187–1195, 2013.

132 Siddharth Bhela, Vassilis Kekatos, and Sriharsha Veeramachaneni. Enhancing observability in distribution grids using smart meter data. *IEEE Trans. Smart Grid*, 9(6):5953–5961, 2017.

133 Wei Gu, Guannan Lou, Wen Tan, and Xiaodong Yuan. A nonlinear state estimator-based decentralized secondary voltage control scheme for autonomous microgrids. *IEEE Trans. Power Syst.*, 32(6):4794–4804, 2017.

134 Matlab codes for the microgrids dynamic modeling, stability and control book. www.wiley.com/go/shafiee/microgrids. Accessed: 2023-07-07.

135 Yoash Levron, Josep M Guerrero, and Yuval Beck. Optimal power flow in microgrids with energy storage. *IEEE Trans. Power Syst.*, 28(3):3226–3234, 2013. doi: 10.1109/TPWRS.2013.2245925.

6

Advances in Microgrid Control

6.1 Introduction

Microgrid (MG) controllers are typically designed using reduced-order linearized models that are centered around the system's operating points for different control layers. However, in an MG that employs converter-based distributed energy resources (DERs), its main dynamic modes tend to vary stochastically due to changes in system parameters. Thus, a fixed linearized time-invariant model may not adequately capture the MG's behavior, and its assumptions may not be valid under such circumstances. Furthermore, associated designed controllers using these simple models may not work well for certain control objectives (e.g. harmonic mitigation) under system nonlinearities, uncertainties, and constraints.

To address this issue, advanced modeling (as presented in Chapters 2 and 3) and control techniques have been proposed to enhance the accuracy and performance of MG controllers. Furthermore, classic control synthesis techniques like proportional-integral (PI) control design methods may not be effective for certain control objectives, prompting the need for advanced control techniques. Different advanced control synthesis methods such as nonlinear control, optimal control, robust control, and artificial intelligence-based control have been proposed and applied to MGs over the past decade. These techniques are better equipped to handle the MG's nonlinear dynamics and system uncertainties and can offer superior performance compared to the classical controllers.

Besides advanced control methods, improved or modified control strategies and new concepts have also been introduced in the literature for some control objectives in the MG control hierarchy. Virtual dynamic control (VDC) is a control synthesis technique that is used to regulate the behavior of power electronics-based systems by controlling the output power of the power converters. This technique is typically implemented in the primary control layer of MGs and allows for the dynamic shaping of the system's behavior to achieve specific control objectives. Virtual impedance and virtual synchronous generator (VSG) are examples of VDCs.

As MGs are considered to be cyber-physical systems (CPSs), it is crucial to ensure effective communication-based control on the upper layers of its control hierarchy. Various advancements and considerations have been proposed in this field, including addressing concerns around communication bandwidth, impairments, and cyber-security. Communication delays and packet loss can affect the performance of the control system, while cyberattacks can compromise the security of the MGs. Techniques such as event-triggered control and consensus-based control have been suggested to reduce the required communication bandwidth for control. Specific detection and mitigation techniques can also be used to enhance the resilience of the control system against cyberattacks. Implementing these techniques is critical to improving the overall effectiveness and reliability of MGs as CPSs.

Microgrids: Dynamic Modeling, Stability and Control, First Edition. Qobad Shafiee, Mobin Naderi, and Hassan Bevrani.

This chapter explores the recent developments in MG control, including cutting-edge methodologies and innovative concepts. The control advancements are divided into three main sections: advanced control synthesis, virtual dynamic control, and resilient and cybersecure control. Advanced control techniques are introduced as alternatives to classical control synthesis methods, and two examples of model predictive control (MPC) are presented to illustrate their applications in MGs. The first example demonstrates the use of MPC for voltage and power control in dc MGs connected to constant power loads (CPLs), while the second example shows the application of MPC for seamless mode transition in ac MGs. Virtual dynamic control is then introduced, along with an example of applying a virtual inertia control loop to dc MGs. The third section introduces the concept of CPSs in MGs and addresses the challenges and advances of such systems. Examples are provided for cyber-security and event-triggered control of ac MGs.

6.2 Advanced Control Synthesis

Classic linear control techniques, such as PI control and proportional control (e.g. conventional droop control), are widely used in power converter-based systems and MGs (see Chapter 5) due to their simplicity and ease of implementation. However, these control techniques are limited in their adaptability to some characteristics of complex MGs. Advanced control synthesis techniques are more sophisticated than linear control methods. They often use model-based control methods and optimization to handle the nonlinearity, uncertainty, and constraints of the systems. They offer higher accuracy, flexibility, and improved MG performance.

The choice of control technique depends on the specific requirements and characteristics of the system. If the system has a high degree of nonlinearity and uncertainty, advanced control synthesis techniques may be necessary to achieve an optimal performance. It is also important to consider the cost and complexity associated with implementing advanced control synthesis techniques. They often require additional sensors, communication infrastructure, and computational resources, which can increase the overall cost of the system. Furthermore, it is worth noting that in practical cases, the simplest designs that meet the MG requirements are prioritized. This is because less complexity results in greater reliability and security, fewer operational challenges, and lower maintenance needs. Therefore, a careful analysis of the benefits and drawbacks of different control techniques is necessary to determine the most appropriate choice for a given MG system. Figure 6.1 shows a very simple implementation of advanced state-feedback control techniques in MGs. The controller generates a control signal based on the measured or estimated states and a specific control law.

6.2.1 Advanced Control Techniques

There are several advanced control techniques that have been widely employed in both ac [1] and dc [2] MG applications to improve their efficiency and reliability. Some of these control techniques

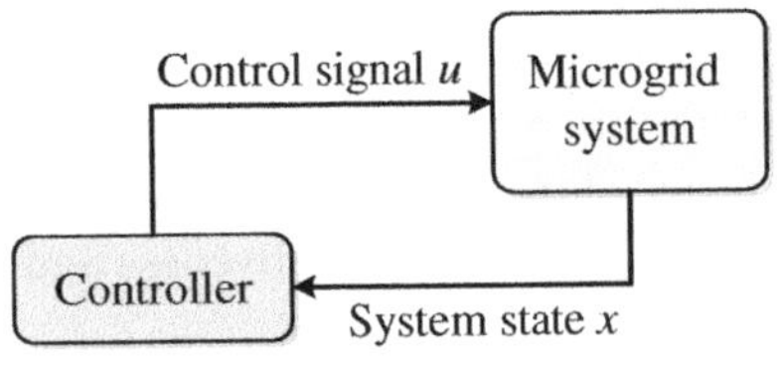

Figure 6.1 Simple implementation of advanced state-feedback control synthesis techniques in microgrids.

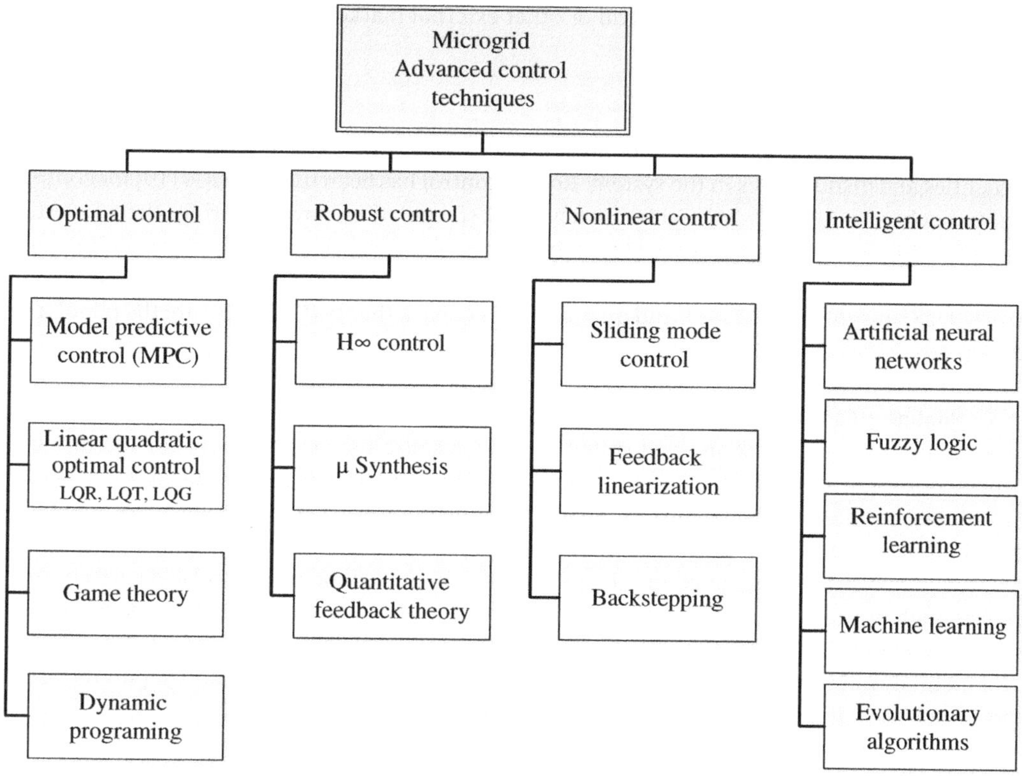

Figure 6.2 Advanced control synthesis techniques used in microgrid applications.

are presented later. A summary of the advanced control techniques used in MG applications is presented in Figure 6.2.

6.2.1.1 Optimal Control

Optimal control is a technique that uses mathematical optimization to find the optimal control actions that minimizes a cost function subject to system constraints. Optimal control has been used in MGs for control objectives such as energy management [3–5], Optimization of economic dispatch [6], energy trade and market operation [7–10], smooth mode transition [11, 12], and voltage and frequency regulation [13–16]. MPC [17], linear quadratic control [18], dynamic programming [3, 4, 6, 15, 16], and game theory [5, 7–10] are the most employed optimal control techniques in MG applications.

Linear quadratic optimal control is a feedback control strategy that minimizes a quadratic cost function subject to linear constraints. Linear quadratic regulator (LQR) can be used to stabilize the voltage and frequency of an MG by adjusting the power output of DERs. On the other hand, linear quadratic tracking (LQT) is a tracking control strategy that minimizes the difference between the actual and desired system outputs. LQT can be used in MGs to track a desired power reference signal or a renewable energy generation profile. By using dynamic programming, the optimization problem can be split into smaller subproblems, and the solutions to these subproblems can be stored and reused to optimize the dispatch strategy. This can result in a more efficient and reliable operation of the MG, reducing costs and improving the resiliency of the system. Game theory can also be used to study the behavior of market participants in MG markets, where the MG operator buys/sells electricity to/from the utility grid or other external markets. MPC is an optimal control method that is widely used in MGs.

6.2.1.2 Robust Control

Robust control is a technique that guarantees stability and performance in the presence of uncertainties and disturbances in the system. Robust control has been used in MGs [19] for control objectives such as voltage [20–24] and frequency [25–28] regulation, power sharing, virtual inertia control [25, 29], fault-tolerant control [30], soft synchronization [31], maintaining stability, and reliability even in fluctuating power conditions. H_∞ control [20–23, 25, 27, 29], LMI-based H_∞ control [24, 28], μ-synthesis [27, 31], and quantitative feedback theory (QFT) [26] are the popularly used methods for MGs.

6.2.1.3 Nonlinear Control

Nonlinear control approaches are well suited to MGs because they can account for the nonlinear interactions and feedback mechanisms between different components of the system. These control strategies offer several benefits for MGs, including improved stability, robustness, and efficiency. Nonlinear control techniques are commonly employed in the primary control layer of MGs to replace the conventional linear control loops as well as in the secondary control layer, for control objectives such as voltage and frequency stabilization [32–38], power sharing [39], power quality improvement [40], and large-signal stability improvement [41, 42]. The most used nonlinear control methods in MGs are based on feedback linearization [32, 33, 36, 37] and Lyapunov-based methods such as sliding mode control [38, 40, 41] and backstepping [34, 35, 42].

6.2.1.4 Intelligent Control

Artificial intelligence (AI) techniques have become increasingly important in the design and operation of MGs [43]. These techniques enable MG operators to optimize the use of renewable energy sources, improve energy efficiency, improve energy management of hybrid energy storage systems, and reduce operating costs. Some of the most commonly used AI techniques in MGs include artificial neural networks [44, 45], fuzzy logic [46–48], evolutionary algorithms [49], machine learning [50], and reinforcement learning [51, 52]. AI techniques can be used for forecasting and prediction, optimization and power management [46], fault detection [44], energy trading [49, 52], and cyber-security [53] in MG applications. These techniques have been used in combination with control techniques such as PI [47], MPC, and optimal control as well as virtual inertia control [48] to tune their design parameters improving the performance and efficiency of MG operation.

Machine learning algorithms can be used to predict energy consumption patterns, forecast renewable energy generation, optimize economic dispatch [50], and cyber-security [54] in MGs. Reinforcement learning algorithms can optimize the operation of microgrids, by learning from the feedback provided by the system and adapting to changing conditions. Artificial neural networks can model complex nonlinear relationships in MGs [45], while fuzzy logic can control the operation of these systems in real time [55], taking into account uncertain or imprecise data. Evolutionary algorithms can optimize the design and operation of MGs by finding the best combinations of energy sources, storage devices, and control strategies. These algorithms are also widely used for tuning the MG controller parameters.

In summary, efficient and reliable operation of MGs requires the use of advanced control techniques. The selection of a suitable control technique depends on the specific control objectives, system requirements, and the availability of system models and measurements. Robust control methods are preferred when the system is subject to uncertainties, while nonlinear techniques are considered when nonlinear dynamics and behaviors pose significant challenges to the MG's performance. Optimal control techniques can be utilized to optimize a performance metric, such as minimizing the cost of operation or maximizing the use of renewable energy sources. Additionally, AI techniques can be used to learn from data and adapt the control strategy to improve the MG's performance in real time. The choice of control technique ultimately depends on the specific needs of the MG and the trade-offs between performance, complexity, and implementation cost.

6.2.2 Model Predictive Control

MPC was first developed in the 1960s and has since been effectively utilized in the chemical process industry, where there is adequate time to perform the necessary calculations. In the 1980s, early applications of MPC in power electronics were observed in high-power systems with low switching frequencies, as higher frequencies were not feasible then due to the extensive computation time required by the control algorithm. However, the advent of fast and powerful microprocessors over the past decade has substantially increased interest in applying MPC to power electronic-based systems and MGs [56].

MPC is a powerful control strategy that utilizes a mathematical model of the system to compute the optimal control signal based on a cost function. Unlike traditional control methods, MPC does not rely on a fixed set of control rules but instead selects the best control input sequence from a range of feasible options over a prediction horizon. This horizon allows the MPC controller to account for future system behavior and make adjustments in real time, resulting in improved performance and efficiency. The *receding horizon* concept is a critical aspect of MPC, which involves continually updating the control input sequence based on the latest system state information.

Figure 6.3 represents the MPC strategy and its main components. As shown in this figure, the MPC strategy can be summarized as follows:

- Use the system model to predict its future output over a prediction horizon (N_p) at upcoming time intervals.
- Minimize an objective function by solving an optimization problem, and compute a control sequence that satisfies the systems constraints.
- Employ a receding horizon strategy where the control sequence is recalculated at each step (see Figure 6.3(a)). Apply the first control signal ($\mathbf{u}(k)$) of the sequence to the system, with the horizon displaced toward the future at each instant. Repeat the previous step with a new state variable.

MPC can employ various types of models depending on the characteristics of the system and the design objectives. The state-space model is a comprehensive description of systems and is often a preferred choice. For example, the discrete-time state-space model below can be used for prediction purposes:

$$\mathbf{x}(k+1) = A\mathbf{x}(k) + B\mathbf{u}(k), \tag{6.1a}$$

$$\mathbf{y}(k) = C\mathbf{x}(k) + D\mathbf{u}(k). \tag{6.1b}$$

A cost function is defined to represent the desired behavior of the system. This function takes into account the references, future states, and future control actions of the system:

$$J = f(\mathbf{x}(k),\ \mathbf{u}(k), \ldots, \mathbf{u}(k+N_c)). \tag{6.2}$$

MPC involves an optimization problem that seeks to minimize the cost function J over the prediction horizon (N_p), while considering the model prediction of the system and the system constraints.

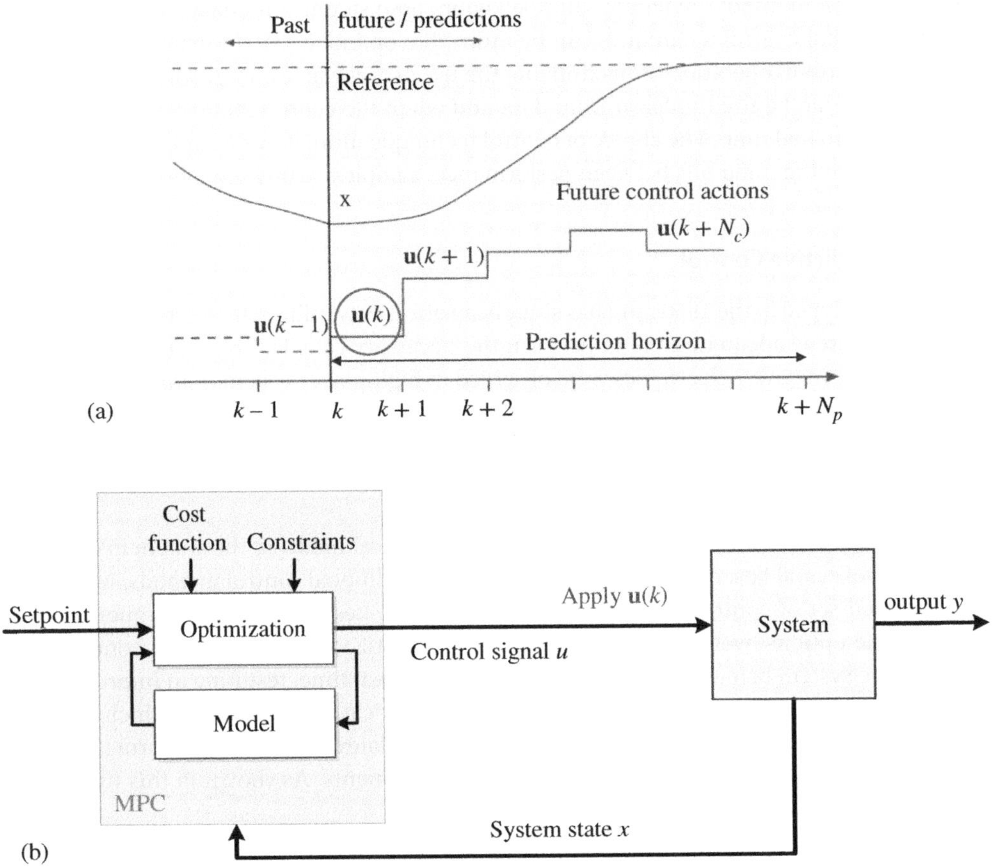

Figure 6.3 MPC strategy: (a) receding horizon concept, (b) main elements of MPC.

The outcome is a sequence of N_c optimal control actions. However, the controller applies only the first element of the sequence

$$\mathbf{u}(k) = [10 \cdots 0] \arg\min_{u} J, \tag{6.3}$$

and solves the optimization problem again at each sampling instant, using the newly measured/estimated data to obtain a new sequence of optimal control actions. This approach is known as a receding horizon strategy.

By continuously solving the optimization problem and applying only the first element of the optimal control sequence, the controller can adapt to changes in the system in real time, while ensuring that the system follows the desired behavior as defined by the cost function J.

6.2.2.1 MPC for Microgrids

The choice of an appropriate MPC technique for a specific MG will depend on a variety of factors, including control objectives, constraints, and system characteristics [57]. For instance, distributed MPC [58] can be utilized to coordinate the operation of multiple DERs without the need for a centralized controller, leading to improved scalability and resilience. In contrast, robust MPC [58] can handle uncertainties and disturbances by designing a controller that maintains desired performance under uncertain conditions. Nonlinear MPC [59] is suitable for systems with nonlinear dynamics, while hybrid MPC [60] can manage systems with both continuous and discrete dynamics. Stochastic MPC [61] is capable of optimizing system performance while accounting for probabilistic uncertainties. Explicit MPC, also known as offline MPC, is useful for very fast system dynamics where online implementation is not feasible. Additionally, finite control set model predictive control (FCS-MPC) is a popular MPC method that is often used for local control of power converters and MGs. This control methodology determines the optimized converter's physical switching states directly.

Given the above-mentioned solutions and capabilities, MPC can well address many challenges of MGs [62]:

- The multivariable nature of MPC can coordinate the operation of different RESs and ESSs in an MG, achieving objectives.
- Stochastic variables can be included in the optimization problem to cope with the intermittence and variability of renewable generation and demand.
- MPC handles logical variables for connection/disconnection of units, changing operation modes, or changing energy prices.
- MPC adapts to sudden changes in an MG such as disconnection or malfunctioning of a certain unit by changing its structure, provided there are degrees of freedom available.
- MPC can solve complex problems in a distributed way, making it useful for IMGs.

MPC has been used in MGs for control objectives like frequency and voltage regulation [17, 63, 64], power sharing and dispatch [65], optimal power flow [66], power quality improvement [62], energy management [67], and demand-side management [62].

The main advantage of using MPC in MGs is that it enables the integration of multiple control objectives into a single framework, while also improving system performance by taking into account the unique characteristics of the system (see examples provided in Sections 6.2.3 and 6.2.4). Additionally, MPC can handle constraints on the system variables and inputs, which is particularly important in MGs where there are various operational constraints. However, a disadvantage of using MPC is that it requires a high computational effort, which may limit its real-time implementation in some cases.

6.2.2.2 Finite Control Set Model Predictive Control

FCS-MPC [68] is a type of MPC that is successfully applied to a wide range of power converters and drive applications [56]. In FCS-MPC, the control input is limited to a finite set of discrete values (switching states), which allows for the implementation of the control algorithm using digital hardware. FCS-MPC is typically utilized in the local control of MGs where the optimal switching states are directly determined for power converters. This control methodology is simple, intuitive, and easy to implement, provided the prediction horizon is small (usually 1 sampling interval). Figure 6.4 shows a block diagram of the FCS-MPC strategy applied to a converter supporting a load or a grid.

FCS-MPC has several advantages over the PWM-based MPC methods (also known as continuous control set MPC), including reduced computational complexity, faster control response, and improved efficiency. FCS-MPC allows easy inclusion of multiple control variables, system constraints, and nonlinearities. Additionally, FCS-MPC can provide fast transient response, good steady-state performance, minimal harmonic distortion, and maximum dc-link utilization, making it a popular choice for many practical applications.

Sections 6.2.3 and 6.2.4 present two case studies that demonstrate the application of MPC in MGs. The first case study focuses on the stability and control of DC MGs with CPLs, while the second

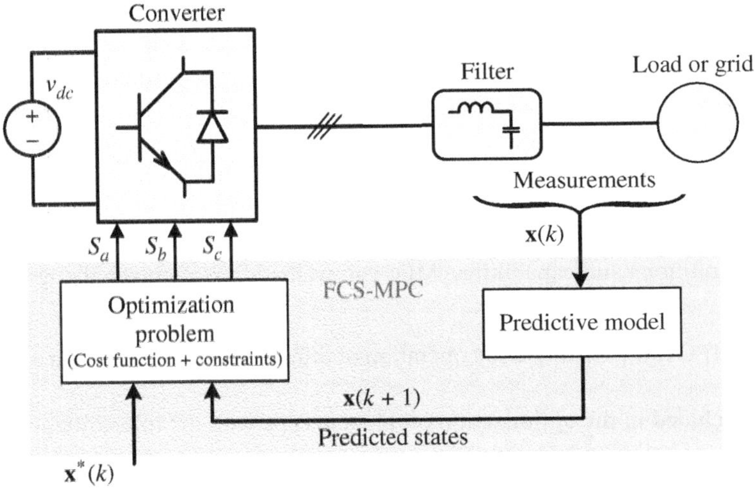

Figure 6.4 Finite control set model predictive control of a power converter connected to a load or a grid.

case study examines the effectiveness of MPC in achieving smooth transitions between different operation modes in MGs.

6.2.3 Model Predictive Control of DC Microgrids with Constant Power Loads

Nowadays, many applications such as aircraft, aerospace, marine, and naval systems utilize the concept of dc MG due to dc system features, e.g. improved efficiency, flexibility, and reliability through providing redundancy [69, 70]. A large majority of end-users in dc MGs are electronic loads that employ converters for power conditioning and voltage regulation. The active regulation capability of CPLs extracts steady power even under dc voltage variation at the dc bus side [71]. A challenging characteristic of CPLs is their input negative impedance feature. This characteristic may lead the system toward instability.

Different solutions have been suggested to deal with this challenge such as (i) passive or active damping of source side impedance [72], (ii) active damping of the CPL's impedance [73], and (iii) utilizing effective (linear and nonlinear) control strategies [74–76]. Passive damping of source side impedance requires installation of a passive component which increases the size and losses of the system [73]. While the active damping one can actively control the power electronic converter at the source side [73, 76–78].

Inspired by the deficiencies of the state-of-the-art dc MG stabilization approaches, direct MPC approaches gain popularity for these systems [79]. Different types of MPC strategies have been recently applied to MGs for various control purposes [79–81], as well as dynamic stabilization of dc MGs [82–84].

This section introduces a decentralized MPC (DMPC) to ensure power sharing and regulate dc bus voltage in dc MGs with CPLs [60]. The proposed method replaces the conventional primary layer of dc MGs, i.e. inner loops and droop control, with a single optimal controller. The proposed solution provides the system with a fast dynamic response and zero steady-state error. The effectiveness of the presented control approach is verified via hardware-in-the-loop real-time experiments, and the results are compared with the conventional primary control.

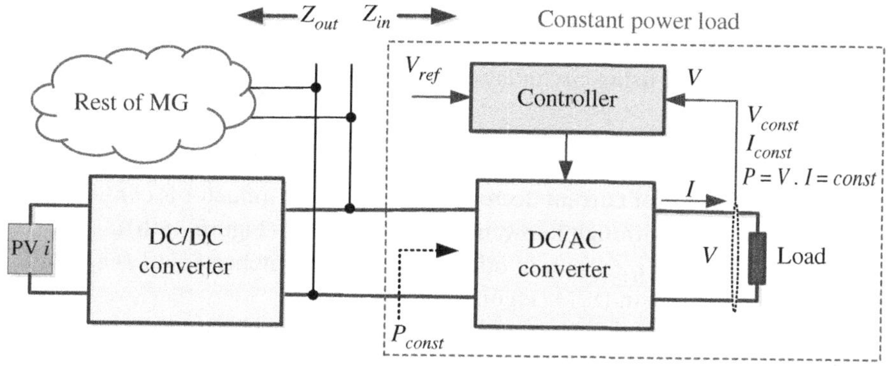

Figure 6.5 Schematic diagram of a dc–dc converter with a CPL.

6.2.3.1 Case Study and Dynamic Modeling

In this study, an islanded dc MG with four PV sources supporting a CPL is considered. In formulation and procedure design, without losing generality, we assume that boost converters are used as dc–dc converters. However, one can use any type of converter in the system. Figure 6.5 shows a general configuration of one DER inside a dc MG, e.g. a PV, connected to a dc–dc converter interfaced with a CPL. State-space model of a dc–dc boost converter with any type of load such as Z considering the inductor current i_L, capacitor voltage v_c and input voltage v_{in}, is expressed as [81].

$$\dot{i}_L(t) = \tilde{d}\left[-\frac{R_L}{L} - \frac{Z}{L(Z+R_c)}\right] i_L(t) - \tilde{d}\frac{Z}{L(Z+R_c)}v_c(t) + \frac{Z}{L(Z+R_c)}(R_c i_L(t) + v_c(t))u(t) + \tilde{d}\frac{v_{in}(t)}{L}, \tag{6.4}$$

$$\dot{v}_c(t) = \tilde{d}\frac{Z}{C_f(Z+R_c)}i_L(t) - \frac{1}{C_f(Z+R_c)}v_c(t) - \frac{Z}{C_f(Z+R_c)}i_L(t)u(t), \tag{6.5}$$

$$u(t) = \begin{cases} 1 & S=1 \\ 0 & S=0 \end{cases}, \tag{6.6}$$

$$\tilde{d}(t) = \begin{cases} 1 & u(t) \geq 0 \text{ and } i_L(t) > 0 \\ 0 & u(t) = 0 \text{ and } i_L(t) = 0 \end{cases}, \tag{6.7}$$

where L is the inductor with R_L representing the inductor equivalent series resistance, C_f is the output capacitor with R_c representing its equivalent series resistance, $u(t)$ is the input of the system, $S = 1$ means the switch is ON and $S = 0$ indicates the switch is OFF. In discontinuous conduction mode (DCM) operation, the inductor current returns to zero in every switching cycle, thus the auxiliary binary variable $\tilde{d}$ is introduced to derive a model that describes the operation of the converter either in DCM (when $\tilde{d} = 0$) or continuous conduction mode (CCM) (when $\tilde{d} = 1$).

Taking into account the inductor current, a dc–dc boost converter can be classified into two models: averaged and automatic models [85, 86]. As the presented MPC controller utilizes an automatic model of the system. Using this model gives the possibility of employing a simple hybrid MPC instead of a nonlinear one. Interested readers can find details of average modeling of dc–dc boost converters for other studies in [60].

Continuous-time model: Unlike the averaged model [60], in the automatic model, the inductor current situation in each sampling time is taken into account in addition to the switching modes.

As shown in Figure 6.6, boost converter has three operational modes for the automatic model in continuous time. The following nonlinear state space equations describe the dynamics of the system based on the automatic model by using circuit laws [84].

(1) When switch S is ON and inductor current increases ($i_{Li}(t) \uparrow$). In this mode, energy is stored in the inductor and its current increases (Figure 6.6(a)).
(2) When switch S is OFF and inductor current decreases ($i_{Li}(t) \downarrow$). The inductor is connected to the output and energy is released through it toward the common bus (Figure 6.6(b)).
(3) When switch S remains OFF and $i_{Li}(t) = 0$, in other words, both switches S and D are OFF. In this case, the converter operates in DCM (Figure 6.6(c)).

The equations of the above-mentioned modes using the circuit law are as follows:

$$\dot{x}(t) = \begin{cases} A_{1i}x(t) + B_i v_{si}(t) & S = 1 \\ A_{2i}x(t) + B_i v_{si}(t) & S = 0 \ \& \ i_{Li}(t) > 0 \\ A_{3i}x(t) & S = 0 \ \& \ i_{Li}(t) = 0 \end{cases}$$
$$y(t) = C_i x(t). \tag{6.8}$$

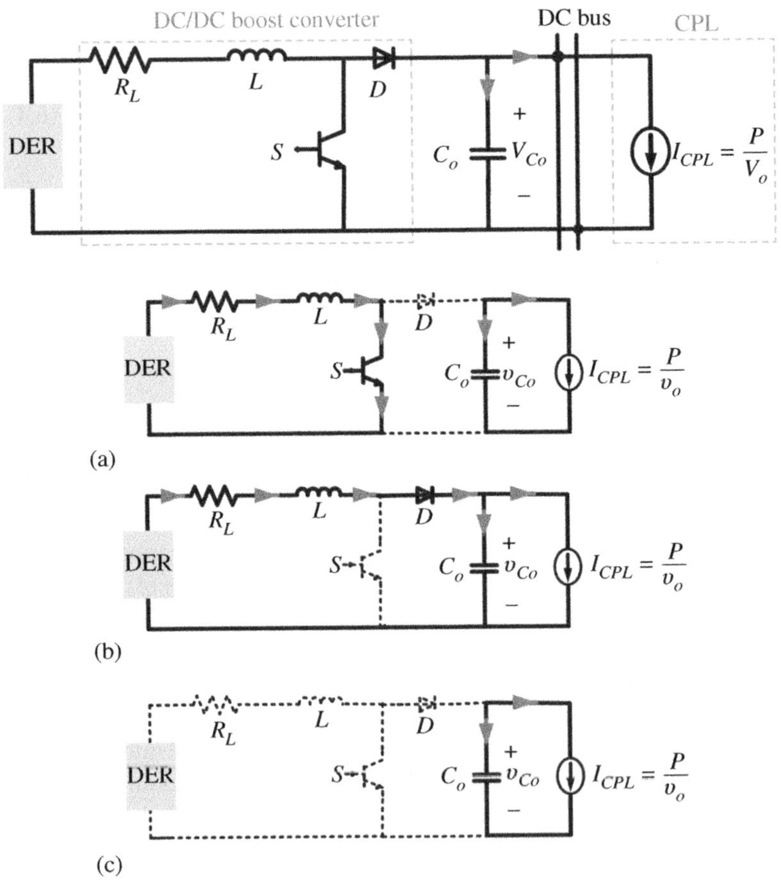

Figure 6.6 Circuit schematic of a DER in a dc MG interfaced with CPL as well as operating modes of a dc–dc boost converter for averaged and automatic models: (a) mode 1, (b) mode 2, and (c) mode 3.

where the matrices $A_{1i}, A_{2i}, A_{3i}, B_i$, and C_i are calculated as follows:

$$A_{1i} = \begin{bmatrix} \frac{-R_{Li}}{L} & 0 \\ 0 & \frac{-P}{C_{oi}v_{oi}^2} \end{bmatrix}, \ A_{2i} = \begin{bmatrix} \frac{-R_{Li}}{L_i} & \frac{-1}{L_j} \\ \frac{1}{C_{oi}} & \frac{-p}{C_{oi}v_{oi}^2} \end{bmatrix}, \ A_{3i} = \begin{bmatrix} 0 & 0 \\ 0 & -\frac{P}{C_{oi}v_{oi}^2} \end{bmatrix}$$

$$B_i = \begin{bmatrix} \frac{1}{L_i} & 0 \end{bmatrix}^T, C_i = \begin{bmatrix} 0 & 1 \end{bmatrix}, \tag{6.9}$$

and $x(t) = [i_L(t)\ v_o(t)]$ is the state vector, inductor current, and capacitor voltage, respectively, $v_{si}(t) = u(t)$ is input voltage of power supply which is equal to nominal voltage of PV array i, for $i = 1, \dots, 4$, $y(t)$ is the output of the system which here is equal to the output voltage of dc bus, i.e. (v_{oi}) which is considered equal to voltage $v_{C_{0i}}$. C_{0i} is the output capacitor filter, R_{Li} is the resistance of the input inductor L_i, and P is the power load.

Discrete-time model: The continuous-time equations of the model given in (6.8) are discretized using the Euler approximation approach. Accordingly, the automatic discrete-time model of the studied system with one unit and CPL:

$$x(k+1) = \begin{cases} G_{1i}x(k) + F_i v_{si}(k), & S = 1 \\ G_{2i}x(k) + F_i v_{si}(k), & S = 0 \ \& \ i_{Li}(k+1) > 0 \\ G_{3i}x(k), & S = 0 \ \& \ i_{Li}(k+1) = 0 \end{cases}$$

$$y(k) = H_i x(k), \tag{6.10}$$

where $G_{1i} = I + A_{1i}T_s$, $G_{2i} = I + A_{2i}T_s$, and $G_{3i} = I + A_{3i}T_s$. Furthermore, I is the identity matrix, $F_i = B_iT_s$ and $H_i = C_i$ with dimension two, and Ts is the sampling time. The discrete automatic model can operate in four different modes as shown in Figure 6.7. A comparison of the averaged and automatic models with details of their modes can be found in [60].

Notice that $v_{si}(k)$ is considered as the input signal in (6.10). Thus, the model should be changed to suit the design purpose as in which an integrator is embedded to avoid steady-state error. An incremental state-space model can also be used if the model input is the control increment $\Delta v_{si}(k)$ instead of the control signal $v_{si}(k)$. This model is called the augmented model, which is used in the design of the predictive controller [60].

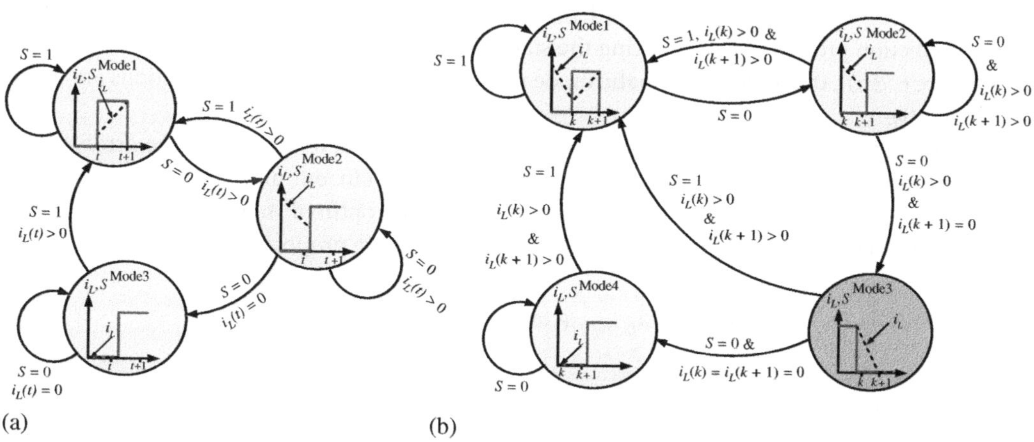

Figure 6.7 Possible modes of the automatic model in a dc–dc boost converter, (a) continuous time (b) discrete time.

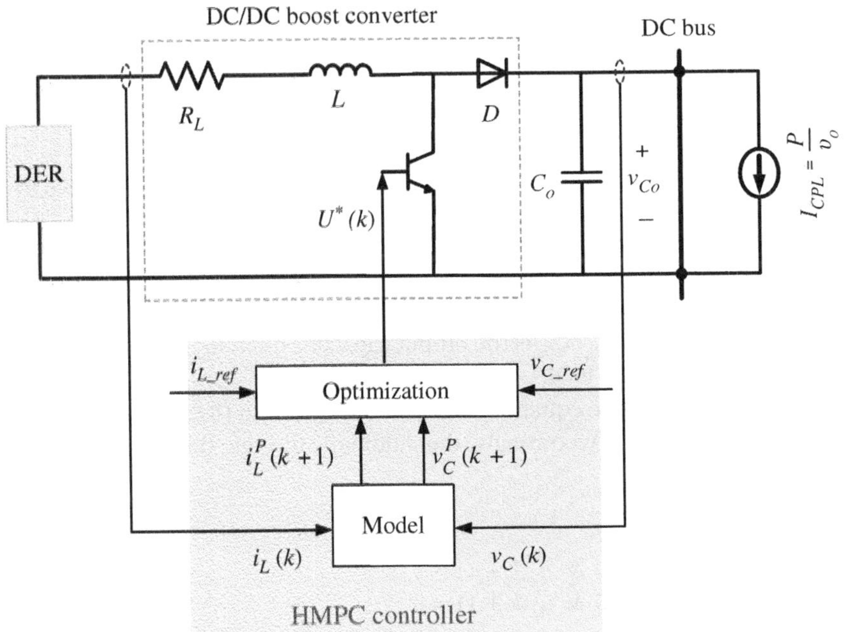

Figure 6.8 Schematic of the DMPC controller for each dc-dc boost converter.

6.2.3.2 Design Methodology

As shown in Figure 6.8, an optimal decentralized MPC is designed for each converter of the dc MG to guarantee appropriate current sharing and control of the dc bus voltage. The controller is implemented locally, and no digital communication is utilized. The DMPC provides optimal switching states for the dc–dc converter by solving an optimization problem using some real measurements, i.e. capacitor voltage and inductor current. The DMPC uses a simple cost function to include the current sharing and voltage control objectives. The following steps, as shown in the flowchart of Figure 6.9, are taken in order to design the controller.

Step 1. Prediction: In this step, the required variables of each converter are measured to be used in the prediction process. After checking the status of switching states and the inductor current (see Figure 6.7(b)), the perfect automatic mode is selected, and the variable predictions are done consequently.

Step 2. Cost function formulation: The cost function $J(k)$ is formulated based on the error of the predicted value (i_{Li}^P and v_{oi}^P) of the state variables (inductive current and output voltage) with their desired reference value over the prediction horizon and optimal state switching ($\Delta u(k)$) of the converter. The cost functions of the MPCs for each DER are proposed as

$$J(k) = \frac{1}{N}\left\{\sum_{1}^{N}\left(\left|i_{Li}^P(k+1) - i_{Li\_ref}\right| + \left|v_{oi}^P(k+1) - v_{oi\_ref}\right|\right) + \lambda|\Delta u(k)|\right\}, \tag{6.11}$$

where N is the prediction horizon, v_{oi_ref} is the voltage reference for DER$_i$ which is equal to v_{dc}^*, and $\lambda > 0$ is the weighting factor, which sets a trade-off between the current/voltage error and

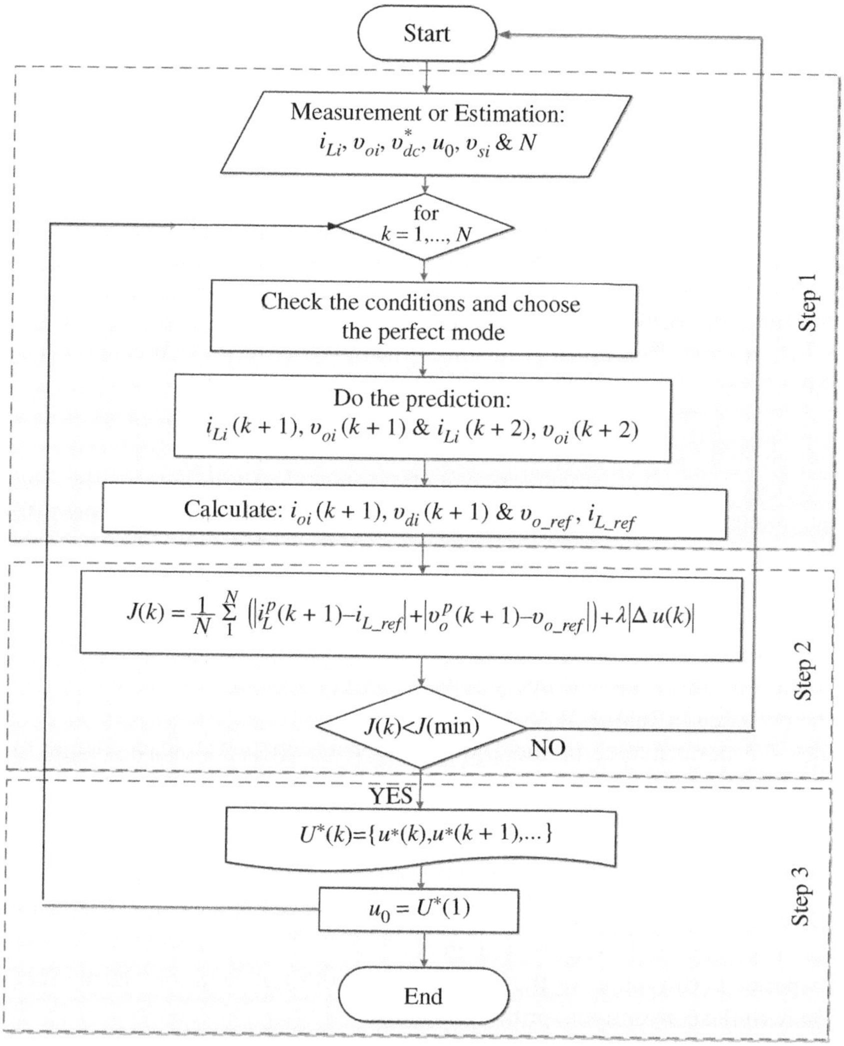

Figure 6.9 Flowchart of the DMPC controller.

the control signal. i_{Li_ref} is the current reference and can be achieved using the power balance equation $P_{in} = P_{out}$ and the desired current as follows:

$$P_{in} = v_{si} i_{Li}, \; P_{out} = P_{Load} \quad \Rightarrow i_{Li\_des} = \frac{P_{Load}}{v_{si}} \tag{6.12}$$

To improve the transient of the output voltage, the proportional term h is multiplied by the voltage error. Hence, the reference inductor current can be obtained as

$$i_{Li\_ref} = i_{Li\_des} + h\,(v_{di} - v_{oi}), \tag{6.13}$$

where $h \in \mathcal{R}^+$ is the small-ripple approximation for regulation of output voltage in steady state. The adjusted voltage for each DER is then obtained using droop control formulation as

$$v_{di}(k+1) = v^*_{dc} - R_{di} i_{oi}(k+1), \tag{6.14}$$

where R_{di} is a droop coefficient. Further, the difference of control signal, i.e. the error between two consecutive switching states is defined as

$$\Delta u(k) = u(k) - u(k-1). \tag{6.15}$$

Step 3. Optimization problem: At each sample time, the following constrained optimization problem must be solved [60]:

$$U^*(k) = \arg\min J(k),$$
$$\textit{Subj. to the augmented model.} \tag{6.16}$$

Minimizing the cost function results in a sequence in the form of U^*, where $U^*(k) = \{u^*(k), u^*(k+1), \dots\}$. There exist 2^N switching sequences and only its first element, i.e. $u^*(k)$ is applied in each sampling time.

6.2.3.3 Real-Time Hardware in the Loop Results

The performance of the DMPC is evaluated using the dc MG system shown in Figure 6.10. The electrical and control parameters of the test system are listed in Table 6.1. As shown in Figure 6.10, the dc MG test system augmented with the control framework is implemented in OPAL-RT OP5600. This allows us to verify the real-time effectiveness of the DMPC. The converter-based sources provide different ratings, i.e. the power rating of the DER_1 and DER_3 is double of the DER_2 and DER_4. The performance of the controller is evaluated under different scenarios, and the results are compared with the conventional primary control, i.e. PI-based inner loops and droop control with the parameters listed in Table 6.1.

Figure 6.11 illustrates the performance of the DMPC under a step power load change in comparison with the conventional primary control. The output power of the CPL connected to the system is stepped up from 400 to 800 W at $t = 4.5s$. Current sharing and the dc-bus voltage are tightly controlled by using the DMPC method, while oscillations can be observed for the conventional droop control method.

Plug and play (PnP) capability of the DMPC controller is studied and shown in Figure 6.12. It is assumed that DER_3 is plugged out from the system at $t = 3$ s and it is plugged in at $t = 6$ s. Figure 6.12 shows a superior performance of the controller for both transient and steady-state response in comparison with the conventional primary control.

6.2.4 Hybrid Fuzzy Predictive Control for Smooth Transition of AC Microgrids

With the expansion of DER-based MGs, the need for dual-mode inverters with the ability to work in both steady (grid-connected and islanded) and transient (transition from grid-connected to islanded and vice versa) modes and smooth transition between their operating modes arises. Hence, providing efficient and comprehensive control approaches to minimize the distortions of voltage and current waveforms during the transition process is a principal need.

In order to achieve a smooth, seamless transition without sudden and harsh oscillations and to reduce waveform distortions during the transition between operating modes, special considerations should be taken into account in the control system of inverters. Beginning with classical control methods, PI control is rarely used for a smooth transition of inverters but for realizing their inner control loops. The reason is that these types of controllers do not achieve proper performance in

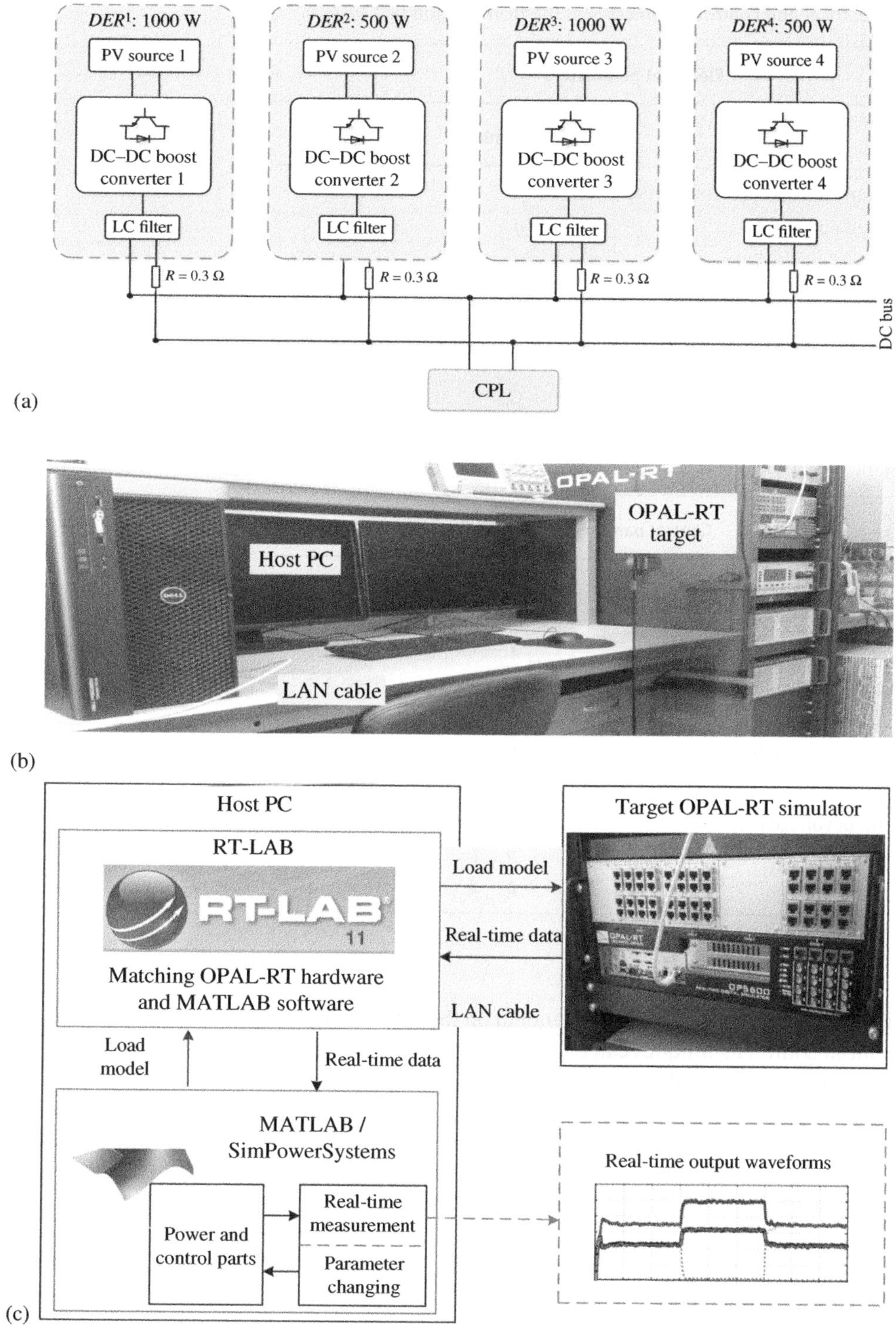

Figure 6.10 Test bench of dc MG and implementation on real-time OPAL-RT: (a) schematic of the system configuration including four DERs, (b) real-time experimental setup including the host PC, the OPAL-RT target and a LAN cable for networking, (c) conceptual diagram of the DMPC real-time process.

Table 6.1 Electrical and control parameters of the dc microgrid test system.

Parameter	Symbol	Value
Electrical parameters		
The dimension of the state variable vector	n	2
The number of outputs	q	1
Input voltage	v_{si}	150 V
Voltage reference	v^*_{dc}	200 V
CPL	P	400 W
Filter inductance	L_i	450 μH
Filter capacitance	C_{oi}	220 μF
Input resistance	R_i	0.01 Ω
Line resistance	R_{Line_i}	0.3 Ω
Control parameters		
Sampling time	T_s	2.5 μs
Prediction horizon	N	3
Weighting factor	λ	0.01
Ripple approximation	h	0.7
Proportional gain of current loop	K_{pI}	3
Integral gain of current loop	K_{iI}	20
Proportional gain of voltage loop	K_{pV}	10
Integral gain of voltage loop	K_{iV}	200
Droop gain for $DER_{1,3}$	R_{d1}, R_{d3}	0.2
Droop gain for $DER_{2,4}$	R_{d2}, R_{d4}	0.4

the transient response, which is highly dependent on the tuning of the controllers' parameters [62]. Using advanced control techniques can overcome some of these problems and enhance the power quality issues in grid-connected converters and MGs.

MPC can be used to enhance the transition conditions between the operating modes [87]. It can be integrated with other control methods such as adaptive control [88, 89] and fuzzy control [90], or be implemented using the FCS concept [91] to improve the smooth transition of inverters and MGs, even under the system's harmonics [92]. Further, adaptive fuzzy control methods [93, 94] and many other advanced control approaches [95] are introduced to improve the transient and steady-state performance of three-phase grid-connected inverters reducing the output distortions.

This section introduces a hybrid fuzzy MPC strategy to improve the behavior of the three-phase inverters under distorted and unbalanced conditions in both steady and transient states, guaranteeing a smooth transition without rapid fluctuations and distortions. The performance of the presented control methodology is verified using some simulation studies, and the results are compared with the conventional MPC in [87].

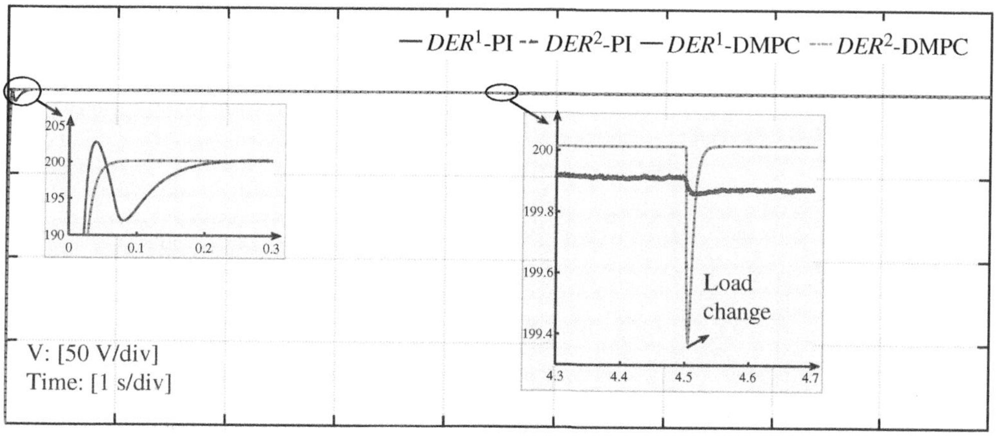

(a)

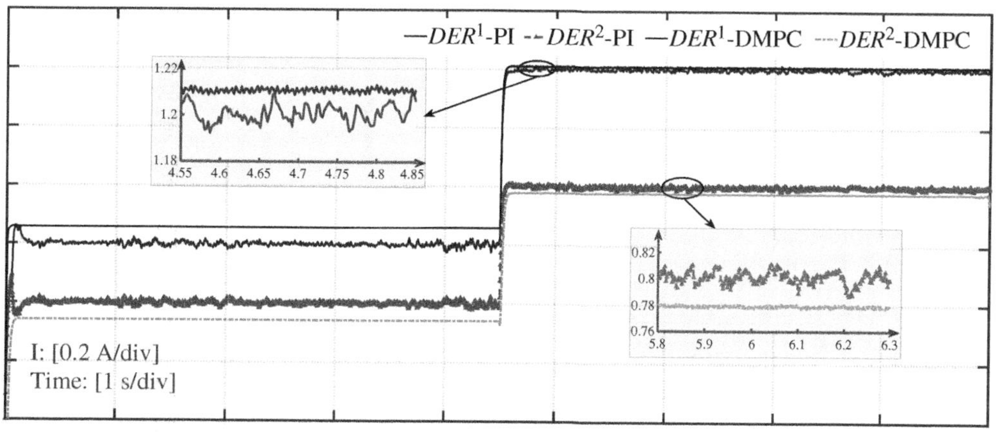

(b)

Figure 6.11 Comparison of the DMPC and the conventional primary control performance under CPL step change: (a) DER_1 and DER_2 voltage, and (b) current sharing among DERs.

6.2.4.1 Case Study and Dynamic Modeling

The general structure of a three-phase grid-connected inverter and its associated control loops is shown in Figure 6.13. The dc link, denoted by C_{dc}, can be charged with renewable energy sources such as solar panels. Here, the energy source is modeled as a dc voltage source with series resistance.

The behavior of the system is defined by the inductor current and the capacitance voltage of the output LC filter as follows [96]:

$$\begin{cases} \frac{d}{dt} i_L(t) = \frac{1}{L}(v_{inv}(t) - v_{out}(t) - i_L(t)R_{esr}) \\ \frac{d}{dt} v_{out}(t) = \frac{1}{C}(i_L(t) - i_{PCC}(t)), \\ v_{inv}(t) = \psi(t) v_{dc}(t) \end{cases} \tag{6.17}$$

where $i_L(t)$ is the inductor current of the inverters' output filter, $i_{PCC}(t)$ refers to the PCC current, or in other words, the output current of the filter, R_{esr} is the equivalent series resistance of the filters' inductance, and $v_{inv}(t)$ and $v_{out}(t)$ are the output voltage of the inverter and the PCC voltage, respectively. Furthermore, $\psi(t)$ is a function based on the switching states of MOSFETs.

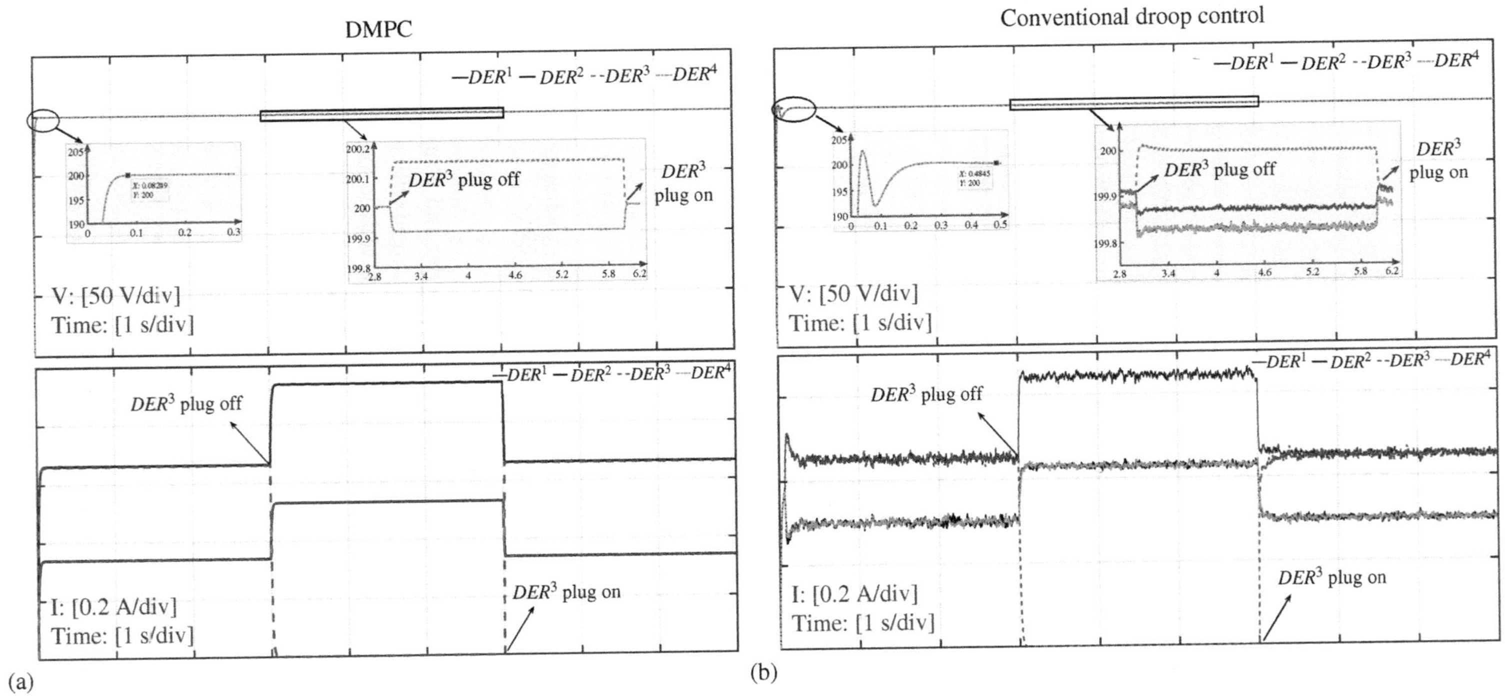

Figure 6.12 PnP capability of the DMPC versus the conventional primary control: (a) the DMPC, and (b) the conventional primary control.

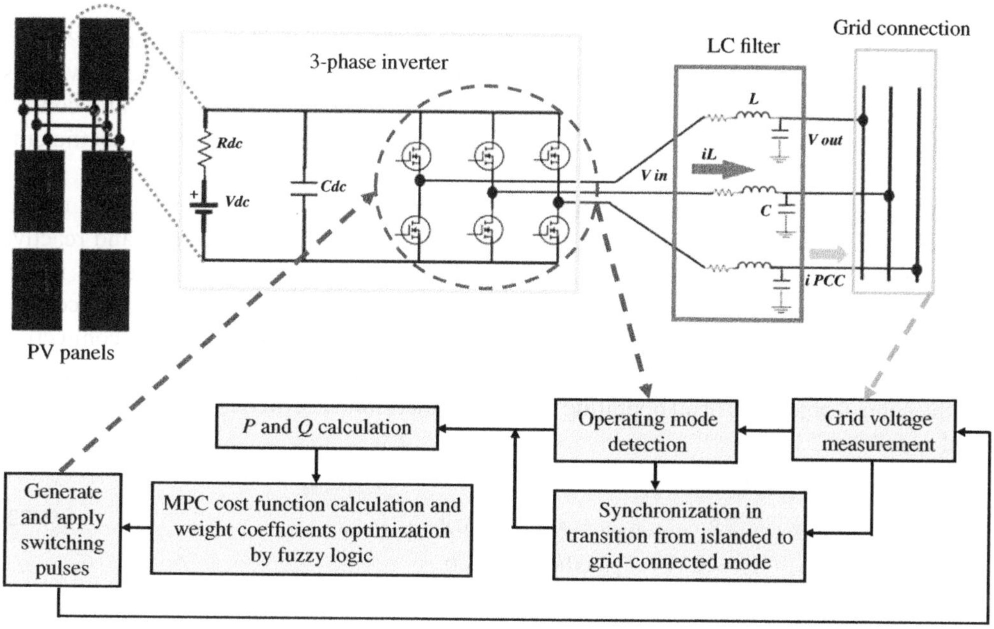

Figure 6.13 The configuration of a three-phase grid-connected PV inverter with its control loops.

Considering the circuit model of Figure 6.13, one can write the state space model of the system in (6.17) as follows:

$$\begin{cases} \dot{x} = \begin{bmatrix} -\frac{R_{esr}}{L} & -\frac{1}{L} \\ \frac{1}{C} & 0 \end{bmatrix} x + \begin{bmatrix} \frac{1}{L} & 0 \\ 0 & -\frac{1}{C} \end{bmatrix} u \\ y = [0 \quad 1]x \end{cases}, \tag{6.18}$$

where $u^T = [v_{inv}(t), i_{PCC}(t)]$ is the input vector, $x^T = [i_L(t), v_{out}(t)]$ is the state vector, which in the three-phase system, the current and voltage are represented as $i_L^T(t) = [i_{L1}(t), i_{L2}(t), i_{L3}(t)]$, and $v_{out}^T(t) = [v_{out_1}(t), v_{out_2}(t), v_{out_3}(t)]$. In order to predict the system states for the MPC controller, the discrete-time model of the system is obtained using the Euler forward method as

$$\begin{cases} i_L(k+1) = \frac{T_S}{L}[v_{inv}(k) - v_{out}(k) - i_L(k)R_{esr}] + i_L(k) \\ v_{out}(k+1) = \frac{T_S}{C}(i_L(k) - i_{PCC}(k)) + v_{out}(k). \end{cases} \tag{6.19}$$

To achieve power control in grid-connected mode, active and reactive power must be predicted. This can be done using the instantaneous grid voltage and current described below:

$$\begin{cases} v_{g1}(t) = V_g \sin \omega t & i_{g1}(t) = I_g \sin(\omega t + \theta) \\ v_{g2}(t) = V_g \sin(\omega t - \frac{2\pi}{3}) & i_{g2}(t) = I_g \sin(\omega t + \theta - \frac{2\pi}{3}) \\ v_{g3}(t) = V_g \sin(\omega t + \frac{2\pi}{3}) & i_{g3}(t) = I_g \sin(\omega t + \theta + \frac{2\pi}{3}). \end{cases} \tag{6.20}$$

Furthermore, the load current in the islanded mode can be calculated as

$$i_{Load}(k+1) = i_L(k+1) - \frac{C}{T_S}(v_{out}(k+1) - v_{out}(k)). \tag{6.21}$$

6.2.4.2 Control System Design

This section introduces Fuzzy MPC (FMPC), a hybrid control approach that combines MPC and fuzzy logic. FMPC adapts to different operating modes by using fuzzy logic to determine the weights of the MPC cost function. This results in a more appropriate response with less cost and control effort. FMPC is an online adaptive system, that continuously adjusts the cost function coefficients during the control process.

The controller design must ensure four operating modes: grid-connected, islanded, and transitions between them. In the grid-connected mode, the objective is to regulate active and reactive power demand, while output voltage regulation is the primary objective in the islanded mode. However, a smooth transition is the main objective in the transient modes. This means accurate and quick synchronization (output voltage phase and amplitude) with the grid during connection and assuring stability and proper load support during disconnection.

Given the control objectives of all operating conditions mentioned earlier, the MPC cost function is defined as follows:

$$\begin{cases} \min\ g = \lambda_P g_P + \lambda_Q g_Q + \lambda_V g_V + \lambda_{V-Sine} g_{V-Sine} + \lambda_{I-Sine} g_{I-Sine} \\ \qquad + \lambda_{V-Unbalance} g_{V-Unbalance} + \lambda_{I-Unbalance} g_{I-Unbalance} \\ \text{Subject to} \quad x(k+1) = x(k) + T_s\left(Ax(k) + Bu(k)\right). \end{cases} \tag{6.22}$$

where

$$\begin{cases} g_P = \frac{1}{P_{rated}} \left| P_{ref}(k+1) - P_{out}(k+1) \right| \\ g_Q = \frac{1}{Q_{rated}} \left| Q_{ref}(k+1) - Q_{out}(k+1) \right| \\ g_V = \frac{1}{V_{rated}} \left| V_{ref}(k+1) - V_{out}(k+1) \right| \\ g_{V-Sine} = \frac{1}{V_{rated}} \left| V_{Ref-Sine}(k+1) - V_{out}(k+1) \right| \\ g_{I-Sine} = \frac{1}{I_{rated}} \left| I_{Ref-Sine}(k+1) - I_{out}(k+1) \right| \\ g_{V-Unbalance} = \frac{1}{V_{rated}} \left(\left(V_A(k+1) - V_B(k+1)\right)^2 + \right. \\ \qquad \left. \left(V_A(k+1) - V_C(k+1)\right)^2 + \left(V_B(k+1) - V_C(k+1)\right)^2 \right) \\ g_{I-Unbalance} = \frac{1}{I_{rated}} \left(\left(I_A(k+1) - I_B(k+1)\right)^2 + \right. \\ \qquad \left. \left(I_A(k+1) - I_C(k+1)\right)^2 + \left(I_B(k+1) - I_C(k+1)\right)^2 \right) \\ \left[\lambda_P, \lambda_Q, \lambda_V, \lambda_{V-Sine}, \lambda_{I-Sine}, \lambda_{V-Unbalance}, \lambda_{I-Unbalance}\right] \in \mathbb{R}. \end{cases} \tag{6.23}$$

The cost function (6.22) comprises seven terms: g_P, g_Q, g_V, g_{V-Sine}, g_{I-Sine}, $g_{V-Unbalance}$, and $g_{I-Unbalance}$. The first three terms aim to minimize the differences between the active and reactive powers and the system's voltage in comparison to their reference values. The remaining four terms aim to minimize voltage distortions, current distortions, voltage unbalances, and current unbalances, respectively, in relation to their reference values. The weight coefficients (i.e. λ_x) correspond to each of the seven variables. Depending on the system's control objective, any of these terms may be used. However, our objective is to ensure appropriate performance in all four steady and transient operation modes, taking into account harmonic and unbalanced distortions.

In addition, in (6.22), the variables P_{rated}, Q_{rated}, V_{rated}, and I_{rated} represent the rated value of the inverter's active power, reactive power, output voltage, and current, respectively. P_{ref}, Q_{ref},

V_{ref}, $V_{Ref-Sine}$, and $I_{Ref-Sine}$ are the active power reference, the reactive power reference, the output voltage reference, the sinusoidal output voltage reference, and the sinusoidal current reference, respectively. V_A, V_B, and V_C refer to the three-phase output voltages, while I_A, I_B, and I_C are the three-phase output currents.

The controller employs a multipurpose automatic adjustment process to determine the priority of the control objectives in each working condition. The following outlines how the weight coefficients of the cost function are continually updated and adapted to the system's operating conditions using fuzzy logic.

Intelligent Adjustment of the Weight Coefficients The weight coefficients of the cost function are critical for the controller's performance. Fuzzy logic is used to intelligently adjust the coefficients based on operating conditions and control objective importance. The adjusted coefficients are used to minimize the cost function in the next cycle of the algorithms' execution. The cost function is divided into seven parts related to control objectives as below:

$$\begin{cases} g_P \leq \psi_P, \quad g_Q \leq \psi_Q, \quad g_V \leq \psi_V, \\ g_{V-Sine} \leq \psi_{V-Sine}, \quad g_{I-Sine} \leq \psi_{I-Sine}, \\ g_{V-Unbalance} \leq \psi_{V-Unbalance}, \quad g_{I-Unbalance} \leq \psi_{I-Unbalance}, \end{cases} \tag{6.24}$$

where ψ_P, ψ_Q, ψ_V, ψ_{V-Sine}, ψ_{I-Sine}, $\psi_{V-Unbalance}$, and $\psi_{I-Unbalance}$ are the acceptable values for the tracking error. If the objective function terms are less than the acceptable values, then the references can be tracked accurately. Selecting smaller values for the parameters in (6.24) results in higher tracking accuracy but a lower probability of obtaining an optimal solution. If all parameters of (6.24) are zero, the optimization problem may have no solution.

For designing the fuzzy system, a combination of human experience and the extraction of proper weight coefficients is used. In other words, relatively suitable weight coefficients for the cost function are selected in all operating modes based on expert experience and the execution of various simulations, and then the fuzzy system chooses the best weight factors among the available values by using its rules [96]. Fuzzy rules are of the Takagi-Sugeno type, meaning their outputs are nonfuzzy numerical functions, and their coefficients are adjusted during the learning process. An example of the fuzzy rules used in the fuzzy system is as follows

$$\text{If} \quad p \in P_1 \ \& \ q \in Q_1 \ \& \ v \in V_1 \ \& \ i \in I_1, \quad \text{then}$$

$$y = f_1(p, q, v, i) = \begin{bmatrix} a_1p + b_1q + c_1v + d_1i + e_1 \\ a_2p + b_2q + c_2v + d_2i + e_2 \\ a_3p + b_3q + c_3v + d_3i + e_3 \\ a_4p + b_4q + c_4v + d_4i + e_4 \\ a_5p + b_5q + c_5v + d_5i + e_5 \\ a_6p + b_6q + c_6v + d_6i + e_6 \\ a_7p + b_7q + c_7v + d_7i + e_7 \end{bmatrix}, \tag{6.25}$$

where p, q, v, and i are the active power, reactive power, voltage, and output current of the inverter, respectively. The number of fuzzy rules is 81. Their output is defuzzified by the weighted average method and forms the weight coefficients of the cost function terms. Each line of function $f_1(p, q, v, i)$, is respectively, associated with the cost function weight coefficients, i.e. λ_P, λ_Q, λ_V, λ_{V-Sine}, λ_{I-Sine}, $\lambda_{V-Unbalance}$, and $\lambda_{I-Unbalance}$.

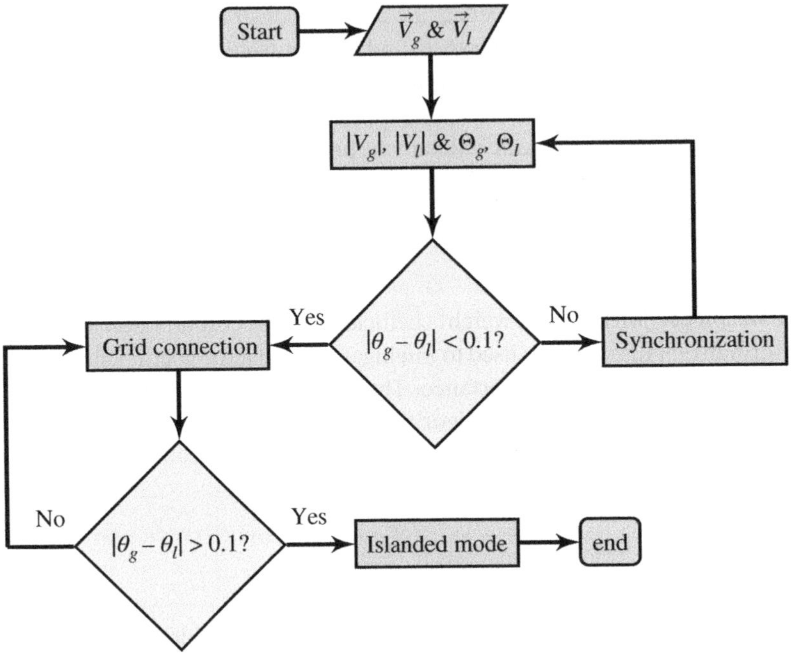

Figure 6.14 The flowchart of the synchronization and phase adjustment algorithm.

The Synchronization and Phase Adjustment Algorithm The algorithm for synchronizing inverter voltage with the grid voltage (shown in Figure 6.14) calculates the phase and amplitude of both voltages and compares the load voltage phase with the grid voltage phase as the reference. If the phase difference is within the acceptable range (usually 0.1 degree), power transfer starts, otherwise, the system waits to reduce the phase difference. The algorithm adjusts the load frequency by increasing it if the grid phase voltage leads, or decreasing it if the inverter phase voltage leads, until the phase difference is reduced, allowing the system to connect to the grid.

The Control Method Flowchart The control method flowchart is depicted in Figure 6.15. The process begins with identifying the operating mode and determining the cost function terms based on measured output voltage and current, as well as calculated active and reactive power. Fuzzy logic selects the initial weight factors, some of which may be zero in certain operating conditions where some control objectives are less important. Next, the cost function is calculated, and the corresponding switching pulse is generated and applied to the inverter. This cycle repeats until a stable control condition is reached, minimizing the cost function. Once the cost function is minimized, the weight coefficients are optimal, and the switching pulse remains unchanged unless the operating mode changes.

6.2.4.3 Simulation Results

The FMPC performance is validated using the test system in Figure 6.13 with the parameters listed in Table 6.2. The simulations are performed in the MATLAB/Simulink environment in two different cases, in the presence of unbalanced load and nonlinear load. The performance of the FMPC

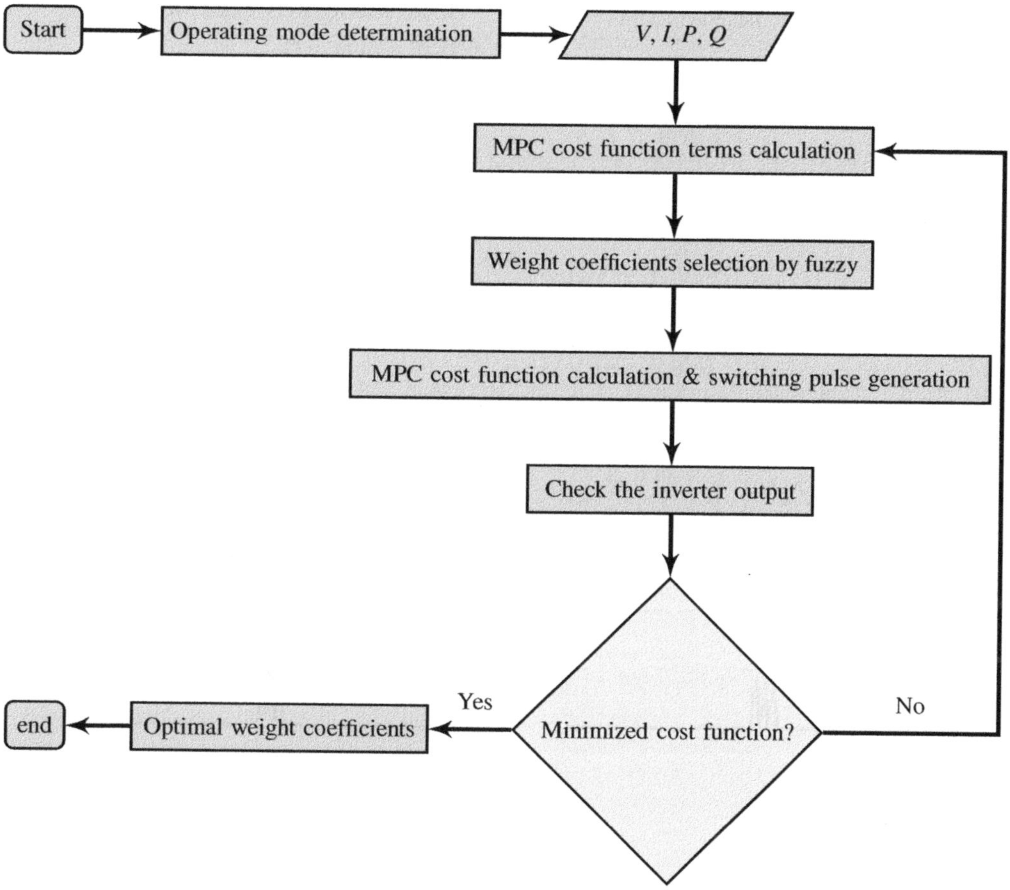

Figure 6.15 The flowchart of the Fuzzy MPC method.

Table 6.2 Electrical parameters of the studied test system.

Parameters	Values
DC voltage source V_{dc}	100 V
DC voltage source series resistance R_{dc}	1 Ω
DC-link capacitance C_{dc}	1 μF
MOSFET resistance R_{on}	0.1 Ω
Internal DIODE resistance R_d	0.01 Ω
Snubber resistance R_s	100 $K\Omega$
Snubber capacitance C_s	∞F
Switching frequency f	1620 Hz
Filter series resistance R_{esr}	1 Ω
Filter inductance L	10 mH
Filter capacitance C	20 μF

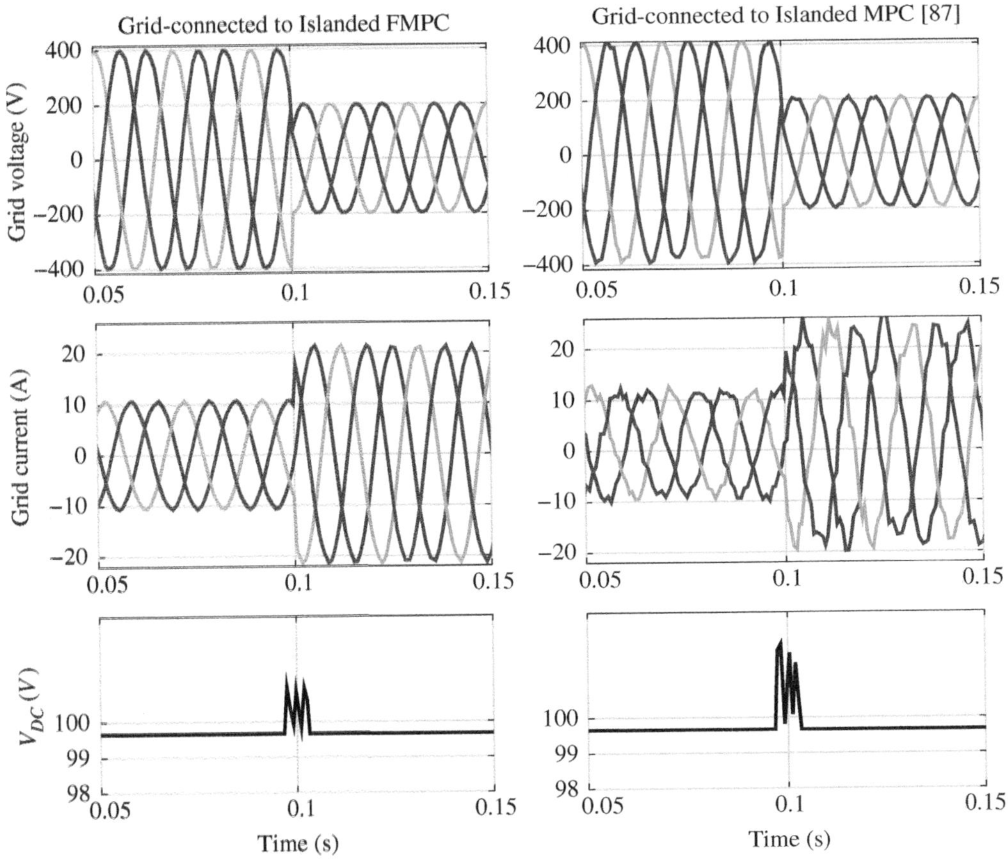

Figure 6.16 Performance of the FMPC and MPC [87] controllers in transition from grid-connected to islanded mode, under unbalanced load.

method in the four operating modes (grid-connected, grid-connected to islanded, islanded, and islanded to grid-connected) of the inverter is compared to the conventional MPC method presented in [87].

Figure 6.16 represents the FMPC performance under unbalanced loads in transition from grid-connected to islanded mode, compared to the conventional MPC. Simulation results show that the FMPC controller outperforms the MPC controller in all operating modes of the inverter, as evidenced by much purer and less distorted output voltage and current waveforms. Additionally, the short-term overshoot observed on the DC link voltage during the inverter transition modes is much lower for the FMPC controller. Therefore, it could be said that the both steady and transient performances of the three-phase inverter under a three-phase unbalanced load are guaranteed.

The same scenario was repeated with nonlinear loads, and results are shown in Figure 6.17 in transition from islanded to grid-connected mode. The simulation results are interpreted as in the previous scenario. Numerical simulations confirm the superior performance of FMPC over conventional methods in terms of various control criteria such as voltage and current THDs, unbalances, DC link voltage overshoot, and peak inrush current. This was mainly due to the proper adjustment of the waiting coefficients of the MPC controller by fuzzy logic. The numerical results are not provided here to save space.

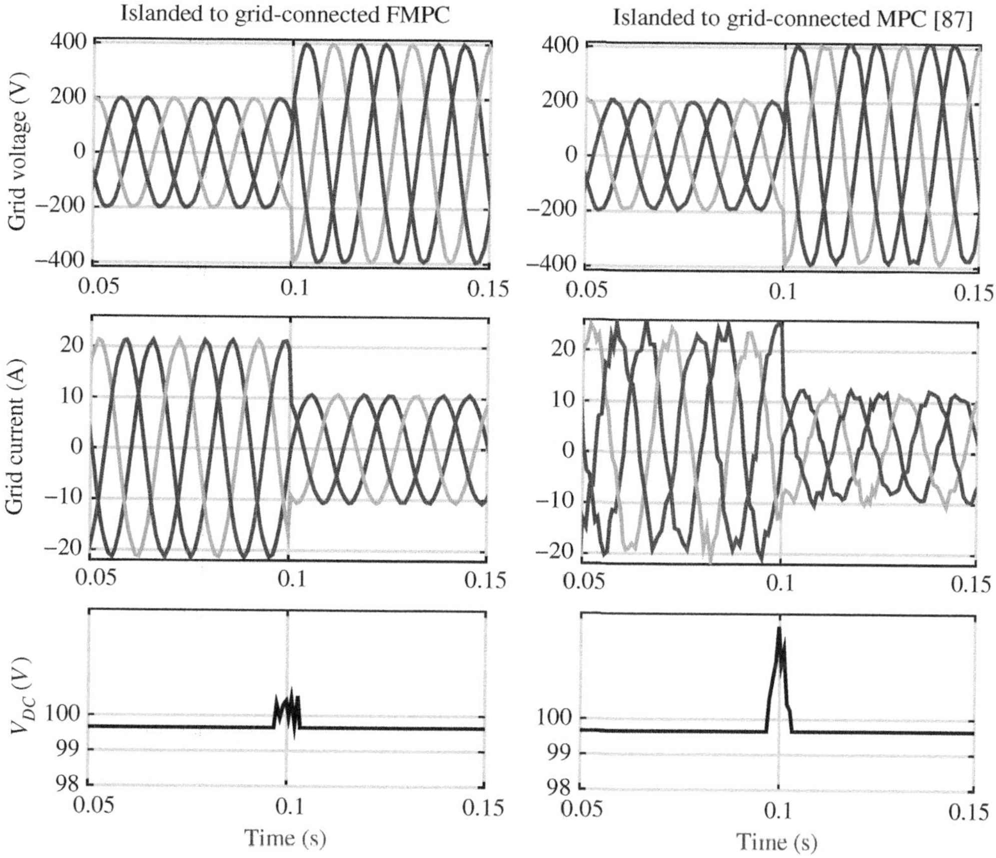

Figure 6.17 Performance of the FMPC and MPC [87] controllers in transition from islanded to grid-connected mode, under nonlinear load.

6.3 Virtual Dynamic Control

VDC is a new concept used to shape various desired power grid dynamics. With this method, power converters are controlled by flexibly adjusting their output active and reactive powers to match specific characteristics or improve system dynamics. This section will introduce the concept, and structure, and provide examples of this control methodology.

6.3.1 Concept and Structure

Power converter-based DERs and MGs, unlike traditional synchronous generators (SGs), provide no or little physical (rotating) inertia and damping support to the power grid. This results in a lack of inertia within the system, leading to a large rate of change of frequency (RoCoF) and low-frequency oscillations in the grid. It may also affect power grid dynamics, voltage regulation, and other control and operation issues [97]. While conventional droop mechanisms, as described in Chapter 5, can be used to share power demand among DERs in an MG, they do not provide inertial support to the power grid, which limits their ability to address this issue. However, the impact of the low-pass filter

(LPF) time constant in droop control has been investigated and demonstrated as a form of virtual inertia for MG frequency, which can be adjusted by tuning the cutoff frequency. Nevertheless, this approach is not as flexible as virtual dynamic methods.

An effective solution to address the challenge of lack of physical inertia and damping in power converter-based systems is to provide virtual ones. This can be achieved by using a stable primary source, e.g. a fast dynamic energy storage to simulate the kinetic energy, coupled with a power converter controlled by a proper control mechanism. This concept is known as *virtual synchronous generator* (VSG) or *virtual synchronous machine* (VSM). The VSG or VSM emulates the behavior of an SG by presenting its inertia and damping characteristics for short time intervals. Therefore, the concept of VSG offers a practical solution to enhance the dynamic stability of MGs and power grids which include MGs [98–100].

The use of power converters for dynamics emulation is not limited to providing virtual inertia and damping characteristics. This concept has been successfully applied to address other power grid/MG dynamic issues, including frequency and active power control [25, 99], voltage regulation [101], reactive power control [102], power sharing [99, 103], and power quality [104]. This means that a power converter can be controlled as a dynamic emulator to shape various desirable power grid dynamics in different directions. To achieve this, the power converter must be controlled in a specific way by flexibly shaping its output active and reactive powers to follow each characteristic or improve any required dynamic.

This general concept is known as *virtual dynamic control*. The VDC can be considered as a reference model-based control system from a control theory perspective, which is used to shape the dynamic behavior of a given power grid based on a desirable dynamic model [105].

The conceptual structure of VDC is shown in Figure 6.18. As can be seen in this figure, depending on the control objective, the VDC strategy performs the required magnitude (v_{ref}) and phase (θ_{ref}) references for the modulation block to ensure that the output power of the converter (ΔP, ΔQ) is properly controlled to shape the required dynamics. The VDC may emulate

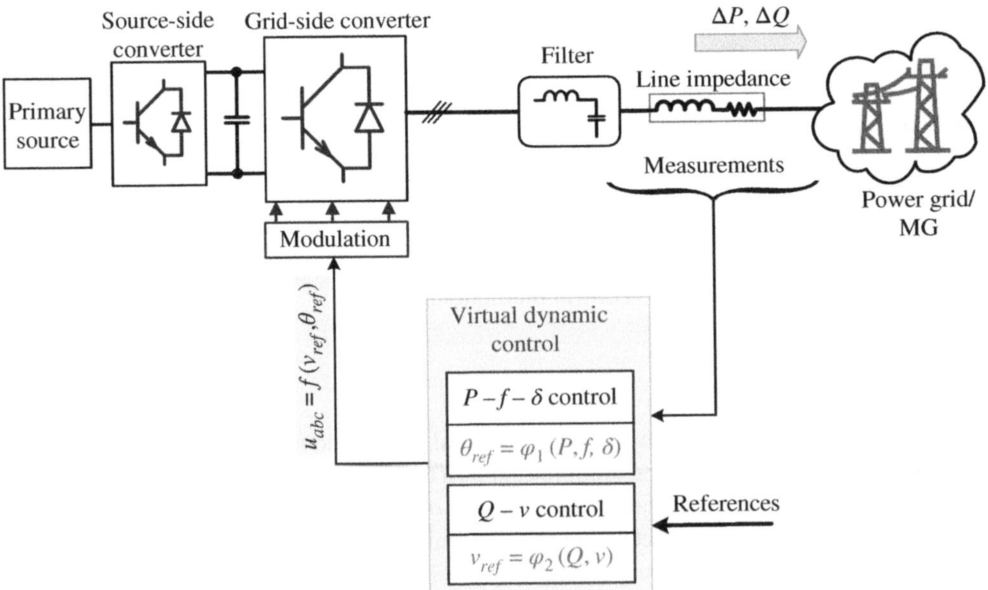

Figure 6.18 A conceptual scheme for virtual dynamic control.

a combination of various control loops (e.g. inertia, damping, droop mechanism, etc.) to achieve desirable system dynamics.

Other promising techniques to emulate virtual dynamics include *virtual impedance*, *virtual oscillator control* (VOC) [106], and *synchronverter* [107]. The VOC digitally emulates a nonlinear oscillator circuit with converter output current as an input, and the oscillator dynamic states are used as the modulation reference to generate PWM switching signals for the converter. While, the synchronverter employs a nonlinear stabilizer to perform self-synchronization and represents the behavior of a synchronous machine, improving the dynamic response. The virtual impedance loop, which is considered a special case for emulating the physical output impedance of power converters, was described in detail in Chapter 5.

The remainder of this section will focus on VSG as a fundamental method for virtual dynamic emulation, followed by an example of implementing VDC on a dc MG. Interested readers can learn more about the other promising techniques in [106–109].

6.3.2 Virtual Synchronous Generator (VSG)

The intrinsic kinetic energy (rotor inertia) and damping property of the bulk SGs play a significant role in the grid stability. As explained previously, the VSG control mechanism mimics these SG properties. In general, the VSG control block embeds a dynamic function similar to the SGs' fundamental swing equation, which determines output power based on grid measurements and their rate of change. Unlike a real SG, the parameters of a VSG can be tuned to improve the dynamic response of the system. Compared to other control techniques, VSG offers several advantages, including improved grid stability, reduced RoCoF, and improved power quality. It can also enable DERs to provide ancillary services to the grid, such as frequency regulation and voltage support, which are traditionally provided by SGs.

Figure 6.19 represents a general VSG scheme for a grid-forming converter in an MG. Comparing the droop mechanism in Figure 5.15, the $\omega - P$ droop control loop is replaced by the VSG control mechanism, while the $v - Q$ droop is intake. The VSG concept actually performs a new grid-forming control strategy with an enhanced $\omega - P$ droop mechanism to provide inertia support. The VSG control loop is composed of two main blocks: virtual governor, and virtual inertia. The virtual governor block mimics the behavior of SG governors, whereas the virtual inertia block emulates SG rotor inertia using the well-known swing equation. The virtual governor of the VSG mechanism can be expressed as

$$P_{in} = P^* - k_\omega(\omega_m - \omega^*) \tag{6.26}$$

where P_{in} is the virtual shaft power defined by the governor, P^* is the reference active power, k_ω is the frequency droop coefficient, ω_m represents the virtual rotor angular frequency, and ω^* represents the nominal angular frequency. The virtual governor expression is indeed the $\omega - P$ droop control mechanism introduced in 5.19a.

The rotor inertia of SGs is emulated using the well-known swing equation as

$$P_{in} - P_{out} = J\omega_m \frac{d\omega_m}{dt} + D(\omega_m - \omega_g), \tag{6.27}$$

where P_{out} represents the measured output power, ω_g is the angular frequency of the grid or the point where the voltage sensor is installed, and J and D are the virtual inertia and virtual damping factor, respectively. Substituting (6.26) in (6.27), one can write the complete VSG equation as

$$P^* - k_\omega(\omega_m - \omega^*) - P_{out} = J\omega_m \frac{d\omega_m}{dt} + D(\omega_m - \omega_g). \tag{6.28}$$

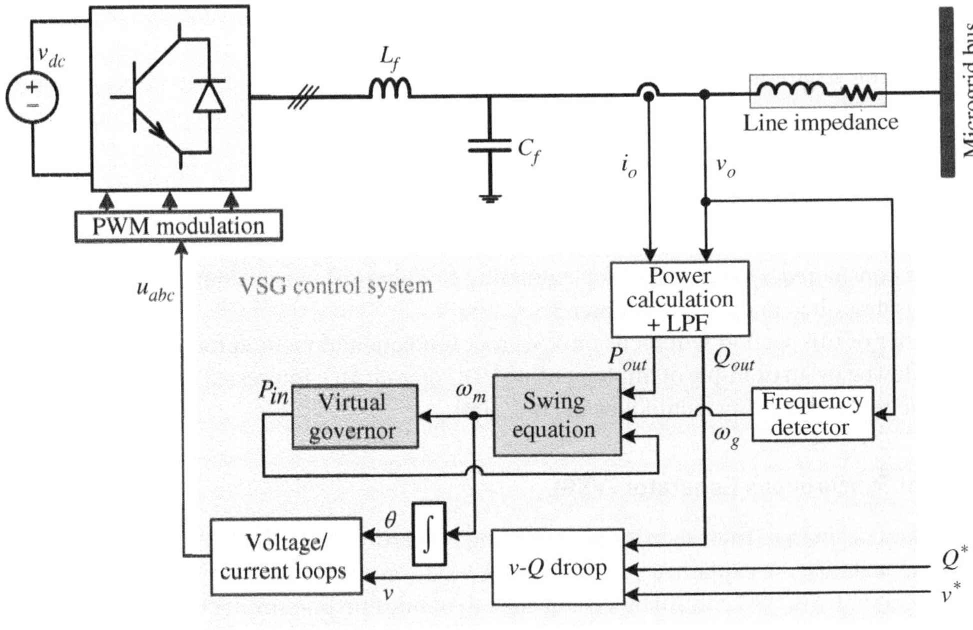

Figure 6.19 A general VSG control scheme for a grid-forming power converter in an MG.

One may consider $L = 0$ and $D = 0$ in (6.28), then the equation is simplified as [110]

$$\omega_m = \omega^* - \frac{1}{k_\omega}(P_{out} - P^*), \tag{6.29}$$

which is equivalent to the $\omega - P$ droop control mechanism in (5.16a) with $k_p = \frac{1}{k_\omega}$. This demonstrates that conventional droop control is a special case of VSG.

The grid frequency ω_g, as shown in Figure 6.19, is detected by a frequency detection method, e.g. PLL (see Section 5.3.3). To avoid using the synchronisation algorithm, the ω_g can be replaced with the nominal frequency ω^* in islanded operation mode.

A typical control scheme of a VSG system is shown in Figure 6.20 [105]. The figure represents the detailed implementation of the VSG's two main blocks described in (6.26) and (6.27) and the other required control blocks for a VSG control. A power converter with the VSG control mechanism can operate in islanded mode, in parallel with other converters, and even directly connected to the power grid, with no control changes required during operation mode transitions [97]. Therefore, the VSG concept performs a new type of grid-forming control strategy with an improved $\omega - P$ control loop, capable of providing inertia support.

Choosing the appropriate value for the VSG parameters, J and D, is a challenging task. These parameters must be chosen so that the VSG exchanges its maximum active power when the maximum specified frequency variation and RoCoF occur. The bigger J and D, the more power is injected/absorbed for the same frequency deviation and RoCoF. Large values of J and D with specific power ratings, on the other hand, result in a small frequency excursion. A slower RoCoF can be achieved with a larger virtual inertia J during contingencies. It is worth noting that the amount of virtual inertia is limited by the energy storage and power converter capacity as well as the PLL accuracy [97].

Section 2.5.1.1 presents a small-signal model for VSG control strategy, which is useful for small-signal stability analysis, sensitivity analysis, and linear control design methods to add

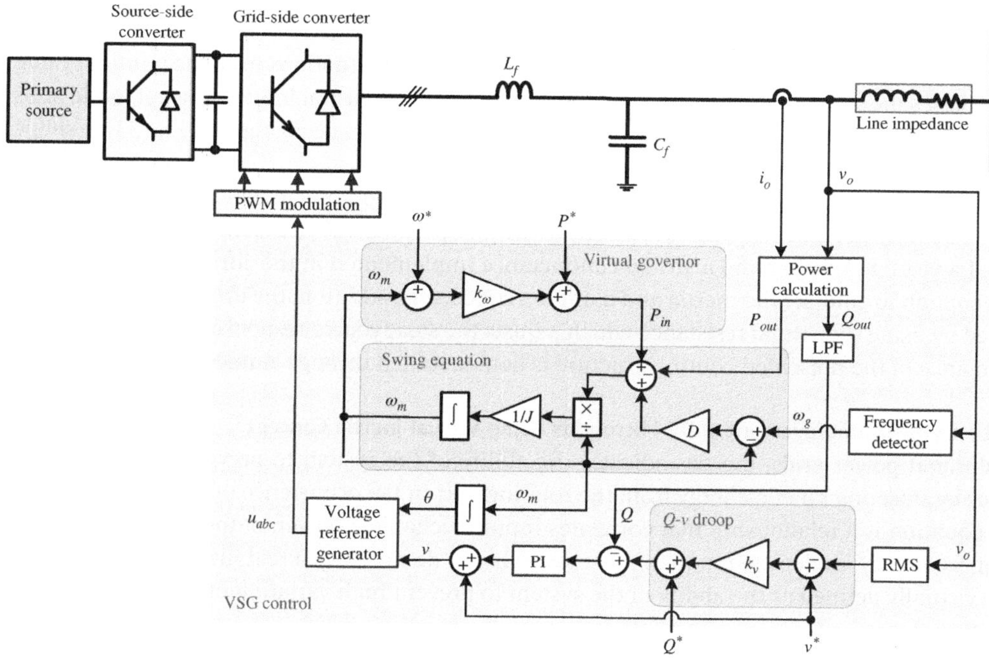

Figure 6.20 A general VSG control scheme for a grid-forming power converter in an MG.

ancillary control loops. Readers are encouraged to refer to Chapter 4 for stability analysis of VDC techniques and their parameter impacts, as the small-signal model presented in Section 2.5.1.1 can be used for this purpose.

6.3.2.1 VSG Applications

Among the VDC methods, VSG is the most developed and various inertia and damping emulation. Methods proposed in the literature [111] to cope with this issue. Small-signal modeling and sensitivity analysis for VSGs are studied in [112] in order to systematically design its parameters. An adaptive virtual inertia method is developed in [113] to improve MG dynamic frequency regulation. In this flexible scheme, a relatively small inertia is emulated to enable faster system dynamics when the frequency starts to return to the nominal frequency, while a relatively large inertia is implemented to limit the RoCoF if the frequency deviates from the rated value once an external disturbance occurs. A self-adaptive VSG control method based on RoCoF [114] and a self-tuning-based algorithm [115] have been used to reduce the frequency deviation and output power fluctuation of MGs. Similar methods to design the VSG parameters are addressed in [97] in order to improve the frequency response. In [103], the authors present a virtual capacitor control to enhance the reactive power sharing in VSG-based MGs. Moreover, the VSG concept has been recently used to enhance power quality in MGs. The virtual dynamic or virtual inertia concept has been recently applied to dc MGs as well [116–118].

Advanced control methods have been also introduced to improve the VSG performance. For example, optimal LQR [119], MPC [120, 121], Fuzzy-Neural-Network [122], and FMPC [123].

It is worth noting that VSG mechanisms have been also proposed to improve the dynamic performance of various applications such as PV systems [124], wind power generation [125], stand-alone gas-engine generators [126], HVDC systems [127, 128], and vehicle-to-grid services [129].

6.3.3 Virtual Dynamic Control of DC Microgrids

Lack of system inertia in dc MGs causes an increase in voltage fluctuations from the nominal value and decreases the stability of the system. The virtual dynamic or virtual inertia concept for dc MGs [116–118] originated from the concept of VSG [130, 131] and virtual dc machine [132, 133] to shape and improve the dynamics of these systems.

This section addresses a virtual inertia control strategy based on linear feedback to improve the dynamic behavior of dc MGs interfaced with CPLs [134]. The presented control scheme is composed of a virtual capacitor and a virtual conductance implemented in the current control loop to be fast enough to emulate the inertia and damping concept. In addition, the droop characteristic is modeled by using the virtual resistance which adjusts the steady-state response of the system. The performance of the presented control structure is demonstrated through numerical simulations.

6.3.3.1 Dynamic Improvement of DC Microgrids Using Virtual Inertia Concept

In traditional power grids, inertia indicates the ability of the system to prevent high-frequency changes by injecting kinetic energy from the rotating part in the occurrence of a fault. In SGs, the swing equation is a relationship that correlates input mechanical power, output electrical power, and the frequency, as depicted in (6.27). For dc MGs, we do not have a real inertia concept, but it can be virtually defined as the ability of the system to prevent high variation of the dc bus voltage due to turbulence.

In order to study the virtual inertia concept, a simplified dc MG is shown in Figure 6.21. The figure includes a dc source connected to the dc load bus through a bidirectional buck converter and a CPL controlled by a buck converter. The primary control is used to handle the dc source including internal voltage and current controllers as well as a v-i droop mechanism and the virtual

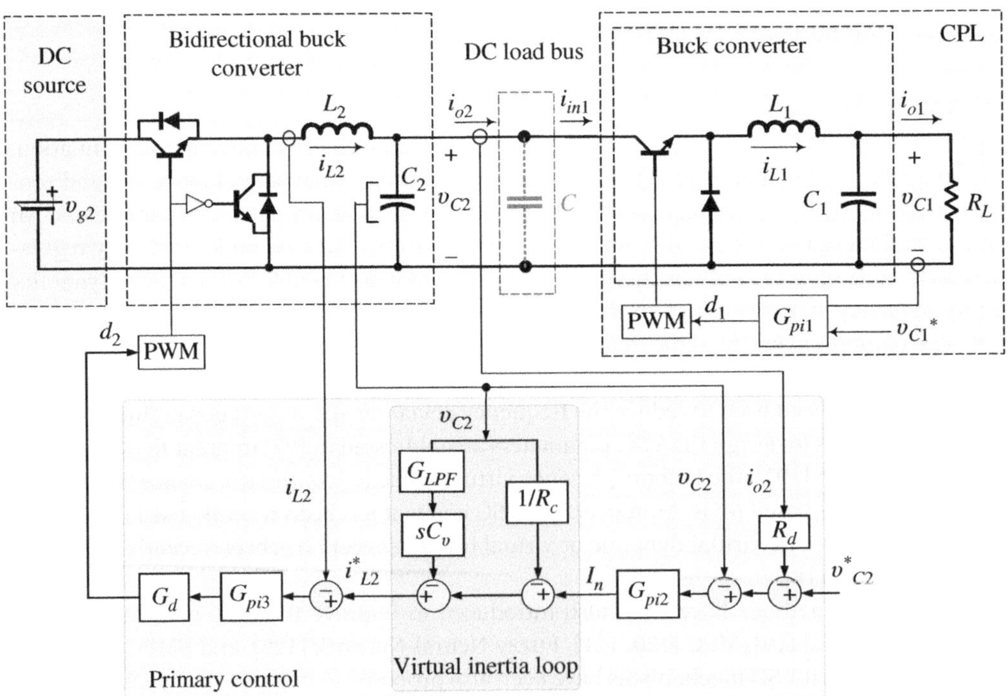

Figure 6.21 Simplified structure of a typical dc MG with CPL equipped with the virtual control loop.

inertia loop. By imitating the swing equation relation (6.27), the balance current equation in the node connected to the dc load bus capacitor is obtained as

$$i_{o2} - i_{in1} = C\frac{dv_{C2}}{dt}, \tag{6.30}$$

where i_{o2}, i_{in1}, v_{C2}, and C are the injected current into the dc load bus from the source-side converter, the load current, the dc load bus voltage, and the dc load bus capacitor, respectively.

Comparing (6.27) with (6.30), one can write that the dc link capacitor in a dc MG shows similar behavior as inertia in SGs. When turbulence occurs in the load current, the dc link capacitor with rapid active power injection prevents the big changes in the dc link voltage. Thus, to increase the inertia in dc MGs, the size of the dc link capacitor has to be increased. Accordingly, instead of increasing the capacity of the dc link capacitor physically, its size can be increased virtually using a virtual capacitor created by a control loop.

Assume that the variables in (6.30) are written as the sum of steady-state variables and their small perturbations, then the corresponding small-signal equation can be expressed as follows:

$$\Delta i_{o2} - \Delta i_{in1} = C\frac{d\Delta v_{C2}}{dt}. \tag{6.31}$$

In such a case, the amount of current to be injected to the load bus can be obtained as

$$\Delta i^*_{o2} = i^*_{o2} - I_{o2} = -\left(C_v\frac{d\Delta v_{C2}}{dt} + \frac{1}{R_c}\Delta v_{C2}\right) \tag{6.32}$$

where i^*_{o2}, I_{o2}, C_v, and $\frac{1}{R_c}$ are the output current reference, the nominal output current, the virtual capacitor and the virtual conductance analogous to the damping factor of SGs, respectively. The negative sign in (6.32) shows negative feedback. By substituting (6.32) into (6.31), assuming that the injected current Δi_{o2} to the load bus follows the reference current Δi^*_{o2}, one can write

$$-\Delta i_{in1} = (C + C_v)\frac{d\Delta v_{C2}}{dt} + \left(\frac{1}{R_c}\right)\Delta v_{C2}. \tag{6.33}$$

The obtained Eq. (6.33) is quite similar to the conventional swing Eq. (6.27) for SGs. According to (6.33), increasing the virtual inertia and the virtual damping factor will result in decreasing the rate of change of voltage (RoCoV) and dc voltage deviation.

The general scheme of the presented virtual inertia loop is illustrated in Figure 6.21, where it is implemented in the primary control layer providing a reference for the inner current loop. The inner loop mechanism applied to the system consists of two modules: voltage controller (G_{pi2}) and current controller (G_{pi3}) with their designed set points. The droop control is used to reach proper load power sharing in the dc MG using the proportional droop gain R_d. The voltage and current controllers employ conventional proportional-integral controllers due to zero steady-state error and easy adjustment. However, the pure differentiator of the virtual capacitance in the introduced method may bring undesired high-frequency noises to the system. To avoid this issue, a first-order low-pass filter (G_{LPF}) is constructed in series with the virtual capacitance feedback. Meanwhile, the PWM delay is expressed as a first-order transfer function that depends on the sampling time. It should be noted that the presented control approach can easily be implemented to all types of dc–dc converters. Here, without loss of generality, a multilevel structure of dc MGs interfaced with a bidirectional buck converter is utilized.

The precise small-signal model of the system considering the introduced virtual inertia loop is presented in [134], where stability analysis is studied by analyzing the system's eigenvalues. Using this small signal model, the impact of virtual controller parameters, the LPF time constant (τ), and the CPL on the system stability, is investigated. It is shown that by selecting proper virtual capacitance ($C_v < 0.02$), the system stability is guaranteed, while larger amounts of C_v move the dominant eigenvalues of the system toward an unstable region. By increasing $\frac{1}{R_c}$ (decreasing R_c), the damping ratio of dominant poles increases and the response speed is decreased. It is also shown that increasing load power of CPL deteriorates the damping and causes instability, whereas selecting the proper time constant ($\tau > 2.5$ ms) moves the unstable poles toward the left side s-plane, hence the system stability is guaranteed.

6.3.3.2 Case Study and Simulation Results

The effectiveness of the virtual control loop is verified using the dc MG test system shown in Figure 6.22. The test system utilizes a three-level configuration. In the first level, namely the source stage, a unidirectional boost converter is employed for PV in order to absorb the maximum available power, while a bidirectional boost converter is used for the battery to provide constant output voltage. The second level consists of an interfacing bidirectional buck converter which establishes the local output bus. Finally, the third level is a common load bus where the CPL is connected. The parameters of the system are listed in Table 6.3. The following scenarios are studied to verify the performance of the virtual inertia loop.

The operation of different elements of the test system's unit 1, i.e. PV output power, battery output power, and CPL power under PV output power reference change, is shown in Figure 6.23. As this figure depicts, the PV output power follows its reference value and feeds the load power together with the battery. When the output power of PV is more than the load power, the difference in their power is stored in the battery ($t > 0.75$ s).

Figure 6.24 shows the performance of the virtual inertia loop in regulating the dynamic response of dc load bus voltage for a +20% change in voltage reference from $V^*_{C2} = 100$ V to $V^*_{C2} = 120$ V. DC bus voltage for various values of C_v while other system parameters are constant is shown in Figure 6.24(a). As the C_v increases, the dc load bus voltage smoothly changes and the RoCoV

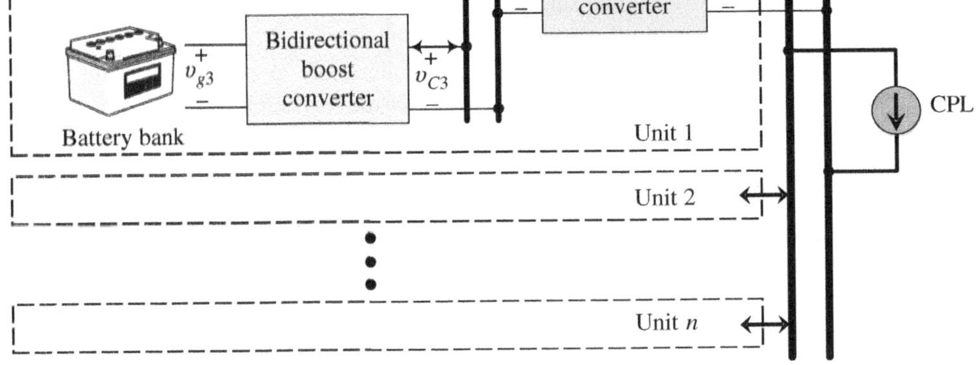

Figure 6.22 Structure of a typical dc MG with CPL as test system.

Table 6.3 Electrical parameters of the dc MG test system in Figure 6.22.

System	Parameters	Values	Parameters	Values
PV system	L4	5 mH	Vg4	55 V
	C4	0.5 mF	VC4	200 V
Battery system	L3	4 mH	Vg3	48 V
	C3	0.8 mF	VC3	200 V
Interface converter	L2	1 mH	Vg2	200 V
	C2	1 mF	VC2	100 V
Load	L1	10 mH	Load power	1 kW
	C1	0.1 mF	VC1	50 V

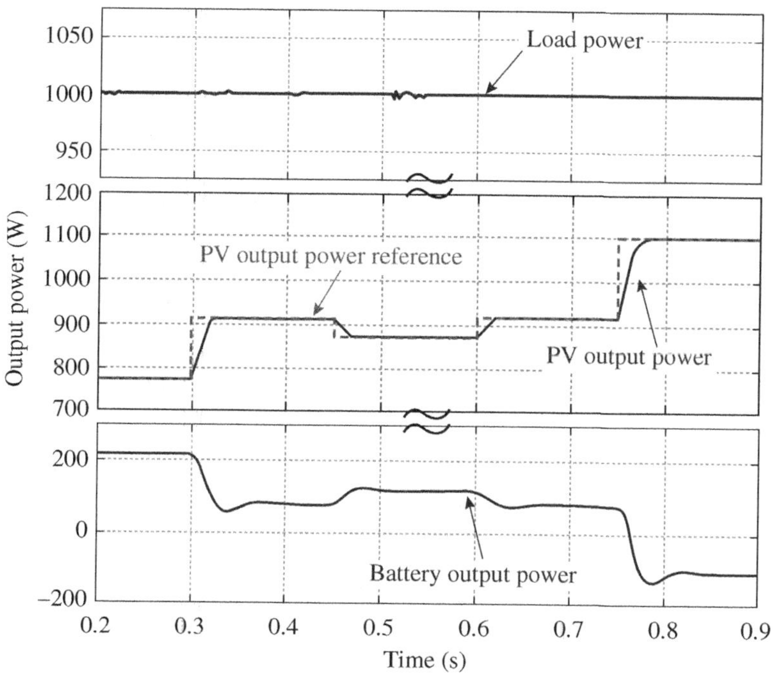

Figure 6.23 Output power of different stage unit 1 when the PV output power reference changes.

decreases. This means that the inertia of the dc MG is increased. However, by increasing C_v the system damping decreases and is prone to cause the DC bus voltage oscillation. This issue can be diminished by increasing the damping coefficient ($\frac{1}{R_c}$) that effectively eliminates the low-frequency voltage oscillations. Figure 6.24(b) shows dc bus voltage for various values of R_c, while other system parameters are constant. By decreasing R_c (increasing damping coefficient), the deviation of dc bus output voltage decreases. One can observe that virtual conductance $\frac{1}{R_c}$ can reduce the deviations of the dc bus output voltage with no influence on the RoCoV when voltage reference changes.

The virtual inertia control behavior is also studied under load power variation. Figure 6.25 shows the dc load bus voltage for a +50% step change in load power from the initial value of $P_o = 1$ kW, to

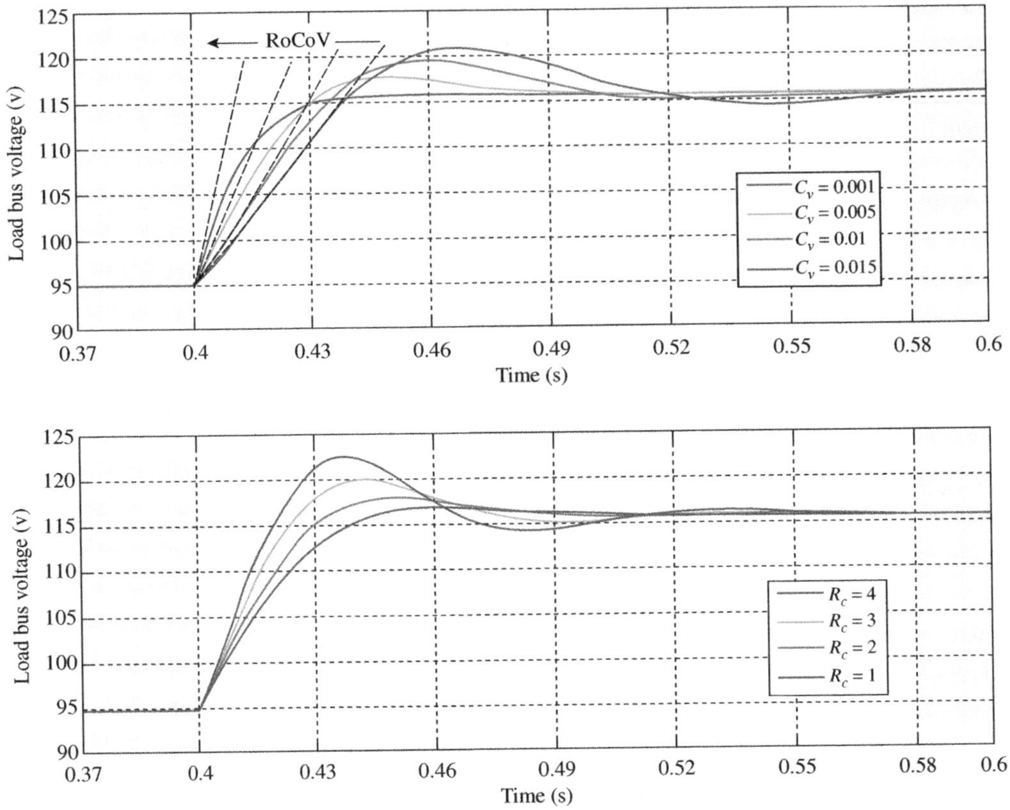

Figure 6.24 Performance of the virtual inertia controller in regulating dc load bus voltage after a step change in voltage reference: (a) varying C_v for $R_c = 2$ and CPL=1 kW, (b) varying R_c for $C_v = 0.005$ and CPL=1 kW.

the final value of $P_o = 1.5$ kW at $t = 0.4$ s. In Figure 6.25 (similar to Figure 6.24), when C_v and/or $\frac{1}{R_c}$ increase (decrease), the RoCoV and voltage deviation will decrease (increase), respectively. Nevertheless, the parameters of inertia response have no influence on the steady-state performance.

6.4 Resilient and Cybersecure Control

Integration of different components and technologies in an MG introduces various cyber-security threats and operational challenges. Therefore, it is crucial to implement robust and resilient control systems that can ensure the safe and secure operation of the MG. This section addresses these new challenges in MGs.

6.4.1 Microgrid as a Cyber-Physical System

CPSs are the integrations of computation, communication, and control that achieve the desired performance of physical systems. Figure 6.26 represents the general scheme and components of a CPS. As shown in this figure, sensors and actuators are the boundaries between the cyber and physical

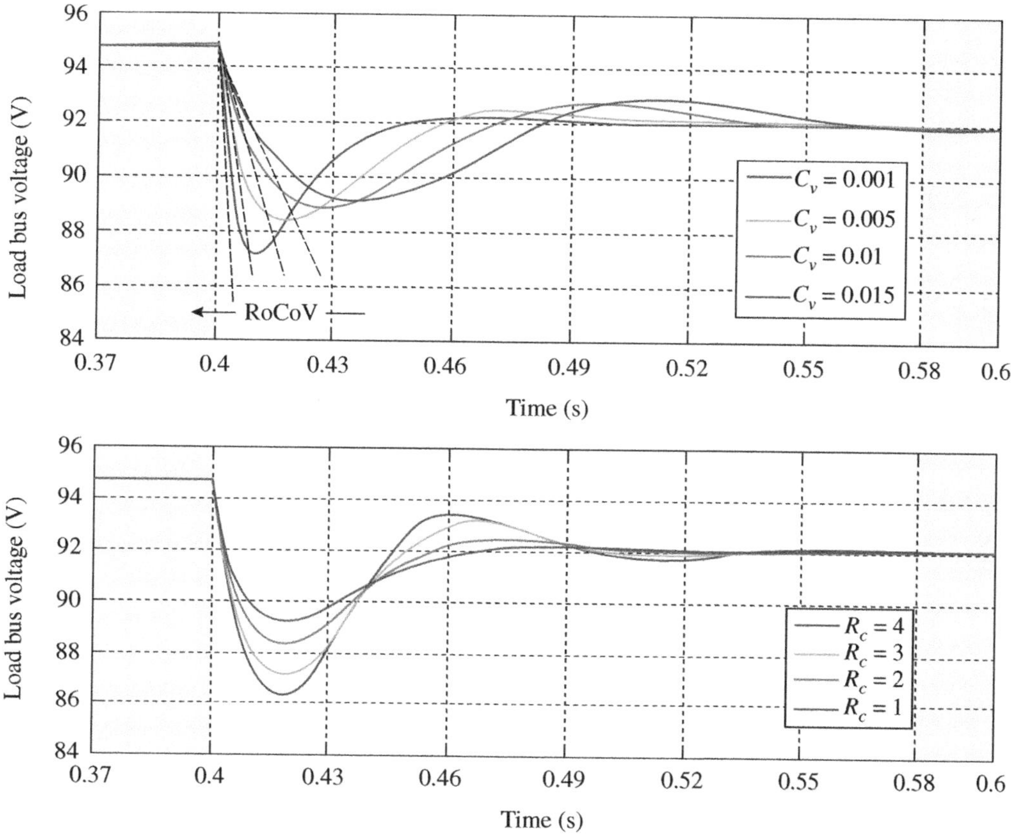

Figure 6.25 Performance of the virtual inertia controller in regulating dc load bus voltage after a step change in load power: (a) varying C_v for $R_c = 2$ and CPL=1 kW, (b) varying R_c for $C_v = 0.005$ and CPL=1 kW.

layers. Because of using processing units in control of these devices, they can also be categorized as a part of the cyber layer.

Smart grid [135] and MG [136] can be described as a CPS because they integrate both physical and cyber components to manage and control the distribution of electrical power. On the physical side, MGs consist of DERs, power converters, LC filters, sensors, loads, and the electrical network. On the cyber side, MGs use advanced control synthesis and communication technologies to monitor and control the system. The integration of cyber and physical components is what makes MGs a CPS. Please refer to Figure 5.9 to see the details of the physical and cyber layers in MGs.

MGs, as CPSs, face several challenges that need to be addressed to ensure a reliable and secure operation. Here are some of the main challenges:

- *Cyber-security*: MGs are vulnerable to various cyber-security threats, such as hacking, viruses, malware, and communication network (CN) attacks. These threats can affect the availability, integrity, and confidentiality of the MGs, leading to disruptions in the MG's operation, and causing safety hazards, equipment damage, and financial losses. Cyber-security for MGs will be further elaborated in the Section 6.4.3 supported by some simple examples.
- *Bandwidth and latency of communication network*: MGs generate a vast amount of data from various sources, such as sensors, smart meters, and control systems. This data needs to be transmitted

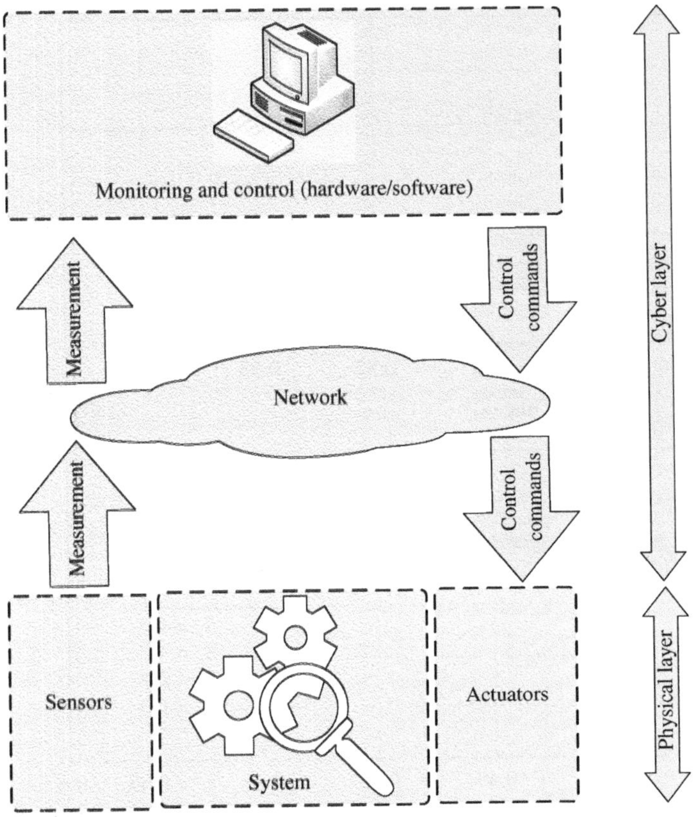

Figure 6.26 General scheme and components of a cyber-physical system.

over the CN to the (centralized or/and distributed) controllers for real-time analysis and control. However, the CN's bandwidth and latency limitations can affect the data transmission quality, resulting in delays and data loss. The provided example in Section 6.4.4 addresses this challenge.

- *Control system complexity*: The integration of physical and cyber components in MGs can make control systems more complex, requiring advanced modeling, control, and optimization techniques. This complexity can increase the risk of control errors and system failures if not managed properly.
- *Communication network reliability*: The communication network that connects physical and cyber components must be reliable and resilient to ensure effective control and coordination even under adverse conditions. Communication networks can be affected by various factors, such as physical disturbances, cyberattacks, and power outages.

6.4.2 Communication Requirements

In a typical MG, data is transmitted between nodes through communication links using various protocols. The nodes act as points where controllers or sensors add data to the link. Data transmission in MGs can be classified according to the type of protocol used, whether it is wired or wireless, and other factors that identify the source and destination [137].

Traditionally, wired communication technologies such as RS-232/422/485, IEEE 802.3, and Power Line Communication have been preferred for transmitting data through the electrical grid due to their superior performance over wireless technologies like WLAN, IEEE 802.15, ZigBee, WiFi, and cellular networks. However, wired systems have higher installation and maintenance costs and may have limitations in terms of adding new devices. On the other hand, wireless systems are user-friendly, but they may experience signal conflicts and power limitations. Furthermore, wireless systems are more vulnerable to cyberattacks since intruders do not need physical connections to access links [138].

When DERs are located close to each other, a local area network (LAN) is employed, whereas if the DERs are far apart, a wide area network (WAN) is required. LAN systems use media access control (MAC) addresses, while WAN systems use Internet Protocol (IP) addresses. In a LAN, each DER can be identified by its MAC address, and data can be transmitted through broadcasting. In contrast, each DER in a WAN adds a source and destination IP address and sends data through the TCP/IP or UDP/IP protocol since broadcasting is not allowed in WAN [138]. Thus, routing methods must be used to transmit data across the network.

From the cybersecurity point of view, using TCP/IP-based protocols offers some benefits such as data encryption and digital signature; however, it requires stronger processors and increases time delay. Since each MG has a unique communication infrastructure, the communication method depends on various factors, including the data type being sent, and the distance between sender and recipient. Figure 6.27 illustrates a typical communication structure between two controllers over a WAN.

6.4.3 Cybersecurity

The growing use of communication infrastructure and information technology systems in MGs has made them vulnerable to cyber-security threats. Cyber-security in MGs refers to the protection of MG systems and data from cyber-attacks.

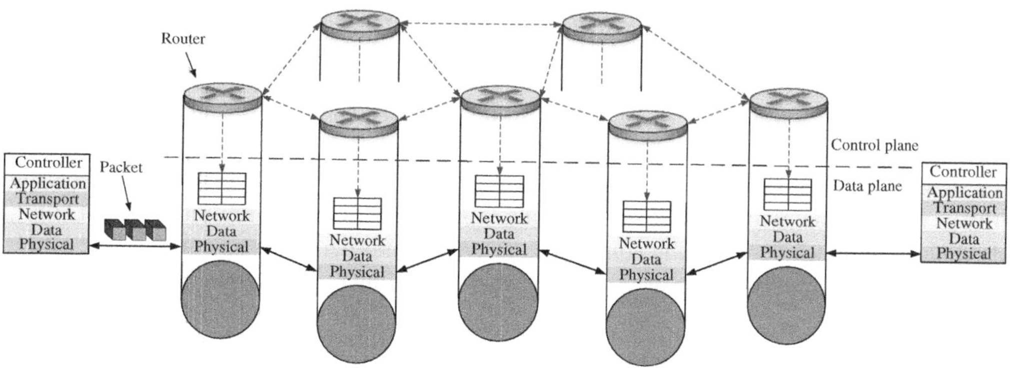

Figure 6.27 A typical scheme for communication of two controllers over WAN.

The vulnerabilities of MGs to cyberattacks can arise from various sources, such as software or hardware vulnerabilities and poor security policies [139]. Some of the most common types of cyber threats on MGs include:

- *Malware attacks*: Malware is a type of software that can be designed to target the control systems of MGs, infecting them and causing significant disruptions to the power supply.
- *Physical attacks*: Physical attacks can also be a significant threat. An attacker could destroy or damage critical components of the MG, resulting in a loss of power supply.
- *Unauthorized access*: One of the biggest vulnerabilities of MGs is unauthorized access to the control system. If an attacker gains access to the control system, they could take control of the MG and disrupt the power supply, causing significant damage to the grid and the connected devices.
- *Network cyber attacks*: Data flow and network infrastructure are susceptible to various vulnerabilities that can compromise the security of an MG. Data is collected by sensors and transmitted through the network to the monitoring and control unit. Occasionally, there are connections between controllers for data exchange. However, attackers can intercept data packets by "sniffing" on communication networks, enabling them to obtain sensitive information about the system. Attackers can manipulate data in several ways, such as flooding the MG communication network with traffic (denial of service (DoS) attack), adding false data to the networks (false data injection attack), or taking control of communication between two parties, e.g. two controllers or two devices (hijacking or man-in-the-middle attack).

To prevent these vulnerabilities and secure the MG systems, it is crucial to employ several strategies such as implementing security measures in the communication networks, using advanced encryption techniques, and deploying intrusion detection systems and mitigation methodologies.

6.4.3.1 Network/Data Cyber Threats on Microgrids

The MG control system infrastructure comprises the control hierarchy and communication links that transfer essential data, with communication between sensors, actuators, and controllers forming the main parts of the cyber layer. All of this equipment can be considered a target for intruders, with attackers typically using malware to target the controller. In such attacks, the hacker needs to have direct physical access to the system in order to install the malware on the system; however, once the control system is connected to the internet, physical access is no longer necessary for a cyberattack. In this case, intruders attempt to circumvent the system security (e.g. firewall or intrusion detection) via communication links.

Various components within the cyber layer, including sensors, controllers, and communication links, are vulnerable to cyberattacks. This section introduces three well-known data cyberattacks on MGs: false data injection (FDI), denial of service (DoS), and hijacking. The FDI attack affects the integrity of information, while the DoS attack interrupts access to the communicated data [140], and the hijacking attack corrupts the control of the system and perturbs the entire signal [141]. More information about these types of attacks and their modeling for MG applications is provided in the following.

FDI Attack Once the transferred information of the MG via communication links is manipulated by false data, the attack can be modeled as [141]

$$x_i^j = x_i + \eta_i x_i^a, \tag{6.34}$$

where x_i^j is the DER i data transmitted to DER j. x_i^a is the corrupted data injected into the controller i and $\eta_i = 1$ indicates the presence of an attack. This type of attack may cause instability or lead to convergence on incorrect output values.

Hijacking Attack In case of a hijacking attack, the intruder replaces the correct data with malicious data. This can be modeled as [142]

$$x_i^c = \left(1 - \eta_i\right) x_i + \eta_i x_i^a, \tag{6.35}$$

where x_i^c is the manipulated data, x_i^a is the false data replaced with correct data, and $\eta_i = 1$ shows an attack occurred on the DER i.

DoS Attack A DoS attack on the agents' exchanged data can be modeled as

$$x_i^j = x_i \times \delta, \tag{6.36}$$

where $\delta \in \{0, 1\}$, (0) means the system is under DoS attack and (1) shows a safe condition. These attacks may target sensor output data, control signals, or data transmitted via communication links.

There are two more types of attacks on MGs from the perspective of their impact on control parameters: *destabilization attacks* and *deception attacks*. A destabilization cyberattack forces agent output parameters outside the allowable operational limits and moves the system toward instability. On the other hand, a deception attack occurs when an intruder penetrates the control system and changes MG parameters without affecting stability. This type of attack may lead to convergence on incorrect output values.

6.4.3.2 Distributed Secondary Control Under Network Cyber Attacks

In the distributed secondary control framework of MGs, controllers and communication links are prone to cyber threats as shown in Figure 6.28. This section investigates the vulnerability of distributed secondary control of MGs against well-known network/data cyberattacks.

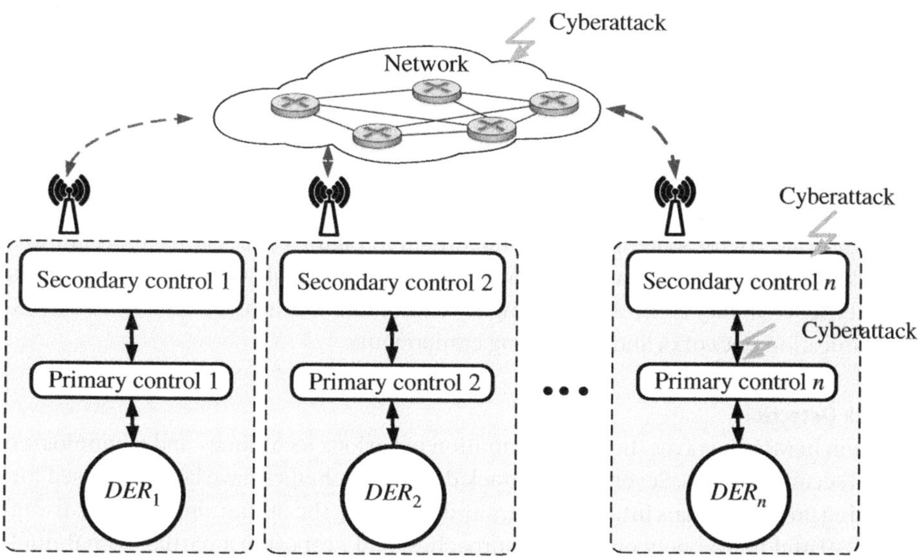

Figure 6.28 Vulnerability of microgrids distributed secondary control under cyber attacks.

As described in Chapter 5, Section 5.6, various distributed secondary control algorithms have been introduced in the literature, mostly based on consensus protocols. Here, the consensus-based distributed secondary control presented in [143] is used here:

$$e_{\omega i} = g_i^f \sum_{j \in N_i} a_{ij}(\omega_i - \omega_j + b_i(\omega_i - \omega_{ref}) + g_i^p \sum_{j \in N_i} a_{ij}\left(m_i P_i - m_j P_j\right), \tag{6.37}$$

where ω_i and ω_j indicate the frequency of the DER i and DER j, ω_{ref} is the reference frequency, N_i is the set of neighboring DERs, a_{ij} is the adjacency matrix coefficient, and b_i is a positive value. g_i^f and g_i^p are positive feedback gains, and $e_{\omega i}$ is a local neighborhood tracking error. The correction term in (6.37) to be added to the droop mechanism is calculated using a control gain (c_ω) as follows:

$$\delta u_{\omega i}(t) = -c_\omega e_{\omega i}(t). \tag{6.38}$$

Considering Eqs. (6.34), (6.4.3.1), and (6.36), one can calculate the amount of error after FDI, DoS and hijacking attack as follows:

$$e_{\omega j}^a = e_{\omega j} - \eta_i \omega, \tag{6.39}$$

$$e_{\omega j}^a = e_{\omega j} + \Delta, \tag{6.40}$$

$$e_{\omega j}^a = e_{\omega i} + \eta_i \omega N_i, \tag{6.41}$$

where $e_{\omega j}^a$ is the calculated error of the compromised agent, $e_{\omega j}^a$ is the error of the neighbors, and Δ shows the impact of the attack on the error magnitude in each agent.

Case Study and Simulation Results The vulnerability of the distributed secondary control under different attacks is studied using the MG test system shown in Figure 6.29, including four dispatchable DERs with all related electrical and control parameters. The MG nominal frequency and voltage are 50 Hz and 220 V, respectively. Regarding the communication graph, DER_1 has bidirectional communication with DER_2 and DER_4, and DER_3 exchanges data with DER_2 and DER_4. DER_1 receives a referenced signal with pinning gain $g_1 = 1$.

Figure 6.30 shows the consensus-based secondary control [143] response to FDI, DoS, and hijacking attacks. The attacks occur on the connection link between DER_2 and DER_3 at $t = 2.5$ s. For the FDI attack, the intruder adds the constant value of 1 Hz to the link data, while for simulating the hijacking attack, the data exchanged between DER_2 and DER_3 is replaced with 51 Hz. The frequency and power rate (mP) of the system units are illustrated in Figure 6.30. As this figure shows, the data cyberattacks affect all the units and lead the system to an unstable condition. Simulation results indicate that the secondary control layer is highly vulnerable to data cyberattacks due to its reliance on communication networks and computing components.

6.4.3.3 Cyberattack Detection

Cyberattack detection in MGs involves the implementation of various techniques and technologies to identify potential security threats. Several cyberattack detection schemes have been proposed for MGs, such as injecting probing signals into controllers and analyzing the output signals [144], using machine learning [54] and neural network [145] approaches, and proposing a partial primal-dual algorithm for distributed detection [146]. In particular, there are various detection strategies for detecting the FDI attacks [147–149].

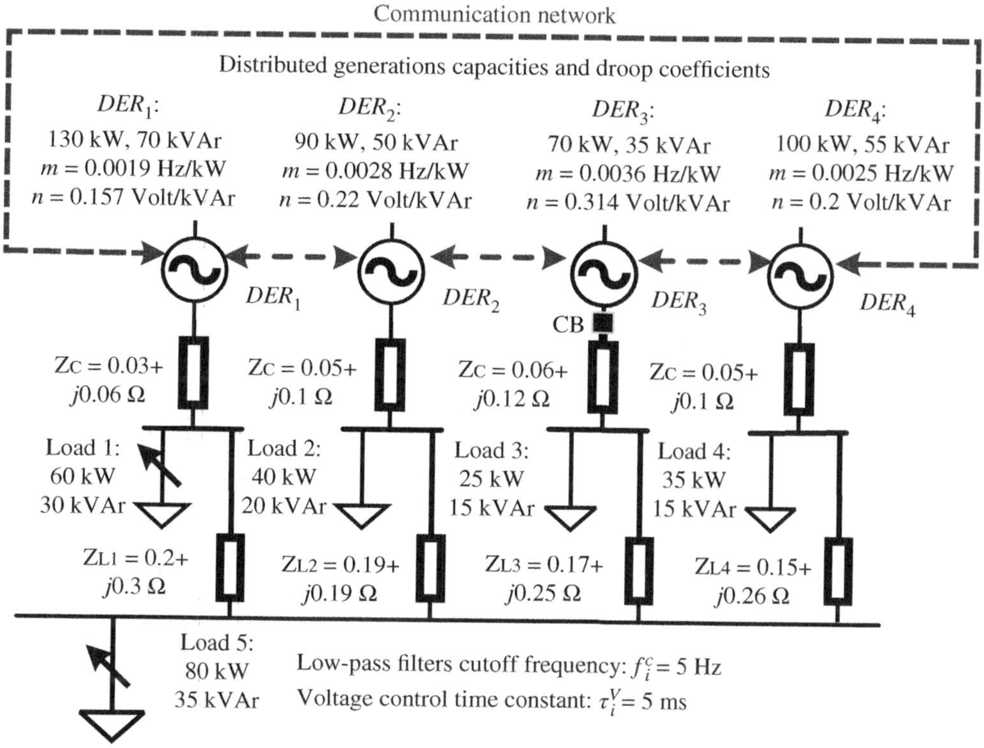

Figure 6.29 MG case study for evaluating vulnerability of distributed secondary control under cyber attacks.

Although detection methods can identify cyber-attacks, they do not provide any correction for their impact. The following provides a simple method to detect deception attacks on secondary control of MGs.

Attack Detection on the Secondary Control of ac MGs A simple distributed detection method is presented to identify deception attacks in ac MGs [150]. The method involves using online estimators to identify system characteristics. These estimators are positioned near the secondary controllers and possess the same input and output as the secondary controllers. By monitoring the behavior of the relevant parameters, this approach can detect any abnormal changes in the system variables.

In a distributed secondary control, the parameters of active power and frequency, or reactive power and voltage, are exchanged between units via communication links, thus they are susceptible to intruder attacks. If an intruder alters these parameters, the estimator will detect unusual changes, leading to the detection of a cyberattack. The study employs the regressive least squares (RLS) method to determine the relationship between these parameters. Figure 6.31 illustrates the RLS estimator based on output measurement.

The online estimator can be similarly applied to monitor the reactive power-voltage parameters. If any of these parameters are attacked, the estimator outputs will be significantly altered. Figure 6.32 demonstrates the effectiveness of the presented detection method in identifying a deception attack on DER_4 frequency measurement. At $t = 3$ s, the power demand of the MG increases, and at $t = 6$ s, a deception attack occurs on the DER_4 frequency communication link. The amount of false data added to the communication link is 0.03 Hz. In the estimators' outputs

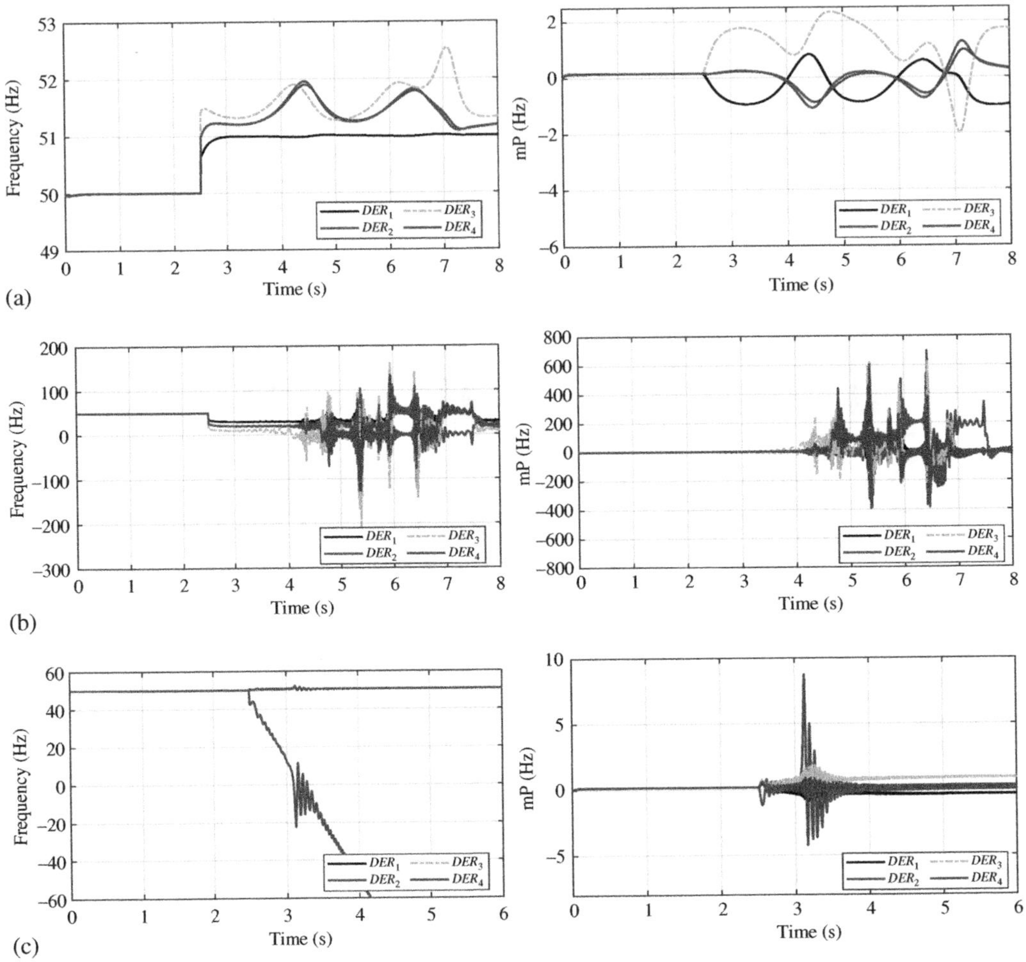

Figure 6.30 Frequency and power rating of the MG test system under: (a) FDI attack, (b) DoS attack, and (c) hijacking attack, on the link between units 2 and 3.

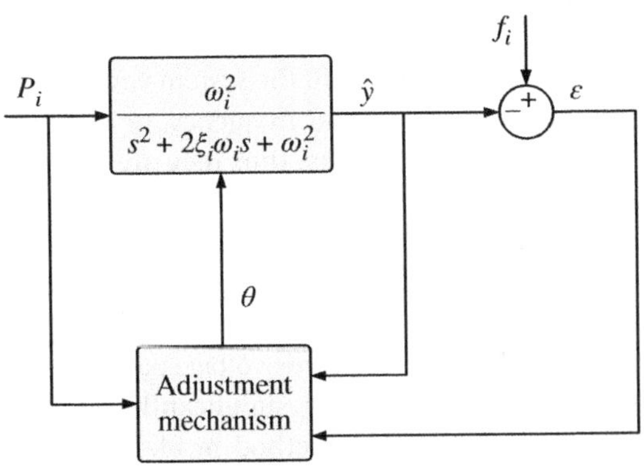

Figure 6.31 Block diagram of an active power/frequency least-squares estimator based on output measurement.

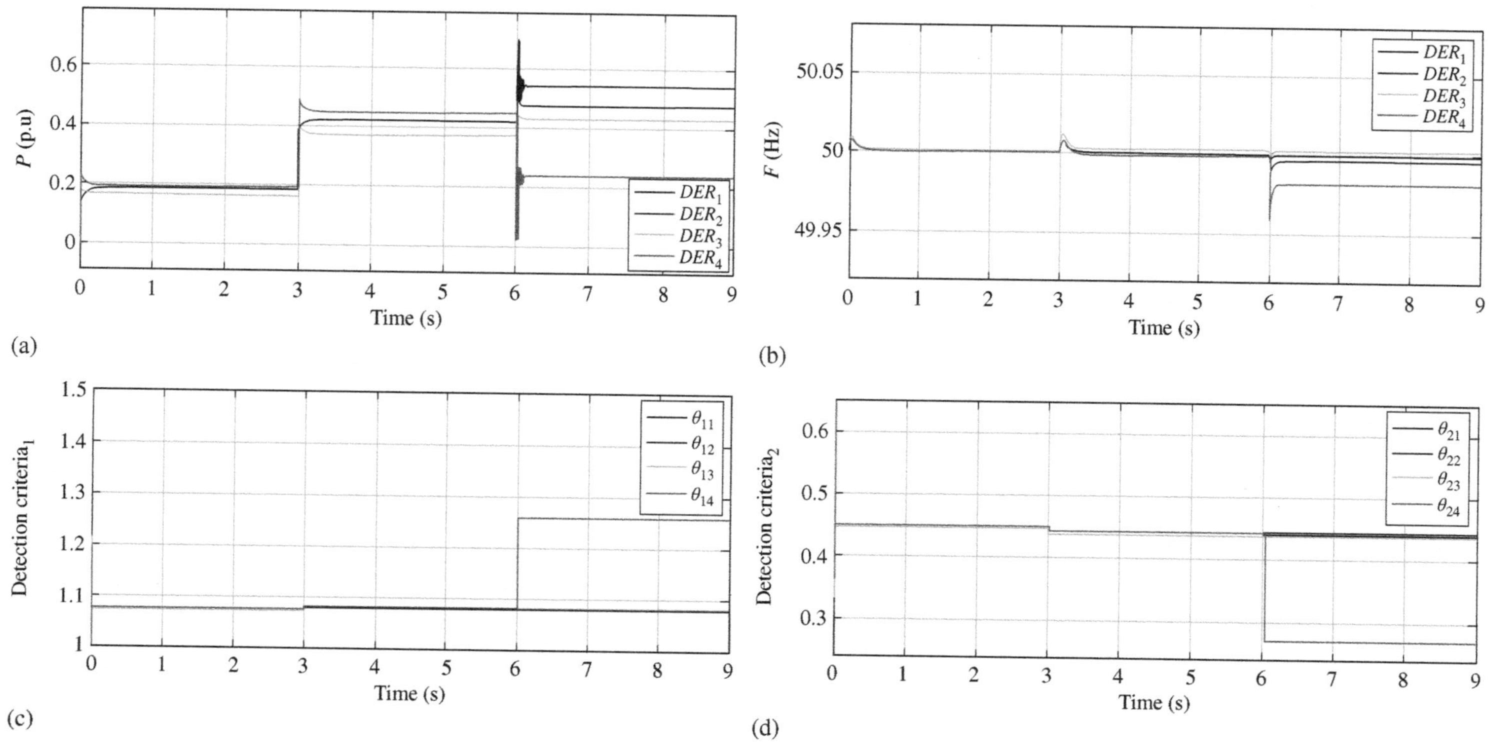

Figure 6.32 Performance of the presented detection approach in detecting deception attack on frequency measurements: (a) DERs output active powers, (b) DERs frequency, (c) and (d) estimators outputs.

shown in Figure 6.32, the first index refers to the parameter number, and the second index represents the DER number. For instance, θ_{11} denotes the first parameter of the first DER. Notably, when the load changes, the estimator output parameters exhibit negligible changes. However, in the presence of a cyberattack, these parameters experience significant variations.

Figure 6.33 illustrates the performance of the detection method against deception attacks on voltage measurements. In this scenario, the value of the injected false data is 0.03 (p.u.). As shown in the figure, the load changes at $t = 3$ s do not influence the parameter estimation. The presented approach demonstrates robust performance in detecting deception attacks on voltage measurements as well.

The simulation results show that the data sent to the secondary layer are disrupted, but the amount of incorrect data is too small to affect the normal operation of the MG. Therefore, the voltage and frequency values remain within the allowable range.

6.4.3.4 Cyberattack Mitigation

Mitigation methods refer to a set of techniques and strategies designed to maintain the stable operation of the system in the presence of cyber threats. To enhance the resilience of MGs against cyberattacks, several approaches can be employed, such as implementing security measures in the communication networks, employing advanced encryption techniques, or deploying intrusion detection systems. Additionally, resilient control techniques can be used to mitigate the effects of cyberattacks on the MG's control system. These techniques involve developing control algorithms that can quickly adapt to the changing conditions of the MG and provide optimal control even in the presence of cyber threats.

Resilient control strategies for MGs typically involve three main steps:

1. Detecting cyberattacks and identifying their impact on the MG (as shown in the provided examples in 6.4.3.2 and 6.4.3.3),
2. Designing control algorithms that can operate under normal and attack conditions, and
3. Deploying these algorithms in the MG control systems.

The mitigation methods often involve the use of advanced control techniques, machine learning algorithms, and probabilistic models to adapt to the changing conditions of the system and detect and mitigate the effects of cyber-attacks.

Various methods have been proposed in the literature to mitigate the impact of cyberattacks on MGs, such as observation-based techniques [151], artificial intelligence-based methods [152], and event-triggered methods [153]. Distributed observers have also been proposed to track reference values and eliminate the effects of cyber-attacks [141]. Some researchers have focused on developing resilient distributed secondary control methods for MGs [154–156].

6.4.4 Event-Triggered Control

From a system scaling standpoint, efficient usage of communication medium is mandatory. In networked control systems, a solution to avoid probable network congestions and reduce the communication burden is to use event-triggered (ET) control strategies to eliminate redundant communications [157, 158].

ET control is a type of control method that is becoming increasingly popular in MGs. In this approach, control actions are only triggered when certain events occur, rather than executing

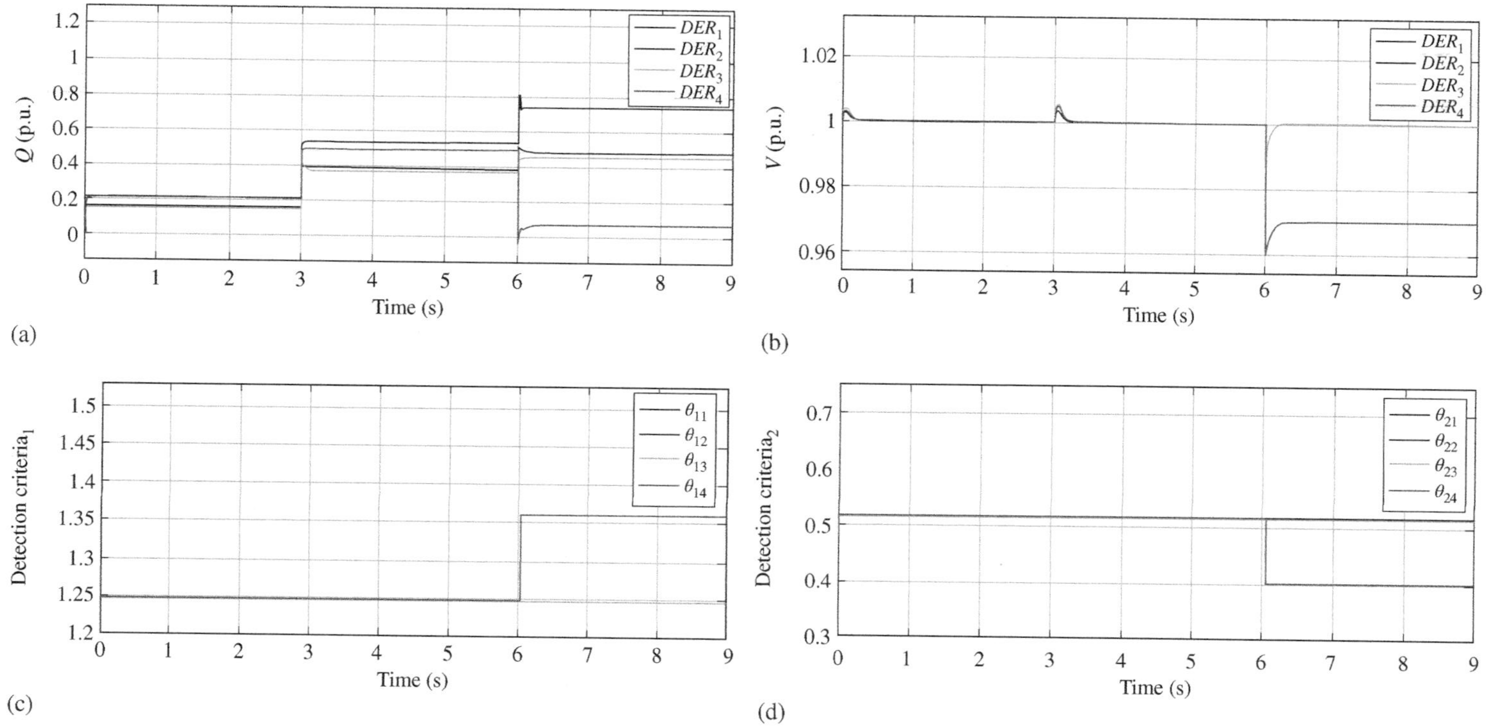

Figure 6.33 Performance of the presented detection approach in detecting deception attack on voltage measurements: (a) DERs output reactive powers, (b) DERs voltage, (c) and (d) estimators outputs.

control actions at fixed time intervals. This results in more efficient use of resources, as control actions are only taken when needed, reducing the computational load and communication bandwidth required by the control system. The events that trigger control actions can be based on various parameters such as changes in the system state, disturbances in the system, or changes in the control objectives. For example, an event may be triggered when the voltage level of the MG exceeds a certain threshold or when there is a sudden increase in the demand for power.

Generally, the ET control rule is updated (i.e. an event is triggered), only if some condition(s) is(are) satisfied. Any ET condition must satisfy two system requirements: Stability and Zeno-freeness. *Zeno behavior* is a phenomenon under which excessive redundant events are triggered over a finite time interval. Because of limited communication and computational capabilities, no control system can be implemented on a digital platform, if the behavior exists [157, 158].

Triggered control of MGs has been introduced in the literature for different objectives [159–169].

6.4.4.1 Event-Triggered Secondary Control of AC Microgrids

As described in Chapter 5, Section 5.6, the secondary control level is mostly realized by using a CN. Therefore, distributed control architectures using sparse CNs are preferred to the centralized ones with complex CNs [170].

Most of the previous works in the context of secondary control are conducted based on the continuous CNs (e.g. see [32, 36, 38, 171–179]), while the realistic data exchange infrastructures are sample-based and have limited bandwidths [180, 181].

This section introduces a secondary voltage, frequency, and active power sharing control for autonomous inverter-based MGs with event-triggered communications [182]. A proportional-integral consensus-based control scheme is introduced which benefits from need-based (event-triggered) data exchange among distributed generators. The effectiveness of the presented controller for various case studies is verified via MATLAB/Simulink-based simulations.

6.4.4.2 Physical and Control Layers

From Figure 6.34, each DER in an MG comprises a voltage source converter (VSC), an LC filter, inner current, and voltage controllers, power measurement filters, and a droop-control unit. Details

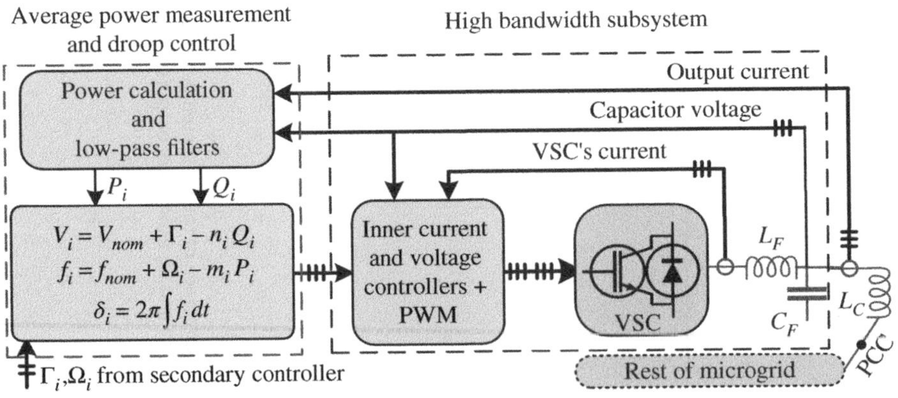

Figure 6.34 Physical and control layers of *i*th DER in a microgrid.

on the design of these parts [183, 184] are provided in Chapter 5. The inner controllers are normally designed such that the subsystem comprising them, the inverter, and LC filter has a high bandwidth but the average power measurement filters are very slow. Therefore, neglecting the dynamics of the mentioned subsystem, one can consider each DER as a droop-controlled voltage source which is connected to the rest of the MG through a connector. The state model of ith DER can then be expressed as

$$\dot{\delta}_i = 2\pi f_i, \tag{6.42a}$$

$$\dot{P}_i = 2\pi f_c(\hat{P}_i - P_i), \tag{6.42b}$$

$$\dot{Q}_i = 2\pi f_c(\hat{Q}_i - Q_i), \tag{6.42c}$$

$$f_i = f_{nom} + \Omega_i - m_i P_i, \tag{6.42d}$$

$$V_i = V_{nom} + \Gamma_i - n_i Q_i. \tag{6.42e}$$

In a Kron-reduced electric network model [185], $\hat{P}_i$ and $\hat{Q}_i$ are as

$$\hat{P}_i = G_i V_i^2 + \sum_j V_i V_j G_{ij} \cos(\delta_i - \delta_j) + \sum_j V_i V_j B_{ij} \sin(\delta_i - \delta_j), \tag{6.43a}$$

$$\hat{Q}_i = -B_i V_i^2 - \sum_j V_i V_j B_{ij} \cos(\delta_i - \delta_j) + \sum_j V_i V_j G_{ij} \sin(\delta_i - \delta_j), \tag{6.43b}$$

where, $G_i = G_{ii} + \sum_j G_{ij}$, $B_i = B_{ii} + \sum_j B_{ij}$.

6.4.4.3 Secondary Control Design

The aim is to shift the droop characteristics along the voltage and frequency axes, so that in steady state one has

$$[V_i, f_i, m_i P_i]^T \rightarrow [V_{ref}, f_{ref}, m_j P_j]^T, \quad \forall i, j. \tag{6.44}$$

This aim ensures proportional active power sharing between the DERs, if one selects $m_i = \overline{\Delta f}/P_i^*$ and $n_i = \overline{\Delta V}/Q_i^*$. As mentioned in Chapter 5, because of the line impedance effect, accurate voltage regulation leads to poor reactive power sharing and vice versa [178]. Here, we focus on the voltage regulation task, and reactive power responses are omitted for brevity. In addition, the ET reactive power controller provided in [161] can be modified slightly, to cover the directed CNs.

A. Event-Triggered Secondary Control: Once the DERs talk to each other via the CN, the aim in (6.44) can be achieved by using a consensus-based secondary control [38, 173–175, 186]. In this way, the following proportional-integral-based (PI-based) correction terms are proposed for ith DER.

$$\Omega_i = K_i^f (b_i^f \Delta f_i + b_i^\Omega \tilde{z}_i^\Omega) + I_i^f \int (b_i^f \Delta f_i + b_i^\Omega \tilde{z}_i^\Omega) dt, \tag{6.45a}$$

$$\Gamma_i = n_i \int \dot{Q}_i dt + K_i^V \left(b_i^V \Delta V_i + (1 - b_i^V) \tilde{z}_i^V\right) + I_i^V \int \left(b_i^V \Delta V_i + (1 - b_i^V) \tilde{z}_i^V\right) dt, \tag{6.45b}$$

$$\Delta f_i = f_{ref} - f_i,\ \ \tilde{z}_i^{\Omega} = \left(\frac{1}{d_i}\sum_j a_{ij}\tilde{\Omega}_j\right) - \tilde{\Omega}_i, \tag{6.45c}$$

$$\Delta V_i = V_{ref} - V_i,\ \ \tilde{z}_i^{V} = \left(\frac{1}{d_i}\sum_j a_{ij}\tilde{V}_j\right) - \tilde{V}_i, \tag{6.45d}$$

where if the DER has direct access to f_{ref} (respective V_{ref}), then $b_i^f = 1$ (respective $b_i^V = 1$), otherwise $b_i^f = 0$ (respective $b_i^V = 0$); if the DER cooperates with its in-neighbors $b_i^{\Omega} = 1$, otherwise $b_i^{\Omega} = 0$. The *local parts* in (6.45) employ the continuous local signals f_i, V_i and $\dot{Q}_i$, while the *neighborhood errors* $\tilde{z}_i^f$, and $\tilde{z}_i^V$ use the latest sampled communication data as follows.

$$\begin{cases} \tilde{\Omega}_i(t) = \Omega_i(t_{k_i}^i), \\ \tilde{V}_i(t) = V_i(t_{k_i}^i), \end{cases} t \in [t_{k_i}^i, t_{k_i+1}^i),\ k_i = 0, 1, \dots . \tag{6.46}$$

The triggering instant $t_{k_i}^i$ is when the signals V_i and Ω_i are sampled and transmitted to the out-neighbors. Please note that over $[t_{k_i}^i, t_{k_i+1}^i)$ the signals $\tilde{\Omega}_i$ and $\tilde{V}_i$ remain constant, while the neighbors' latest received data, i.e. $\tilde{\Omega}_j$ and $\tilde{V}_j$, $\forall j$ may vary.

Triggering Condition In a time-triggered strategy, a high-frequency periodic data sampling and communication should be adopted to implement the controller in (6.45) such that the time interval between any two consecutive triggering instants is equal to a sampling time τ_s, i.e. $t_{k_i+1}^i - t_{k_i}^i = \tau_s, \forall i, k$.

Although the periodic transmission strategy provides a good performance, it leads to many redundant data exchanges, especially in steady state, when the control goal is achieved and not so many communications are required. Thus, an ET control under which every DER decides on its own to sample and transmit data to CN *only when it is needed* seems effective. To this end, an appropriate triggering instant should be designed such that (i) the performance of the system undergoes a little degradation while using $\tilde{\Omega}_i$ and $\tilde{V}_i$ instead of Ω_i and V_i in (5) and (ii) the intervals between consecutive triggering instants have a lower bound τ which is the digital communication platform's (infrastructure's) capability (working frequency).

One way is to control the deviations of the continuous signals Ω_i, V_i from their latest sampled quantities $\tilde{\Omega}_i, \tilde{V}_i$, i.e. $e_i^{\Omega} = \tilde{\Omega}_i - \Omega_i$ and $e_i^V = \tilde{V}_i - V_i$. To this end, one should compare $|e_i^{\Omega}|$ and $|e_i^V|$ with some thresholds, every τ seconds. The employed thresholds should comprise a state-dependent part and a constant part [157]. Accordingly, the following triggering instant is proposed for ith DER.

$$t_{k_i+1}^i = \inf\{t_i > t_{k_i}^i + \tau : \text{(6.48) or (6.49)}\}, \tag{6.47}$$

$$\text{where,}\quad |e_i^{\Omega}| > \sigma b_i^{\Omega}|\tilde{z}_i^{\Omega}| + \gamma_{\Omega}, \tag{6.48}$$

$$|e_i^V| > \sigma(1 - b_i^V)|\tilde{z}_i^V| + \gamma_V, \tag{6.49}$$

where, $\tilde{z}_i^{\Omega} = \frac{1}{d_i}\sum a_{ij}(\tilde{\Omega}_j - \tilde{\Omega}_i)$, $\tilde{z}_i^V = \frac{1}{d_i}\sum a_{ij}(\tilde{V}_j - \tilde{V}_i)$ are the local neighborhood errors; inf denotes the infimum sign.

In (6.47), the triggering conditions in (6.48) and (6.49) are checked every τ seconds (not continuously); therefore, in addition to the communication burden, the computational burden is accordingly reduced. Moreover, if τ is selected greater than or equal to the communication

Algorithm 6.1 The presented event-triggering algorithm.

initialization: ▷ The first triggering instant

1: Sample $\tilde{\Omega}_i \leftarrow \Omega_i$, $\tilde{V}_i \leftarrow V_i$, and $\tilde{t}_i \leftarrow t_i$
2: Store the samples in the Data Store Unit (DSU)
3: Send $\tilde{\Omega}_i$ and $\tilde{V}_i$ to out-neighbors
4: Compute Ω_i and Γ_i in (6.45a) and (6.45b)
5: **while** secondary control is active **do**
6: Compute $e_i^{\Omega} = \tilde{\Omega}_i - \Omega_i$, $e_i^V = \tilde{V}_i - V_i$, and $t_i - \tilde{t}_i$
7: Alert the data receiver
8: **if** $\{t_i - \tilde{t}_i > \tau\}$ and {(6.48) or (6.49) holds} **then** ▷ Trigger
9: Do lines 1 to 3 of the algorithm
10: **end if**
11: **for all** $j \in N_i$ **do**
12: **if** new data received **then** ▷ Data receive
13: Update and store $\tilde{\Omega}_i$ and $\tilde{V}_i$
14: **end if**
15: **end for**
16: Compute Ω_i and Γ_i in (6.45a) and (6.45b)
17: **end while**

** *t_i (respective $\tilde{t}_i$) is the local clock (respective latest triggering instant) of ith DER.*

medium's working sampling time, one can ensure that the Zeno phenomenon (sampling chattering) is canceled out. Note that b_i^{Ω} and b_i^V indicate whether ith DER cooperates with its in-neighbors or not. Therefore, they are included to make (6.48) and (6.49) adaptive to this issue.

The ET controller is explained in Algorithm 6.1. The theoretical analyses and design guidelines for the presented controller are provided in [182].

6.4.4.4 Case Study and Simulation Results

To verify the effectiveness of the controller, an islanded 220 V, 50 Hz three-phase 5-bus MG with four DERs is simulated in MATLAB/Simulink environment. Each DER has the dynamics in (6.42) which feeds its local bus through the connector Z_C. Each feeder line is modeled with a lumped series RL impedance Z_L. The loads are modeled by constant impedance loads. Single-line diagram of the test MG is depicted in Figure 6.35, where the electrical and control requirements of different parts are also given. Note that $\overline{\Delta f} = 0.25$ Hz and $\overline{\Delta V} = 11$ V. The parameters in (8)–(12) are as follows: $\tau^{-1} = 25$ Hz, $K_i^V = K_i^f = 5$, $I_i^V = I_i^f = 10$, $\mathbf{b}_f = \mathbf{b}_V = [1, 0, 0, 0]^T$, $\mathbf{b}_{\Omega} = [0, 1, 1, 1]^T$, $\sigma = 0.49$, $\gamma_{\Omega} = 0.001/2$, and $\gamma_V = 0.05/2$.

Figure 6.36 indicates the performance of the controller over different stages. Prior to the controller activation at $t = 4$ s, the MG is controlled by droop control. At $t = 8$ s, load 5 increases by 33% of total load. Load 5 decreases to its initial value, at $t = 12$ s. Figures 6.36(a)–6.36(c) indicate that for $t < 3$ s the DERs' voltages and frequencies are deviated from the nominal values, and the active powers are shared properly. Activation of the secondary controller at $t = 4$ s restores the voltages and frequencies to the reference values of 220 V and 50 Hz while guaranteeing proportional active power sharing between the DERs. It is shown that after load changes at $t = 8$ s and $t = 12$ s, the voltages and frequencies are restored back to the reference values, and proportional active power sharing still remains. Figure 6.36(d) depicts the triggering (communication) instants for all the DERs. It is indicated that (i) the communication burden is highly reduced, (ii) most of

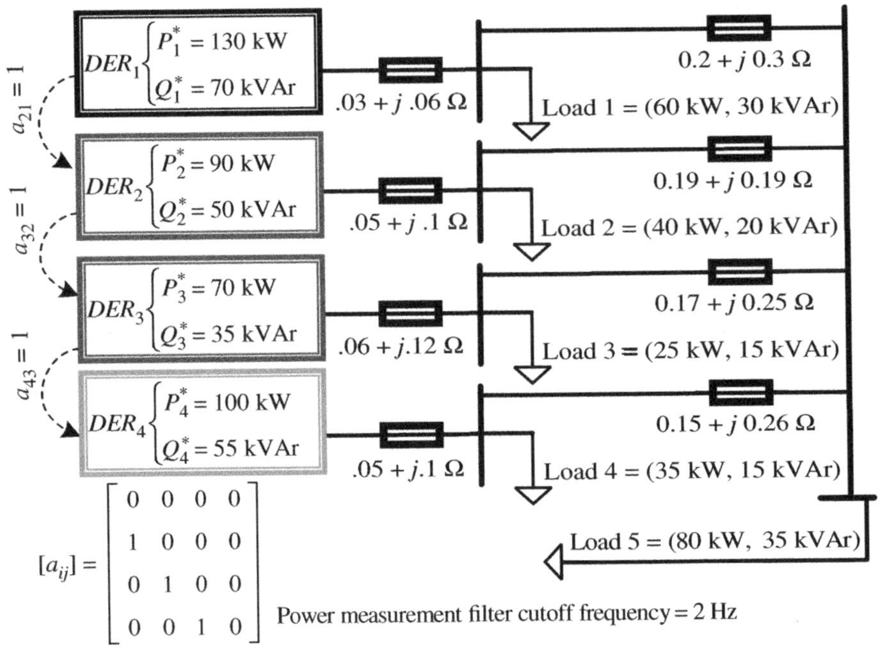

Figure 6.35 Test microgrid system; specifications of different parts are also given.

the communications are triggered during the transient stages, and (iii) the interevent time intervals are always greater than τ, i.e. Zeno behavior does not appear.

To preserve the connectivity of the CN in Figure 6.35, a link is established from DER_1 to DER_4. From $t = 6$ s to $t = 10$ s, the link between DER 3 and DER 4 is interrupted. The circuit breaker CB is intentionally opened at $t = 11$ s, and the 4th DER becomes disconnected (its communication links are also interrupted) until $t = 15$ s when the DER joins the secondary controlled MG again. Note that prior to $t = 15$ s a synchronization process is started, CB is closed, and the communication links associated with the 4th DER are all recovered. Figure 6.37 depicts the performance of the controller for this scenario. Figures 6.37(a)–6.37(c) indicate that (i) the controller is resilient to the link failure, (ii) once the 4th DER leaves the MG, other DERs increase their supplied active powers proportional to their ratings and regulate their voltages and frequencies, and (iii) as soon as the 4th DER connects back to the MG, immediately participates in proportional active power sharing, as well as, voltage and frequency regulation.

Figure 6.38 shows the controller performance for the system with similar parameters as previous cases, but with $K_i^V = K_i^f = 10$, $I_i^V = I_i^f = 20$ and communication delays of 0 and 160 milliseconds. Comparing Figures 6.36(a)–6.36(c) and Figures 6.38(a)–6.38(c), it is obvious that, the more feedback gain, the more synchronization speed, and the better transient response. Comparing Figures 6.38(a)–6.38(c) and Figures 6.38(d)–6.38(f), one can see that the communication delay slows down the synchronization speed and results in some transient oscillations. However, the desired control aim can be reached properly, for a practical range of delays (in the order of tens of milliseconds).

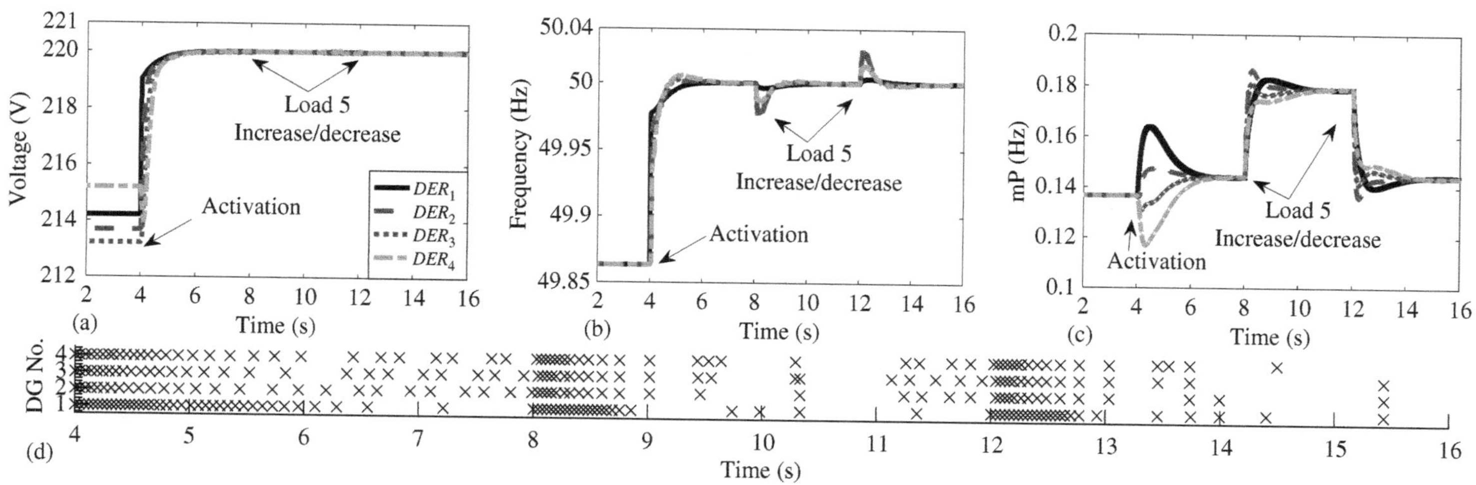

Figure 6.36 Performance of the presented controller; (a) voltages, (b) frequencies, (c) m_iP_is, and (d) communication (triggering) instants.

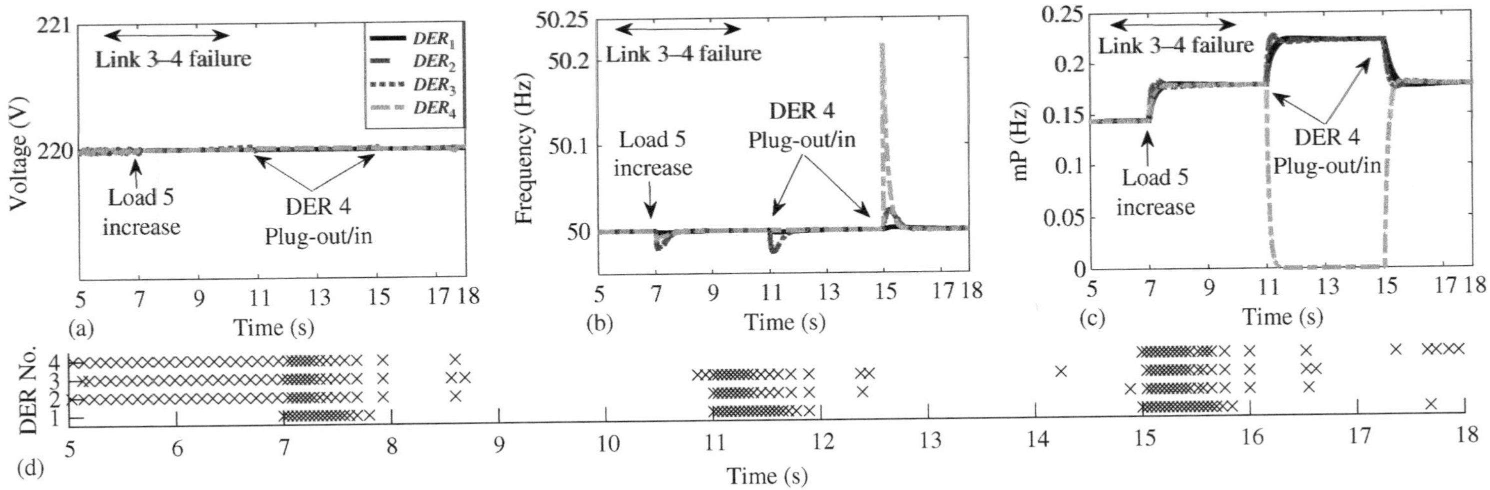

Figure 6.37 Link 3–4 failure and plug-and-play capability with 4th DER; (a) voltages, (b) frequencies, and (c) m_iP_is.

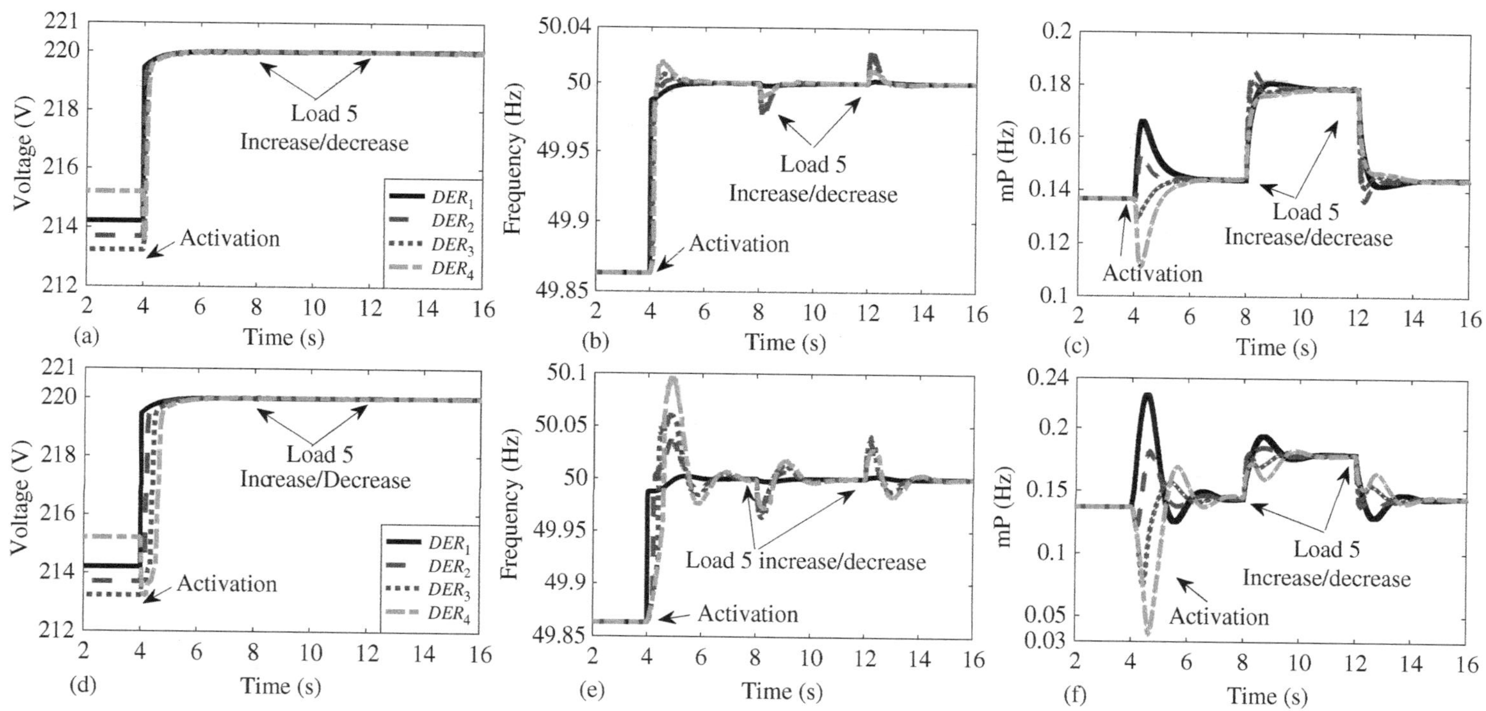

Figure 6.38 Performance of the controller with doubled gains under delay of (a)–(c) 0 s and (d)–(f) 0.16 s; (a) and (d) voltages, (b) and (e) frequencies, and (c) and (f) m_iP_is.

6.5 Summary

The chapter provides an overview of the recent developments in MG control, focusing on advanced control techniques, virtual dynamic control, and resilient and cybersecure control. The advanced control methods are more suitable for dealing with the nonlinear dynamics and uncertainties of MGs, and offer superior performance compared to classical controllers. The virtual dynamic control concept includes various virtual control loops, such as VSG and virtual impedance, used for different purposes in power converter-based systems and MGs such as virtual inertia control, frequency control, power sharing, and power quality improvement. Resilient and cybersecure control methods can address the challenges faced by MGs as cyber-physical systems. The effectiveness of these control methodologies is demonstrated through several simulations and laboratory experiments.

References

1 Tomislav Dragičević, Sergio Vazquez, and Pat Wheeler. Advanced control methods for power converters in DG systems and microgrids. *IEEE Trans. Ind. Electron.*, 68(7):5847–5862, 2021. doi: 10.1109/TIE.2020.2994857.

2 Qianwen Xu, Navid Vafamand, Linglin Chen, Tomislav Dragičević, Lihua Xie, and Frede Blaabjerg. Review on advanced control technologies for bidirectional DC/DC converters in DC microgrids. *IEEE J. Emerging Sel. Top. Power Electron.*, 9(2):1205–1221, 2021. doi: 10.1109/JESTPE.2020.2978064.

3 Peng Zeng, Hepeng Li, Haibo He, and Shuhui Li. Dynamic energy management of a microgrid using approximate dynamic programming and deep recurrent neural network learning. *IEEE Trans. Smart Grid*, 10(4):4435–4445, 2019. doi: 10.1109/TSG.2018.2859821.

4 Qinglai Wei, Derong Liu, Frank L Lewis, Yu Liu, and Jie Zhang. Mixed iterative adaptive dynamic programming for optimal battery energy control in smart residential microgrids. *IEEE Trans. Ind. Electron.*, 64(5):4110–4120, 2017. doi: 10.1109/TIE.2017.2650872.

5 Xin Dong, Xianshan Li, and Shan Cheng. Energy management optimization of microgrid cluster based on multi-agent-system and hierarchical Stackelberg game theory. *IEEE Access*, 8:206183–206197, 2020. doi: 10.1109/ACCESS.2020.3037676.

6 Hang Shuai, Jiakun Fang, Xiaomeng Ai, Yufei Tang, Jinyu Wen, and Haibo He. Stochastic optimization of economic dispatch for microgrid based on approximate dynamic programming. *IEEE Trans. Smart Grid*, 10(3):2440–2452, 2019. doi: 10.1109/TSG.2018.2798039.

7 Mehmet Hazar Cintuglu, Harold Martin, and Osama A Mohammed. Real-time implementation of multiagent-based game theory reverse auction model for microgrid market operation. *IEEE Trans. Smart Grid*, 6(2):1064–1072, 2015. doi: 10.1109/TSG.2014.2387215.

8 Amrit Paudel, Kalpesh Chaudhari, Chao Long, and Hoay Beng Gooi. Peer-to-peer energy trading in a prosumer-based community microgrid: A game-theoretic model. *IEEE Trans. Ind. Electron.*, 66(8):6087–6097, 2019. doi: 10.1109/TIE.2018.2874578.

9 Hao Wang and Jianwei Huang. Incentivizing energy trading for interconnected microgrids. *IEEE Trans. Smart Grid*, 9(4):2647–2657, 2018. doi: 10.1109/TSG.2016.2614988.

10 Kelvin Anoh, Sabita Maharjan, Augustine Ikpehai, Yan Zhang, and Bamidele Adebisi. Energy peer-to-peer trading in virtual microgrids in smart grids: A game-theoretic approach. *IEEE Trans. Smart Grid*, 11(2):1264–1275, 2020. doi: 10.1109/TSG.2019.2934830.

11 Mohyedin Ganjian-Aboukheili, Majid Shahabi, Qobad Shafiee, and Josep M Guerrero. Linear quadratic regulator based smooth transition between microgrid operation modes. *IEEE Trans. Smart Grid*, 12(6):4854–4864, 2021. doi: 10.1109/TSG.2021.3094419.

12 Dibakar Das, Gurunath Gurrala, and U Jayachandra Shenoy. Linear quadratic regulator-based bumpless transfer in microgrids. *IEEE Trans. Smart Grid*, 9(1):416–425, 2018. doi: 10.1109/TSG.2016.2580159.

13 Yousef Khayat, Mobin Naderi, Qobad Shafiee, Yazdan Batmani, Mohammad Fathi, Josep M Guerrero, and Hassan Bevrani. Decentralized optimal frequency control in autonomous microgrids. *IEEE Trans. Power Syst.*, 34(3):2345–2353, 2019. doi: 10.1109/TPWRS.2018. 2889671.

14 Sachidananda Sen and Vishal Kumar. Decentralized output-feedback-based robust LQR V-F controller for PV-battery microgrid including generation uncertainties. *IEEE Syst. J.*, 14(3):4418–4429, 2020. doi: 10.1109/JSYST.2020.2967774.

15 Masoud Davari, Weinan Gao, Zhong-Ping Jiang, and Frank L Lewis. An optimal primary frequency control based on adaptive dynamic programming for islanded modernized microgrids. *IEEE Trans. Autom. Sci. Eng.*, 18(3):1109–1121, 2021. doi: 10.1109/TASE.2020.2996160.

16 Yi Zhang, Qiuye Sun, Jianguo Zhou, Josep M Guerrero, Rui Wang, and Abderezak Lashab. Optimal frequency control for virtual synchronous generator based AC microgrids via adaptive dynamic programming. *IEEE Trans. Smart Grid*, 14(1):4–16, 2023. doi: 10.1109/TSG.2022. 3196412.

17 Alessandra Parisio, Evangelos Rikos, and Luigi Glielmo. A model predictive control approach to microgrid operation optimization. *IEEE Trans. Control Syst. Technol.*, 22(5):1813–1827, 2014. doi: 10.1109/TCST.2013.2295737.

18 Juan F Patarroyo-Montenegro, Fabio Andrade, Josep M Guerrero, and Juan C Vasquez. A linear quadratic regulator with optimal reference tracking for three-phase inverter-based islanded microgrids. *IEEE Trans. Power Electron.*, 36(6):7112–7122, 2021. doi: 10.1109/ TPEL.2020.3036594.

19 Fazel Mohammadi, Behnam Mohammadi-Ivatloo, Gevork B Gharehpetian, Mohd. Hasan Ali, Wei Wei, Ozan Erdinç, and Mohammadamin Shirkhani. Robust control strategies for microgrids: A review. *IEEE Syst. J.*, 16(2):2401–2412, 2022. doi: 10.1109/JSYST.2021.3077213.

20 Wenjie Ma, Sen Ouyang, Jun Zhang, and Weidong Xu. Control strategy based on H_∞ repetitive controller with active damping for islanded microgrid. *IEEE Access*, 7:162157–162168, 2019. doi: 10.1109/ACCESS. 2019.2951753.

21 Marjan Shafiee-Rad, Mahdieh S Sadabadi, Qobad Shafiee, and Mohammad Reza Jahed-Motlagh. Robust performance satisfaction of DC microgrids using a decentralized optimal voltage control strategy. *IEEE Syst. J.*, 16(1):464–474, 2022. doi: 10.1109/JSYST. 2021.3051922.

22 Mohammad Raeispour, Hajar Atrianfar, Hamid Reza Baghaee, and Gevork B Gharehpetian. Robust sliding mode and mixed H_2/H_∞ output feedback primary control of AC microgrids. *IEEE Syst. J.*, 15(2):2420–2431, 2021. doi: 10.1109/JSYST.2020.2999553.

23 Siamak Derakhshan, Marjan Shafiee-Rad, Qobad Shafiee, Mohammad Reza Jahed-Motlagh, Subham Sahoo, and Frede Blaabjerg. Decentralized voltage control of autonomous DC microgrids with robust performance approach. *IEEE J. Emerging Sel. Top. Power Electron.*, 9(5):5508–5520, 2021. doi: 10.1109/JESTPE.2021.3054723.

24 Mahdieh S Sadabadi and Qobad Shafiee. Scalable robust voltage control of DC microgrids with uncertain constant power loads. *IEEE Trans. Power Syst.*, 35(1):508–515, 2020. doi: 10.1109/ TPWRS. 2019.2928512.

25 Abdolwahhab Fathi, Qobad Shafiee, and Hassan Bevrani. Robust frequency control of microgrids using an extended virtual synchronous generator. *IEEE Trans. Power Syst.*, 33(6):6289–6297, 2018. doi: 10.1109/TPWRS.2018.2850880.

26 Ali Rafiee, Yazdan Batmani, Farahnaz Ahmadi, and Hassan Bevrani. Robust load-frequency control in islanded microgrids: Virtual synchronous generator concept and quantitative feedback theory. *IEEE Trans. Power Syst.*, 36(6):5408–5416, 2021. doi: 10.1109/TPWRS.2021.3077768.

27 Hassan Bevrani, Mohammad Ramin Feizi, and Sirwan Ataee. Robust frequency control in an islanded microgrid: H_∞ and μ synthesis approaches. *IEEE Trans. Smart Grid*, 7(2):706–717, 2016. doi: 10.1109/TSG.2015.2446984.

28 Maniza Armin, Mizanur Rahman, Md Mukidur Rahman, Subrata K Sarker, Sajal K Das, Md Rabiul Islam, Abbas Z Kouzani, and M A Parvez Mahmud. Robust extended H_∞ control strategy using linear matrix inequality approach for islanded microgrid. *IEEE Access*, 8:135883–135896, 2020. doi: 10.1109/ACCESS.2020.3009188.

29 Thongchart Kerdphol, Fathin Saifur Rahman, Masayuki Watanabe, and Yasunori Mitani. Robust virtual inertia control of a low inertia microgrid considering frequency measurement effects. *IEEE Access*, 7:57550–57560, 2019. doi: 10.1109/ACCESS.2019.2913042.

30 Mingyu Huang, Li Ding, Wenqu Li, Chao-Yang Chen, and Zhiwei Liu. Distributed observer-based H_∞ fault-tolerant control for DC microgrids with sensor fault. *IEEE Trans. Circuits Syst. Regul. Pap.*, 68(4):1659–1670, 2021. doi: 10.1109/TCSI.2020.3048971.

31 Seyed Abbas Taher, Mahdi Zolfaghari, Changhee Cho, Mehrdad Abedi, and Mohammad Shahidehpour. A new approach for soft synchronization of microgrid using robust control theory. *IEEE Trans. Power Delivery*, 32(3):1370–1381, 2017. doi: 10.1109/TPWRD.2016.2596106.

32 Ali Bidram, Ali Davoudi, Frank L Lewis, and Josep M Guerrero. Distributed cooperative secondary control of microgrids using feedback linearization. *IEEE Trans. Power Syst.*, 28(3):3462–3470, 2013. doi: 10.1109/TPWRS.2013.2247071.

33 Qianwen Xu, Chuanlin Zhang, Changyun Wen, and Peng Wang. A novel composite nonlinear controller for stabilization of constant power load in DC microgrid. *IEEE Trans. Smart Grid*, 10(1):752–761, 2019. doi: 10.1109/TSG.2017.2751755.

34 Tushar Kanti Roy, Md Apel Mahmud, Amanullah Maung Than Oo, Md Enamul Haque, Kashem M Muttaqi, and Nishad Mendis. Nonlinear adaptive backstepping controller design for islanded DC microgrids. *IEEE Trans. Ind. Appl.*, 54(3):2857–2873, 2018. doi: 10.1109/TIA.2018.2800680.

35 Jiarong Wu and Yimin Lu. Adaptive backstepping sliding mode control for boost converter with constant power load. *IEEE Access*, 7:50797–50807, 2019. doi: 10.1109/ACCESS.2019.2910936.

36 Shan Zuo, Ali Davoudi, Yongduan Song, and Frank L Lewis. Distributed finite-time voltage and frequency restoration in islanded AC microgrids. *IEEE Trans. Ind. Electron.*, 63(10): 5988–5997, 2016. doi: 10.1109/TIE.2016.2577542.

37 Md Apel Mahmud, Tushar Kanti Roy, Sajeeb Saha, M Enamul Haque, and Hemanshu R Pota. Robust nonlinear adaptive feedback linearizing decentralized controller design for islanded DC microgrids. *IEEE Trans. Ind. Appl.*, 55(5):5343–5352, 2019. doi: 10.1109/TIA.2019.2921028.

38 Alessandro Pilloni, Alessandro Pisano, and Elio Usai. Robust finite-time frequency and voltage restoration of inverter-based microgrids via sliding-mode cooperative control. *IEEE Trans. Ind. Electron.*, 65(1):907–917, 2018. doi: 10.1109/TIE.2017.2726970.

39 Shivam Chaturvedi, Deepak Fulwani, and Josep M Guerrero. Adaptive-SMC based output impedance shaping in DC microgrids affected by inverter loads. *IEEE Trans. Sustainable Energy*, 11(4):2940–2949, 2020. doi: 10.1109/TSTE.2020.2982414.

40 Quentin Tabart, Ionel Vechiu, Aitor Etxeberria, and Seddik Bacha. Hybrid energy storage system microgrids integration for power quality improvement using four-leg three-level NPC inverter and second-order sliding mode control. *IEEE Trans. Ind. Electron.*, 65(1):424–435, 2018. doi: 10.1109/TIE.2017.2723863.

41 Wentao Jiang, Xinan Zhang, Fanghong Guo, Jiawei Chen, Peng Wang, and Leong Hai Koh. Large-signal stability of interleave boost converter system with constant power load using sliding-mode control. *IEEE Trans. Ind. Electron.*, 67(11):9450–9459, 2020. doi: 10.1109/TIE.2019.2955401.

42 Qianwen Xu, Wentao Jiang, Frede Blaabjerg, Chuanlin Zhang, Xinan Zhang, and Tyrone Fernando. Backstepping control for large signal stability of high boost ratio interleaved converter interfaced DC microgrids with constant power loads. *IEEE Trans. Power Electron.*, 35 (5):5397–5407, 2020. doi: 10.1109/TPEL.2019.2943889.

43 Ebrahim Mohammadi, Mojtaba Alizadeh, Mohsen Asgarimoghaddam, Xiaoyu Wang, and Marcelo Godoy Simões. A review on application of artificial intelligence techniques in microgrids. *IEEE J. Emerging Sel. Top. Ind. Electron.*, 3(4):878–890, 2022. doi: 10.1109/JESTIE.2022.3198504.

44 James J Q Yu, Yunhe Hou, Albert Y S Lam, and Victor O K Li. Intelligent fault detection scheme for microgrids with wavelet-based deep neural networks. *IEEE Trans. Smart Grid*, 10(2):1694–1703, 2019. doi: 10.1109/TSG.2017.2776310.

45 Saman Darvish Kermany, Mahmood Joorabian, Sara Deilami, and Mohammad A S Masoum. Hybrid islanding detection in microgrid with multiple connection points to smart grids using fuzzy-neural network. *IEEE Trans. Power Syst.*, 32(4):2640–2651, 2017. doi: 10.1109/TPWRS.2016.2617344.

46 Rohit R Deshmukh, Makarand S Ballal, and Hiralal M Suryawanshi. A fuzzy logic based supervisory control for power management in multibus DC microgrid. *IEEE Trans. Ind. Appl.*, 56(6):6174–6185, 2020. doi: 10.1109/TIA.2020.3012415.

47 Hassan Bevrani, F Habibi, P Babahajyani, Masayuki Watanabe, and Yasunori Mitani. Intelligent frequency control in an AC microgrid: Online PSO-based fuzzy tuning approach. *IEEE Trans. Smart Grid*, 3(4):1935–1944, 2012. doi: 10.1109/TSG.2012.2196806.

48 Thongchart Kerdphol, Masayuki Watanabe, Komsan Hongesombut, and Yasunori Mitani. Self-adaptive virtual inertia control-based fuzzy logic to improve frequency stability of microgrid with high renewable penetration. *IEEE Access*, 7:76071–76083, 2019. doi: 10.1109/ACCESS.2019.2920886.

49 Ahmed Belgana, Bhaskar P Rimal, and Martin Maier. Open energy market strategies in microgrids: A Stackelberg game approach based on a hybrid multiobjective evolutionary algorithm. *IEEE Trans. Smart Grid*, 6(3):1243–1252, 2015. doi: 10.1109/TSG.2014.2363119.

50 Wei Dong, Qiang Yang, Wei Li, and Albert Y Zomaya. Machine-learning-based real-time economic dispatch in islanding microgrids in a cloud-edge computing environment. *IEEE Internet Things J.*, 8 (17):13703–13711, 2021. doi: 10.1109/JIOT.2021.3067951.

51 Elham Foruzan, Leen-Kiat Soh, and Sohrab Asgarpoor. Reinforcement learning approach for optimal distributed energy management in a microgrid. *IEEE Trans. Power Syst.*, 33(5):5749–5758, 2018. doi: 10.1109/TPWRS.2018.2823641.

52 Guanyu Gao, Yonggang Wen, and Dacheng Tao. Distributed energy trading and scheduling among microgrids via multiagent reinforcement learning. *IEEE Transactions on Neural Networks and Learning Systems*, 1–15, 2022. doi: 10.1109/TNNLS.2022.3170070.

53 Mohammad Reza Habibi, Hamid Reza Baghaee, Tomislav Dragičević, and Frede Blaabjerg. Detection of false data injection cyber-attacks in DC microgrids based on recurrent neural networks. *IEEE Journal of Emerging and Selected Topics in Power Electronics*, 9(5):5294–5310, 2021. doi: 10.1109/JESTPE.2020.2968243.

54 Abdollah Kavousi-Fard, Wencong Su, and Tao Jin. A machine-learning-based cyber attack detection model for wireless sensor networks in microgrids. *IEEE Trans. Ind. Inf.*, 17(1):650–658, 2021. doi: 10.1109/TII.2020.2964704.

55 Diego Arcos-Aviles, Julio Pascual, Luis Marroyo, Pablo Sanchis, and Francesc Guinjoan. Fuzzy logic-based energy management system design for residential grid-connected microgrids. *IEEE Trans. Smart Grid*, 9(2):530–543, 2018. doi: 10.1109/TSG.2016.2555245.

56 Jose Rodriguez and Patricio Cortes. *Predictive control of power converters and electrical drives.* John Wiley & Sons, 2012.

57 Tomislav Dragičević. Model predictive control of power converters for robust and fast operation of AC microgrids. *IEEE Trans. Power Electron.*, 33(7):6304–6317, 2017.

58 Zhuoli Zhao, Juntao Guo, Xi Luo, Chun Sing Lai, Ping Yang, Loi Lei Lai, Peng Li, Josep M Guerrero, and Mohammad Shahidehpour. Distributed robust model predictive control-based energy management strategy for islanded multi-microgrids considering uncertainty. *IEEE Trans. Smart Grid*, 13(3):2107–2120, 2022. doi: 10.1109/TSG.2022.3147370.

59 Josip Vasilj, Damir Jakus, and Petar Sarajcev. Robust nonlinear economic MPC based management of a multi energy microgrid. *IEEE Trans. Energy Convers.*, 36(2):1528–1536, 2021. doi: 10.1109/TEC.2020.3046459.

60 Zeinab Karami, Qobad Shafiee, Yousef Khayat, Meysam Yaribeygi, Tomislav Dragičević, and Hassan Bevrani. Decentralized model predic-tive control of DC microgrids with constant power load. *IEEE Journal of Emerging and Selected Topics in Power Electronics*, 9(1):451–460, 2021. doi: 10.1109/JESTPE.2019.2957231.

61 Najmeh Bazmohammadi, Ahmadreza Tahsiri, Amjad Anvari-Moghaddam, and Josep M Guerrero. Stochastic predictive control of multi-microgrid systems. *IEEE Trans. Ind. Appl.*, 55(5):5311–5319, 2019. doi: 10.1109/TIA.2019.2918051.

62 Carlos Bordons, Félix Garcia-Torres, and Miguel A Ridao. *Model predictive control of microgrids*, volume 358. Springer, 2020.

63 Guannan Lou, Wei Gu, Yinliang Xu, Ming Cheng, and Wei Liu. Distributed MPC-based secondary voltage control scheme for autonomous droop-controlled microgrids. *IEEE Trans. Sustainable Energy*, 8(2):792–804, 2017. doi: 10.1109/TSTE.2016.2620283.

64 Jonggrist Jongudomkarn, Jia Liu, Yuta Yanagisawa, Hassan Bevrani, and Toshifumi Ise. Model predictive control for indirect boost matrix converter based on virtual synchronous generator. *IEEE Access*, 8:60364–60381, 2020. doi: 10.1109/ACCESS.2020.2983115.

65 Tianhao Chen, Omar Abdel-Rahim, Faxiang Peng, and Haoyu Wang. An improved finite control set-MPC-based power sharing control strategy for islanded AC microgrids. *IEEE Access*, 8:52676–52686, 2020. doi: 10.1109/ACCESS.2020.2980860.

66 Ahmed Ouammi, Hanane Dagdougui, Louis Dessaint, and Roberto Sacile. Coordinated model predictive-based power flows control in a cooperative network of smart microgrids. *IEEE Trans. Smart Grid*, 6(5):2233–2244, 2015. doi: 10.1109/TSG.2015.2396294.

67 Daniel E Olivares, Jose D Lara, Claudio A. Cañizares, and Mehrdad Kazerani. Stochastic-predictive energy management system for isolated microgrids. *IEEE Trans. Smart Grid*, 6(6):2681–2693, 2015. doi: 10.1109/TSG.2015.2469631.

68 Samir Kouro, Patricio Cortes, RenÉ Vargas, Ulrich Ammann, and JosÉ Rodriguez. Model predictive control—a simple and powerful method to control power converters. *IEEE Trans. Ind. Electron.*, 56(6):1826–1838, 2009. doi: 10.1109/TIE.2008.2008349.

69 Lexuan Meng, Qobad Shafiee, Giancarlo Ferrari Trecate, Houshang Karimi, Deepak Fulwani, Xiaonan Lu, and Josep M Guerrero. Review on control of DC microgrids and multiple microgrid clusters. *IEEE J. Emerging Sel. Top. Power Electron.*, 5(3):928–948, 2017. doi: 10.1109/JESTPE.2017.2690219.

70 Qobad Shafiee, Tomislav Dragičević, Juan C Vasquez, and Josep M Guerrero. Hierarchical control for multiple DC-microgrids clusters. *IEEE Trans. Energy Convers.*, 29(4):922–933, 2014. doi: 10.1109/TEC.2014.2362191.

71 Ali Emadi, Alireza Khaligh, Claudio H Rivetta, and Geoffrey A Williamson. Constant power loads and negative impedance instability in automotive systems: Definition, modeling, stability, and control of power electronic converters and motor drives. *IEEE Trans. Veh. Technol.*, 55(4):1112–1125, 2006.

72 Mauricio Cespedes, Lei Xing, and Jian Sun. Constant-power load system stabilization by passive damping. *IEEE Trans. Power Electron.*, 26(7):1832–1836, 2011.

73 Amir M Rahimi, Geoffrey A Williamson, and Ali Emadi. Loop-cancellation technique: A novel nonlinear feedback to overcome the destabilizing effect of constant-power loads. *IEEE Trans. Veh. Technol.*, 59(2):650–661, 2010.

74 Zhi Qu, Seyyedmilad Ebrahimi, Navid Amiri, Juri Jatskevich, and Alex Pizniur. Adaptive control method for stabilizing DC distribution systems with constant-power loads based on tunable active damping. In *2018 IEEE 19th Workshop on Control and Modeling for Power Electronics (COMPEL)*, pages 1–6. IEEE, 2018.

75 Fei Gao, Serhiy Bozhko, Alessandro Costabeber, Chintan Patel, Pat Wheeler, Christopher Ian Hill, and Greg Asher. Comparative stability analysis of droop control approaches in voltage-source-converter-based DC microgrids. *IEEE Trans. Power Electron.*, 32(3):2395–2415, 2016.

76 Alexis Kwasinski and Chimaobi N Onwuchekwa. Dynamic behavior and stabilization of DC microgrids with instantaneous constant-power loads. *IEEE Trans. Power Electron.*, 26(3):822–834, 2010.

77 Xiaogang Feng, Jinjun Liu, and Fred C Lee. Impedance specifications for stable DC distributed power systems. *IEEE Trans. Power Electron.*, 17(2):157–162, 2002.

78 Xinan Zhang, D Mahinda Vilathgamuwa, King-Jet Tseng, Bikramjit Singh Bhangu, and Chandana Jayampathi Gajanayake. Power buffer with model predictive control for stability of vehicular power systems with constant power loads. *IEEE Trans. Power Electron.*, 28(12): 5804–5812, 2012.

79 Long Cheng, Pablo Acuna, Ricardo P Aguilera, Jiuchun Jiang, Shaoyuan Wei, John E Fletcher, and Dylan D C Lu. Model predictive control for DC–DC boost converters with reduced-prediction horizon and constant switching frequency. *IEEE Trans. Power Electron.*, 33(10):9064–9075, 2017.

80 Kun Liu, Tao Liu, Zhiyuan Tang, and David J Hill. Distributed MPC-based frequency control in networked microgrids with voltage constraints. *IEEE Trans. Smart Grid*, 10(6):6343–6354, 2019.

81 Zeinab Karami, Qobad Shafiee, and Hassan Bevrani. Model predictive and SDRE control of DC microgrids with constant power loads: A comparative study. In *2018 Smart Grid Conference (SGC)*, pages 1–6, 2018. doi: 10.1109/SGC.2018.8777914.

82 Ming Yu, Yi Wang, and Yonggang Li. Hierarchical control of DC microgrid based on model predictive controller. In *IECON 2016-42nd Annual Conference of the IEEE Industrial Electronics Society*, pages 4139–4144. IEEE, 2016.

83 Shunlong Xiao, Mohammad B Shadmand, and Robert S Balog. Model predictive control of multi-string PV systems with battery back-up in a community DC microgrid. In *2017 IEEE Applied Power Electronics Conference and Exposition (APEC)*, pages 1284–1290. IEEE, 2017.

84 Petros Karamanakos, Tobias Geyer, and Stefanos Manias. Direct model predictive current control strategy of DC–DC boost converters. *IEEE J. Emerging Sel. Top. Power Electron.*, 1(4):337–346, 2013.

85 Mohammad Mehdi Mardani, Mohammad Hassan Khooban, Ali Masoudian, and Tomislav Dragičević. Model predictive control of DC–DC converters to mitigate the effects of pulsed power loads in naval DC microgrids. *IEEE Trans. Ind. Electron.*, 66(7):5676–5685, 2018.

86 Xinan Zhang, Benfei Wang, Ujjal Manandhar, Hoay Beng Gooi, and Gilbert Foo. A model predictive current controlled bidirectional three-level DC/DC converter for hybrid energy storage system in DC microgrids. *IEEE Trans. Power Electron.*, 34(5):4025–4030, 2018.

87 Xiao Li, Haiyu Zhang, Mohammad B Shadmand, and Robert S Balog. Model predictive control of a voltage-source inverter with seamless transition between islanded and grid-connected operations. *IEEE Trans. Ind. Electron.*, 64(10):7906–7918, 2017. doi: 10.1109/TIE.2017.2696459.

88 Yun Yang, Siew-Chong Tan, and Shu Yuen Ron Hui. Adaptive reference model predictive control with improved performance for voltage-source inverters. *IEEE Trans. Control Syst. Technol.*, 26(2):724–731, 2018. doi: 10.1109/TCST.2017.2670529.

89 Hosein Gholami-Khesht, Pooya Davari, and Frede Blaabjerg. An adaptive model predictive voltage control for LC-filtered voltage source inverters. *Appl. Sci.*, 11(2):1–13, 2021. ISSN 1454-5101. doi: 10.3390/app11020704.

90 Badii Gmati, Imed Jlassi, Sejir Khojet El Khil, and Antonio J Marques Cardoso. Open-switch fault diagnosis in voltage source inverters of PMSM drives using predictive current errors and fuzzy logic approach. *IET Power Electron.*, 14(6):1059–1072, 2021.

91 Bingtao Zhang, Weimin Wu, Yong Yang, Ning Gao, Jianming Chen, Eftichios Koutroulis, Henry Shu-Hung Chung, Marco Liserre, and Frede Blaabjerg. A novel simplified finite control set repeat-model predictive control for grid-connected inverters. *IEEE Trans. Ind. Electron.*, 70(11):11324–11333, 2022. doi: 10.1109/TIE.2022.3231252.

92 Hector A Young, Marcelo A Perez, Jose Rodriguez, and Haitham Abu-Rub. Assessing finite-control-set model predictive control: A comparison with a linear current controller in two-level voltage source inverters. *IEEE Ind. Electron. Mag.*, 8(1):44–52, 2014. doi: 10.1109/MIE.2013.2294870.

93 Myada Shadoul, Hassan A Yousef, Rashid Al Abri, and Amer Al-Hinai. Adaptive fuzzy approximation control of PV grid-connected inverters. *Energies*, 14(4):942, 2021.

94 Myada Shadoul, Hassan Yousef, Rashid Al Abri, and Amer Al-Hinai. Adaptive interval type-2 fuzzy tracking control of PV grid-connected inverters. *IEEE Access*, 9:130853–130861, 2021. doi: 10.1109/ACCESS.2021.3114311.

95 Xiaodong Liang and Chowdhury Andalib Bin-Karim. Harmonics and mitigation techniques through advanced control in grid-connected renewable energy sources: A review. *IEEE Trans. Ind. Appl.*, 54(4):3100–3111, 2018. doi: 10.1109/TIA.2018.2823680.

96 Sahba Moslemimoghadam, Qobad Shafiee, Mobin Naderi, and Hassan Bevrani. A hybrid fuzzy-predictive control for smooth transition between operating modes of voltage source inverters. In *2022 8th International Conference on Control, Instrumentation and Automation (ICCIA)*, pages 1–6, 2022. doi: 10.1109/ICCIA54998.2022.9737176.

97 Hassan Bevrani, Bruno François, and Toshifumi Ise. *Microgrid dynamics and control*. John Wiley & Sons, Hoboken, NJ, 2017.

98 Hassan Bevrani, Toshifumi Ise, and Yushi Miura. Virtual synchronous generators: A survey and new perspectives. *Int. J. Electric. Power Energy Syst.*, 54:244–254, 2014. ISSN 0142-0615. doi: 10.1016/j.ijepes.2013.07.009 URL https://www.sciencedirect.com/science/article/pii/S0142061513003062.

99 Jia Liu, Yushi Miura, and Toshifumi Ise. Comparison of dynamic characteristics between virtual synchronous generator and droop control in inverter-based distributed generators. *IEEE Trans. Power Electron.*, 31(5):3600–3611, 2016. doi: 10.1109/TPEL.2015.2465852.

100 Salvatore D'Arco and Jon Are Suul. Equivalence of virtual synchronous machines and frequency-droops for converter-based microgrids. *IEEE Trans. Smart Grid*, 5(1):394–395, 2014. doi: 10.1109/TSG.2013.2288000.

101 Lin-Yu Lu and Chia-Chi Chu. Consensus-based secondary frequency and voltage droop control of virtual synchronous generators for isolated AC micro-grids. *IEEE J. Emerging Sel. Top. Circuits Syst.*, 5(3):443–455, 2015.

102 Aazim Rasool, Shah Fahad, Xiangwu Yan, Jia Jiaoxin, Haaziq Rasool, and Mohsin Jamil. A virtual parallel inductor approach for mitigating reactive power sharing error in a VSG controlled microgrid. *IEEE Syst. J.*, 17(1):1363–1374, 2023. doi: 10.1109/JSYST.2022.3180998.

103 Haizhen Xu, Xing Zhang, Fang Liu, Rongliang Shi, Changzhou Yu, and Renxian Cao. A reactive power sharing strategy of VSG based on virtual capacitor algorithm. *IEEE Trans. Ind. Electron.*, 64(9):7520–7531, 2017. doi: 10.1109/TIE.2017.2686374.

104 Guannan Lou, Quan Yang, Wei Gu, Xiangjun Quan, Josep M Guerrero, and Shanlin Li. Analysis and design of hybrid harmonic suppression scheme for VSG considering non-linear loads and distorted grid. *IEEE Trans. Energy Convers.*, 36(4):3096–3107, 2021. doi: 10.1109/TEC.2021.3063607.

105 H Bevrani, T Kato, T Ise, and K Inoue. *Grid connected converters: Modeling, stability and control*. Elsevier Science, 2022. ISBN 9780323999021.

106 Brian B Johnson, Sairaj V Dhople, Abdullah O Hamadeh, and Philip T Krein. Synchronization of parallel single-phase inverters with virtual oscillator control. *IEEE Trans. Power Electron.*, 29 (11):6124–6138, 2014. doi: 10.1109/TPEL.2013.2296292.

107 Qing-Chang Zhong and George Weiss. Synchronverters: Inverters that mimic synchronous generators. *IEEE Trans. Ind. Electron.*, 58(4):1259–1267, 2011. doi: 10.1109/TIE.2010.2048839.

108 Brian B Johnson, Mohit Sinha, Nathan G Ainsworth, Florian Dörfler, and Sairaj V Dhople. Synthesizing virtual oscillators to control islanded inverters. *IEEE Trans. Power Electron.*, 31(8):6002–6015, 2016. doi: 10.1109/TPEL.2015.2497217.

109 Krishnakumar R Vasudevan, Vigna K Ramachandaramurthy, Thanikanti Sudhakar Babu, and Aref Pouryekta. Synchronverter: A comprehensive review of modifications, stability assessment, applications and future perspectives. *IEEE Access*, 8:131565–131589, 2020. doi: 10.1109/ACCESS.2020.3010001.

110 Jiefeng Hu, Yinghao Shan, Ka Wai Cheng, and Syed Islam. Overview of power converter control in microgrids—challenges, advances, and future trends. *IEEE Trans. Power Electron.*, 37(8):9907–9922, 2022. doi: 10.1109/TPEL.2022.3159828.

111 Jia Liu, Yushi Miura, Hassan Bevrani, and Toshifumi Ise. A unified modeling method of virtual synchronous generator for multi-operation-mode analyses. *IEEE J. Emerging Sel. Top. Power Electron.*, 9(2):2394–2409, 2021. doi: 10.1109/JESTPE.2020.2970025.

112 Heng Wu, Xinbo Ruan, Dongsheng Yang, Xinran Chen, Wenxin Zhao, Zhipeng Lv, and Qing-Chang Zhong. Small-signal modeling and parameters design for virtual synchronous generators. *IEEE Trans. Ind. Electron.*, 63(7):4292–4303, 2016. doi: 10.1109/TIE.2016.2543181.

113 Xiaochao Hou, Yao Sun, Xin Zhang, Jinghang Lu, Peng Wang, and Josep M Guerrero. Improvement of frequency regulation in VSG-based AC microgrid via adaptive virtual inertia. *IEEE Trans. Power Electron.*, 35(2):1589–1602, 2020. doi: 10.1109/TPEL.2019.2923734.

114 Dongdong Li, Qianwei Zhu, Shunfu Lin, and X Y Bian. A self-adaptive inertia and damping combination control of VSG to support frequency stability. *IEEE Trans. Energy Convers.*, 32(1):397–398, 2017. doi: 10.1109/TEC.2016.2623982.

115 Miguel A Torres L, Luiz A C Lopes, Luis A Morán T, and José R Espinoza C. Self-tuning virtual synchronous machine: A control strategy for energy storage systems to support dynamic frequency control. *IEEE Trans. Energy Convers.*, 29(4):833–840, 2014. doi: 10.1109/TEC.2014.2362577.

116 Pedro José dos Santos Neto, Tárcio André dos Santos Barros, João Pedro Carvalho Silveira, Ernesto Ruppert Filho, Juan Carlos Vasquez, and Josep M Guerrero. Power management strategy based on virtual inertia for DC microgrids. *IEEE Trans. Power Electron.*, 35(11):12472–12485, 2020. doi: 10.1109/TPEL.2020.2986283.

117 Xiaorong Zhu, Fanqi Meng, Zhiyun Xie, and Ying Yue. An inertia and damping control method of DC–DC converter in DC microgrids. *IEEE Trans. Energy Convers.*, 35(2):799–807, 2020. doi: 10.1109/TEC.2019.2952717.

118 Gang Lin, Junjie Ma, Yong Li, Christian Rehtanz, Jiayan Liu, Ziya Wang, Pengcheng Wang, and Fengjian She. A virtual inertia and damping control to suppress voltage oscillation in islanded DC microgrid. *IEEE Trans. Energy Convers.*, 36(3):1711–1721, 2021. doi: 10.1109/TEC.2020.3039364.

119 Uros Markovic, Zhongda Chu, Petros Aristidou, and Gabriela Hug. LQR-based adaptive virtual synchronous machine for power systems with high inverter penetration. *IEEE Trans. Sustainable Energy*, 10(3):1501–1512, 2019. doi: 10.1109/TSTE.2018.2887147.

120 Changming Zheng, Tomislav Dragičević, and Frede Blaabjerg. Model predictive control-based virtual inertia emulator for an islanded alternating current microgrid. *IEEE Trans. Ind. Electron.*, 68(8):7167–7177, 2021. doi: 10.1109/TIE.2020.3007105.

121 Bo Long, Yong Liao, Kil To Chong, José Rodríguez, and Josep M Guerrero. MPC-controlled virtual synchronous generator to enhance frequency and voltage dynamic performance in islanded microgrids. *IEEE Trans. Smart Grid*, 12(2):953–964, 2021. doi: 10.1109/TSG.2020.3027051.

122 Yu Wang and Rong-Jong Wai. Adaptive fuzzy-neural-network power decoupling strategy for virtual synchronous generator in micro-grid. *IEEE Trans. Power Electron.*, 37(4):3878–3891, 2022. doi: 10.1109/TPEL.2021.3120519.

123 Bo Long, Yong Liao, Kil To Chong, José Rodríguez, and Josep M Guerrero. Enhancement of frequency regulation in AC microgrid: A fuzzy-MPC controlled virtual synchronous generator. *IEEE Trans. Smart Grid*, 12(4):3138–3149, 2021. doi: 10.1109/TSG.2021.3060780.

124 Xiangjun Quan, Ruiyang Yu, Xin Zhao, Yang Lei, Tianxiang Chen, Chengjing Li, and Alex Q Huang. Photovoltaic synchronous generator: Architecture and control strategy for a grid-forming PV energy system. *IEEE J. Emerging Sel. Top. Power Electron.*, 8 (2):936–948, 2020. doi: 10.1109/JESTPE.2019.2953178.

125 Yiwei Ma, Wenchao Cao, Liu Yang, Fei Wang, and Leon M Tolbert. Virtual synchronous generator control of full converter wind turbines with short-term energy storage. *IEEE Trans. Ind. Electron.*, 64(11):8821–8831, 2017. doi: 10.1109/TIE.2017.2694347.

126 Htar Su Hlaing, Jia Liu, Yushi Miura, Hassan Bevrani, and Toshifumi Ise. Enhanced performance of a stand-alone gas-engine generator using virtual synchronous generator and energy storage system. *IEEE Access*, 7:176960–176970, 2019. doi: 10.1109/ACCESS.2019.2957890.

127 Andres E. Leon and Juan Manuel Mauricio. Virtual synchronous generator for VSC-HVDC stations with DC voltage control. *IEEE Trans. Power Syst.*, 38(1):728–738, 2023. doi: 10.1109/TPWRS.2022.3164027.

128 Yijia Cao, Weiyu Wang, Yong Li, Yi Tan, Chun Chen, Li He, Ulf Häger, and Christian Rehtanz. A virtual synchronous generator control strategy for VSC-MTDC systems. *IEEE Trans. Energy Convers.*, 33(2):750–761, 2018. doi: 10.1109/TEC.2017.2780920.

129 Jon Are Suul, Salvatore D'Arco, and Giuseppe Guidi. Virtual synchronous machine-based control of a single-phase bi-directional battery charger for providing vehicle-to-grid services. *IEEE Trans. Ind. Appl.*, 52(4):3234–3244, 2016. doi: 10.1109/TIA.2016.2550588.

130 Wenhua Wu, Yandong Chen, An Luo, Leming Zhou, Xiaoping Zhou, Ling Yang, Yanting Dong, and Josep M Guerrero. A virtual inertia control strategy for DC microgrids analogized with virtual synchronous machines. *IEEE Trans. Ind. Electron.*, 64(7):6005–6016, 2017. doi: 10.1109/TIE.2016.2645898.

131 Zhehan Yi, Xiaoying Zhao, Di Shi, Jiajun Duan, Yingmeng Xiang, and Zhiwei Wang. Accurate power sharing and synthetic inertia control for DC building microgrids with guaranteed performance. *IEEE Access*, 7:63698–63708, 2019. doi: 10.1109/ACCESS.2019.2915046.

132 Soumya Samanta, Jyoti Prakash Mishra, and Binoy Krishna Roy. Virtual DC machine: An inertia emulation and control technique for a bidirectional DC–DC converter in a DC microgrid. *IET Electr. Power Appl.*, 12(6):874–884, 2018. doi: 10.1049/iet-epa.2017.0770 URL https://ietresearch.onlinelibrary.wiley.com/doi/abs/10.1049/iet-epa.2017.0770.

133 Na Zhi, Ke Ding, Liang Du, and Hui Zhang. An SOC-based virtual DC machine control for distributed storage systems in DC microgrids. *IEEE Trans. Energy Convers.*, 35(3):1411–1420, 2020. doi: 10.1109/TEC.2020.2975033.

134 Mehran Jami, Qobad Shafiee, and Hassan Bevrani. Dynamic improvement of DC microgrids with CPLs using virtual inertia concept. *Int. J. Ind. Electron. Control Optim.*, 4(1):67–76, 2021. doi: 10.22111/ieco.2020.33152.1245.

135 Mehmet Hazar Cintuglu, Osama A Mohammed, Kemal Akkaya, and A Selcuk Uluagac. A survey on smart grid cyber-physical system testbeds. *IEEE Commun. Surv. Tutorials*, 19(1):446–464, 2017. doi: 10.1109/COMST.2016.2627399.

136 Ge Cao, Wei Gu, Chenxiao Gu, Wanxing Sheng, Jing Pan, Rui Li, and Lijing Sun. Real-time cyber-physical system co-simulation testbed for microgrids control. *IET Cyber-Phys. Syst.: Theory Appl.*, 4(1):38–45, 2019.

137 Ioan Serban, Sandra Cespedes, Corneliu Marinescu, Cesar A Azurdia-Meza, Juan S Gomez, and Doris Sáez Hueichapan. Communication requirements in microgrids: A practical survey. *IEEE Access*, 8:47694–47712, 2020.

138 Abedalsalam Bani-Ahmed, Luke Weber, Adel Nasiri, and Hossein Hosseini. Microgrid communications: State of the art and future trends. In *2014 International Conference on Renewable Energy Research and Application (ICRERA)*, pages 780–785. IEEE, 2014.

139 Zhiyi Li, Mohammad Shahidehpour, and Farrokh Aminifar. Cybersecurity in distributed power systems. *Proc. IEEE*, 105(7):1367–1388, 2017.

140 Shichao Liu, Zhijian Hu, Xiaoyu Wang, and Ligang Wu. Stochastic stability analysis and control of secondary frequency regulation for islanded microgrids under random denial of service attacks. *IEEE Trans. Ind. Inf.*, 15(7):4066–4075, 2018.

141 Shankar Abhinav, Hamidreza Modares, Frank L Lewis, Frank Ferrese, and Ali Davoudi. Synchrony in networked microgrids under attacks. *IEEE Trans. Smart Grid*, 9(6):6731–6741, 2017.

142 Subham Sahoo, Sukumar Mishra, Seyed Mahdi Fazeli, Furong Li, and Tomislav Dragičević. A distributed fixed-time secondary controller for DC microgrid clusters. *IEEE Trans. Energy Convers.*, 34(4):1997–2007, 2019.

143 Mengxuan Shi, Xia Chen, Jianyu Zhou, Yin Chen, Jinyu Wen, and Haibo He. PI-consensus based distributed control of AC microgrids. *IEEE Trans. Power Syst.*, 35(3):2268–2278, 2019.

144 Yan Li, Peng Zhang, Liang Zhang, and Bing Wang. Active synchronous detection of deception attacks in microgrid control systems. *IEEE Trans. Smart Grid*, 8(1):373–375, 2016.

145 Mohammad Reza Habibi, Hamid Reza Baghaee, Tomislav Dragičević, and Frede Blaabjerg. Detection of false data injection cyber-attacks in DC microgrids based on recurrent neural networks. *IEEE J. Emerging Sel. Top. Power Electron.*, 9(5):5294–5310, 2020.

146 Lin-Yu Lu, Hao Jan Liu, Hao Zhu, and Chia-Chi Chu. Intrusion detection in distributed frequency control of isolated microgrids. *IEEE Trans. Smart Grid*, 10(6):6502–6515, 2019.

147 Subham Sahoo, Sukumar Mishra, Jimmy Chih-Hsien Peng, and Tomislav Dragicevic. A stealth cyber attack detection strategy for DC microgrids. *IEEE Trans. Power Electron.*, 34(8):8162–8174, 2018.

148 Omar Ali Beg, Taylor T Johnson, and Ali Davoudi. Detection of false-data injection attacks in cyber-physical DC microgrids. *IEEE Trans. Ind. Inf.*, 13(5):2693–2703, 2017.

149 Subham Sahoo, Jimmy Chih-Hsien Peng, Annavaram Devakumar, Sukumar Mishra, and Tomislav Dragičević. On detection of false data in cooperative DC microgrids—a discordant element approach. *IEEE Trans. Ind. Electron.*, 67(8):6562–6571, 2019.

150 Abbas Ahmadi and Qobad Shafiee. An estimation based detection method for deception cyber attack in AC microgrids. In *2021 11th Smart Grid Conference (SGC)*, pages 1–5. IEEE, 2021.

151 Yajie Jiang, Yun Yang, Siew-Chong Tan, and Shu Yuen Hui. Distributed sliding mode observer-based secondary control for DC microgrids under cyber-attacks. *IEEE J. Emerging Sel. Top. Circuits Syst.*, 11(1):144–154, 2020.

152 Dillip Kumar Mishra, Prakash Kumar Ray, Li Li, Jiangfeng Zhang, M J Hossain, and Asit Mohanty. Resilient control based frequency regulation scheme of isolated microgrids considering cyber attack and parameter uncertainties. *Appl. Energy*, 306:118054, 2022.

153 Subham Sahoo, Yongheng Yang, and Frede Blaabjerg. Resilient synchronization strategy for AC microgrids under cyber attacks. *IEEE Trans. Power Electron.*, 36(1):73–77, 2020.

154 Ali Bidram, Binod Poudel, Lakshmisree Damodaran, Rafael Fierro, and Josep M Guerrero. Resilient and cybersecure distributed control of inverter-based islanded microgrids. *IEEE Trans. Ind. Inf.*, 16(6):3881–3894, 2019.

155 Mahdieh S Sadabadi, Subham Sahoo, and Frede Blaabjerg. A fully resilient cyber-secure synchronization strategy for AC microgrids. *IEEE Trans. Power Electron.*, 36(12):13372–13378, 2021.

156 Quan Zhou, Mohammad Shahidehpour, Ahmed Alabdulwahab, and Abdullah Abusorrah. A cyber-attack resilient distributed control strategy in islanded microgrids. *IEEE Trans. Smart Grid*, 11(5):3690–3701, 2020.

157 Lei Ding, Qing Long Han, Xiaohua Ge, and Xian Ming Zhang. An overview of recent advances in event-triggered consensus of multiagent systems. *IEEE Trans. Cybernet.*, 48(4):1110–1123, 2018.

158 Xian Ming Zhang, Qing Long Han, and Bao Lin Zhang. An overview and deep investigation on sampled-data-based event-triggered control and filtering for networked systems. *IEEE Trans. Ind. Inf.*, 13(1):4–16, 2017.

159 M Tahir and S K Mazumder. Self-triggered communication enabled control of distributed generation in microgrids. *IEEE Trans. Ind. Inf.*, 11(2):441–449, 2015.

160 W Meng, X Wang, and S Liu. Distributed load sharing of an inverter-based microgrid with reduced communication. *IEEE Trans. Smart Grid*, 9(2):1354–1364, 2017.

161 Y Fan, G Hu, and M Egerstedt. Distributed reactive power sharing control for microgrids with event-triggered communication. *IEEE Trans. Control Syst. Technol.*, 25(1):118–128, 2017.

162 R Han, N L Diaz Aldana, L Meng, J M Guerrero, and Q Sun. Droop-free distributed control with event-triggered communication in DC micro-grid. In *Proceedings of IEEE Applied Power Electronics Conference and Exposition (APEC)*, pages 1160–1166, Tampa, FL, USA, March 2017.

163 R Han, L Meng, J M Guerrero, and J C Vasquez. Distributed nonlinear control with event-triggered communication to achieve current-sharing and voltage regulation in DC microgrids. *IEEE Trans. Power Electron.*, 33(7):6416–6433, 2018.

164 J Zhou, H Zhang, Q Sun, D Ma, and B Huang. Event-based distributed active power sharing control for interconnected AC and DC microgrids. *IEEE Trans. Smart Grid*, 9(6):6815–6828, 2018.

165 S Sahoo and S Mishra. An adaptive event-triggered communication-based distributed secondary control for DC microgrids. *IEEE Trans. Smart Grid*, 9(6):6674–6683, 2018.

166 M Chen, X Xiao, and J M Guerrero. Secondary restoration control of islanded microgrids with a decentralized event-triggered strategy. *IEEE Trans. Ind. Inf.*, 14(9):3870–3880, 2018.

167 Babak Abdolmaleki, Qobad Shafiee, Mohammad Mehdi Arefi, and Tomislav Dragičević. An instantaneous event-triggered Hz–Watt control for microgrids. *IEEE Trans. Power Syst.*, 34(5):3616–3625, 2019.

168 Shengxuan Weng, Dong Yue, Chunxia Dou, Jing Shi, and Chongxin Huang. Distributed event-triggered cooperative control for frequency and voltage stability and power sharing in isolated inverter-based microgrid. *IEEE Trans. Cybernet.*, 49(4):1427–1439, April 2019.

169 Babak Abdolmaleki and Qobad Shafiee. Online Kron reduction for economical frequency control of microgrids. *IEEE Trans. Ind. Electron.*, 67(10):8461–8471, 2020. doi: 10.1109/TIE.2019.2950860.

170 M Yazdanian and A Mehrizi-Sani. Distributed control techniques in microgrids. *IEEE Trans. Smart Grid*, 5(6):2901–2909, 2014.

171 A Bidram, A Davoudi, F L Lewis, and Z Qu. Secondary control of microgrids based on distributed cooperative control of multi-agent systems. *IET Gener. Transm. Distrib.*, 7(8):822–831, 2013.

172 Ali Bidram, Frank L Lewis, and Ali Davoudi. Distributed control systems for small-scale power networks: Using multiagent cooperative control theory. *IEEE Control Syst. Mag.*, 34(6):56–77, 2014.

173 N M Dehkordi, N Sadati, and M Hamzeh. Fully distributed cooperative secondary frequency and voltage control of islanded microgrids. *IEEE Trans. Energy Convers.*, 32(2):675–685, 2017.

174 Fanghong Guo, Changyun Wen, Jianfeng Mao, and Yong Duan Song. Distributed secondary voltage and frequency restoration control of droop-controlled inverter-based microgrids. *IEEE Trans. Ind. Electron.*, 62(7):4355–4364, 2015.

175 N M Dehkordi, N Sadati, and M Hamzeh. Distributed robust finite-time secondary voltage and frequency control of islanded microgrids. *IEEE Trans. Power Syst.*, 32(5):3648–3659, 2017.

176 Nima Mahdian Dehkordi, Hamid Reza Baghaee, Nasser Sadati, and Josep M Guerrero. Distributed noise-resilient secondary voltage and frequency control for islanded microgrids. *IEEE Trans. Smart Grid*, 10(4):3780–3790, 2019.

177 Lin Yu Lu and Chia Chi Chu. Consensus-based droop control synthesis for multiple DICs in isolated micro-grids. *IEEE Trans. Power Syst.*, 30(5):2243–2256, 2015.

178 J W Simpson-Porco, Q Shafiee, F Dörfler, J C Vasquez, J M Guerrero, and F Bullo. Secondary frequency and voltage control of islanded microgrids via distributed averaging. *IEEE Trans. Ind. Electron.*, 62(11):7025–7038, 2015.

179 Babak Abdolmaleki, Qobad Shafiee, and Hassan Bevrani. Kron reduction and L_2-stability for plug-and-play frequency control of microgrids. In *Proceedings of 2018 Smart Grid Conference (SGC)*, pages 1–6, Sanandaj, Iran, November 2018.

180 Q Yang, J A Barria, and T C Green. Communication infrastructures for distributed control of power distribution networks. *IEEE Trans. Ind. Inf.*, 7(2):316–327, 2011.

181 G Xie, H Liu, L Wang, and Y Jia. Consensus in networked multi-agent systems via sampled control: Fixed topology case. In *Proceedings of 2009 American Control Conference*, pages 3902–3907, St. Louis, MO, USA, June 2009.

182 Babak Abdolmaleki, Qobad Shafiee, Ali Reza Seifi, Mohammad Mehdi Arefi, and Frede Blaabjerg. A zeno-free event-triggered secondary control for AC microgrids. *IEEE Trans. Smart Grid*, 11(3):1905–1916, 2020. doi: 10.1109/TSG.2019.2945250.

183 Amirnaser Yazdani and Reza Iravani. *Voltage-sourced converters in power systems: Modeling, control, and applications*. John Wiley & Sons, 2010.

184 X Wu, C Shen, and R Iravani. Feasible range and optimal value of the virtual impedance for droop-based control of microgrids. *IEEE Trans. Smart Grid*, 8(3):1242–1251, 2017.

185 Johannes Schiffer, Romeo Ortega, Alessandro Astolfi, Jörg Raisch, and Tevfik Sezi. Conditions for stability of droop-controlled inverter-based microgrids. *Automatica*, 50(10):2457–2469, 2014.

186 Ali Bidram, Frank L Lewis, and Ali Davoudi. Synchronization of nonlinear heterogeneous cooperative systems using input–output feedback linearization. *Automatica*, 50(10):2578–2585, 2014.

Part II

Interconnected Microgrids

7

Interconnected Microgrids: Opportunities and Challenges

7.1 Introduction

Microgrids (MGs) are fundamental components of smart grids, which try to aggregate nearby distributed energy resources (DERs) and loads in distribution networks. This reconfiguration results in an independent management of MGs in the grid-connected mode to maximize MG owners' revenue, which increases the electric energy and also leads to an economical operation. Nevertheless, such an operation should be in a coordinated manner with the distribution network operator. MGs are also able to be operated/controlled fully autonomous due to existing DERs and energy storage systems as independent energy sources, which can benefit from enhanced reliability, resilience, and security [1]. The transition between grid-connected and islanded operating modes makes MG-oriented modern active distribution grids more flexible than traditional power systems with unilateral power flow. The potential for much more flexibility has been revealed once the idea of interconnected microgrids (IMGs) was presented [2].

IMGs provide a new operation mode in addition to islanded and grid-connected modes. Figure 7.1 shows a general schematic of IMGs. Each MG can be connected to other MGs through an interlinking device (ILD) and connected to the main grid through a circuit breaker (CB) or another connection device. Interlinking lines are shown by their impedances, e.g., Z_{ij} is the impedance of the interlinking line between MG_i and MG_j. The IMG global controller coordinates the power sharing/exchange among IMGs and can be implemented in the form of a central, distributed, or even a decentralized control scheme in special cases [3, 4].

In the IMG operation mode, the new grid is larger with higher capacity than the individual MGs. Therefore, the flexibility of supplying consumers and optimized operating DERs increases. Moreover, reliability, resiliency, sustainability, and supply security improve with respect to the islanded MGs [5]. This new operating mode permits IMGs to exchange power in order to support the frequency/voltage of a critical MG in an emergency condition as well as supply power deficits in a planned manner, etc. [5–8]. The IMGs are also able to be connected with the utility grid to profit from power selling and purchasing when it is required. The idea of MGs interconnection can also be beneficial to divide an active distribution network into some financially independent MGs; however, they are connected physically and have power transfers. In such a new configuration and correlated electricity market, the energy consumption will be more economical.

The IMGs are also called MG clusters, multi-MGs, multiple MGs, community MG, and networked MGs in the literature. All the terms are appropriate to refer to a set of MGs operating and being controlled together. However, the term "network" may lead to a misunderstanding of the

Microgrids: Dynamic Modeling, Stability and Control, First Edition. Qobad Shafiee, Mobin Naderi, and Hassan Bevrani.

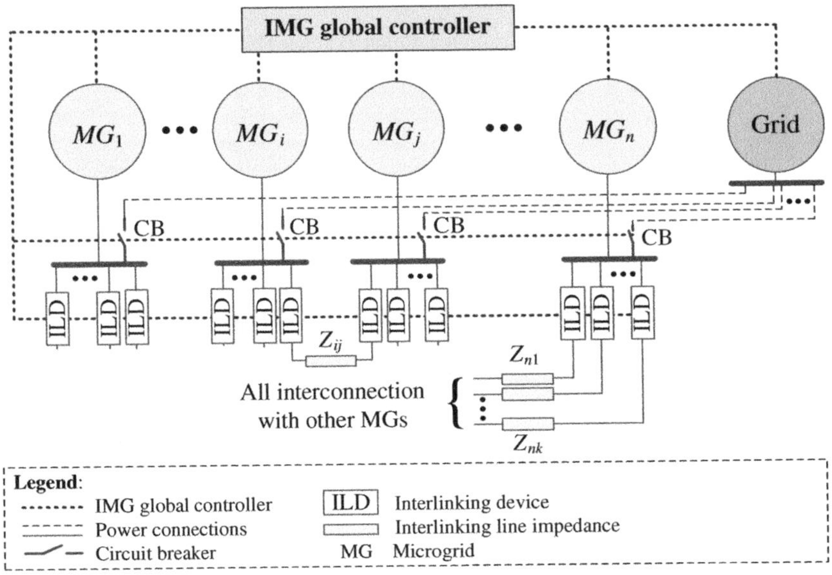

Figure 7.1 Multi MGs interconnected through interlinking lines and devices, and coordinated by a global controller.

necessity of digital communication and cybernetworks. The term "interconnected microgrids" is used in the present book, which focuses on the physical interconnection among MGs.

Dynamic modeling is the starting point for analyzing stability and synthesizing a large group of controllers. According to applications and uses, various modeling methods are represented for modern DER-integrated power systems focusing on the generation side such as DER type [9], power converters [10], grid topology [11], load types [12], DER/MG control loops [1, 13], and reduced order models [14–18]. Note that modeling the IMGs requires an individual modeling procedure to find subsystem models, which is done using some known concepts of the mentioned modeling methods. In addition, special considerations on the large scale of the IMGs and simplification procedures based on application are required.

Although the detailed models are investigated to recognize the most significant participants in the stability of different structures of the IMGs [19–21], simplified/reduced-order models are addressed to investigate and facilitate the stability analysis [22–24], analyze the stability of a designed controller for IMGs [25–28], and design the secondary controller for load-frequency control objectives [28–31]. This book mostly focuses on dynamic modeling, which can be realized in short-term studies, e.g. in few seconds to minute-timescale, and not on the static modeling, which is applied in 15-minute time slots or larger. Such modeling methods are suitable for operation, power/energy management, and planning studies of MGs/IMGs.

Due to the widespread system of IMGs and the possible presence of several types of DERs, converters, loads, and control loops, a stable operation is challenging to obtain and ensure in all possible operational points. In general, stability study fields include phenomena stability assessment, component influences on stability, stability analysis of designed controllers, and stability improvement, which are investigated for different IMG architectures.

Several control objectives can be followed in the field of IMGs. Emergency conditions, mostly over-generation/load, are presented as a dominant control objective [3, 4, 32–36]. The distribution network is reconfigured, and the IMGs are formed to improve reliability, controllability, availability, and supply security [37–40]. Different multilevel control approaches of power exchange/sharing among the IMGs are presented to enhance certain control indices [27, 41–46]. Similar control methods are investigated to optimize power management and consider economic aspects [26, 47–50], improve stability indices, e.g. parameters margins, time-delay stability, and dynamic stability [20, 25, 26, 51–55], enhance communication indices, e.g. communication burden, and delay and packet loss [56–64]. The last research field is load frequency control, where authors have tried to use multiarea power system control and analysis methods for the IMGs [30, 31, 43, 65, 66].

7.2 An Overview

Basically, MGs have been proposed as a solution for sustainable and the efficient operation of DERs [1]. Other benefits of MGs are improved reliability, reduced distribution network losses, and reduced voltage sags with respect to the traditional distribution network as well as dispersed DER-integrated networks. Increasing penetration of renewable energy sources is another considerable MGs advantage. Challenges in stability, control, power energy management, protection, and low inertia of MGs are still being investigated by many researchers.

MGs usually are operated in the grid-connected mode, which switch to the islanded mode when an event such as a fault occurs in the upstream grid. In the grid-connected operation, the flexibility of MGs is greater than the islanded mode in terms of energy generated by their DERs and supplying local loads. In this mode, DERs can produce their maximum power because the upstream grid, like an infinite bus, is extremely receptive to their energy production. Moreover, supplying local MG loads will not be critical due to their support from two energy sources, i.e. the upstream grid and MG's DERs.

On the other hand, the flexibility in production and consumption is greatly reduced in the islanded mode. The amount of DER generations in each period of operation should only be equal to the total consumption of local loads, MG losses, and storage capacity. If renewable energy sources can produce more than the total consumption within hours of islanded operation, there is no choice to use. In contrast, local MG loads may not be provided within other hours of operation due to deficiency of DERs' generation, and thus load shedding process will inevitably take place [1]. Such load shedding process at some times and the lack of maximum production of renewable energy sources at some other times are evidence of low MG flexibility in the islanded mode. In the same way, less reliability and economic efficiency of MG power and energy management in an isolated/islanded mode can be understood.

One of the basic solutions to solve the mentioned challenges in the MG islanded mode is interconnecting MGs to each other [2]. In fact, a new operating mode will be defined called interconnected MGs, multi MGs, multiple MGs, clusters of MG, or networked MGs in different literatures. In this mode, MGs are separate from the main network while connected to each other and have the ability to exchange power. Figure 7.2 shows a simple schematic of the IMGs. Each MG has its own production units and local loads. Several energy resources and storages can be integrated into each MG, including photovoltaic, wind turbine, micro-turbine, battery energy storage, and fuel cell. The MGs can exchange electrical power with other MGs using the ILDs. The ILD can be a back-to-back converter (BTBC) that connects ac lines or it can be used to consider its dc link as a dc line, which can reduce power losses. Another option for interface devices is an instantaneous static switch (SS) or a programmable CB, in which case the communication lines can only be ac. However, dc SSs are also able to be used to transmit power using dc lines.

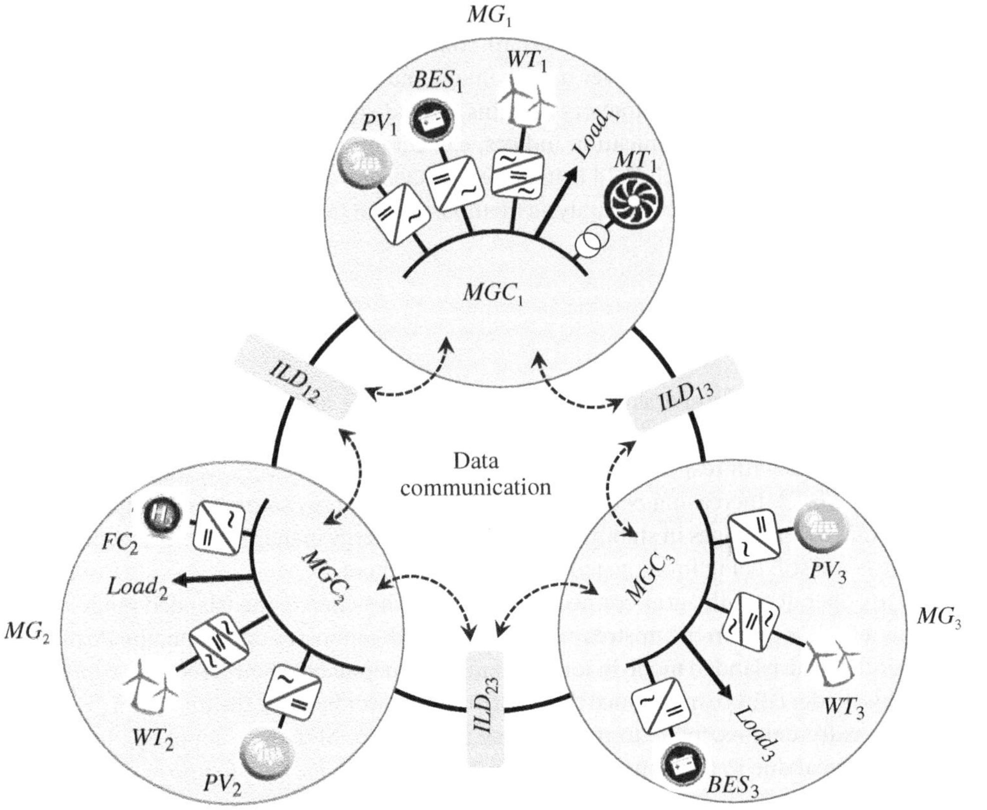

Figure 7.2 A simple schematic of an IMG with three MGs to show main components and relationships.

In order to control MG power exchanges by a specific interface, e.g. ILD_{12} in Figure 7.2, the minimum information exchange required is the information communicated between that interface and the higher layer of microgrid controllers (MGCs) connected to that device, i.e. MGC_1 and MGC_2 in the example. In the case of using a distributed control architecture, it is necessary to communicate information between the desired interface and other interfaces connected to the two MGs, i.e. interfaces ILD_{23} and ILD_{13} in Figure 7.2. If central control architecture is used, information exchanges among all interfaces and all MGCs are required.

By connecting two or more MGs to each other and providing more parallel paths to supply loads, reliability increases. Similarly, the flexibility of DERs and MG load supply are improved. Furthermore, the efficiency of power sharing/exchange increases with an increasing number of acceptable combinations of DERs to supply the load. Figure 7.3 shows a typical power generation in the three IMGs shown in Figure 7.2 to support the power deficiencies of the overloaded MGs. MG_1 load increases at t_0 and required power is provided by MG_2. A pricing limit is considered, which causes the MG_2 priority to support MG_1 with respect to MG_3. MG_2 local load increases at t_0 and thus it has no reserve power after that. Therefore, the next MG_1 load increases at t_3 cannot be compensated by MG_2. However, MG_3 supports the additional demand of MG_1 with a higher energy price and under its generation limit. As a result, IMGs can be a good option to generally improve the operation of islanded MGs and a better option for the operation of distribution networks.

Nevertheless, IMGs have their own challenges that need to be addressed. Dynamic interference between MGs and relevant stability issues, increase of uncertainties associated with each MG,

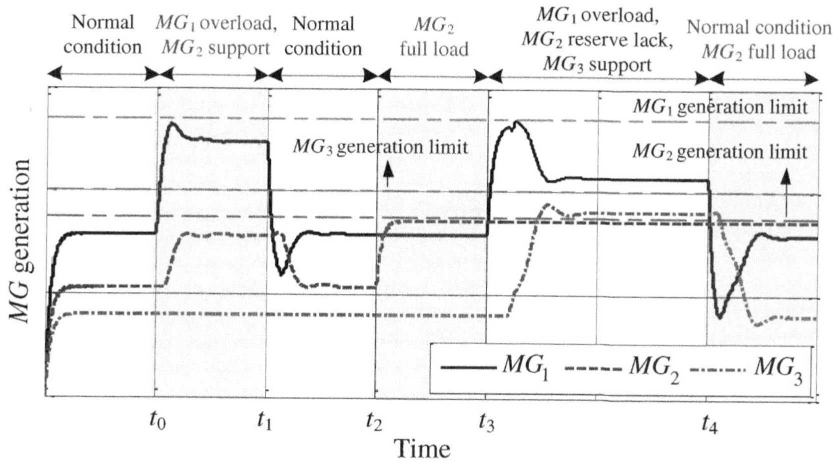

Figure 7.3 A typical power generation in the three IMGs shown in Figure 7.2 to support power deficiency of the overloaded MG.

control of power exchange/sharing between MGs, and interferences between power exchange controllers and internal MGCs are some of the IMG challenges, which will be discussed later.

7.3 Architectures of Interconnected Microgrids

Different types of MG structures, i.e. ac and dc types, different types of interlinking lines, i.e. ac and dc types, different types of ILDs including CBs, power converters, e.g. BTBCs, and SSs, methods of power exchange control, as well as communication methods will result in different IMG architectures [7, 8]. Figure 7.4 shows various structures of IMGs with regard to ILD and MG types including hybrid ac/dc MGs, two ac IMGs, multiple ac IMGs, multiple ac–dc IMGs, and multiple dc IMGs. Figure 7.4(a) shows simply hybrid ac/dc MGs, which are connected via an ac–dc converter (ADC). The stability analysis of such MGs is less studied [67–69]. However, control and operation challenges and different control approaches of this architecture are studied and fully reviewed [70–73].

Figure 7.4(b) shows the architecture of two ac IMGs, where CBs, BTBCs, and SSs can be used as the ILD. This architecture is widely investigated due to its simple structure, being the first step in IMG formation, and being a good case study for many IMG challenges. According to the ILD type, this architecture can be categorized into three groups. The first group is named two IMGs via CBs (CB-2IMGs) and consists of two MGs with the same rated voltage and frequency whether they are islanded or grid-connected. Numerous studies have been conducted on the dynamics, modeling, stability, and control of CB-IMGs. [29–32, 37, 40, 47, 74–76]. The second is named two IMGs via BTBCs (BTBC-2IMGs), which uses BTBC to improve power flow controllability and earn supplementary objectives [39, 59, 65, 77–81]. In the third group, an SS is used, which typically supports overloaded MG by emergency connect/reconnect of the IMGs [3].

Multiple ac interconnected MGs (MIMGs) shown in Figure 7.4(c) have attracted more interest due to the existing structure of the ac distribution networks, which can be restructured into MIMG to improve stability, controllability, resiliency, reliability, and power quality. The same classification can also be considered for MIMG as CB-MIMG [20–25, 27, 28, 34, 36, 38, 43, 45, 48–51, 54–58, 60,

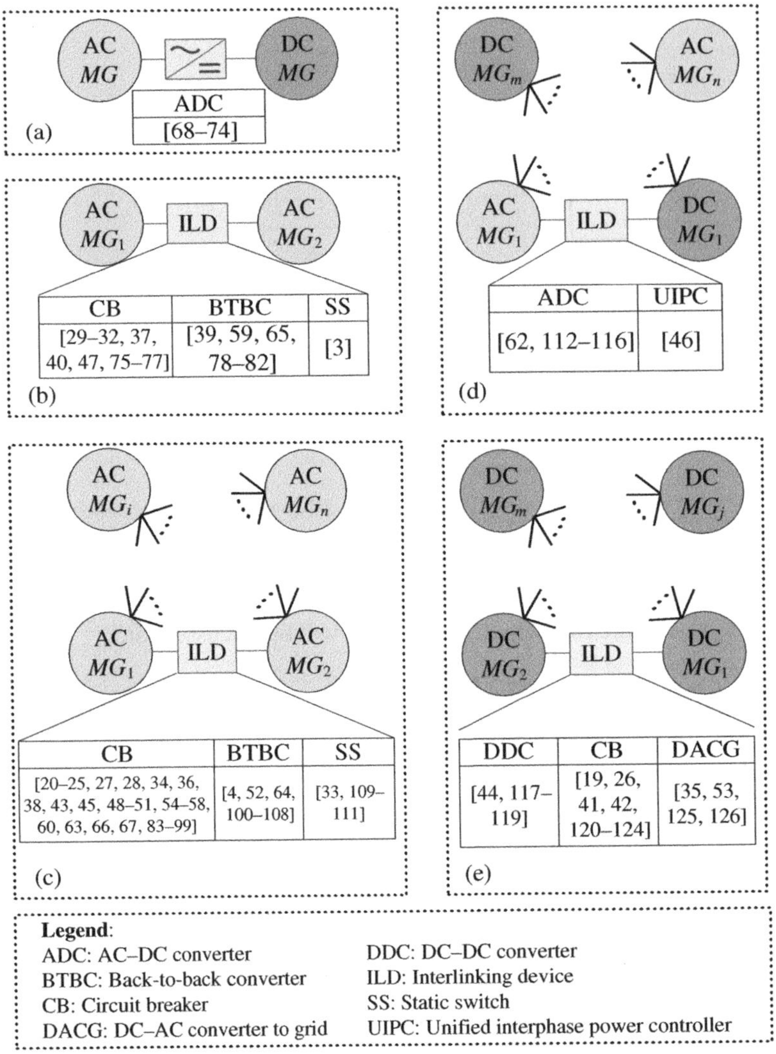

Figure 7.4 General architectures of interconnected microgrids: (a) hybrid ac/dc microgrids, (b) two ac IMGs, (c) multiple ac IMGs, (d) multiple ac/dc IMGs, and (e) multiple dc IMGs.

63, 66, 82–97], BTBC-MIMG [4, 52, 64, 98–104], and SS-MIMG [33, 105–107]. Note that CB-MIMG is an economical choice; however, it is difficult to control due to operating in one frequency as synchronous IMGs. In fact, control challenges such as synchronization and power sharing should be fully solved to have a reliable structure. In BTBC-MIMG, BTBCs enhance the controllability. This is because BTBCs are controllable agents in addition to MGs and increase the controllable actuators with respect to CB-MIMG. Nevertheless, BTBCs are more expensive and cause less tendency to be used for interconnecting MGs. The SS-MIMG are more similar to CB-MIMG, which can provide better time responses and smaller transients during switching.

Another architecture is multiple ac–dc IMGs shown in Figure 7.4(d), where ILDs are usually AC/DC power converters. Studies around this architecture are mostly dealing with hierarchical/distributed coordination control of active power sharing/exchange among the IMGs [46, 62, 108–112] and the topology of the power exchange interface [46, 110–112]. The last architecture includes multiple dc MGs shown in Figure 7.4(e), where they are interconnected together through dc–dc converters [44, 113–115], and dc CBs [19, 26, 41, 42, 116–120], as well as connected in parallel to utility grid via dc–ac converters [35, 53, 121, 122].

7.4 Benefits, Challenges, and Research Fields

Areas of research in the operation of IMGs and some of correlated papers are summarized in Table 7.1. Also, the research issues in the literature are emphasized. Before conducting extensive research, such as the studies presented in Table 7.1, the benefits of IMGs were primarily discussed in theory without substantiating them through mathematical models or demonstrating them through simulation and laboratory results. Some benefits can be summarized as follows:

- Improving reliability,
- Increasing resiliency, i.e. recovery of MGs after a severe incident,
- Increment of consumption and production flexibilities, i.e. increasing efficiency and effectiveness of production and finding the best options for load supply,
- Economic power management and energy scheduling,
- Delaying load shedding process compared to individual MGs [2, 5].

These benefits have been fully proven in the literature. However, there are many challenges related to the operation mode of IMGs that are still being studied, and some are listed below.

- Measurement-based and data-driven modeling and control
- Reduced-order models for the large-scale IMGs suitable for applications such as frequency dynamics and control studies [28]

Table 7.1 Overview of research areas in interconnected microgrids.

IMG research fields	Main research problem	References
Basic studies	Advantages, disadvantages, opportunities, and challenges	[2, 5]
	Basic simulation and dynamics study	[40, 86]
Modeling	The detailed modeling of CB-IMGs	[27, 45, 106, 107]
	The detailed modeling of BTBC-IMGs	[100, 101]
	Module-based simplified modeling	[51, 55, 93, 96]
	Reduced-order models	[24]
Stability	Small-signal stability and dominant low-frequency modes	[24, 27, 100, 106, 107]
	Stability improvement	[55, 79]
	Power generation uncertainty analysis	[22, 23]
	Power and frequency oscillation damping	[24, 52]
	Grid-connected IMG stability analysis	[100]
Control	Power sharing control enhancement	[3, 98, 99, 108, 109, 117]
	DC IMG control	[116, 117]
	AC-DC IMG control	[98, 108, 109]
Operation	Power/energy management and operation	[124–127]
	Ancillary services for voltage regulation in distribution networks	[129]

- Low-frequency oscillation in hybrid MGs [68]
- Inter-MG power oscillations [52]
- Power exchange/sharing methods among the IMGs [3, 98, 99, 108, 109, 117]
- Uncertainties impact on the IMG stability [22]
- Inter-MG dynamic modes [28]
- Stability analysis of the dc IMGs considering time-delay caused by wide-area measurements [51, 118]
- Frequency stability of CB-MIMG [123]
- BTBC dc voltage stability in BTBC-MIMG [102]
- IMG stability assessment using eigenvalue analysis, stability margin criteria, Lyapunov methods, and time-domain simulations (TDS) [22, 23, 27, 55, 68, 82, 102]
- stability improvement solutions [20, 81, 95]
- power/energy management [124–128].

These challenges and issues regarding modeling, stability, and control of IMGs are taken into account in the next chapters. Studies and investigations concerning IMG operation and power/energy management are not within the scope of the book's research focus. They are briefly covered in Section 7.5. Nevertheless, research gaps in the main fields of the book are taken into consideration afterward.

7.5 Operation of Interconnected Microgrids

Most of the previous research works have been related to the operation of IMGs, which are well reviewed in the literature [130–133]. This field can be found in the literature through topics of energy management, energy trading, and MG cluster operation. Generally, in energy management problems, a static optimization problem is solved in which a cost function is minimized, and some constraints are considered to realize the limitations of IMG operation. In such problems, researchers consider power exchange between the IMGs without dynamic control challenges such as differences in CB-IMG frequencies, and suppose the control and stability issues have been solved. The time intervals studied in operation problems are in the range of a few minutes to an hour and hence are much larger than the time intervals studied in stability and control issues, which are in the fraction of a second to a few seconds. However, slow-dynamic control constraints are taken into account in energy management, e.g. bus voltage amplitude and its acceptable range. Note that the IMG operation is done in the global control level of the hierarchical control architecture in which the secondary controllers of grid-forming (GFO) DERs are fed with desired frequency and voltage values in a grid-isolated IMG operating mode. However, it generally determines the active and reactive power references in a grid-connected IMG operation or for grid-following (GFL) DERs in the grid-isolated IMG operating mode.

Many methods are used to manage and schedule energy resources or demand side in an IMG operation process. An intelligent agent-based demand response is presented to comfort energy trading among the IMGs [134]. In order to minimize the total power generation and exchange costs, a peer-to-peer architecture is used to exchange energy between the IMGs [135]. A distributed control architecture based on noncooperative game theory is proposed to coordinate energy scheduling among the IMGs in an electricity market [136]. In order to investigate the cooperative behavior of the IMGs in the concept of smart grids [137, 138] and analyze the price and cost of IMG development [139], methods of cooperative game theory have been applied. A multiagent consensus algorithm is addressed to solve the real-time cooperative power dispatch problem of the IMGs [140]. A system-of-systems control architecture is investigated to optimize IMG energy management.

Convex programming is a global optimizer searching and finding method, which is used to minimize the energy scheduling of the IMGs [127, 128]. The authors have studied a set of MGs connected to the main grid, where the problem of power dispatching among the IMGs has been studied considering full coordination between them [128]. However, the problem of energy consumption planning under an indefinite demand of neighboring MGs is solved in [127]. IMG energy management has been also investigated in the form of an MG community using a hierarchical energy management system [124]. In [125], a different operation problem is considered in the case of emergency conditions, where a radial distribution network is considered that suffers many faults when exposed to natural disasters. In order to increase the resilience of such a network and recover as many sensitive loads as possible after occurring faults and power outages, the formation of independent MGs has been proposed. In order to recover the maximum number of sensitive loads, the problem of MG formation has been solved through the mixed integer linear programming optimization method.

Another important field of IMG operation is the electricity market with the goal of voltage control in distribution networks [129]. MGs can play the role of virtual power plants in the future network of IMGs. Each MGC bids energy and ancillary services to other IMGs based on the aggregation of internal bids from the DERs in the MG, e.g. renewable energy resources, controllable loads, micro-generators, and storage devices [141]. The distribution network operator can run a day-ahead market for reactive power, which is required for the voltage maintained within acceptable boundaries.

7.6 Vacancies for Future Research

7.6.1 IMG Dynamic Modeling

IMGs are large-scale and complex systems, whose modeling results in high-order models even using model-based (MOB) and module-based (MDB) simplification techniques. Therefore, measurement-based (MEB) and data-driven reduced-order modeling approaches are necessary to be used to achieve very low-order models based on measured data of just a specific variable, e.g. frequency. System identification methods, e.g. fast Fourier transform, Prony analysis [142, 143], discrete wavelet transform [144], and vector-fitting technique [145] as well as machine learning algorithms, e.g. reinforcement learning [146, 147], Q-learning [148], and decision tree [149], can be employed to analyze the measured data. Artificialintelligence modeling techniques are also used [150, 151]. In order to earn very low-order models, the aggregation methods can be applied to simplified MOB and MDB models [28]. On the other hand, many IMG models, reviewed in Chapter 8, are obtained without an acceptable validation. To achieve standard IMG models used in real and laboratory case studies, secure and verified models are needed.

7.6.2 IMG Stability Analysis

CB-MIMG is the most common structure of IMGs. Due to the weakness of these grids and their low inertia, their voltage and frequency are notably unstable and challenging to control. Recently, powerful control methods have been proposed for these IMGs using a distributed pinning consensus-based secondary controller. Therefore, many previous stability analyses of

CB-MIMG belong to poor control methods and should be re-established for new control methods. Although an asymptotic stability proof based on Lyapunov theory is theoretically represented [25, 38, 48, 60, 63], small-signal stability is not assessed to identify the impact of IMG power exchange controller parameters as well as critical dynamic modes are not clearly determined. In spite of the existing small-signal stability analyses for different IMG types, more analytical large-signal stability assessment methods are needed to investigate various IMG operating points, e.g. equal-area criterion [152, 153].

Another gap is related to the analysis of IMG elements. GFO DERs as the most important control actuators are well taken into account. However, GFL DERs [38] are not considered in terms of IMG stability and control. Different load types are another group of elements, which are not studied, particularly in ac IMGs. Furthermore, harmonic modeling and stability analysis [154] of IMGs, which dominantly are integrated with power electronics devices, are necessary to be paid attention to achieve a much more complete analysis of IMG stability.

7.6.3 IMG Control

Although the PI controllers are extensively used in different control layers of the IMGs, advanced control methods and strategies such as adaptive control methods, robust control theory, and artificial intelligence-based control types can be used for different objectives. Some other ideas for future works in control of the IMGs are listed below.

- The control of CB-IMGs and SS-IMGs is more challenging due to direct interconnection by CBs/SSs. Therefore, it should be studied from different aspects, e.g. cooperation of generation units, smooth synchronization, and islanding controls.
- Developing CB-IMG controllers to dampen the inter-MG oscillation modes in an efficient way.
- Investigating the low inertia of the IMGs and the role of virtual synchronous generator control in improving it.
- Reconfiguring the MG load shedding control considering IMG power exchange/sharing.
- A complete technical-economic comparison between CB-IMGs and BTBC-IMGs in terms of stability margins and controllability to make a decision on selecting these IMG types based on the applications and necessities.
- Eliminating communication delay issues using advanced control methods and communication architectures, e.g. robust and predictive controllers, and developed distributed communication methods.
- Introducing more objectives for deciding the power exchange among MGs and realizing the corresponding controllers, e.g. cybersecurity with respect to control communication and power quality.

7.7 Summary

This chapter introduces the concept of interconnecting MGs to maximize the benefits of adjacent isolated MGs, such as flexibility, resiliency, and reliability, as well as to divide an active distribution network into financially independent MGs during faults and emergency conditions. The basic concepts and a literature review of IMGs are presented in terms of dynamic modeling, stability, control, and operation. Different architectures of IMGs are discussed, including hybrid ac/dc IMGs, multiple ac IMGs, multiple ac/dc IMGs, and multiple dc IMGs, which can be connected/disconnected through CBs, BTBCs, and other methods. While each IMG architecture has its own benefits, there are some challenges in modeling, stability, and control, such as finding simplified models for large-scale systems for specific goals like frequency control and analysis, synchronization stability, and power-sharing/exchanging control in CB/BTBC IMGs.

Some of these challenges will be addressed in more detail in the next chapters. For example, Chapter 8 focuses on detailed and simplified modeling of IMGs, while Chapter 9 covers stability and sensitivity analysis of different IMG architectures. Finally, vacancies and gaps in the three main research fields of this book are presented for future studies of IMGs.

References

1 Hassan Bevrani, Bruno François, and Toshifumi Ise. *Microgrid dynamics and control*. John Wiley & Sons, Hoboken, NJ, 2017.

2 Robert H Lasseter. Smart distribution: Coupled microgrids. *Proc. IEEE*, 99(6):1074–1082, 2011. doi: 10.1109/JPROC.2011.2114630.

3 Ehsan Pashajavid, Arindam Ghosh, and Firuz Zare. A multimode supervisory control scheme for coupling remote droop-regulated microgrids. *IEEE Trans. Smart Grid*, 9(5):5381–5392, 2018.

4 Mobin Naderi, Yousef Khayat, Qobad Shafiee, Tomislav Dragicevic, Frede Blaabjerg, and Hassan Bevrani. An emergency active and reactive power exchange solution for interconnected microgrids. *IEEE J. Emerging Sel. Top. Power Electron.*, 9(5):5206–5218, 1–13, 2019.

5 Mohammad Shahidehpour, Zhiyi Li, Shay Bahramirad, Zuyi Li, and Wei Tian. Networked microgrids: Exploring the possibilities of the iit-bronzeville grid. *IEEE Power Energy Mag.*, 15(4):63–71, 2017.

6 Mahamad Nabab Alam, Saikat Chakrabarti, and Arindam Ghosh. Networked microgrids: State-of-the-art and future perspectives. *IEEE Trans. Ind. Inf.*, 15(3):1238–1250, 2018.

7 Zhirong Xu, Ping Yang, Chengli Zheng, Yujia Zhang, Jiajun Peng, and Zhiji Zeng. Analysis on the organization and development of multi-microgrids. *Renewable Sustainable Energy Rev.*, 81:2204–2216, 2018.

8 Eduard Bullich-Massagué, Francisco Díaz-González, Mònica Aragüés-Peñalba, Francesc Girbau-Llistuella, Pol Olivella-Rosell, Andreas Sumper. Microgrid clustering architectures. *Appl. Energy*, 212:340–361, 2018.

9 Tiankui Sun, Jian Lu, Zhimin Li, David Lee Lubkeman, and Ning Lu. Modeling combined heat and power systems for microgrid applications. *IEEE Trans. Smart Grid*, 9(5):4172–4180, 2017.

10 Joan Rocabert, Alvaro Luna, Frede Blaabjerg, and Pedro Rodriguez. Control of power converters in AC microgrids. *IEEE Trans. Power Electron.*, 27(11):4734–4749, 2012.

11 Nagaraju Pogaku, Milan Prodanovic, and Timothy C Green. Modeling, analysis and testing of autonomous operation of an inverter-based microgrid. *IEEE Trans. Power Electron.*, 22(2):613–625, 2007.

12 Francis K Tuffner, Kevin P Schneider, Jacob Hansen, and Marcelo A Elizondo. Modeling load dynamics to support resiliency-based operations in low-inertia microgrids. *IEEE Trans. Smart Grid*, 10(3):2726–2737, 2018.

13 Magdi Sadek Mahmoud, S Azher Hussain, and Mohammad Ali Abido. Modeling and control of microgrid: An overview. *J. Franklin Inst.*, 351(5):2822–2859, 2014.

14 Ling Luo and Sairaj V Dhople. Spatiotemporal model reduction of inverter-based islanded microgrids. *IEEE Trans. Energy Convers.*, 29(4):823–832, 2014.

15 Valerio Mariani, Francesco Vasca, Juan C Vásquez, and Josep M Guerrero. Model order reductions for stability analysis of islanded microgrids with droop control. *IEEE Trans. Ind. Electron.*, 62(7):4344–4354, 2014.

16 Md Rasheduzzaman, Jacob A Mueller, and Jonathan W Kimball. Reduced-order small-signal model of microgrid systems. *IEEE Trans. Sustainable Energy*, 6(4):1292–1305, 2015.

17 Petr Vorobev, Po-Hsu Huang, Mohamed Al Hosani, James L Kirtley, and Konstantin Turitsyn. High-fidelity model order reduction for microgrids stability assessment. *IEEE Trans. Power Syst.*, 33(1):874–887, 2017.

18 Iraklis P Nikolakakos, H H Zeineldin, Mohamed Shawky El-Moursi, and James L Kirtley. Reduced-order model for inter-inverter oscillations in islanded droop-controlled microgrids. *IEEE Trans. Smart Grid*, 9(5):4953–4963, 2017.

19 Renke Han, Michele Tucci, Andrea Martinelli, Josep M Guerrero, and Giancarlo Ferrari-Trecate. Stability analysis of primary plug-and-play and secondary leader-based controllers for DC microgrid clusters. *IEEE Trans. Power Syst.*, 34(3):1780–1800, 2018.

20 Jinghan He, Xiaoyu Wu, Xiangyu Wu, Yin Xu, and Josep M Guerrero. Small-signal stability analysis and optimal parameters design of microgrid clusters. *IEEE Access*, 7:36896–36909, 2019.

21 Ouyang Jianna, Liang Shuo, Chen Weidong, and Guo Min. Modeling and analysis of microgrid cluster simulation based on RTDS. In *2nd IEEE Conference Energy Internet and Energy Syst. Integration (EI2)*, pages 1–5. IEEE, 2018.

22 Yan Li, Peng Zhang, and Peter B Luh. Formal analysis of networked microgrids dynamics. *IEEE Trans. Power Syst.*, 33(3):3418–3427, 2018.

23 Yan Li, Peng Zhang, and Meng Yue. Networked microgrid stability through distributed formal analysis. *Appl. Energy*, 228:279–288, 2018.

24 Iraklis P Nikolakakos, Hatem H Zeineldin, Mohamed Shawky El-Moursi, and Nikos D Hatziargyriou. Stability evaluation of interconnected multi-inverter microgrids through critical clusters. *IEEE Trans. Power Syst.*, 31(4):3060–3072, 2016.

25 Jingang Lai, Xiaoqing Lu, Xinghuo Yu, and Antonello Monti. Cluster-oriented distributed cooperative control for multiple AC microgrids. *IEEE Trans. Ind. Inf.*, 15(11):5906–5918, 2019.

26 Shivraman Mudaliyar, Bhanu Duggal, and Sukumar Mishra. Distributed tie-line power flow control of autonomous DC microgrid clusters. *IEEE Trans. Power Electron.*, 35(10):11250–11266, 2020.

27 Zhuoli Zhao, Ping Yang, Yuewu Wang, Zhirong Xu, and Josep M Guerrero. Dynamic characteristics analysis and stabilization of PV-based multiple microgrid clusters. *IEEE Trans. Smart Grid*, 10(1):805–818, 2017.

28 Mobin Naderi, Qobad Shafiee, Hassan Bevrani, and Frede Blaabjerg. Low-frequency small-signal modeling of interconnected AC microgrids. *IEEE Trans. Power Syst.*, 36(4):2786–2797, 2020.

29 A Hasib Chowdhury and Md Asaduz-Zaman. Load frequency control of multi-microgrid using energy storage system. In *8th International Conference on Electrical and Computer Engineering*, pages 548–551, 2014.

30 Meysam Gheisarnejad and Mohammad Hassan Khooban. Secondary load frequency control for multi-microgrids: HiL real-time simulation. *Soft Comput.*, 23(14):5785–5798, 2019.

31 Deepak Kumar Lal, Ajit Kumar Barisal, and M Tripathy. Load frequency control of multi area interconnected microgrid power system using grasshopper optimization algorithm optimized fuzzy PID controller. In *2018 Recent Advances on Engineering, Technology and Computational Sciences (RAETCS)*, pages 1–6. IEEE, 2018.

32 Mir Nahidul Ambia, Ke Meng, Weidong Xiao, and Zhao Yang Dong. Comprehensive solution of networked microgrid towards enhanced overload resiliency. In *2018 International Conference on Power System Technology (POWERCON)*, pages 1736–1742. IEEE, 2018.

33 Munira Batool, Farhad Shahnia, and Syed M Islam. Multi-level supervisory emergency control for operation of remote area microgrid clusters. *J. Mod. Power Syst. Clean Energy*, 7(5):1210–1228, 2019.

34 Moein Choobineh, Deborath Silva-Ortiz, and Salman Mohagheghi. An automation scheme for emergency operation of a multi-microgrid industrial park. *IEEE Trans. Ind. Appl.*, 54(6):6450–6459, 2018.

35 Md Tanvir Arafat Khan, Aranya Chakrabortty, Iqbal Husain, and Rafael Cisneros. Coordinated control of energy storage in networked microgrids under unpredicted load demands. In *2019 American Control Conference (ACC)*, pages 1573–1578. IEEE, 2019.

36 Yiping Shen, Yongming Xu, Hongwei Yao, Jianxun Shi, Feiwei Li, Zhen Wen, Chongbiao Zhang, and Wei Liu. Distributed cluster control for multi-microgrids using pinning-based group consensus of multi-agent system. In *5th IEEE International Conference on Cloud Computing and Intelligence Systems (CCIS)*, pages 1077–1080. IEEE, 2018.

37 Chunxia Dou, Yuanzhe Li, Dong Yue, Zhanqiang Zhang, and Bo Zhang. Distributed cooperative control method based on network topology optimisation in microgrid cluster. *IET Renew. Power Gener.*, 14(5):939–947, 2020.

38 Wei Liu, Wei Gu, Yinliang Xu, Yong Wang, and Kaifeng Zhang. General distributed secondary control for multi-microgrids with both PQ-controlled and droop-controlled distributed generators. *IET Gener. Transm. Distrib.*, 11(3):707–718, 2017.

39 Inam Ullah Nutkani, Poh Chiang Loh, Peng Wang, Tseng King Jet, and Frede Blaabjerg. Intertied AC–AC microgrids with autonomous power import and export. *Int. J. Electr. Power Energy Syst.*, 65:385–393, 2015.

40 Fan Zhang, Reza Jamalzadeh, and Mingguo Hong. A dynamic simulation study of networked microgrids in active distribution system. In *2016 IEEE Power & Energy Society Innovative Smart Grid Technologies Conference (ISGT)*, pages 1–5. IEEE, 2016.

41 Ying Han, Yuchen Pu, Qi Li, Wangxuan Fu, Weirong Chen, Zhiyu You, and Hong Liu. Coordinated power control with virtual inertia for fuel cell-based DC microgrids cluster. *Int. J. Hydrogen Energy*, 44(46):25207–25220, 2019.

42 Subham Sahoo, Sukumar Mishra, Seyed Mahdi Fazeli, Furong Li, and Tomislav Dragičević. A distributed fixed-time secondary controller for DC microgrid clusters. *IEEE Trans. Energy Convers.*, 34(4):1997–2007, 2019.

43 Kevin P Schneider, Nikitha Radhakrishnan, Yingying Tang, Francis K Tuffner, Chen-Ching Liu, Jing Xie, and Dan Ton. Improving primary frequency response to support networked microgrid operations. *IEEE Trans. Power Syst.*, 34(1):659–667, 2018.

44 Umamaheswararao Vuyyuru, Suman Maiti, and Chandan Chakraborty. Active power flow control between DC microgrids. *IEEE Trans. Smart Grid*, 10(5):5712–5723, 2019.

45 Xiaoyu Wu, Yin Xu, Xiangyu Wu, Jinghan He, Josep M Guerrero, Chen-Ching Liu, Kevin P Schneider, and Dan T Ton. A two-layer distributed control method for islanded networked microgrid systems. *IEEE Trans. Smart Grid*, 11(2):942–957, 2019.

46 Mahdi Zolfaghari, Mehrdad Abedi, and Gevork B Gharehpetian. Power exchange control of clusters of multiple AC and DC microgrids interconnected by UIPC in hybrid microgrids. In *2019 24th Electrical Power Distribution Conference (EPDC)*, pages 22–26. IEEE, 2019.

47 Diambomba Hyacinthe Tungadio, Ramesh C Bansal, and Mukwanga Willy Siti. Optimal control of active power of two micro-grids interconnected with two AC tie-lines. *Electr. Power Compon. Syst.*, 45(19):2188–2199, 2017.

48 Xiaoqing Lu, Jingang Lai, and Xinghuo Yu. A novel secondary power management strategy for multiple AC microgrids with cluster-oriented two-layer cooperative framework. *IEEE Trans. Ind. Inf.*, 17(2):1483–1495, 2020.

49 A G Madureira, J C Pereira, N J Gil, J A Peças Lopes, G N Korres, N D Hatziargyriou. Advanced control and management functionalities for multi-microgrids. *Eur. Trans. Electr. Power*, 21(2):1159–1177, 2011.

50 Eric Harmon, Utku Ozgur, Mehmet Hazar Cintuglu, Ricardo de Azevedo, Kemal Akkaya, and Osama A Mohammed. The internet of microgrids: A cloud-based framework for wide area networked microgrids. *IEEE Trans. Ind. Inf.*, 14(3):1262–1274, 2017.

51 Ran Hao, Qian Ai, Ziqing Jiang, and Yuchao Zhu. A novel adaptive control strategy of interconnected microgrids for delay-dependent stability enhancement. *Int. J. Electr. Power Energy Syst.*, 99:566–576, 2018.

52 Md Jahangir Hossain, Md Apel Mahmud, Federico Milano, Seddik Bacha, and Ahmad Hably. Design of robust distributed control for interconnected microgrids. *IEEE Trans. Smart Grid*, 7(6):2724–2735, 2016.

53 David G Wilson, Rush D Robinett III, Wayne W Weaver, and Steven F Glover. Nonlinear power flow control for networked AC/DC microgrids, May 26 2020. US Patent 10,666,054.

54 Li Yang, Zhijian Hu, Shiwei Xie, Shunfei Kong, and Weiwei Lin. Adjustable virtual inertia control of supercapacitors in PV-based AC microgrid cluster. *Electr. Power Syst. Res.*, 173:71–85, 2019.

55 Yun Zhang, Le Xie, and Qifeng Ding. Interactive control of coupled microgrids for guaranteed system-wide small signal stability. *IEEE Trans. Smart Grid*, 7(2):1088–1096, 2016.

56 Yan Li, Yanyuan Qin, Peng Zhang, and Amir Herzberg. SDN-enabled cyber-physical security in networked microgrids. *IEEE Trans. Sustainable Energy*, 10(3):1613–1622, 2018.

57 Yaran Li, Ke Meng, and Zhao Yang Dong. Distributed consensus control with event-triggered communication for multi-microgrid cluster. In *2019 IEEE Milan PowerTech*, pages 1–6. IEEE, 2019.

58 Qiang Li, Mengkai Gao, Houfei Lin, Ziyu Chen, and Minyou Chen. Mas-based distributed control method for multi-microgrids with high-penetration renewable energy. *Energy*, 171:284–295, 2019.

59 Inam Ullah Nutkani, Poh Chiang Loh, and Frede Blaabjerg. Power flow control of intertied AC microgrids. *IET Power Electron.*, 6(7):1329–1338, 2013.

60 Lingyu Ren, Yanyuan Qin, Yan Li, Peng Zhang, Bing Wang, Peter B Luh, Song Han, Taofeek Orekan, and Tao Gong. Enabling resilient distributed power sharing in networked microgrids through software defined networking. *Appl. Energy*, 210:1251–1265, 2018.

61 David Rua, L F Moura Pereira, Nuno Gil, and J A Peças Lopes. Impact of multi-microgrid communication systems in islanded operation. In *2nd IEEE PES International Conference and Exhibition on Innovative Smart Grid Technologies*, pages 1–6. IEEE, 2011.

62 Zhaojian Wang, Laijun Chen, Feng Liu, Peng Yi, Ming Cao, Sicheng Deng, and Shengwei Mei. Asynchronous distributed power control of multi-microgrid systems. *IEEE Trans. Control Netw. Syst.*, 7(4):1960–1973, 2020.

63 Shengxuan Weng, Yusheng Xue, Jianbo Luo, and Yanman Li. Distributed secondary control for islanded microgrids cluster based on hybrid-triggered mechanisms. *Processes*, 8(3):370, 2020.

64 Ramon Zamora and Anurag K Srivastava. Multi-layer architecture for voltage and frequency control in networked microgrids. *IEEE Trans. Smart Grid*, 9(3):2076–2085, 2018.

65 Mojtaba Khederzadeh, Hamed Maleki, and Vahid Asgharian. Frequency control improvement of two adjacent microgrids in autonomous mode using back to back voltage-sourced converters. *Int. J. Electr. Power Energy Syst.*, 74:126–133, 2016.

66 Vladimir Toro and Eduardo Mojica-Nava. Droop-free control for networked microgrids. In *2016 IEEE Conference on Control Applications (CCA)*, pages 374–379. IEEE, 2016.

67 Moudud Ahmed, Arash Vahidnia, Lasantha Meegahapola, and Manoj Datta. Small signal stability analysis of a hybrid AC/DC microgrid with static and dynamic loads. In *2017 Australasian Universities Power Engineering Conference (AUPEC)*, pages 1–6. IEEE, 2017.

68 Moudud Ahmed, Lasantha Meegahapola, Arash Vahidnia, and Manoj Datta. Analysis and mitigation of low-frequency oscillations in hybrid AC/DC microgrids with dynamic loads. *IET Gener. Transm. Distrib.*, 13(9):1477–1488, 2019.

69 Zhiyi Li and Mohammad Shahidehpour. Small-signal modeling and stability analysis of hybrid AC/DC microgrids. *IEEE Trans. Smart Grid*, 10 (2):2080–2095, 2017.

70 Moudud Ahmed, Lasantha Meegahapola, Arash Vahidnia, and Manoj Datta. Stability and control aspects of microgrid architectures–a comprehensive review. *IEEE Access*, 8:144730–144766, 2020.

71 Sarmad Majeed Malik, Xin Ai, Yingyun Sun, Chen Zhengqi, and Zhou Shupeng. Voltage and frequency control strategies of hybrid AC/DC microgrid: A review. *IET Gener. Transm. Distrib.*, 11(2):303–313, 2017.

72 Saroja Kanti Sahoo, Avinash Kumar Sinha, and N K Kishore. Control techniques in AC, DC, and hybrid AC–DC microgrid: A review. *IEEE J. Emerging Sel. Top. Power Electron.*, 6(2):738–759, 2017.

73 Ajay Gupta, Suryanarayana Doolla, and Kishore Chatterjee. Hybrid AC–DC microgrid: Systematic evaluation of control strategies. *IEEE Trans. Smart Grid*, 9(4):3830–3843, 2017.

74 Kun Liu, Tao Liu, and D J Jill. Frequency control in networked microgrids with voltage-sensitive loads. In *Proceedings of IREP Symposium*, pages 1–6, 2017.

75 Ali Akhavan, Hamid Reza Mohammadi, Juan C Vasquez, and Josep M Guerrero. Coupling effect analysis and control for grid-connected multi-microgrid clusters. *IET Power Electron.*, 13(5):1059–1070, 2020.

76 Kishan Veerashekar, Stefan Eichner, and Matthias Luther. Modelling and transient stability analysis of interconnected autonomous hybrid microgrids. In *2019 IEEE Milan PowerTech*, pages 1–6. IEEE, 2019.

77 Megha Goyal and Arindam Ghosh. Microgrids interconnection to support mutually during any contingency. *Sustainable Energy Grids Netw.*, 6:100–108, 2016.

78 Inam Ullah Nutkani, Poh Chiang Loh, and Frede Blaabjerg. Distributed operation of interlinked AC microgrids with dynamic active and reactive power tuning. *IEEE Trans. Ind. Appl.*, 49(5):2188–2196, 2013.

79 Julius Susanto, Farhad Shahnia, Arindam Ghosh, and Sumehda Rajakaruna. Interconnected microgrids via back-to-back converters for dynamic frequency support. In *IEEE Power Engineering Conference (AUPEC)*, pages 1–6, September 2014.

80 Sandeep Bala and Giri Venkataramanan. Autonomous power electronic interfaces between microgrids. In *2009 IEEE Energy Conversion Congress and Exposition, ECCE'2009*, pages 3006–3013. IEEE, 2009.

81 Raouia Aouini, Bogdan Marinescu, Khadija Ben Kilani, and Mohamed Elleuch. Stability improvement of the interconnection of weak AC zones by synchronverter-based HVDC link. *Electr. Power Syst. Res.*, 142:112–124, 2017.

82 Tong Huang, Sicun Gao, and Le Xie. Transient stability assessment of networked microgrids using neural lyapunov methods. *arXiv preprint arXiv:2012.01333*, 2020.

83 Seyed Ali Arefifar, Martin Ordonez, and Yasser Abdel-Rady I Mohamed. Voltage and current controllability in multi-microgrid smart distribution systems. *IEEE Trans. Smart Grid*, 9(2):817–826, 2016.

84 Nuno José Gil and J A Peças Lopes. Hierarchical frequency control scheme for islanded multi-microgrids operation. In *2007 IEEE Lausanne Power Tech*, pages 473–478. IEEE, 2007.

85 Edward J Ng and Ramadan A El-Shatshat. Multi-microgrid control systems (MMCS). In *IEEE PES General Meeting*, pages 1–6. IEEE, 2010.

86 Fan Zhang, Huiying Zhao, and Mingguo Hong. Operation of networked microgrids in a distribution system. *CSEE J. Power Energy Syst.*, 1(4):12–21, 2015.

87 Abdollah Kavousi-Fard, Mengqi Wang, and Wencong Su. Stochastic resilient post-hurricane power system recovery based on mobile emergency resources and reconfigurable networked microgrids. *IEEE Access*, 6:72311–72326, 2018.

88 F O Resende, Nuno José Gil, and J A Peças Lopes. Service restoration on distribution systems using multi-microgrids. *Eur. Trans. Electr. Power*, 21(2):1327–1342, 2011.

89 Farhad Shahnia, Ruwan P S Chandrasena, Sumedha Rajakaruna, and Arindam Ghosh. Primary control level of parallel distributed energy resources converters in system of multiple interconnected autonomous microgrids within self-healing networks. *IET Gener. Transm. Distrib.*, 8(2):203–222, 2014.

90 Mohammadhassan A Sofla and Roger King. Control method for multi-microgrid systems in smart grid environment—stability, optimization and smart demand participation. In *2012 IEEE PES Innovative Smart Grid Technologies (ISGT)*, pages 1–5. IEEE, 2012.

91 Cherry Yuen, Alexandre Oudalov, and Adrian Timbus. The provision of frequency control reserves from multiple microgrids. *IEEE Trans. Ind. Electron.*, 58(1):173–183, 2010.

92 Yun Zhang and Le Xie. A transient stability assessment framework in power electronic-interfaced distribution systems. *IEEE Trans. Power Syst.*, 31(6):5106–5114, 2016.

93 Zhikang Shuai, Yelun Peng, Xuan Liu, Zuyi Li, Josep M Guerrero, and Z John Shen. Dynamic equivalent modeling for multi-microgrid based on structure preservation method. *IEEE Trans. Smart Grid*, 2018. doi: 10.1109/TSG.2018.2844107.

94 Sergio Rivera, Amro Farid, and Kamal Youcef-Toumi. A multi-agent system coordination approach for resilient self-healing operations in multiple microgrids. In *Industrial Agents*, pages 269–285. Elsevier, 2015.

95 Francis K Tuffner, Nikitha Radhakrishnan, Yingying Tang, and Kevin P Schneider. Grid friendly appliance controllers to increase the dynamic stability of networked resiliency-based microgrids. In *2018 IEEE/PES Transmission and Distribution Conference and Exposition (T&D)*, pages 1–5. IEEE, 2018.

96 Yun Zhang and Le Xie. Online dynamic security assessment of microgrid interconnections in smart distribution systems. *IEEE Trans. Power Syst.*, 30(6):3246–3254, 2014.

97 Xiangyu Wu, Yin Xu, Jinghan He, Xiaojun Wang, Juan C Vasquez, and Josep M Guerrero. Pinning-based hierarchical and distributed cooperative control for AC microgrid clusters. *IEEE Trans. Power Electron.*, 35(9):9867–9887, 2020.

98 Pan Wu, Wentao Huang, Nengling Tai, and Shuo Liang. A novel design of architecture and control for multiple microgrids with hybrid AC/DC connection. *Appl. Energy*, 210:1002–1016, 2018.

99 Hyeong-Jun Yoo, Thai-Thanh Nguyen, and Hak-Man Kim. Multi-frequency control in a stand-alone multi-microgrid system using a back-to-back converter. *Energies*, 10(6):1–18, 2017.

100 Ritwik Majumder and Gargi Bag. Parallel operation of converter interfaced multiple microgrids. *Int. J. Electr. Power Energy Syst.*, 55:486–496, 2014.

101 Mobin Naderi, Yousef Khayat, Qobad Shafiee, Tomislav Dragicevic, Hassan Bevrani, and Frede Blaabjerg. Interconnected autonomous AC microgrids via back-to-back converters—Part I: Small-signal modeling. *IEEE Trans. Power Electron.*, 35(5):4728–4740, 2019.

102 Mobin Naderi, Yousef Khayat, Qobad Shafiee, Tomislav Dragičević, Hassan Bevrani, and Frede Blaabjerg. Interconnected autonomous AC microgrids via back-to-back converters-Part II: Stability analysis. *IEEE Trans. Power Electron.*, 35(11):11801–11812, 2020.

103 S M Ferdous, Farhad Shahnia, and G M Shafiullah. Power sharing and control strategy for microgrid clusters. In *9th International Conference on Power and Energy Systems (ICPES)*, pages 1–5. IEEE, 2019.

104 Blessy John, Arindam Ghosh, Megha Goyal, and Firuz Zare. A DC power exchange highway based power flow management for interconnected microgrid clusters. *IEEE Syst. J.*, 13(3):3347–3357, 2019.

105 Mohammad S Golsorkhi, David John Hill, and Hamid Reza Karshenas. Distributed voltage control and power management of networked microgrids. *IEEE J. Emerging Sel. Top. Power Electron.*, 6(4):1892–1902, 2018.

106 Farhad Shahnia. Stability and eigenanalysis of a sustainable remote area microgrid with a transforming structure. *Sustainable Energy Grids Netw.*, 8:37–50, 2016.

107 Farhad Shahnia and Ali Arefi. Eigenanalysis-based small signal stability of the system of coupled sustainable microgrids. *Int. J. Electr. Power Energy Syst.*, 91:42–60, 2017.

108 Jianguo Zhou, Huaguang Zhang, Qiuye Sun, Dazhong Ma, and Bonan Huang. Event-based distributed active power sharing control for interconnected AC and DC microgrids. *IEEE Trans. Smart Grid*, 5(6):6815–6828, 2018.

109 Liang Che, Mohammad Shahidehpour, Ahmed Alabdulwahab, and Yusuf Al-Turki. Hierarchical coordination of a community microgrid with AC and DC microgrids. *IEEE Trans. Smart Grid*, 6(6):3042–3051, 2015.

110 Xiaoping Zhou, Leming Zhou, Yandong Chen, Josep M Guerrero, An Luo, Wenhua Wu, and Ling Yang. A microgrid cluster structure and its autonomous coordination control strategy. *Int. J. Electr. Power Energy Syst.*, 100:69–80, 2018.

111 Lei Qiao, Xialin Li, Di Huang, Li Guo, Yutao Xu, Zhukui Tan, and Chengshan Wang. Coordinated control for medium voltage DC distribution centers with flexibly interlinked multiple microgrids. *J. Mod. Power Syst. Clean Energy*, 7(3):599–611, 2019.

112 Li Guo, Pengfei Li, Xialin Li, Di Huang, and Jiebei Zhu. Flexible control of DC interlinked multiple mgs cluster. *IET Gener. Transm. Distrib.*, 13(11):2088–2101, 2018.

113 Mashood Nasir, Zheming Jin, Hassan A Khan, Nauman Ahmad Zaffar, Juan C Vasquez, and Josep M Guerrero. A decentralized control architecture applied to DC nanogrid clusters for rural electrification in developing regions. *IEEE Trans. Power Electron.*, 34(2):1773–1785, 2018.

114 Srayashi Konar and Arindam Ghosh. Interconnection of islanded DC microgrids. In *2015 IEEE PES Asia-Pacific Power and Energy Engineering Conference (APPEEC)*, pages 1–5. IEEE, 2015.

115 Mahesh Kumar, Suresh Chandra Srivastava, Sri Niwas Singh, and Mylavarapu Ramamoorty. Development of a control strategy for interconnection of islanded direct current microgrids. *IET Renew. Power Gener.*, 9(3):284–296, 2014.

116 Wayne W Weaver, Rush Robinett, Gordon G Parker, and David G Wilson. Distributed control and energy storage requirements of networked DC microgrids. *Control Eng. Pract.*, 44:10–19, 2015.

117 Qobad Shafiee, Tomislav Dragičević, Juan C Vasquez, and Josep M Guerrero. Hierarchical control for multiple DC-microgrids clusters. *IEEE Trans. Energy Convers.*, 29(4):922–933, 2014. doi: 10.1109/TEC.2014.2362191.

118 Chaoyu Dong, Qingbin Gao, Qian Xiao, Xiaodan Yu, Libor Pekař, and Hongjie Jia. Time-delay stability switching boundary determination for DC microgrid clusters with the distributed control framework. *Appl. Energy*, 228:189–204, 2018.

119 Yilin Li, Ping Dong, Mingbo Liu, and Guokang Yang. A distributed coordination control based on finite-time consensus algorithm for a cluster of DC microgrids. *IEEE Trans. Power Syst.*, 34(3):2205–2215, 2018.

120 Sucheng Liu, Rui Liu, Jiazhu Zheng, and Xiaodong Liu. Predictive function control in tertiary level for power flow management of DC microgrid clusters. *Electron. Lett.*, 56(13):675–676, 2020.

121 David G Wilson, Wayne W Weaver, Rush D Robinett, and Steve F Glover. Nonlinear power flow control design for networked AC/DC based microgrid systems. In *2018 American Control Conference (ACC)*, pages 5698–5705. IEEE, 2018.

122 Jianjun Ma, Miao Zhu, Xu Cai, and Yun Wei Li. Configuration and operation of DC microgrid cluster linked through DC–DC converter. In *2016 IEEE 11th Conference on Industrial Electronics and Applications (ICIEA)*, pages 2565–2570. IEEE, 2016.

123 Hêmin Golpîra, Hossein Seifi, Arturo Román Messina, and Mahmoud-Reza Haghifam. Maximum penetration level of micro-grids in large-scale power systems: Frequency stability viewpoint. *IEEE Trans. Power Syst.*, 31(6):5163–5171, 2016.

124 Peigen Tian, Xi Xiao, Kui Wang, and Ruoxing Ding. A hierarchical energy management system based on hierarchical optimization for microgrid community economic operation. *IEEE Trans. Smart Grid*, 7(5):2230–2241, 2016.

125 Chen Chen, Jianhui Wang, Feng Qiu, and Dongbo Zhao. Resilient distribution system by microgrids formation after natural disasters. *IEEE Trans. Smart Grid*, 7(2):958–966, 2016.

126 Mehmet Hazar Cintuglu and Osama A Mohammed. Behavior modeling and auction architecture of networked microgrids for frequency support. *IEEE Trans. Ind. Inf.*, 13(4):1772–1782, 2017.

127 Mohammad Fathi and Hassan Bevrani. Adaptive energy consumption scheduling for connected microgrids under demand uncertainty. *IEEE Trans. Power Delivery*, 28(3):1576–1583, 2013.

128 Mohammad Fathi and Hassan Bevrani. Statistical cooperative power dispatching in interconnected microgrids. *IEEE Trans. Sustainable Energy*, 4(3):586–593, 2013.

129 Adam Hirsch, Yael Parag, and Josep Guerrero. Microgrids: A review of technologies, key drivers, and outstanding issues. *Renewable Sustainable Energy Rev.*, 90:402–411, 2018.

130 Bin Zhou, Jianting Zou, Chi Yung Chung, Huaizhi Wang, Nian Liu, Nikolai Voropai, and Daosen Xu. Multi-microgrid energy management systems: Architecture, communication, and scheduling strategies. *J. Mod. Power Syst. Clean Energy*, 9(3):463–476, 2021.

131 Hualei Zou, Shiwen Mao, Yu Wang, Fanghua Zhang, Xin Chen, and Long Cheng. A survey of energy management in interconnected multi-microgrids. *IEEE Access*, 7:72158–72169, 2019.

132 Yimy E García Vera, Rodolfo Dufo-López, and José L Bernal-Agustín. Energy management in microgrids with renewable energy sources: A literature review. *Appl. Sci.*, 9(18):3854, 2019.

133 F Bandeiras, E Pinheiro, M Gomes, P Coelho, and J Fernandes. Review of the cooperation and operation of microgrid clusters. *Renewable Sustainable Energy Rev.*, 133:110311, 2020.

134 H S V S Kumar Nunna and Suryanarayana Doolla. Demand response in smart distribution system with multiple microgrids. *IEEE Trans. Smart Grid*, 3(4):1641–1649, 2012.

135 Javier Matamoros, David Gregoratti, and Mischa Dohler. Microgrids energy trading in islanding mode. In *2012 IEEE Third International Conference on Smart Grid Communications (SmartGridComm)*, pages 49–54. IEEE, 2012.

136 Wei Liu, Wei Gu, Jianhui Wang, Wenwu Yu, and Xinze Xi. Game theoretic non-cooperative distributed coordination control for multi-microgrids. *IEEE Trans. Smart Grid*, 9(6):6986–6997, 2018.

137 Walid Saad, Zhu Han, and H Vincent Poor. Coalitional game theory for cooperative micro-grid distribution networks. In *2011 IEEE International Conference on Communications Workshops (ICC)*, pages 1–5. IEEE, 2011.

138 Luis M Camarinha-Matos. Collaborative smart grids–a survey on trends. *Renewable Sustainable Energy Rev.*, 65:283–294, 2016.

139 Chiara Lo Prete and Benjamin F Hobbs. A cooperative game theoretic analysis of incentives for microgrids in regulated electricity markets. *Appl. Energy*, 169:524–541, 2016.

140 Xingli Zhai and Ning Wang. Multi-agent consensus algorithm-based optimal power dispatch for islanded multi-microgrids. *Processes*, 7(10):679, 2019.

141 A G Madureira and J A Peças Lopes. Ancillary services market framework for voltage control in distribution networks with microgrids. *Electr. Power Syst. Res.*, 86:1–7, 2012.

142 Formerly TP462. Identification of electromechanical modes in power systems. *IEEE Task Force Report*, 2012.

143 Peter W Sauer, Mangalore A Pai, and Joe H Chow. *Power system dynamics and stability: With synchrophasor measurement and power system toolbox*. John Wiley & Sons, 2017.

144 Hassan Bevrani, Masayuki Watanabe, and Yasunori Mitani. *Power system monitoring and control*. John Wiley & Sons, 2014.

145 Eleftherios O Kontis, Theofilos A Papadopoulos, Andreas I Chrysochos, and Grigoris K Papagiannis. Measurement-based dynamic load modeling using the vector fitting technique. *IEEE Trans. Power Syst.*, 33(1):338–351, 2017.

146 Mehdi Ahrarinouri, Mohammad Rastegar, and Ali Reza Seifi. Multiagent reinforcement learning for energy management in residential buildings. *IEEE Trans. Ind. Inf.*, 17(1):659–666, 2020.

147 Nathan Tsang, Collin Cao, Serena Wu, Zilin Yan, Ashkan Yousefi, Alexander Fred-Ojala, and Ikhlaq Sidhu. Autonomous household energy management using deep reinforcement learning. In *2019 IEEE International Conference on Engineering, Technology and Innovation (ICE/ITMC)*, pages 1–7. IEEE, 2019.

148 Heeyun Lee, Changbeom Kang, Yeong-Il Park, Namwook Kim, and Suk Won Cha. Online data-driven energy management of a hybrid electric vehicle using model-based Q-learning. *IEEE Access*, 8:84444–84454, 2020.

149 Geoffrey K F Tso and Kelvin K W Yau. Predicting electricity energy consumption: A comparison of regression analysis, decision tree and neural networks. *Energy*, 32(9):1761–1768, 2007.

150 C H Lu. Wavelet fuzzy neural networks for identification and predictive control of dynamic systems. *IEEE Trans. Ind. Electron.*, 58(7):3046–3058, 2010.

151 Eleftherios O Kontis, Theofilos A Papadopoulos, Mazheruddin H Syed, Efren Guillo-Sansano, Graeme M Burt, and Grigoris K Papagiannis. Artificial-intelligence method for the derivation of generic aggregated dynamic equivalent models. *IEEE Trans. Power Syst.*, 34(4):2947–2956, 2019.

152 Mostafa Farrokhabadi, Claudio A Cañizares, John W Simpson-Porco, Ehsan Nasr, Lingling Fan, Patricio A Mendoza-Araya, Reinaldo Tonkoski, Ujjwol Tamrakar, Nikos Hatziargyriou, Dimitris Lagos, et al. Microgrid stability definitions, analysis, and examples. *IEEE Trans. Power Syst.*, 35(1):13–29, 2019.

153 Xiongfei Wang, Mads Graungaard Taul, Heng Wu, Yicheng Liao, Frede Blaabjerg, and Lennart Harnefors. Grid-synchronization stability of converter-based resources—an overview. *IEEE Open J. Ind. Appl.*, 1:115–134, 2020.

154 Xiongfei Wang and Frede Blaabjerg. Harmonic stability in power electronic-based power systems: Concept, modeling, and analysis. *IEEE Trans. Smart Grid*, 10(3):2858–2870, 2018.

8

Modeling of Interconnected Microgrids

8.1 Introduction

In order to interconnect ac MGs, back-to-back converters (BTBCs), circuit breakers (CBs), and static switches can be used as interlinking devices. CBs and instantaneous static switches are used as interlinking devices in [1–7] that can only network ac MGs with the same nominal voltages and frequencies. In such IMGs (named CB-IMGs) a synchronization algorithm is required to interconnect MGs. Power exchange/sharing can be controlled by changing the droop coefficients of primary controllers [1], adding a signal to secondary controllers [5, 7], or employing hierarchical control [2]. More flexible power exchange can be achieved by networking ac MGs using BTBCs [8–12]. Moreover, independent control of frequency and voltage of each MG [10, 13], power quality improvement by reactive/harmonic power interchange control [12], and integration of multiple ac MGs with different nominal voltages/frequencies [9, 10] can be addressed for IMGs via BTBCs (called BTBC-IMGs). In addition, several applications of BTBCs for bidirectional power flow between MGs and the utility grid are reported [14, 15].

Dynamic modeling is the starting point of analyzing stability and designing a large group of controllers, i.e. model-based controllers. According to applications, many modeling methods are represented for modern DER-integrated power grids focusing on the generation side, e.g. heat and power systems [16]. Other aspects of modeling such power grids are taken into account including power converters modeling [17], grid topology considerations [18], load types [19], DER/MG control loops [20, 21], and reducing the model order of inverter-based DERs and MGs [22–26]. However, IMGs require an individual modeling procedure using some known concepts of the mentioned modeling methods as well as special considerations on the large scale of IMGs and model simplification based on the application. Modeling of large-scale IMGs comprises high-burden computations unlike small-scale system modeling, which should be considered. Moreover, the IMG models should be simplified to facilitate different required analyses, e.g. stability analysis, controller design, and verification.

Small-signal modeling is a well-known method to understand the behavior of systems around an equilibrium point. It is engaged for autonomous MGs performance and stability analysis as well as controller design [18, 26–29]. In the case of IMGs, the presented small-signal models are generally an expansion of [18]. A special configuration of MGs is considered in [8], in which any MG has a static synchronous compensator (STATCOM) to be coordinated with the BTBC exchanging power. Then, an eigenvalue analysis and a robust distributed controller design are done based on the small-signal model, while BTBC modeling has not been addressed. Another modeling method of BTBC-IMGs refers to multiple MGs connected to an ideal main grid, where a lower challenging structure of IMGs is concerned with respect to the grid-isolated IMGs [14]. Detailed small-signal

Microgrids: Dynamic Modeling, Stability and Control, First Edition. Qobad Shafiee, Mobin Naderi, and Hassan Bevrani.

modeling is developed for CB-IMGs and is simplified using a truncation order reduction model [3]. Another detailed model for fully PV-based CB-IMGs is obtained in order to find oscillating modes and increase their damping [5]. An aggregation-based simplified model has been presented recently, which finds a reduced-order model for CB-IMGs in low-frequency and medium-frequency dynamics [30]. More CB-IMGs are analyzed in order to define a suitable range of electrical and control parameters such as interlinking line length and droop gains [3, 5, 6],[31], as well as to analyze power exchange control performance [2, 7, 32].

Although the detailed models are investigated to recognize the most significant participants in the stability of different structures of IMGs [6, 31, 33–38], simplified/reduced-order models are proposed to facilitate the stability analysis [3, 14, 32, 39–45], analyze the closed-loop stability [5, 46–48], and synthesize the secondary controller for load-frequency control [48–51]. Note that dynamic modeling realized in short-term studies, in few seconds to minute-timescale, is of interest in this book, not the static modeling in 15-minute time slots or larger one-hour time slots, which is usually suitable for the studies of operation, energy management, and planning of MGs and IMGs.

8.2 Interconnection Method

A simple schematic of ac IMGs in a general form is shown in Figure 8.1, and the BTBCs are typically considered as interlinking devices. The control requirements and considerations are represented in Chapter 10, Section 10.5. In order to have a comprehensive and easily generalized model of IMGs, similar modules are modeled in the same format. Then, the modeling process is divided into two main steps: (i) modeling of each module separately and (ii) interconnecting all the modules using the interconnection method.

Figure 8.2 shows the modeling process including the two steps. The main modules consist of islanded ac MGs, ac ILs, and interlinking devices, which are considered BTBCs here. Note that the comprehensive modeling of both BTBC-IMGs and CB-IMGs is represented in Sections 8.4.1 and 8.4.2. Once all modules are modeled separately in the first step, the interconnection method is applied to build the overall model in the second step. In the interconnection method, each module, i.e. an MG, a BTBC, or an interlinking line is represented individually by a state-space model. The connections with other modules typically for MG_i, MG_j, IL_{ij}, IL_{ji}, and $BTBC_{ij}$ are considered in the model input/output as shown in Figure 8.1. The interconnection process can be completed without more analytical calculation and only using RCT functions in MATLAB, which strongly reduces calculation burden/error. The RCT functions in the proposed modeling calculations and their applications are listed in Table 8.1. The functions facilitate interconnecting a lot of subsystems in order to model large-scale systems, e.g. IMGs.

In the case of IMGs modeling by the interconnection method (i) any number of separated modeled modules are transformed into a system matrix using the *pck* function and are considered as subsystems using the *systemnames* function, (ii) all inputs to the modules are specified using the *input_to* function, (iii) the desired inputs and outputs for the overall model of IMGs are specified by the *inputvar* and *outputvar* functions, (iv) the overall model of IMGs is calculated by the *sysic* function, and finally (v) it is transformed to the state-space representation by the *unpck* function to use in eigenvalue analysis. The details of the RCT functions in the general process of interconnecting the subsystems are presented in [52].

Thus, modeling of any number of ac IMGs with different structures can be realized only by interconnecting the modules using the interconnection method. Moreover, the interconnection method is used within the modeling of each main module to further facilitate overall IMG modeling. Note

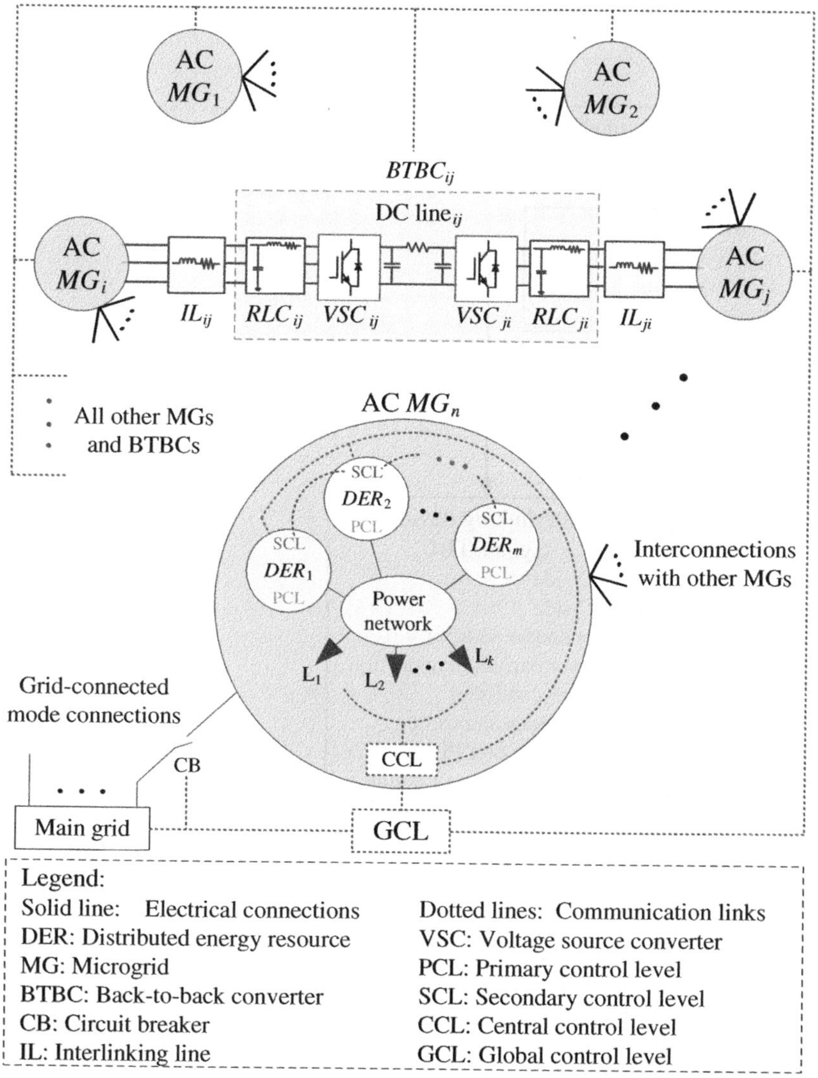

Figure 8.1 Interconnected ac microgrids using back-to-back converters, dc and ac interlinking lines for physical interconnection, and the four-level hierarchical control for control interconnection.

that the analytical calculation of the IMGs model is also presented just for the matrices in Section 8.4.1, which has more calculation burden than the interconnection method.

8.3 Module Modeling

In this section, the modeling of each IMG module is addressed. MGs, interlinking lines, and interlinking devices, i.e. BTBCs and CBs, are the main modules to be modeled. Generally, each main module is modeled separately using a state-space representation as

$$\dot{X}_M^k = A_M^k X_M^k + B_M^k U_M^k, \tag{8.1}$$

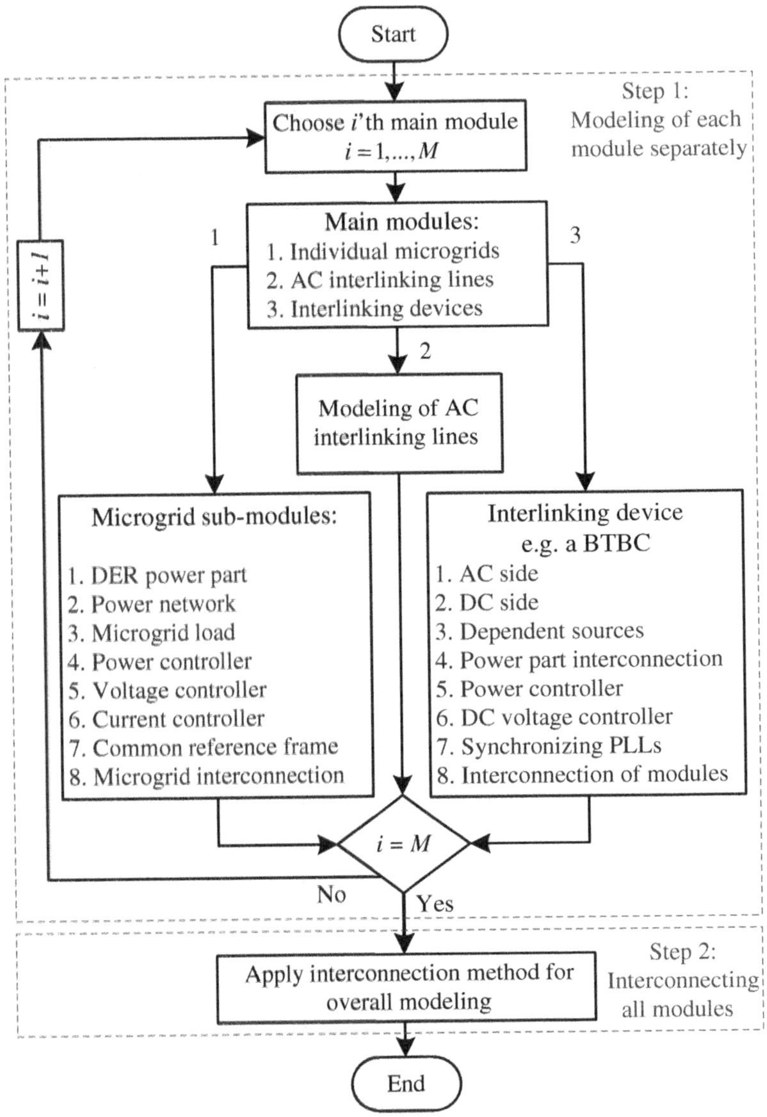

Figure 8.2 Process of the proposed interconnection modeling method including two generic steps: modeling of each module separately and interconnecting all modules.

$$Y_M^k = C_M^k X_M^k + D_M^k U_M^k, \tag{8.2}$$

where X_M^k is the state vector of the k'th IMG module, i.e.

$$X_M^k \subseteq \left\{ X_{MG}^1, \dots, X_{MG}^n, \dots, X_{IL}^{ij}, \dots, X_I D^{ij}, \dots \right\}, \tag{8.3}$$

where U_M^k is the input vector of the k'th module determined based on all electrical, control, or communication interconnections from other modules. Y_M^k includes all required connections to other modules, and the matrices which can be calculated for each module using the governing relationship, e.g. electrical circuit laws and block diagram rules. In cases where the module is large and formed from a number of subsystems, the interconnection method is also required to calculate the matrices.

Table 8.1 Applications of robust control toolbox functions for system interconnection.

Function	Application
pck	Transformation from the state-space representation to a system matrix used in system interconnection calculation
unpck	Reverse transformation from the system matrix to a state-space representation used in eigenvalue analysis
sysic	Forming interconnection of large-scale linear/linearized systems using the following subfunctions
systemnames	Specifying all subsystems modeled separately
inputvar	Specifying desired inputs for the main calculating system
input_to	Specifying the inputs to each subsystem from external input signals or other subsystems' outputs
outputvar	Specifying desired outputs for the main calculating system

8.3.1 Microgrid Modeling

Modeling of individual MGs is presented in Chapter 2 in detail. Therefore, we do not repeat those contents here. Nevertheless, in order to have a successful control and a good performance of CB-IMGs, coordination control is necessary, which leads to some modifications in the secondary control layer of individual MGs. In fact, power sharing of CB-IMGs is possible when there is a communication among IMGs in addition to the communication among DERs of each MG. In order to solve such a problem, usually a distributed secondary control architecture based on the pinning consensus is employed [2, 34, 46, 53–56] in which DERs are classified into two types, leader and follower. Leader DERs establish communication among IMGs to coordinate them for common power sharing. On the other hand, they have another communication with their followers in each MG to apply the power sharing and frequency/voltage restoration objectives in a distributed manner. Therefore, a modified secondary control architecture required for CB-IMGs is presented here. Note that concepts, architecture classifications, and basic issues of the MG secondary control level have been addressed in [57].

Note that BTBC-IMGs also require coordination control. In fact, MGs have their set points to generate a scheduled power and supply their individual load. On the other hand, BTBCs have their set points to exchange a certain power among MGs. Nevertheless, BTBC power set points should be coordinated with MGs capacities and excess powers. Otherwise, BTBC set-points cannot be satisfied. Therefore, coordination control among BTBC set-points and MGs capacities and also operation constraints should be employed. However, for such a coordination control, MG control architecture including primary and secondary control levels does not need to be restructured. In fact, the coordination control of BTBC-IMGs should be realized at the global control level, which is only a static control process to coordinate setpoints of MGs and BTBCs.

Figure 8.3 shows the general structure of ac MGs consisting of m grid-forming (GFO) and k grid-following (GFL) DERs. Two assumptions are considered in relation to the power part: (i) the structure of single-PCC is selected [23, 58], where all DERs are connected by the coupling lines, (ii) all loads including local DER loads and common loads are integrated into a lumped load at the PCC. Note that at least a GFO DER of each MG is necessary to be operated in islanded and interconnected modes to control frequency and voltage. However, GFL DERs are operated to generate

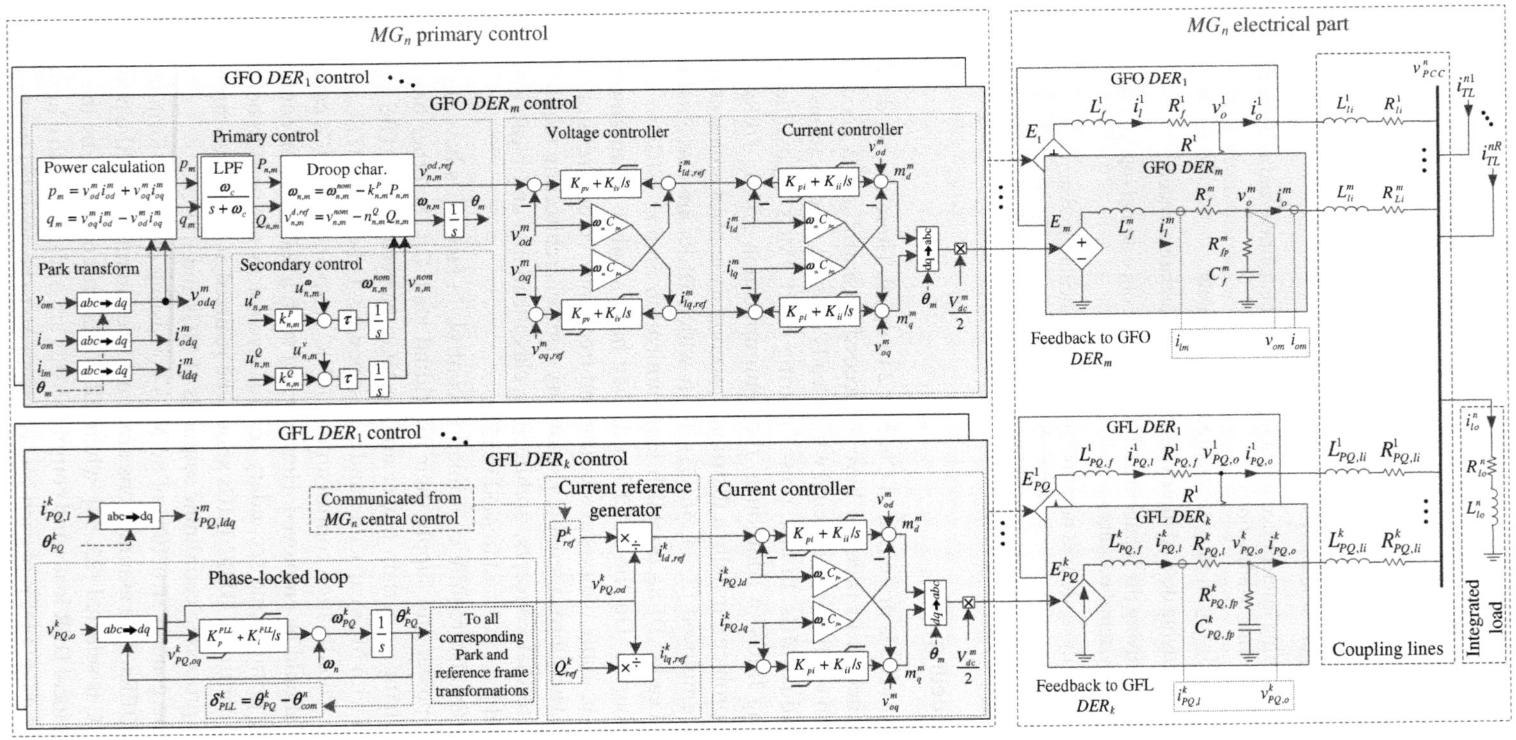

Figure 8.3 The structure of the studied autonomous ac microgrids including both GFO and GFL DERs, and considering a simplified network into DER coupling lines and an integrated load.

scheduled power. Therefore, the operation of GFL DERs in islanded and interconnected modes is based on the need and generally is not necessary. Accordingly, GFL DERs are taken into account in some studies of this book and they are not considered in some others.

8.3.1.1 Modeling of Secondary Control for CB-IMGs

Figure 8.4 displays a general structure of synchronous IMGs, including autonomous ac MGs, interlinking lines, as well as CBs as reconfiguration devices. The MGs have the general structures shown in Figure 8.3. The IMGs are controlled using a hierarchical control strategy including local (primary), supplementary (secondary), central, and global control layers. Here, the primary and secondary control levels of the hierarchical control strategy [20] are considered for being modeled. Individual MG control is achieved through the primary control level, utilizing voltage/frequency restoration and intra-MG power-sharing functions from the secondary control level. Furthermore, coordination control among IMGs is established through the pinning consensus-based function of the secondary control level. [2, 34, 46, 53–56].

Although the grid-connected operating mode is not of interest in this book, it is worth mentioning that its required control should be accomplished in the global control layer. Thus, the connections with the main grid are opened. Another application of the outermost control layer is scheduling for connection/disconnection of the CBs among IMGs based on power/energy management plans. This application can almost be considered static, which does not require to be dynamically modeled. The central controller produces the global set-points such as the nominal frequency and voltage of the GFO DERs and the power references of the GFL DERs. Load-shedding procedures

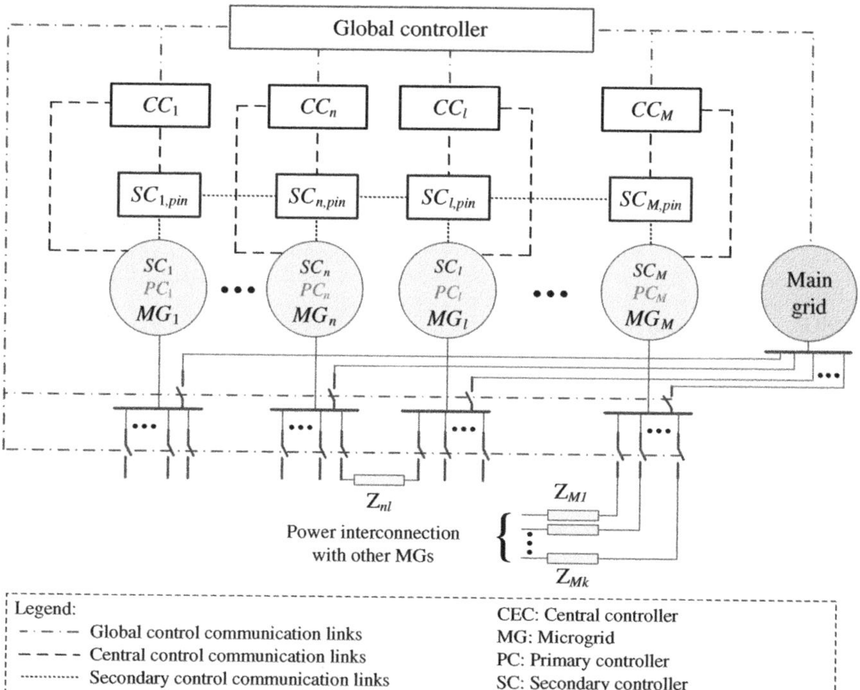

Figure 8.4 Synchronous ac microgrids interconnected through ac lines and circuit breakers, and controlled using hierarchical control strategy based on a pinning consensus secondary control level.

and other emergency plans are also done in this control layer. In a normal operation of the IMGs, which the modeling process is presenting for, all duties of the central-emergency control layer can be considered static in the presented dynamic modeling.

According to Figure 8.3, the secondary control is used only for GFO DERs, for which the frequency and voltage control is the main duty. Hence, in this section, GFO DERs are called in brief DERs. In the secondary control level, DERs should communicate some of their own features to regulate MG voltage and frequency and improve power sharing. One of the most common secondary control methods is the distributed control, which communicates the lowest required data leading to decreasing communication failures [57]. In synchronous CB-IMGs, generally, the pinning consensus protocol is used, which is a two-layer communication infrastructure including intra-MG and inter-MG communication networks [2, 34, 46, 53–56]. In each MG, one of the DERs is selected as the head (pinning) DER. Therefore, DER_{pin} is as a leader for other follower DERs within the MG to determine the nominal voltage/frequency in the intra-MG layer. It is also a communicating node in the inter-MG layer to share the information with other MG leaders and thus share the active and reactive powers at IMG level. Figure 8.5(a), shows a simple schematic of the pinning consensus-based distributed secondary control communications including typical inter-MG and intra-MG/inter-DER communication links.

As shown in Figure 8.3, the secondary controller for both frequency and voltage loops of all DERs is selected as an integral type, where the delay parameters are τ and T for follower and leader DERs, respectively. The secondary controller outputs are the nominal frequency and voltage, which also have the same form for all DERs. Nevertheless, the secondary control inputs of the leader and follower DERs are different. In fact, the leader DERs receive the rated MG values (ω_{rated} and v_{rated}) and follower DERs receive the MG leader values (ω_{pin} and v_{pin}). Therefore, the governing relationships of these two DER types are different.

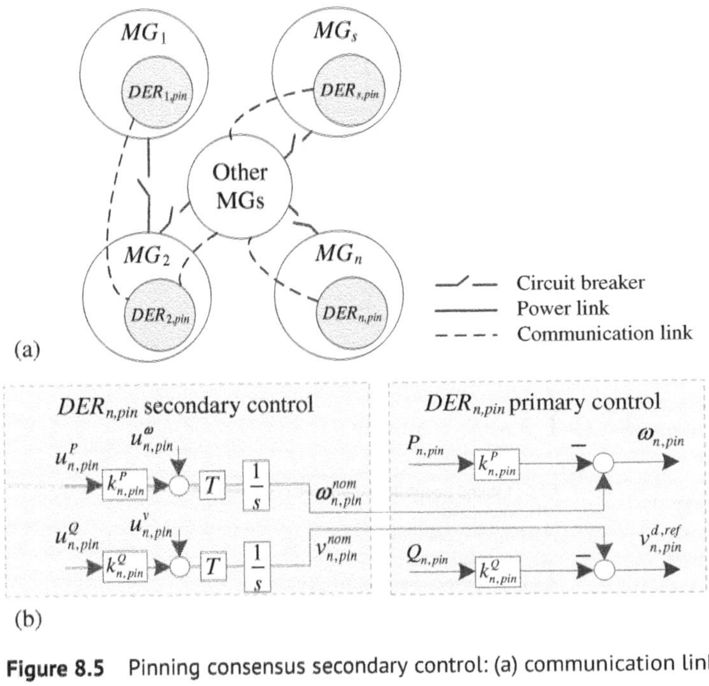

Figure 8.5 Pinning consensus secondary control: (a) communication links, (b) control loops.

According to Figure 8.5(b), for the typical leader/pinning DER of MG_n, the dynamic relations are given as follows:

$$\dot{\omega}_{n,pin}^{nom} = T\left(u_{n,pin}^{\omega} + k_{n,pin}^{P} u_{n,pin}^{P}\right), \tag{8.4a}$$

$$\dot{v}_{n,pin}^{nom} = T\left(u_{n,pin}^{v} + k_{n,pin}^{Q} u_{n,pin}^{Q}\right), \tag{8.4b}$$

where the input signals are as follows:

$$u_{n,pin}^{\omega} = \sum_{l\in\tilde{N}_n} \tilde{a}_{nl}\left(\omega_{l,pin} - \omega_{n,pin}\right) + \tilde{a}_{n0}\left(\omega_{rated} - \omega_{n,pin}\right), \tag{8.5a}$$

$$u_{n,pin}^{P} = \left(1/k_{n,pin}^{P}\right) \sum_{l\in\tilde{N}_n} \tilde{a}_{nl}\left(k_{l,pin}^{P} P_{l,pin} - k_{n,pin}^{P} P_{n,pin}\right), \tag{8.5b}$$

$$u_{n,pin}^{v} = \sum_{l\in\tilde{N}_n} \tilde{a}_{nl}\left(v_{l,pin} - v_{n,pin}\right) + \tilde{a}_{n0}\left(v_{rated} - v_{n,pin}\right), \tag{8.5c}$$

$$u_{n,pin}^{Q} = \left(1/k_{n,pin}^{Q}\right) \sum_{l\in\tilde{N}_n} \tilde{a}_{nl}\left(k_{l,pin}^{Q} Q_{l,pin} - k_{n,pin}^{Q} Q_{n,pin}\right). \tag{8.5d}$$

The inter-MG cyber network can be defined as $\tilde{\chi}(\tilde{\vartheta}, \tilde{\varepsilon}, \tilde{A})$ with leader node set $\tilde{\vartheta} = \{\vartheta_{1,pin}, \dots, \vartheta_{n,pin}, \dots, \vartheta_{M,pin}\}$, set of edges $\tilde{\varepsilon} \subseteq \tilde{\vartheta} \times \tilde{\vartheta}$, i.e. communication links among IMGs, and adjacency matrix $\tilde{A} = (\tilde{a}_{nl})_{M\times M}$, where $\tilde{a}_{ll} = 0$ and $\tilde{a}_{nl} \geq 0$. $\tilde{a}_{nl} = 1$ if and only if $\left(\vartheta_{n,pin}, \vartheta_{l,pin}\right) \in \tilde{\varepsilon}$. The neighbors of MG_n is expressed in the set $\tilde{N}_n = \left\{ \vartheta_{l,pin} \in \tilde{\vartheta} \middle| \left(\vartheta_{n,pin}, \vartheta_{l,pin}\right) \in \tilde{\varepsilon} \right\}$. The access of leader DERs to the rated MG values is specified by the leader-adjacency matrix $\tilde{B} = diag\{\tilde{a}_{10}, \dots, \tilde{a}_{n0}, \dots, \tilde{a}_{M0}\}$, where $\tilde{a}_{n0} = 1$ if the rated values are available for the MG_n head, otherwise, $\tilde{a}_{n0} = 0$.

According to Figure 8.3, for the typical follower $DER_{n,m}$, the dynamic relations are as follows:

$$\dot{\omega}_{n,m}^{nom} = \tau\left(u_{n,m}^{\omega} + k_{n,m}^{P} u_{n,m}^{P}\right), \tag{8.6a}$$

$$\dot{v}_{n,m}^{nom} = \tau\left(u_{n,m}^{v} + k_{n,m}^{Q} u_{n,m}^{Q}\right), \tag{8.6b}$$

where the input signals are

$$u_{n,m}^{\omega} = \sum_{j\in N_{n,m}} a_{mj}^{n}\left(\omega_{n,j} - \omega_{n,m}\right) + a_{m0}^{n}\left(\omega_{n,pin} - \omega_{n,m}\right), \tag{8.7a}$$

$$u_{n,m}^{P} = \left(1/k_{n,m}^{P}\right) \sum_{j\in N_{n,m}} a_{mj}^{n}\left(k_{n,j}^{P} P_{n,j} - k_{n,m}^{P} P_{n,m}\right) + a_{m0}^{n}\left(k_{n,pin}^{P} P_{n,pin} - k_{n,m}^{P} P_{n,m}\right), \tag{8.7b}$$

$$u_{n,m}^{v} = \sum_{j\in N_{n,m}} a_{mj}^{n}\left(v_{n,j} - v_{n,m}\right) + a_{m0}^{n}\left(v_{n,pin} - v_{n,m}\right), \tag{8.7c}$$

$$u_{n,m}^{Q} = \left(1/k_{n,m}^{Q}\right) \sum_{j\in N_{n,m}} a_{mj}^{n}\left(k_{n,j}^{Q} Q_{n,j} - k_{n,m}^{Q} Q_{n,m}\right) + a_{m0}^{n}\left(k_{n,pin}^{Q} Q_{n,pin} - k_{n,m}^{Q} Q_{n,m}\right). \tag{8.7d}$$

The inter-DER cyber network in each MG_n can be defined as $\chi_n(\vartheta_n, \varepsilon_n, A_n)$, which describes the communication interactions among $S-1$ follower DERs, where S indicates the total number of DERs of each MG. The node set $\vartheta_n = \{\vartheta_{n,1}, \dots, \vartheta_{n,pin-1}, \vartheta_{n,pin+1}, \dots, \vartheta_{n,S_n}\}$ represents the follower DERs, the set of edges $\varepsilon_n \subseteq \vartheta_n \times \vartheta_n$, expresses the communication links among the DERs, and $A_n = (a_{mj}^{n})_{(S_n-1)\times(S_n-1)}$ is an adjacency matrix, where $a_{mm}^{n} = 0$ and $a_{mj}^{n} \geq 0$. $a_{mj}^{n} = 1$ if and only if $(\vartheta_{n,m}, \vartheta_{n,j}) \in \varepsilon_n$. The neighbor set of MG_n can be expressed by $N_{n,m} = \left\{ \vartheta_{n,j} \in \vartheta_n \middle| (\vartheta_{n,m}, \vartheta_{n,j}) \in \varepsilon_n \right\}$. The access of each DER to the MG_n leader values is specified by the leader-adjacency matrix $B_n = diag\{a_{10}^{n}, \dots, a_{m0}^{n}, \dots, a_{S0}^{n}\}$, where $a_{m0}^{n} = 1$ if the leader DER values are available for the $DER_{n,m}$, otherwise, $a_{m0}^{n} = 1$.

Table 8.2 Input vectors of the state space representation of the leader and follower DERs' secondary controllers.

DER type	U_{SC1}	U_{SC2}	U_{SC3}
Leader $DER_{n,pin}$	—	Column vector of $\left\{ Y_{PC}^{l,pin} \middle| \left(\vartheta_{n,pin}, \vartheta_{l,pin}\right) \in \tilde{\varepsilon} \right\}$	$\mathbf{Y}_{PC}^{n,pin}$
Follower $DER_{n,m}$	$\mathbf{Y}_{PC}^{n,m}$	Column vector of $\left\{ Y_{PC}^{n,j} \middle| \left(\vartheta_{n,m}, \vartheta_{n,j}\right) \in \varepsilon_n \right\}$	$\mathbf{Y}_{PC}^{n,pin}$

By specifying the two integrator outputs as the state variables, one can find a state space representation for the secondary controller of each DER as follows:

$$\begin{aligned} \dot{\boldsymbol{X}}_{SC} &= \boldsymbol{A}_{SC}\boldsymbol{X}_{SC} + \boldsymbol{B}_{SC1}\boldsymbol{U}_{SC1} + \boldsymbol{B}_{SC2}\boldsymbol{U}_{SC2} + \boldsymbol{B}_{SC3}\boldsymbol{U}_{SC3}, \\ \boldsymbol{Y}_{SC} &= \boldsymbol{C}_{SC}\boldsymbol{X}_{SC}, \end{aligned} \tag{8.8}$$

where $\boldsymbol{Y}_{SC} = \boldsymbol{X}_{SC}$ and equal $[\omega_{n,pin}^{nom}\ v_{n,pin}^{nom}]^T$ for leader DERs, or $[\omega_{n,m}^{nom}\ v_{n,m}^{nom}]^T$ for follower DERs. Input vectors of the leader and follower DERs are tabulated in Table 8.2. $\boldsymbol{A}_{SC} = \mathbf{0}_2$ and $\boldsymbol{C}_{SC} = \mathbf{I}_2$. Obviously, the size and content of the input matrices are strongly related to the inter-MG and inter-DER communication network configurations.

8.3.1.2 Other MG Modules

Modeling of each module is realized by finding its state space representation as (8.1), where the state, input, and output vectors of the module are presented in Table 8.3 for all MG modules. The matrices A_m, B_m, C_m, and D_m can easily be found by considering the vectors expressed in Table 8.3 and using the circuit and control laws in Figure 8.3. Furthermore, they are given in Part I, Chapter 2 in detail.

8.3.1.3 Overall MG Model

To find the state space representation of overall MG, the interconnections among all modeled modules should be considered. The interconnections are based on the MG power and control configurations and can be calculated using the advantageous Robust Control Toolbox functions based on the interconnection method.

Figure 8.6 shows the process of the interconnection method for ac MG modules presented in Figure 8.3. An expandable MG model can be obtained according to the interconnections details of all MG modules shown in Figure 8.7. Note that the interconnections should be exactly specified based on module inputs and outputs and considering all connections among them. Hence, each ac MG with any number of DERs, lines, and loads can be represented as the state space model:

$$\begin{aligned} \dot{\boldsymbol{X}}_{MG}^{n} &= \boldsymbol{A}_{MG}^{n}\boldsymbol{X}_{MG}^{n} + \boldsymbol{B}_{MG}^{n}\boldsymbol{U}_{MG}^{n}, \\ \boldsymbol{Y}_{MG}^{n} &= \boldsymbol{C}_{MG}^{n}\boldsymbol{X}_{MG}^{n} + \boldsymbol{D}_{MG}^{n}\boldsymbol{U}_{MG}^{n}, \end{aligned} \tag{8.9}$$

where the state vector can be organized as

$$\begin{aligned} X_{MG}^{n} = \Bigg[& X_{DER}^{DB1} \cdots \overbrace{X_{SC}^{m}, X_{PC}^{m}, X_{VC}^{m}, X_{CC}^{m}, X_{PD}^{m}}^{X_{DER}^{DB,m}} \; X_{DER}^{PQ1} \cdots \\ & \overbrace{X_{PLL}^{k}, X_{CC}^{k}, X_{PD}^{k}}^{X_{DER}^{PQ,k}} \; X_{CL}^{1} \cdots X_{CL}^{m+k} \; X_{ML} \Bigg], \end{aligned} \tag{8.10}$$

Table 8.3 Variables, input, and output vectors of state space representation of microgrid modules related to Figure 8.3.

Module	State variables (X_m)	Inputs (U_m)	Outputs (Y_m)
GFO DER$_m$			
Output filter	$[i_l^m\ v_o^m]^T$	$[E_m\ \omega_{com}\ i_o^m]^T$	$= X_m$
Current controller	Integrator outputs	$[i_l^m\ v_o^m\ i_{l,ref}^m]^T$	E_m
Voltage controller	Integrator outputs	$[v_o^m\ v_{o,ref}^m]^T$	$i_{l,ref}^m$
Power controller	$[\delta_m\ P_m\ Q_m]^T$	$[v_o^m\ i_o^m\ \omega_{com}]^T$	a)
GFL DER$_k$			
Output filter	$[i_{PQ,l}^k\ v_{PQ,o}^k]^T$	b)	$= X_m$
Current controller	Integrator outputs	$[i_{PQ,l}^k\ v_{PQ,o}^k]^T$	$E_{PQ,k}$
PLL	Integrator outputs	$[v_{PQ,o}^k\ \omega_{com}]^T$	δ_{PLL}
Power sections			
Coupling lines[c)]	$i_{PQ,o}^k$	d)	$= X_m$
PCC voltage	—	e)	v_{PCC}^n
Integrated load	i_{lo}^n	$[v_{PCC}^n\ \omega_{com}]^T$	$= X_m$

a) $[\omega_m\ v_{o,ref}^m\ \delta_m]^T$,
b) $[E_{PQ}^k\ \omega_{com}\ i_{PQ,o}^k]^T$,
c) $[v_{PCC}^n\ v_{PQ,o}^k\ \omega_{com}]^T$,
d) For instance, DER_k,
e) Currents of coupling lines, tie-lines and load.

and it consists of $15m + 10k + 2$ state variables for the MG structure shown in Figure 8.3, where m is the number of GFO DERs and k is the number of GFL DERs. Interlinking line currents to MG_n are the disturbance inputs, i.e. $\boldsymbol{U}_{MG}^n = [\boldsymbol{i}_{IL,dq}^{n1} \cdots \boldsymbol{i}_{IL,dq}^{nM}]^T$ and the MG_n output can be considered as $\boldsymbol{Y}_{MG}^n = [\Delta\omega_{com}^n\ \Delta\boldsymbol{v}_{pcc,dq}^n\ \boldsymbol{Y}_{SC}^{n,pin}]^T$. Finally, the matrices can be numerically calculated employing the Robust Control Toolbox functions.

8.3.2 Interlinking Line Modeling

Interlinking lines of IMGs usually have low/medium lengths and can be categorized into medium-length lines. Therefore, they are modeled as series RL branches like coupling lines of DERs. In BTBC-IMGs, the current direction of the interlinking line is considered from a BTBC to the connected MGs. However, in CB-IMGs, the current direction is fully arbitrary, but it should be taken into account when modeling the overall interconnections of the IMGs. In accordance with the reference frame, their equations are presumed to be expressed in the CRF of the connected MG for BTBC-IMGs and in the global CRF for CB-IMGs. Since the dynamics are completely similar to the dynamics of the DER coupling lines, (2.49) can be rewritten and a state space representation for IL_{ij} can be derived as:

$$\begin{aligned}\dot{X}_{IL}^{ij} &= A_{IL}^{ij}X_{IL}^{ij} + B_{ILM}^{ij}U_{ILM}^{ij} + B_{ILB}^{ij}U_{ILB}^{ij},\\ Y_{IL}^{ij} &= C_{IL}^{ij}X_{IL}^{ij},\end{aligned} \tag{8.11}$$

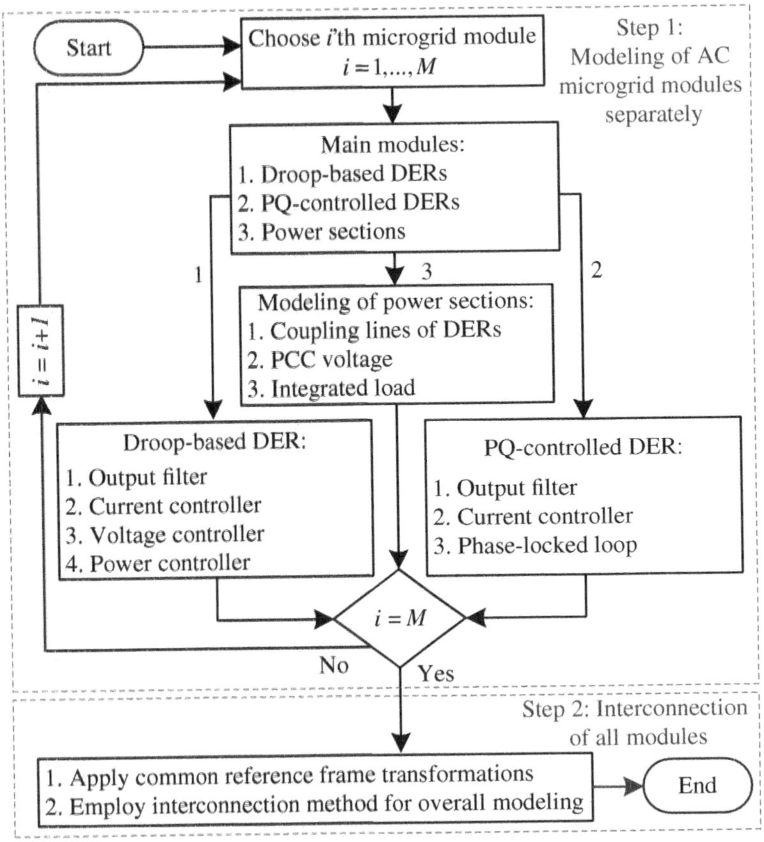

Figure 8.6 Process of the interconnection modeling method [38] for finding the detailed model of autonomous ac microgrids in two generic steps.

where $X_{IL}^{ij} = Y_{IL}^{ij} = \Delta i_{IL,dq}^{ij}$, $U_{ILB}^{ij} = \Delta v_{fc}^{ij}$, $U_{ILM}^{ij} = [\Delta v_{pcc,dq}^{i} \;\; \Delta\omega_{com}^{i}]^T$, and $B_{ILB}^{ij} = \left(1/L_{IL}^{ij}\right) I_2$,

$$A_{IL}^{ij} = \begin{bmatrix} -\frac{R_{IL}^{ij}}{L_{IL}^{ij}} & \omega_{com0}^{i} \\ -\omega_{com0}^{i} & -\frac{R_{IL}^{ij}}{L_{IL}^{ij}} \end{bmatrix},$$

$$B_{ILM}^{ij} = \begin{bmatrix} \frac{-1}{L_{IL}^{ij}} & 0 & i_{IL,q0}^{ij} \\ 0 & \frac{-1}{L_{IL}^{ij}} & -i_{IL,d0}^{ij} \end{bmatrix}.$$

8.3.3 Back-to-Back Converter Modeling

According to Figure 8.8(a), a BTBC can be controlled by two controllers in order to exchange power between two ac MGs. The power controller receives the active and reactive power references from a higher-level controller and tries to exchange scheduled powers by controlling the VSC_i current. On the other hand, the dc voltage controller stabilizes the dc side voltage (V_{dc}^{j}) by controlling the VSC_j current. In addition, two phase-locked loops (PLLs) are required for the VSCs to synchronize with the MGs [38, 59]. The power part comprises two AC sides and one common dc link. A precise dynamic model of a BTBC consists of ac and dc sides, a power controller, a dc voltage controller, and PLLs [60, 61].

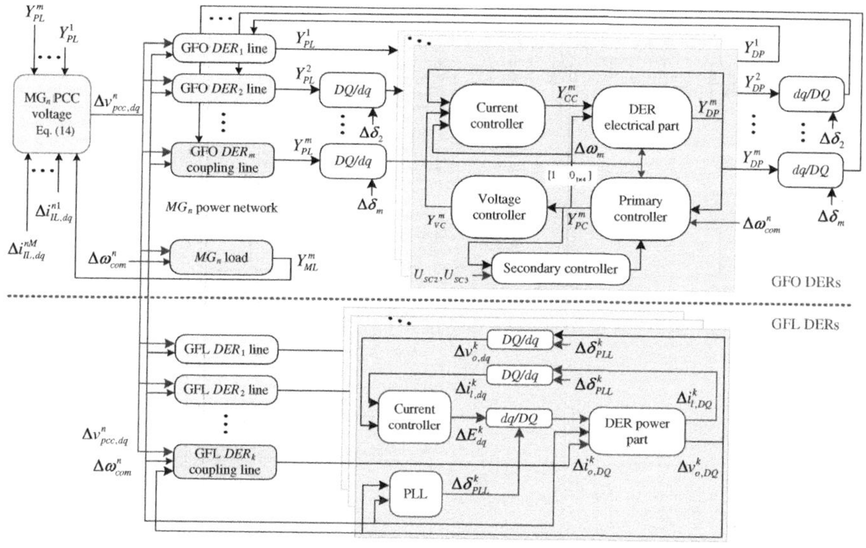

Figure 8.7 All power and control interconnections of the proposed generalizable modeling for autonomous ac MG_n.

8.3.3.1 AC Side of the BTBC

Figure 8.8(b) shows the circuit model of the BTBC power part. Each BTBC ac side is an RLC filter, where its dynamics are presented in MG modeling in Part I. Hence, a similar representation to (2.12) can be given for the BTBC ac side as follows:

$$\begin{aligned}\dot{X}^i_{Bac} &= A^i_{Bac}X^i_{Bac} + B^i_{Bac1}U^i_{Bac1} + B^i_{Bac2}U^i_{Bac2},\\ Y^i_{Bac} &= C^i_{Bac}X^i_{Bac},\end{aligned} \tag{8.12}$$

where $X^i_{Bac} = Y^i_{Bac} = [\Delta i^i_{fc,dq}\ \ \Delta v^i_{fc,dq}]^T$, $U^i_{Bac1} = [\Delta E^i_{Cdq}\ \ \Delta\omega^i_B]^T$, $U^i_{Bac2} = \Delta I^{ij}_{IL,dq}$, and the matrices are

$$A^i_{Bac} = \begin{bmatrix} -R^i_{fc}/L^i_{fc} & \omega^i_{B0} & -1/L^i_{fc} & 0\\ -\omega^i_{B0} & -R^i_{fc}/L^i_{fc} & 0 & -1/L^i_{fc}\\ 1/C^i_{fc} & 0 & 0 & \omega^i_{B0}\\ 0 & 1/C^i_{fc} & -\omega^i_{B0} & 0 \end{bmatrix},$$

$$B^i_{Bac1} = \begin{bmatrix} \frac{1}{L^i_{fc}} & 0 & i^i_{fc,q0}\\ 0 & \frac{1}{L^i_{fc}} & -i^i_{fc,d0}\\ 0 & 0 & v^i_{fc,q0}\\ 0 & 0 & -v^i_{fc,d0} \end{bmatrix}, B^i_{Bac2} = \begin{bmatrix} 0_2\\ -\left(1/C^i_{fc}\right)I_2 \end{bmatrix}.$$

8.3.3.2 DC Side of the BTBC

The drawn currents by ac sides in order to form the dependent voltage sources, i.e. E^i_C and E^j_C, are modeled by current sources I^i_C and I^j_C shown in Figure 8.8(b). In addition to the VSC's capacitors and series resistors, the dc line is modeled by a resistor on the BTBC dc side. One can easily find a state space representation as follows:

$$\dot{X}^{ij}_{Bdc} = A^{ij}_{Bdc}X^{ij}_{Bdc} + B^{ij}_{Bdc}U^{ij}_{Bdc},$$

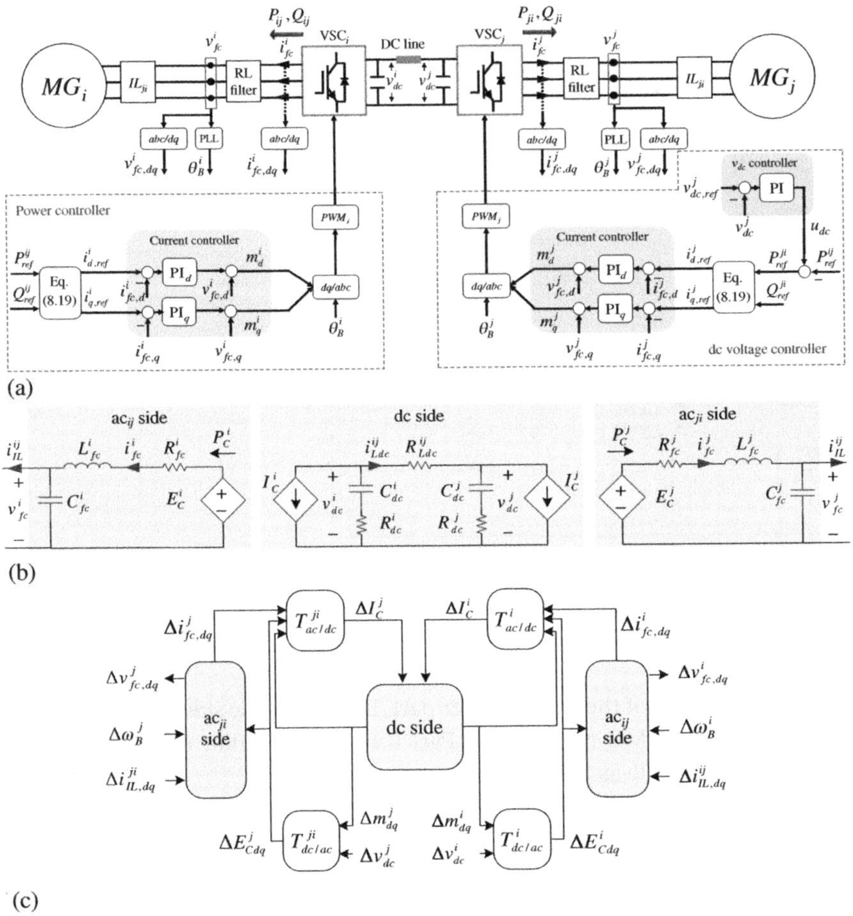

Figure 8.8 Interlinking back-to-back converter: (a) control and operation block diagram, (b) power part averaging model, (c) power part interconnection.

$$Y^{ij}_{Bdc} = C^{ij}_{Bdc} X^{ij}_{Bdc}, \tag{8.13}$$

where $X^{ij}_{Bdc} = [\Delta v^i_{dc}\ \Delta v^j_{dc}]^T$, $U^{ij}_{Bdc} = [\Delta I^i_C\ \Delta I^j_C]^T$, and

$$A^{ij}_{Bdc} = \begin{bmatrix} -1/\left(R^{eq}_{dc} C^i_{dc}\right) & 1/\left(R^{eq}_{dc} C^i_{dc}\right) \\ 1/\left(R^{eq}_{dc} C^j_{dc}\right) & -1/\left(R^{eq}_{dc} C^j_{dc}\right) \end{bmatrix},$$

$$B^{ij}_{Bdc} = \begin{bmatrix} \left(R^i_{dc}/R^{eq}_{dc}\right) - 1 & -\left(R^j_{dc}/R^{eq}_{dc}\right) \\ -\left(R^i_{dc}/R^{eq}_{dc}\right) & \left(R^j_{dc}/R^{eq}_{dc}\right) - 1 \end{bmatrix},$$

where $R^{eq}_{dc} = R^i_{dc} + R^j_{dc} + R^{ij}_{Ldc}$, and $C^{ij}_{Bdc} = I_2$.

8.3.3.3 Dependent Current and Voltage Sources

As shown in Figure 8.8(b), dependent current and voltage sources are as transformations from ac/dc side to dc/ac side, which correlate both sides as follows:

$$I^i_C = P^i_C / v^i_{dc},$$

$$E^i_C = (1/2)\, m_i v^i_{dc}, \tag{8.14}$$

where P^i_C is the produced ac power of VSC_i and m_i is the PWM control signal shown in Figure 8.8(a). By linearizing (8.14), the transformations can be calculated as follows:

$$\begin{aligned}
\Delta I^i_C &= T^i_{ac/dc}\left[\Delta v^i_{dc}\ \Delta E^i_{Cdq}\ \Delta i^i_{fc,dq}\right]^T,\\
\Delta E^i_{Cdq} &= T^i_{dc/ac}\left[\Delta v^i_{dc}\ \Delta m^i_{dq}\right]^T,
\end{aligned} \tag{8.15}$$

where

$$T^i_{ac/dc} = \frac{3}{2v^i_{dc0}}\left[-\frac{2}{3}I^i_{C0}\ E^i_{Cd0}\ E^i_{Cq0}\ i^i_{fc,d0}\ i^i_{fc,q0}\right], \tag{8.16}$$

$$T^i_{dc/ac} = \frac{1}{2}\begin{bmatrix} m^i_{d0} & v^i_{dc0} & 0 \\ m^i_{q0} & 0 & v^i_{dc0} \end{bmatrix}. \tag{8.17}$$

8.3.3.4 BTBC Power Part Interconnection

In order to simplify the BTBC modeling, the power part interconnections among ac and dc sides and independent voltage and current sources are fulfilled before the control modeling. Figure 8.8(c) displays this interconnection, which consists of the BTBC power modules and their interconnections with each other and the outside. The state space representation is calculated readily using the interconnection method, which can be expressed as:

$$\begin{aligned}
\dot{X}^{ij}_{BP} &= A^{ij}_{BP}X^{ij}_{BP} + B^{ij}_{BP}U^{ij}_{BP},\\
Y^{ij}_{BP} &= C^{ij}_{BP}X^{ij}_{BP} + D^{ij}_{BP}U^{ij}_{BP},
\end{aligned} \tag{8.18}$$

where X^{ij}_{BP} is a 10×1 vector, including the state variables of the ac and dc sides as $X^{ij}_{BP} = [X^i_{Bac}\ X^j_{Bac}\ X^{ij}_{Bdc}]$ and

$$\begin{aligned}
U^{ij}_{BP} &= \left[\Delta m^i_{dq}\ \Delta\omega^i_B\ \Delta m^j_{dq}\ \Delta\omega^j_B\ \Delta i^{ij}_{IL,dq}\ \Delta i^{ji}_{IL,dq}\right]^T,\\
Y^{ij}_{BP} &= \left[\Delta i^i_{fc,dq}\ \Delta i^j_{fc,dq}\ \Delta v^j_{dc}\ \Delta v^i_{fc,dq}\ \Delta v^j_{fc,dq}\right]^T.
\end{aligned}$$

The matrices A^{ij}_{BP}, B^{ij}_{BP}, C^{ij}_{BP}, and D^{ij}_{BP} are calculated numerically for an input data using the RCT functions.

8.3.3.5 Power Controller

Here, a current control method is employed to control the BTBC power exchange. In this method, the power references P^{ij}_{ref} and Q^{ij}_{ref} are provided by the GCL. Then, the current references $i^{ij}_{dq,ref}$ are calculated as follows [60]:

$$\begin{aligned}
i^{ij}_{d,ref} &= 2P^{ij}_{ref}/3v^i_{fc,d},\\
i^{ij}_{q,ref} &= -2Q^{ij}_{ref}/3v^i_{fc,d}.
\end{aligned} \tag{8.19}$$

It is easy to obtain a state space model similar to the PI current controller model of DERs and active load, explained in Section 2.4.6.4, by considering the integrators outputs of the PI controllers as the state variables (X^i_{Bpc}) as follows:

$$\begin{aligned}
\dot{X}^i_{Bpc} &= A^i_{Bpc}X^i_{Bpc} + B^i_{Bpc}U^i_{Bpc},\\
Y^i_{Bpc} &= C^i_{Bpc}X^i_{Bpc} + D^i_{Bpc}U^i_{Bpc},
\end{aligned} \tag{8.20}$$

where $U^i_{Bpc} = [\Delta V^i_{PCC,dq}\ \Delta i^i_{fc,dq}\ \Delta i^i_{dq,ref}\ \Delta\omega^i_B]^T$, $Y^i_{Bpc} = \Delta m^i_{dq}$, the matrices are $A^i_{Bpc} = 0_2$, $B^i_{Bpc} = [B^i_{Bpc1}\ B^i_{Bpc2}\ B^i_{Bpc3}\ B^i_{Bpc4}]$, $C^i_{Bpc} = (2/V_{dc})\, I_2$, $D^i_{Bpc} = [D^i_{Bpc1}\ D^i_{Bpc2}\ D^i_{Bpc3}\ D^i_{Bpc4}]$ and the sub-matrices are $B^i_{Bpc1} = 0_2$, $B^i_{Bpc2} = -K^i_{Bi}I_2$, $B^i_{Bpc3} = K^i_{Bi}I_2$, $B^i_{Bpc4} = 0_{2\times1}$, $D^i_{Bpc1} = (2/V_{dc})\, I_2$, $D^i_{Bpc3} = \left(2K^i_{Bp}/V_{dc}\right) I_2$, and

$$D^i_{Bpc2} = \frac{2}{V_{dc}} \begin{bmatrix} -K^i_{Bp} & -L^i_{fc}\omega^i_B \\ L^i_{fc}\omega^i_B & -K^i_{Bp} \end{bmatrix},$$

$$D^i_{Bpc4} = \frac{2}{V_{dc}} \begin{bmatrix} -L^i_{fc} i^i_{fc,q0} \\ L^i_{fc} i^i_{fc,d0} \end{bmatrix}.$$

K^i_{Bp} and K^i_{Bi} are the proportional and integral gains of the PI current controller and V_{dc} is the nominal dc link voltage.

8.3.3.6 DC Voltage Controller

In order to have a stable BTBC operation, the VSC_j needs to exchange the same active power with the VSC_i by applying $-P^{ij}_{ref}$ as its active power reference. In addition, the dc voltage should be controlled through the same control signal. On the other hand, although the VSC_j's reactive power reference is free from the VSC_i's Q^{ij}_{ref}, usually references with equal magnitudes are of interest. Therefore, power references of the VSC_j are expressed as:

$$P^{ji}_{ref} = u_{dc} - P^{ij}_{ref},\ Q^{ji}_{ref} = -Q^{ij}_{ref}, \tag{8.21}$$

where u_{dc} is the control effort of the V^j_{dc} controller shown in Figure 8.8(a). Thus the current references can be calculated by (8.19) and a state space can be represented for VSC_j's current controller similar to (8.20) as follows:

$$\begin{aligned} \dot{X}^j_{Bvc} &= A^j_{Bvc}X^j_{Bvc} + B^j_{Bvc}U^j_{Bvc}, \\ Y^j_{Bvc} &= C^j_{Bvc}X^j_{Bvc} + D^j_{Bvc}U^j_{Bvc}, \end{aligned} \tag{8.22}$$

where X^j_{Bvc} is a 2×1 vector of the integrator outputs of the PI controllers. Since the current controller structure is the same for the power controller and the dc voltage controller, input vector, output vector, and all matrices and sub-matrices are equal to those of the BTBC power controller expressed below (8.20) with superscript j.

The V^j_{dc} controller, which is usually a PI controller is considered using state space representation as follows:

$$\begin{aligned} \dot{X}^j_{DVC} &= B^j_{DVC}U^j_{DVC}, \\ Y^j_{DVC} &= C^j_{DVC}X^j_{DVC} + D^j_{DVC}U^j_{DVC}, \end{aligned} \tag{8.23}$$

where X^j_{DVC} is the integrator output, $U^j_{DVC} = \Delta v^j_{dc}$, $Y^j_{DVC} = \Delta i^j_{d,ref}$, $B^j_{DVC} = K^{DVC}_i$, $C^j_{DVC} = [2/3v^i_{fc,d}\ \ 0]^T$ and $D^j_{DVC} = [2K^{DVC}_p/3v^i_{fc,d}\ \ 0]^T$. K^{DVC}_p and K^{DVC}_i are the proportional and integral gains of the V^j_{dc} PI controller.

8.3.3.7 Synchronizing PLLs

As mentioned earlier, the BTBC needs two PLLs to synchronize ac sides with the MG's PCC voltages through ILs. The typical PLL structure is indicated in Figure 8.9. An LPF is used to clear the $v^i_{pcc,q}$ signal, and a PI controller is used to enforce it to zero in order to lock θ^i_B on the phase of v^i_{pcc}. Note that before enabling the BTBC to exchange the power, $v^i_{fc} = v^i_{pcc}$. Therefore, the v^i_{pcc} is measured by

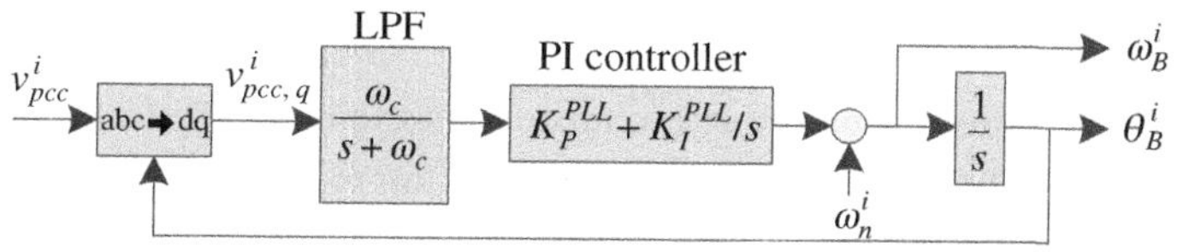

Figure 8.9 The structure of phase-locked loop typically used for VSC_i.

the v^i_{fc} transducer. In order to find the small-signal model of the PLL, PI and LPF integrator outputs are considered state variables. Note that the third state variable is $\delta^i_B = \theta^i_B - \theta^i_{com}$ to state all BTBC AC side dynamics into the related CRF, e.g. VSC_i dynamics state into the MG_i CRF. Therefore, the state space representation is as follows:

$$\begin{aligned}\dot{X}^i_{PLL} &= A^i_{PLL}X^i_{PLL} + B^i_{PLL}U^i_{PLL},\\ Y^i_{PLL} &= C^i_{PLL}X^i_{PLL},\end{aligned} \tag{8.24}$$

where $U^i_{PLL} = [\Delta v^i_{pcc,q}\ \Delta\omega^i_{com}]^T$, $Y^i_{PLL} = [\Delta\omega^i_B\ \Delta\delta^i_B]$ and

$$A^i_{PLL} = \begin{bmatrix} 0 & K_I^{PLL} & 0 \\ 0 & -\omega_c & 0 \\ 1 & K_P^{PLL} & 0 \end{bmatrix}.$$

$$B^i_{PLL} = \begin{bmatrix} 0 & \omega_c & 0 \\ 0 & 0 & -1 \end{bmatrix}^T, C^i_{PLL} = \begin{bmatrix} 1 & K_P^{PLL} & 0 \\ 0 & 0 & 1 \end{bmatrix},$$

8.3.3.8 Complete Interconnection of BTBC Modules

As the final step, the interconnections among all power and control modules represented by (8.18), (8.20), (8.22), (8.23), and (8.24) should be considered as shown in Figure 8.10. The $BTBC_{ij}$ is modeled as:

$$\begin{aligned}\dot{X}^{ij}_B &= A^{ij}_BX^{ij}_B + B^{ij}_BU^{ij}_B,\\ Y^{ij}_B &= C^{ij}_BX^{ij}_B + D^{ij}_BU^{ij}_B,\end{aligned} \tag{8.25}$$

where X^{ij}_B is a 21×1 state vector as

$$X^{ij}_B = [X^{ij}_{BP}\ X^i_{Bpc}\ X^j_{Bcc}\ X^j_{DVC}\ X^i_{PLL}\ X^j_{PLL}]^T.$$

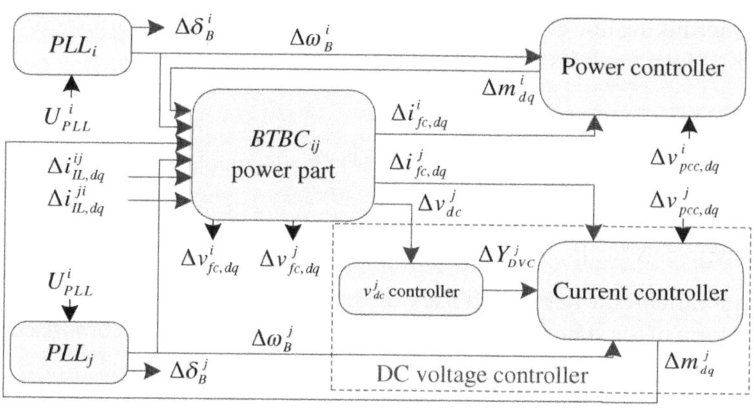

Figure 8.10 Interconnections of power and control modules of the $BTBC_{ij}$.

By specifying the input and output vectors as $U_B^{ij} = [\Delta v^i_{pcc,dq}\ \Delta i^{ij}_{IL,dq}\ \Delta v^j_{pcc,dq}\ \Delta i^{ji}_{IL,dq}\ \Delta\omega^i_{com}\ \Delta\omega^j_{com}]^T$ and $Y_B^{ij} = \left[\Delta v^i_{fc,dq}\ \Delta v^j_{fc,dq}\ \Delta v^j_{dc}\ \Delta\delta^i_B\ \Delta\delta^j_B\right]^T$, considering the interconnections shown in Figure 8.10 among all the BTBC modules, and using RCT functions in the interconnection method, the matrices can be calculated numerically for certain input data.

8.3.4 Circuit Breaker Modeling

The CBs and instantaneous SSs have fast responses in the range of microseconds to milliseconds. On the other hand, the time range of the dynamic studies, which are taken into account in this book, corresponds to a fraction of a second to a few seconds. Therefore, it is a logical assumption to consider that interlinking CBs switch as quickly as they are neglected in modeling the much slower studied dynamics. In other words, CB dynamics are finished much sooner than the times that the studied dynamics will start.

8.4 Overall IMG Modeling

In this section, comprehensive closed-loop models of both CB-IMGs and BTBC-IMGs are presented, which are useful for stability analysis and parameter behaviour identification. The models are validated using a measurement-based method done by Prony analysis. The reduced order models are also represented for CB-IMGs, which can be used in the frequency controller synthesis.

8.4.1 Comprehensive Modeling of CB-IMGs

By modeling all modules of CB-IMGs including ac MGs controlled by distributed pinning consensus-based secondary control architecture and interlinking lines separately, small-signal modeling of various structures of IMGs is possible. The main method, employed to calculate the overall IMG model, is the interconnection method, which is a low-calculation burden numerical method. Nevertheless, in this section, the state matrix of the IMG closed-loop model is analytically obtained to show the impact of each module on the final state matrix and, in turn, IMG stability.

Figure 8.11 indicates the details of interconnections among the IMGs shown in Figure 8.4. Note that the CRF zone consists of all interlinking lines and it is extended to the MG power part except the DER power part. The PCC voltage of each MG is considered as an input for each interlinking line to the MG. Inversely, the interlinking line current is considered as the input of the corresponding MGs. Thus, the overall closed-loop model is represented in the state space form as:

$$\dot{\mathbf{X}}_{IMG} = \mathbf{A}_{IMG}\mathbf{X}_{IMG}, \tag{8.26}$$

where the integrated state vector consisting of the main modules is as:

$$\mathbf{X}_{IMG} = [\overbrace{\mathbf{X}^1_{MG} \cdots \mathbf{X}^n_{MG} \cdots \mathbf{X}^M_{MG}}^{MGs} \cdots \overbrace{\mathbf{X}^{nl}_{IL}}^{ILs} \cdots]^T, \tag{8.27}$$

including $(15m + 2)\,n + 2p$ state variables, where n and p are the number of MGs and interlinking lines, respectively. The general form of the state matrix is analytically obtained, which is comprised of four main sub-matrices as follows:

$$\mathbf{A}_{IMG} = \begin{bmatrix} \mathbf{A}_{MGs} & \mathbf{A}_{MG,IL} \\ \mathbf{A}_{IL,MG} & \mathbf{A}_{ILs} \end{bmatrix}, \tag{8.28}$$

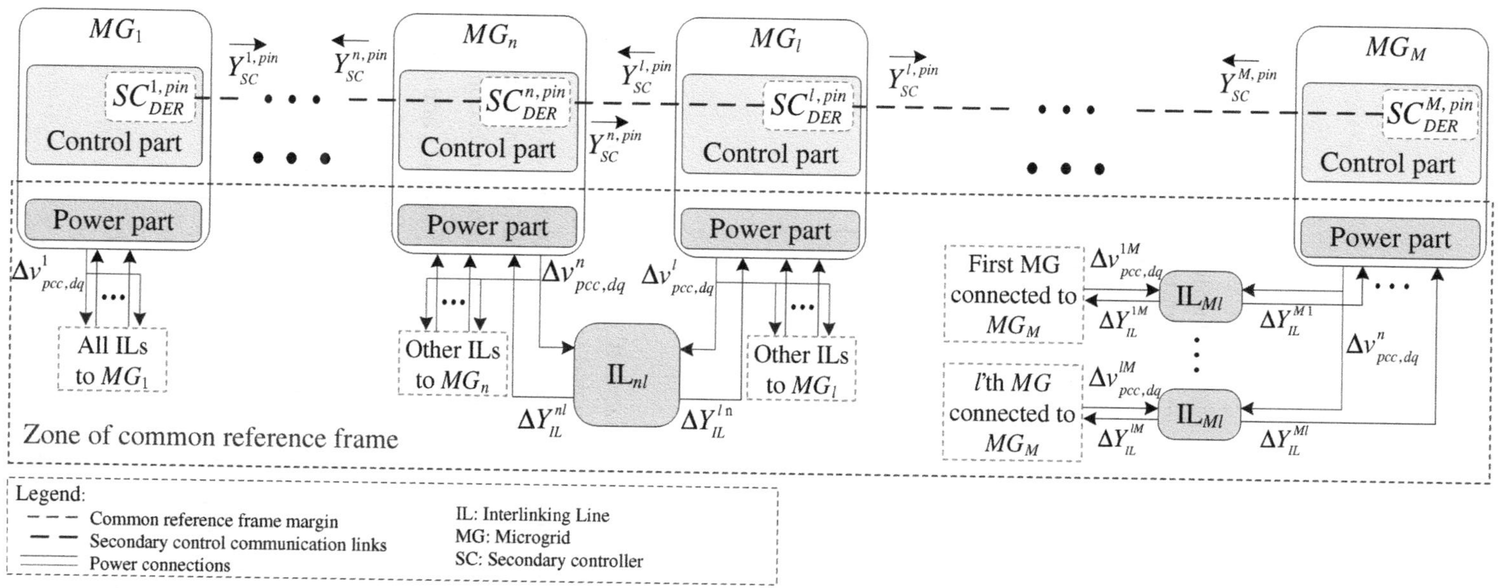

Figure 8.11 Specifying interconnections among synchronous microgrids through interlinking lines including all inputs and outputs of the main modules.

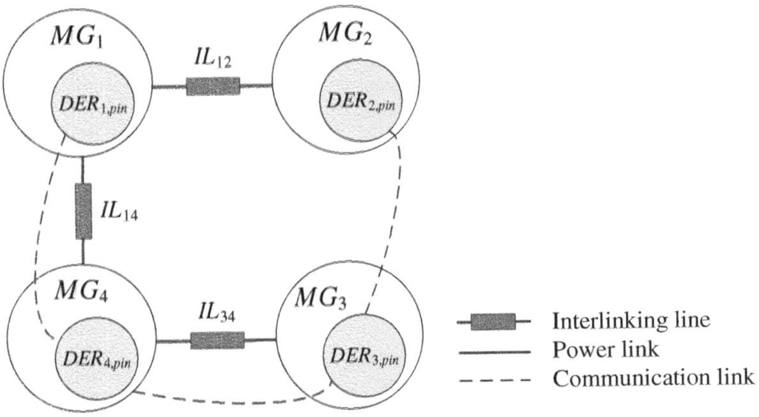

Figure 8.12 A typical system of four interconnected microgrids via three interlinking lines and an inter-MG communication network.

where

$$\mathbf{A}_{MGs} = \begin{bmatrix} \mathbf{A}_{MG}^{1} & \cdots & \mathbf{B}_{SC}^{1n} \\ \vdots & \ddots & \vdots \\ \mathbf{B}_{SC}^{n1} & \cdots & \mathbf{A}_{MG}^{n} \end{bmatrix}, \mathbf{A}_{MG,IL} = \begin{bmatrix} \mathbf{B}_{MG_1}^{IL_1} & \cdots & \mathbf{B}_{MG_1}^{IL_p} \\ \vdots & \ddots & \vdots \\ \mathbf{B}_{MG_n}^{IL_1} & \cdots & \mathbf{B}_{MG_n}^{IL_p} \end{bmatrix},$$

$$\mathbf{A}_{IL,MG} = \begin{bmatrix} \mathbf{B}_{IL_1}^{MG_1} & \cdots & \mathbf{B}_{IL_1}^{MG_n} \\ \vdots & \ddots & \vdots \\ \mathbf{B}_{IL_p}^{MG_1} & \cdots & \mathbf{B}_{IL_p}^{MG_n} \end{bmatrix}, \mathbf{A}_{ILs} = \begin{bmatrix} \mathbf{A}_{IL}^{1} & \cdots & 0 \\ \vdots & \ddots & \vdots \\ 0 & \cdots & \mathbf{A}_{IL}^{p} \end{bmatrix}. \tag{8.29}$$

$\mathbf{A}_{MG}^{n}$ is given in (8.9) and it is always nonzero. $\mathbf{B}_{SC}^{1n}$ shows the secondary control communication structure from MG_1 to MG_n, which is obtained from $\mathbf{B}_{SC2}$ in (8.8). It is nonzero if there exists a communication link between these two MGs. $\mathbf{B}_{MG_n}^{IL_p}$ indicates interconnections from MG_n to IL_p state variables, which is obtained as a sub-matrix of $\mathbf{B}_{MG}^{n}$ in (8.9), and it is nonzero if the MG_n is connected to another MG through the IL_p. $\mathbf{B}_{IL_p}^{MG_n}$ indicates interconnections from IL_p to MG_n state variables, which is obtained as a sub-matrix of the typical $\mathbf{B}_{IL}^{nl}$ in (8.11), and it is nonzero if $\mathbf{B}_{MG_n}^{IL_p} \neq 0$. Finally, $\mathbf{A}_{IL}^{p}$ is the same as typical $\mathbf{A}_{IL}^{nl}$ in (8.9). For the typical four-IMG case shown in Figure 8.12, $\mathbf{A}_{4\text{-}IMG}$ is obtained as follows:

$$\mathbf{A}_{4-IMG} = \left[\begin{array}{cccc|ccc} \mathbf{A}_{MG}^{1} & 0 & 0 & \mathbf{B}_{SC}^{14} & \mathbf{B}_{MG_1}^{IL_{12}} & \mathbf{B}_{MG_1}^{IL_{14}} & 0 \\ 0 & \mathbf{A}_{MG}^{2} & \mathbf{B}_{SC}^{23} & 0 & \mathbf{B}_{MG_2}^{IL_{12}} & 0 & 0 \\ 0 & \mathbf{B}_{SC}^{32} & \mathbf{A}_{MG}^{3} & \mathbf{B}_{SC}^{34} & 0 & 0 & \mathbf{B}_{MG_3}^{IL_{34}} \\ \mathbf{B}_{SC}^{41} & 0 & \mathbf{B}_{SC}^{43} & \mathbf{A}_{MG}^{4} & 0 & \mathbf{B}_{MG_4}^{IL_{14}} & \mathbf{B}_{MG_4}^{IL_{34}} \\ \hline \mathbf{B}_{IL_{12}}^{MG_1} & \mathbf{B}_{IL_{12}}^{MG_2} & 0 & 0 & \mathbf{A}_{IL}^{12} & 0 & 0 \\ \mathbf{B}_{IL_{14}}^{MG_1} & 0 & 0 & \mathbf{B}_{IL_{14}}^{MG_4} & 0 & \mathbf{A}_{IL}^{14} & 0 \\ 0 & 0 & \mathbf{B}_{IL_{34}}^{MG_3} & \mathbf{B}_{IL_{34}}^{MG_4} & 0 & 0 & \mathbf{A}_{IL}^{34} \end{array}\right].$$

Therefore, the overall IMG state matrix $\boldsymbol{A}_{IMG}$ can be achieved analytically using (8.28) and (8.29). Nevertheless, it is easier calculate numerically for interconnecting a desired number of modules using the Robust Control Toolbox functions in MATLAB. Note that the main requirement in this method is to specify the interconnections exactly in each IMG case study as generally shown in Figure 8.11.

8.4.2 Comprehensive Modeling of BTBC-IMGs

By modeling all BTBC-IMG modules, including ac MGs, BTBCs and ac/dc ILs, modeling of various structures of BTBC-IMGs with any number of MGs and interconnections is possible using the interconnection method. Figure 8.13 indicates the IMG interconnection focusing on the interconnection between MG_i and MG_j including $BTBC_{ij}$, IL_{ij}, IL_{ji}, and all modeling requirements (see Figure 8.1). In fact, MG_i, $i = 1, \dots, n$ can be interlinked to all other MGs, e.g. MG_j,$j = 1, \dots, n$ and $j \neq i$ through corresponding BTBCs and ILs, which shows the generality of the interconnection method.

Note that the MGs have independent CRFs, where their functional zones are shown in Figure 8.13. All ILs and BTBC ac sides connected to each MG are covered by its CRF. Therefore, the CRF to individual reference frames and inverse transformations are used for power interconnections between the MG and IL with the BTBC. The small-signal model can be represented in a free motion state-space form as:

$$\dot{X}_{IMG} = A_{IMG}X_{IMG}, \tag{8.30}$$

where X_{IMG} consists of the state variables of all participating modules in the IMGs and can be expressed as

$$X_{IMG} = [\overbrace{X^1_{MG} \dots X^n_{MG}}^{MGs} \dots \overbrace{X^{ij}_{IL} \dots}^{ILs} \dots \overbrace{X^{ij}_{B} \dots}^{BTBCs}]^T,$$

including $(13m + 2)\,n + 2p + 21q$ state variables, where n, p, and q are the number of MGs, ILs, and BTBCs, respectively.

For each IMG case with certain data, A_{IMG} can be calculated numerically by considering the modules' connections as generally shown in Figure 8.13, and applying the interconnection method.

8.5 Model Validation

A model validation based on the Prony analysis method is presented in this section. Fundamentals of the Prony analysis method are represented in Chapter 2, Section 2.3.4. Participation matrix is another basic concept required for validating the detailed models of CB-IMGs and BTBC-IMGs, which are already presented. Section 4.3.2 of Chapter 4 includes participation matrix principles. In the next sections, the model validation procedure is first represented, and then CB-IMGs and BTBC-IMGs models are validated according to the Prony analysis-based model validation.

8.5.1 Model Validation Procedure

Figure 8.14 shows a block diagram of the proposed validation method. The measured waveforms of the real-time simulated IMGs are analyzed by the Prony method. In order to take precise results from the Prony analysis, a perturbed form of the measured waveform must be found by subtracting

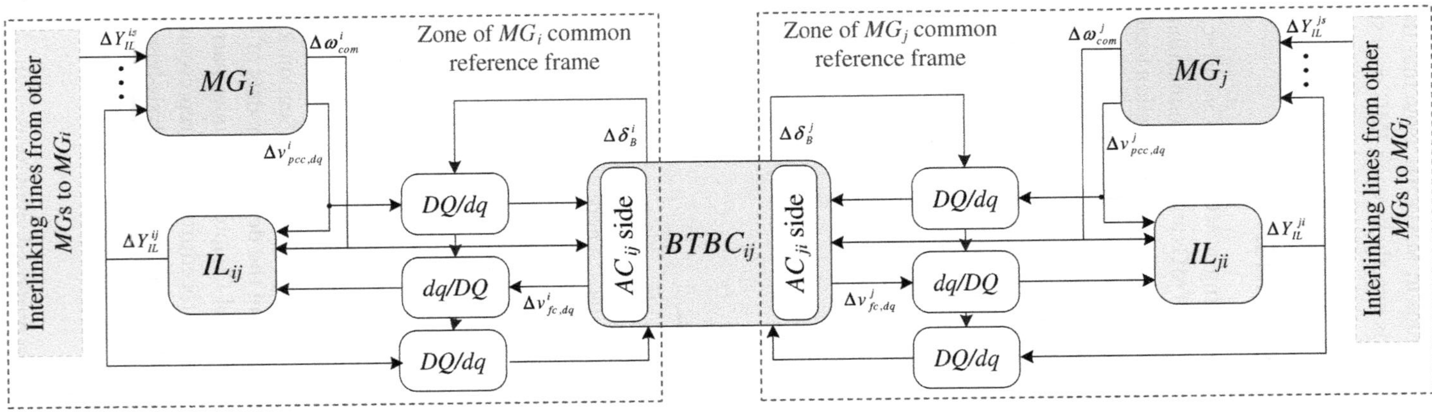

Figure 8.13 Complete interconnection of BTBC-IMGs focusing on the connection between MG_i and MG_j through IL_{ij}, IL_{ji} and $BTBC_{ij}$.

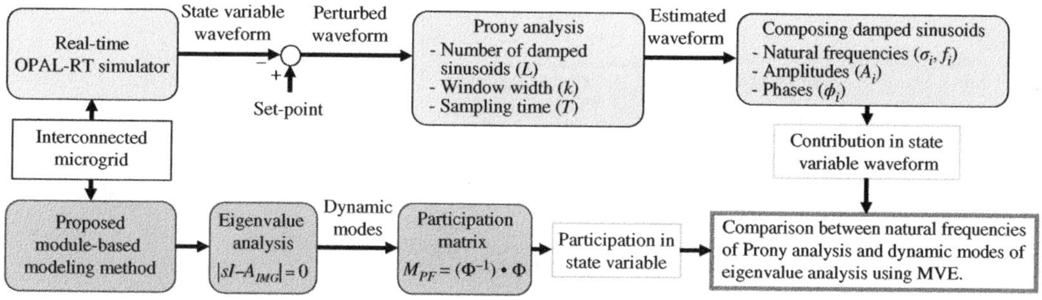

Figure 8.14 Block diagram of the proposed model validation by comparing the outputs of the eigenvalue and Prony analyses.

from the set point. In addition, the Prony analysis parameters should be chosen carefully including a number of damped sinusoids (L), data window width (equivalent to k), and sampling time (T). As a result, all composing damped sinusoids of the perturbed waveform are estimated in the form of natural frequencies ($\gamma_i = \sigma_i \pm j2\pi f_i$), amplitudes ($A_i$) and phases ($\phi_i$).

In order to realize the proposed validation, the dominant participating eigenvalues in the state variable and their participation factors should be equal or near to the natural frequencies and their corresponding amplitudes of the Prony estimated state variable, respectively. The model validation error (MVE), which shows the similarity percentage, is as:

$$MVE_i = \frac{|\lambda_i - \gamma_i|}{|\lambda_i|} \times 100, \tag{8.31}$$

where MVE_i is the relative error percentage that indicates the difference between estimated frequency γ_i by Prony analysis and correlated eigenvalue λ_i calculated by eigenvalue analysis. A similar relation can also be used for the difference between per-unit contribution in waveform $A_i\,(pu)$ (instead of γ_i in (8.31)) and correlated normalized participation factor $M_{PF,n}$ (instead of λ_i in (8.31)). Note that the values are per-unit in this case. Therefore, a relative error for MVE_{PF} is not required.

8.5.2 Real-Time Simulator

The simulations are done using OPAL-RT simulator OP5600 to show the ability to perform the study IMG simulations in real-time. Figure 8.15(a) shows the experimental setup including the host PC, the OPAL-RT target, and a LAN cable for networking. Figure 8.15(b) shows the conceptual diagram of the real-time simulation process. RT-LAB software is used as the interface between MATLAB and a real-time OPAL-RT simulator. The OP5600 is configured with 4 activated Intel Xeon E5, 3.2 GHz processing cores, and works under the Linux operating system. Moreover, it provides user-programmable I/O management, handled by a fast Xilinx Spartan-3 FPGA. Version 11.2.1.91 of RT-LAB software is used as the interface between MATLAB and OPAL-RT digital simulator.

The MATLAB/SimPowerSystems model is loaded on the OPAL-RT through the RT-LAB and the real-time data comes back to MATLAB conversely. The model is split into two subsystems, i.e. a subsystem including all permanent power and control parts during the real-time simulation

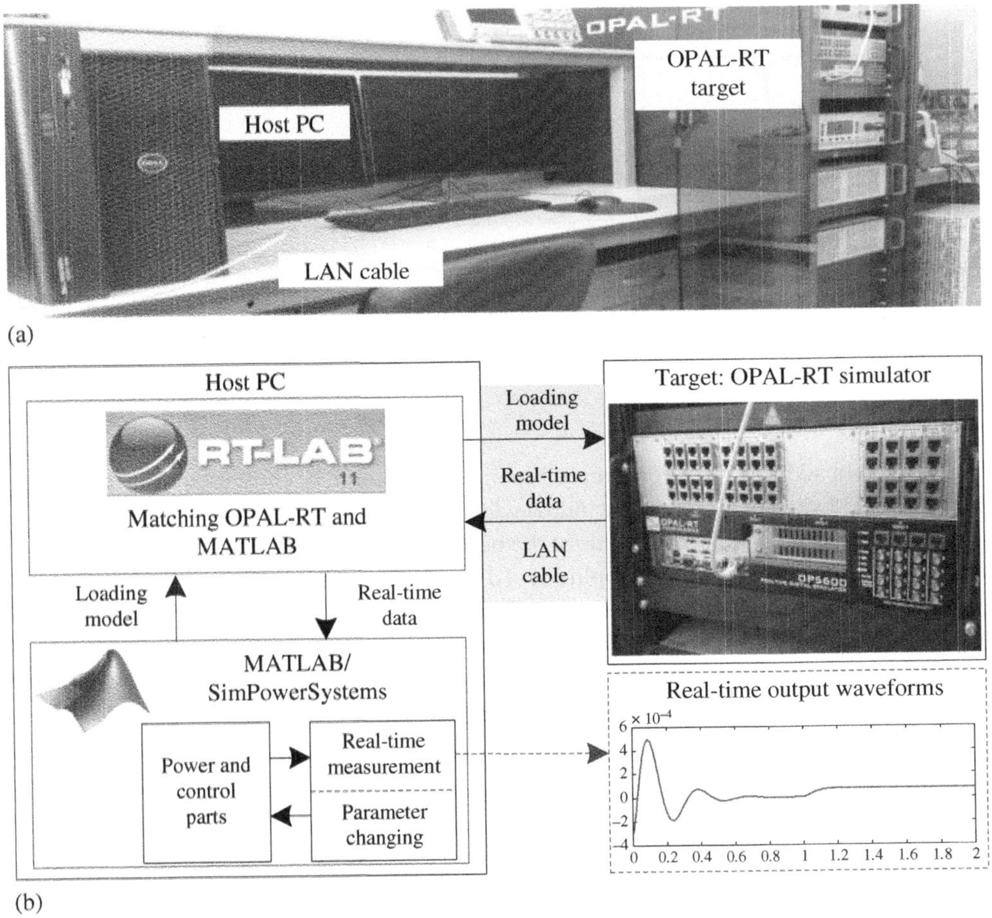

Figure 8.15 (a) Real-time setup including the host PC, the OPAL-RT target and a LAN cable, (b) conceptual diagram of the real-time simulation process.

and a subsystem including real-time displays and changeable parameters. After model loading and real-time data receiving, real-time waveforms can be measured in the SimPowerSystems environment.

8.5.3 Validation of CB-IMG Modeling

8.5.3.1 Case Study Information

In order to verify the proposed modeling method, a typical two IMGs presented in [62], which is based on the general structure shown in Figure 8.4 is implemented in an OPAL-RT simulator. Table 8.4 comprises the electrical and control parameters used to find the numerical form of A_{IMG} for the two IMGs. Moreover, initial values are obtained from the real-time simulation of the IMGs based on the information in Table 8.4.

8.5.3.2 Prony Analysis Results

Figure 8.16 shows the perturbed form of three measured waveforms from the real-time simulated IMGs and their estimation using the Prony analysis, which is applied at the time interval of 0 to 1 s, i.e. $k.T = 1$ s. For a precise estimation in each case, L and T are changed and the convenient

Table 8.4 Power and control information of two interconnected ac microgrids.

Parameter	Value			
RMS line-line voltage (V)	400			
DC voltage (V)	780			
Angular frequency (rad/s)	100π			
Interlinking line impedance (Ω)	3.6 + j1.23			
MGs	MG_1		MG_2	
MG P_{max} (kW)	3		3	
MG Q_{max} (kVAr)	1.5		1.5	
Initial load	$0.6P_{max}$ $0.6Q_{max}$		$0.5P_{max}$ $0.5Q_{max}$	
Controller coefficients	K_P		K_I	
PI voltage controller	1		200	
DERs	DER_1	DER_2	DER_3	DER_4
DER ratings (S_{max}) (kVA)	1.12	2.24	1.12	2.24
Series filter inductance (mH)	1.8	1.8	1.8	1.8
Series filter resistance (Ω)	0.05	0.05	0.05	0.05
Shunt filter capacitance (μF)	3.7	7.5	3.7	7.5
Line inductance (mH)	15.6	7.8	15.6	7.8
Line resistance (Ω)	3.6	1.8	3.6	1.8
P-f droop coefficients (10^{-3}rad/s · W)	1.6	0.8	1.6	0.8
Q-V droop coefficients (10^{-2}V/VAR)	6.5	3.2	6.5	3.2
Leader/follower secondary control delay (T/τ) (s)	0.2	10	0.2	10

values are found by trial and error, which are indicated in Table 8.5. MG_1 and MG_2 are operated in 60% and 50% of their nominal load, respectively. As the second set-point, the MG_1 load increases to 90% of its nominal value at $t = 1$ s. This recent change shows only the different dynamic responses of two different set points and initial conditions. Note that a state variable waveform can show all participating natural frequencies only when the input or the initial condition is applied in the direction of all its eigenvectors. In this situation, the Prony analysis can correctly estimate all natural frequencies. By comparing several initial conditions, the best identification can be achieved by the zero initial condition.

According to Figure 8.16, ΔI_{ILd}, $\Delta f_2^{MG_2}$, and $\Delta\delta_2^{MG_1}$ are estimated well by the Prony analysis. In fact, the real-time simulated waveform is fitted by the Prony estimated waveform in all three cases.

8.5.3.3 Comparison Results

The estimation results of the Prony analysis method are classified in Table 8.5 including the natural frequencies and their related contribution in the waveforms. On the other hand, the results of the eigenvalue analysis and participation matrix are depicted in Table 8.5 comprising the eigenvalues/dynamic modes and their participation in the state variables.

Table 8.5 Comparison between outputs of the Prony and eigenvalue analyses.

State variable	Number of damped sinusoids L	Sampling time T(s)	Mode number	Prony analysis outputs: Natural frequencies (γ)	Prony analysis outputs: Contribution in waveform (per-unit)	Eigenvalues analysis outputs: Dynamic mode (λ)	Eigenvalues analysis outputs: Contribution in state variable/ participation factors (per-unit)
ΔI_{IL}	6	0.004	1	$-4.74 \pm j18.47$	0.134	$-6.75 \pm j20.95$	0.114
			2	-59.76	0.36	-73.49	0.33
			3	$-453.61 \pm j511.31$	0.34	$-509.69 \pm j261.80$	0.14
						$-635.07 \pm j312.67$	0.132
			4	-452	0.16	-467	0.14
						$-166.26 \pm j19.38$	0.137
						Other	0.01
Δf_2^{MG2}	9	0.004	1	$-5.35 \pm j22.35$	0.218	$-6.75 \pm j20.95$	0.334
						$-8.26 \pm j28.82$	0.206
			2	-22.40	0.531	-19.94	0.312
						-22.12	0.025
			3	-81.22	0.175	-73.49	0.122
				-305.72	0.002	Other	0.01
				$-146.69 \pm j73.18$	0.01		
				$-605 \pm j219.91$	0.064		
$\Delta \delta_2^{MG1}$	6	0.001	1	$-6.52 \pm j21.39$	0.76	$-8.27 \pm j28.68$	0.806
			2	-79.88	0.208	-73.49	0.152
			3	$-140.58 \pm j217.93$	0.024	$-145.17 \pm j186.91$	0.012
						$-143.71 \pm j192.51$	0.016
				-3419.19	0.006	Other	0.01

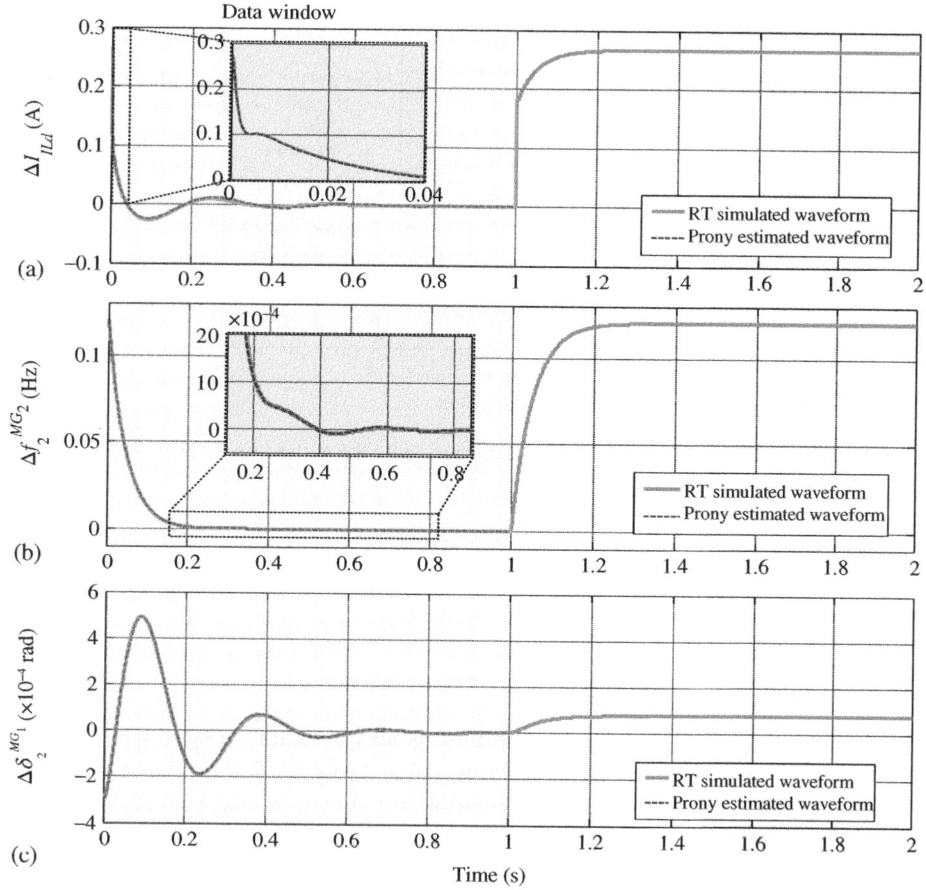

Figure 8.16 Simulated waveforms and their Prony estimation: (a) direct component of interlinking line current, (b) frequency of DER_2 in MG_2, (c) angle difference of DER_2 voltage in MG_1 from the common reference frame.

According to the comparison results in all three cases, the natural frequencies and dynamic modes are not exactly the same. However, there is an appropriate similarity between the frequency ranges and even the real and imaginary values in some cases. These similarities are shown by placing similar natural frequencies and dynamic modes in the same rows. Note that the lack of complete equivalency can be due to the deletion of some dynamics during linearization or the lack of some exciting natural frequencies. Moreover, Figure 8.17 shows MVE for the listed variables in Table 8.5 including dynamic modes and their normalized participation factor (PF in the legend). The third mode of ΔI_{IL}, which is calculated by averaging, has the maximum MVE. This mode is medium frequency, thus it is considerably affected by the linearization process. The maximum MVE_{PF} is 32%, which belongs to the first mode of $\Delta f_2^{MG_1}$. The low-frequency modes, which can be observed in the power controller variables, e.g. $\Delta f_2^{MG_2}$ and $\Delta\delta_2^{MG_1}$, are more taken into account in the stability studies [26] and generally are validated with MVE less than 20%.

Note that in order to simplify the model and use it in stability analysis, one can reduce the main model to the low-frequency modes [48]. As can be seen in the first two rows of $\Delta f_2^{MG_2}$ and $\Delta\delta_2^{MG_1}$ in Table 8.5, these low-frequency modes are validated with an appropriate MVE shown in Figure 8.17. Therefore, an important point is that Prony analysis can be used to find a simplified model of

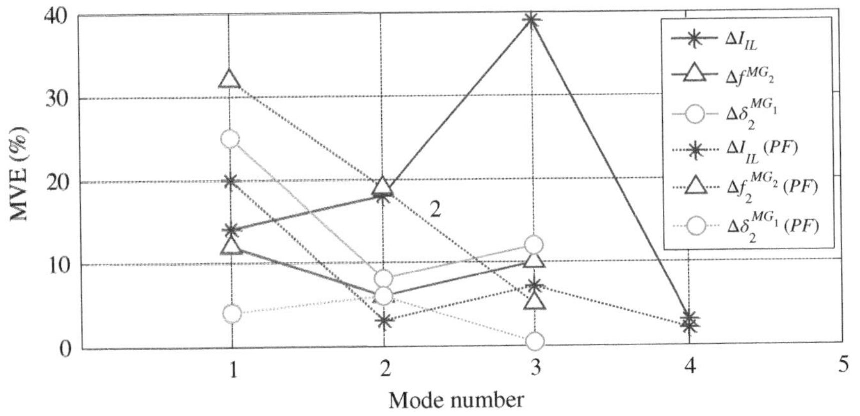

Figure 8.17 Model validation error (MVE) for listed variables in Table 8.5 (solid line for MVE and dotted line for MVE_{PF}).

large-scale systems such as IMGs by measuring a remarkable characteristic, e.g. frequency, and analyzing it by Prony tool. Such a measurement-based modeling method and correlated data-driven control methods will be studied in future works.

8.5.4 Validation of BTBC-IMG Modeling

The introduced Prony analysis-based validation method is also used for BTBC-IMGs. The two BTBC-IMGs introduced in [63] are simulated in real time using the OPAL-RT simulator. The waveforms of OPAL-RT output and Prony analysis estimation are shown in Figure 8.18 for $\Delta\delta_2^{MG_1}$, $\Delta P_1^{MG_2}$, and ΔX_{DVC}, where generally $\Delta x = x_{ref} - x$, Δx is the perturbed form of the state variable (x) and x_{ref} is the reference value. The Prony method estimates the perturbed form of waveforms for a certain set-point, which is 850 W active power flow from MG_2 to MG_1 at $t = 1$ s due to a power deficiency in MG_1. The data window shows the time interval of Prony estimation. All three waveforms show appropriate estimation by Prony analysis.

Comparative results for the estimated state variables are shown in Table 8.6, where the eigenvalues and participation factors are also calculated for the two BTBC-IMGs. The MVE is less than 10 % in all cases except $\lambda = -21.46$, which is not an effective dominant mode based on its participation factor, i.e. 0.02. Note that the presented Prony-based validation is able to be employed for all state variables.

8.6 Reduced-Order Models

The detailed models of IMGs are high-order, particularly for large-scale IMGs comprising many MGs to share power. Such models are hard to use in applicable cases. On the other hand, the models according to the application can focus on certain dynamics, i.e. special frequency range. For instance, in the stability analysis of interconnected systems, the low-frequency dynamics are taken into account in order to investigate power/frequency oscillations. Therefore, reduced-order models of IMGs are needed in different applications.

In Chapter 3, two reduced-order models of ac MGs are addressed including a simplified low-frequency model and a very simplified single-order model. In fact, the single-order model of ac

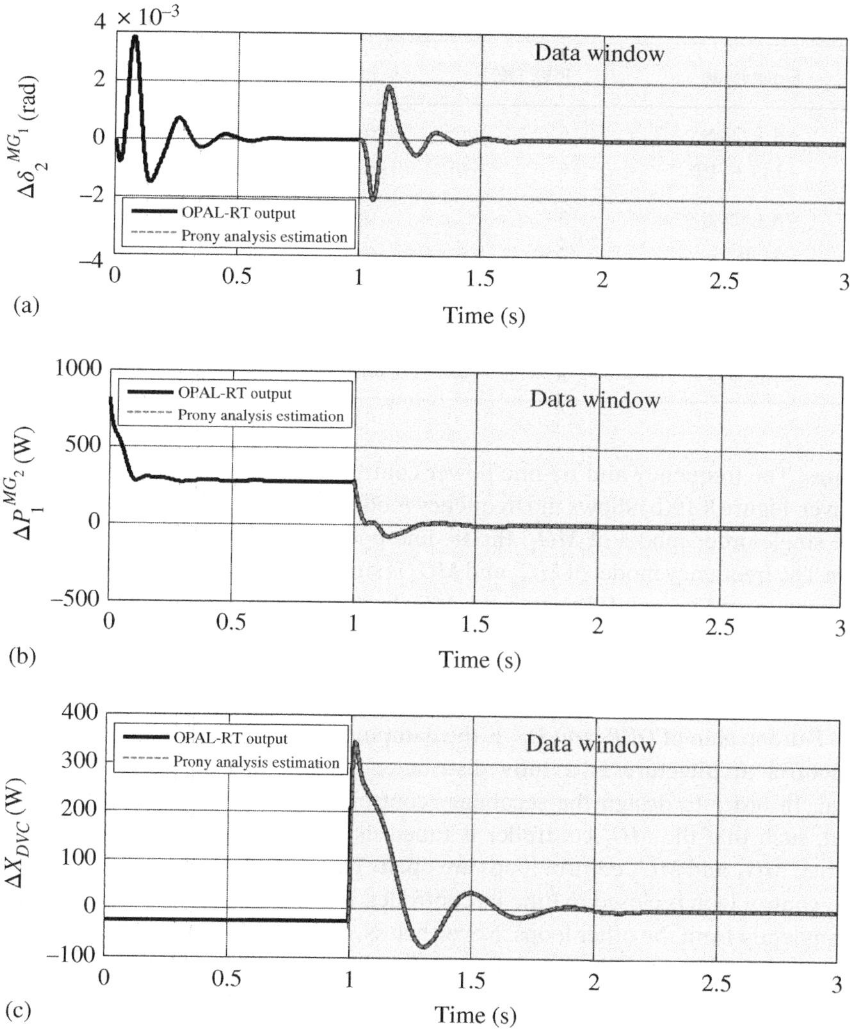

Figure 8.18 Prony analysis-based estimation for perturbed form of OPAL-RT output waveforms: (a) the angle difference of the DER_2 reference frame from the common reference frame of MG_1, (b) DER_1 active power of MG_2, (c) integrator output of the dc voltage controller.

MGs is obtained using the low-frequency model and according to the virtual swing equation, which is applied to the aggregated oscillatory mode. Hence, each MG is modeled by just a single-order transfer function. Such a simple model can be very useful in the small-signal stability and frequency control of IMGs, where the fast dynamics are not of interest and the focus is on the slow dynamics. Both CB-IMGS and BTBC-IMGs models can be simplified using the single-order model of individual IMGs. However, in this section, CB-IMGs are considered to find the simplified model.

8.6.1 Simplified Model Application in CB-IMG Frequency Control

The proposed single-order model facilitates the high-level control studies of large-scale IMGs, e.g. the frequency control. Figure 8.19(a) shows three different autonomous ac MGs interconnected

Table 8.6 Prony analysis-based model validation.

State variable	Eigenvalue	MVE (%)	Participation factor	MVE (%)
$\Delta\delta_2^{MG_1}$	$-8 \pm j28.8$	5	0.96	8
	$-145 \pm j106$	8	0.03	3
$\Delta P_1^{MG_2}$	$-8 \pm j28.8$	3	0.66	6
	-21.46	7	0.02	21
	-20	0	0.29	9
ΔX_{DVC}	$-2 \pm j13.8$	6	0.45	3
	$-108 \pm j87$	8	0.55	3

through three tie-lines. The frequency and tie-line power control among the MGs is done in the secondary control layer. Figure 8.19(b) shows the frequency model of MG_1 within the studied three IMGs including the single-order model of MG_1, the tie-line power (ΔP_{TL1}) model [64], and the secondary controller. The frequency model of MG_2 and MG_3 is similar to the MG_1 frequency model with individual parameters. In Figure 8.19(b), β_1 is the MG_1 frequency bias calculated as

$$\beta_1 = \sum_{i=1}^{m} \left(1/m_p^i\right) + D_{L1}, \tag{8.32}$$

where m_p^i is the $\omega - P$ droop gain of DER_i and D_{L1} is the damping coefficient of the MG_1 load.

The secondary control architecture is a fully distributed control and the controllers are proportional-integral. In order to design the secondary controllers, the sequential loop closing method [65] is used, such that the MG_1 controller is tuned using a PID tuning block in MATLAB/SIMULINK when MG_2 and MG_3 control loops are open. Then, the MG_2 controller is tuned and finally, the MG_3 control loop is closed to tune its controller. Note that in this method, the first loop is closed independently from the other loops. Nevertheless, during closing the next loops and designing the corresponding PI controllers, the previous loops are closed and their interactions are considered in the design process. Hence, the frequency control is done uniformly.

8.6.2 Simplified Model of MGs and CB-IMGs

In order to find the overall simplified model of CB-IMGs, the interconnection model can also be used. Although the model is of low order and it can be analytically found for a few IMGs, it is better to use the interconnection method to generalize the model for any number of IMGs. Hence, the model of each MG should be achieved including the single-order model, secondary controller, and interlinking line power model as shown in Figure 8.19(b). Then, the interconnection method can be applied to any number of IMGs with different connection structures.

For each typical MG_n, three state variables $\Delta\omega_{MG,n}$, $\Delta x_{SC,n}$, and $\Delta P_{IL,n}$ are considered as shown in Figure 8.19(b), where $\Delta x_{SC,n}$ is the integrator output of the secondary PI controller, and it cannot be seen. According to system block diagram and signal flow principles, a state space representation can be found for MG_n as follows:

$$\dot{X}_{MG,ro} = A_{MG,ro}X_{MG,ro} + B_{MG,ro}U_{MG,ro} \tag{8.33a}$$

$$Y_{MG,ro} = C_{MG,ro}X_{MG,ro} + D_{MG,ro}U_{MG,ro}, \tag{8.33b}$$

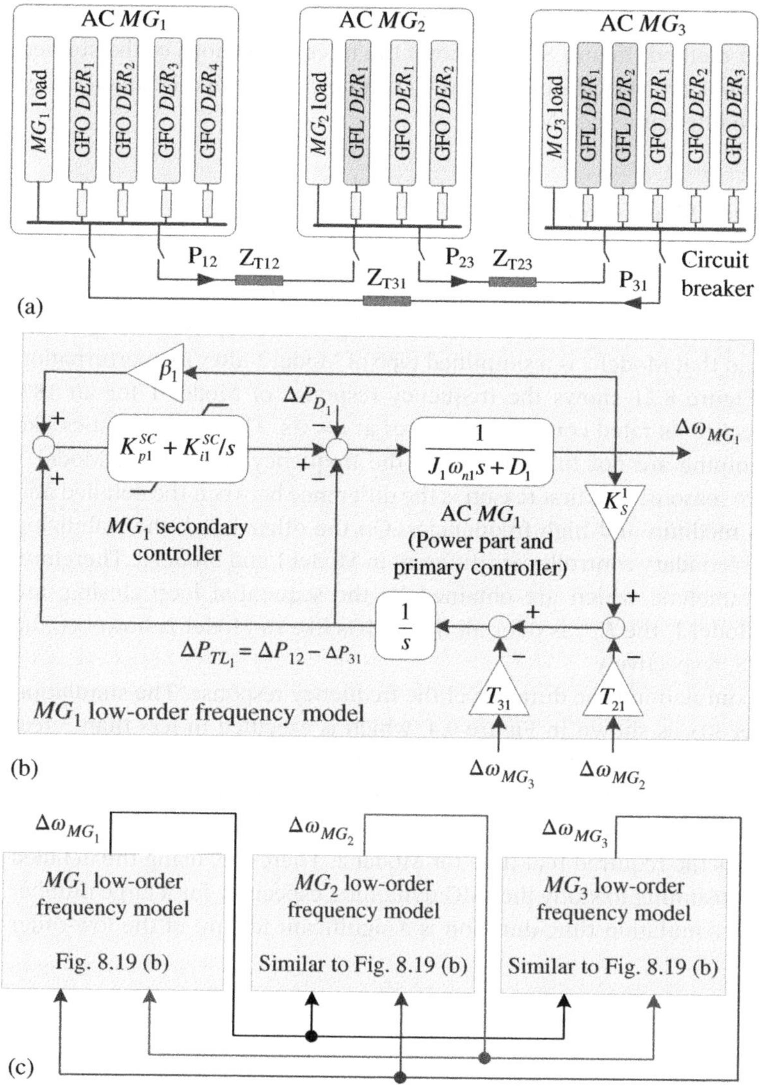

Figure 8.19 Case study for frequency control: (a) Three interconnected ac microgrids through circuit breakers and tie-lines, (b) low-order frequency model of MG_1 including the single-order model, controller, and tie line power, (c) low-order frequency model of the studied system.

where the state vector is as $X_{MG,ro} = [\Delta\omega_{MG,n}\ \Delta x_{SC,n}\ \Delta P_{IL,n}]^T$, the input vector is as $U_{MG,ro} = [\Delta\omega_{MG,1} \dots \Delta\omega_{MG,M}]^T$ including all angular frequencies of the MGs connected to MG_n, and M indicates the number of MGs. Note that $T_{kn} = 0$ if MG_n and MG_k are not connected. Finally, the matrices are given as follows:

$$A_{MG,ro} = \begin{bmatrix} \frac{1}{J_n\omega_n}\left(K_{pn}^{SC}\beta_n - D_n\right) & \frac{1}{J_n\omega_n} & \frac{K_{pn}^{SC}}{J_n\omega_n} \\ K_{in}^{SC}\beta_n & 0 & K_{in}^{SC} \\ K_S^n & 0 & 0 \end{bmatrix}, B_{MG,ro} = \begin{bmatrix} 0 & \dots & 0 \\ 0 & \dots & 0 \\ T_{1n} & \dots & T_{Mn} \end{bmatrix},$$
$$C_{MG,ro} = \begin{bmatrix} 1 & 0 & 0 \end{bmatrix}, D_{MG,ro} = 0_{1\times(M-1)}.$$

The state-space representation of IMGs with any number of MGs and interconnections is easy using (8.33) and the interconnection method. Figure 8.19(c) shows the interconnections of the studied three IMGs. The interconnections are very simple, and they can be summarized in angular frequency connections among the interconnected MGs.

8.6.3 Comparing Detailed and Single-Order Models

Figure 8.20 shows a comparison between two different models of the studied IMGs including Model 1 with the detailed MG_1 model and the single-order models of MG_2 and MG_3 and Model 2 with the single-order models of all the three MGs shown in Figure 8.19(c). According to the DLFMs, it is easy to understand that Model 2 is a simplified type of Model 1 after the perturbation and aggregation processes. Figure 8.21 shows the frequency response of Model 1 for an 18% MG_1 load increase with respect to its rated controllable power at $t = 0$ s. The characteristics like overshoot and oscillation damping are not fully similar to the frequency response of Model 1 shown in Figure 9.4 due to two reasons. The first reason is the difference between the detailed and simplified models of MG_1 in medium and high frequencies. On the other hand, the stabilizing parameter ranges of the MG_1 secondary controller are different in Model 1 and Model 2. Therefore, the secondary controller parameters, which are obtained by the sequential loop closing, are different for the models. In Model 1, the K_p^{SC} is 0 for all three MGs like in Model 2; however, the K_i^{SC} is −3.11, −4.71, and −8.98, respectively.

The important point is the simulation time duration of the frequency response. The simulation time of Model 2 is selected as 30 s as shown in Figure 9.4, which is executed in less than 3 real seconds. Nevertheless, the simulation time of Model 1 is selected as 3 s, which is executed in 510 real seconds using the same processor Intel Core i7-7500U, 2.7 GHz with Turbo boost up to 3.5 GHz. In other words, for the same simulation time of both the models, e.g. 30 s, the required real time for Model 1 is 5100/3 = 1700 times the required real time for Model 2. Therefore, using the detailed MG model is hard and time-consuming to study the IMG dynamics, especially for a large number of IMGs. In contrast, very low simulation time duration is a significant feature of the low-order frequency model of the IMGs.

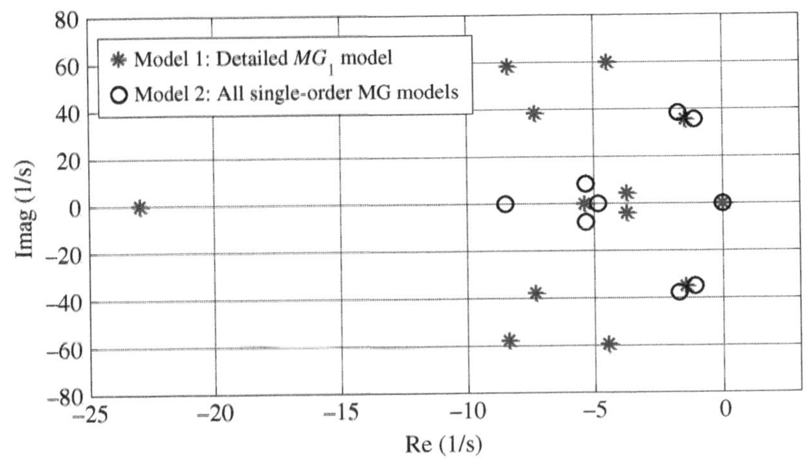

Figure 8.20 The comparison between two models of the studied IMGs including Model 1 with detailed MG_1 model and single-order model of MG_2 and MG_3, Model 2 with single-order model of the all three MGs.

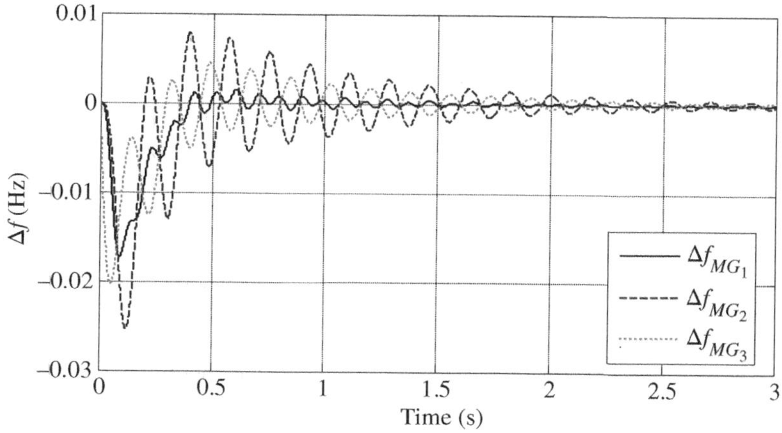

Figure 8.21 Frequency response of Model 1 when MG_1 load increases as 18% rated controllable power at $t = 0$ s.

8.7 Summary

This chapter focuses on extending the interconnection method presented in Chapters 2 and 3 to obtain small-signal models of complex MG subsystems and different structures of MGs, specifically CB-IMGs, and BTBC-IMGs, which are very large-scale systems. The method involves modeling each subsystem (module) separately and interconnecting all modules according to their electrical, communication, and control links. For CB-IMG modeling, the most important module is the secondary control, where requirements of interconnections among MGs are provided. For BTBC-IMG modeling, BTBCs are vital, including power connection and required control loops. Therefore, modeling of both modules is presented in detail, while modeling of other modules can be found in Chapters 2 and 3. In BTBC-IMG modeling, it is essential to consider several common reference frames due to separated frequency zones caused by BTBC performance.

Small-signal models of both IMG architectures are validated by comparing some state variables with the Prony analysis outputs obtained from real-time large-signal equivalent variables. An OPAL-RT simulator is used to find the nonlinear models. The validated closed-loop IMG models are useful for small-signal stability analysis and sensitivity analysis of parameters, which are discussed in Chapter 9. However, according to the presented general rules, researchers can find some open-loop models for their own case studies, excluding controller models, in order to design them based on model-based synthesis methods. Finally, a very low-order module-based model is proposed for frequency studies of CB-IMGs in which each MG model order is reduced to one. The single-order MG model has already been attained in Chapter 3 and has been validated compared to the detailed MG model.

References

1 Ehsan Pashajavid, Arindam Ghosh, and Firuz Zare. A multimode supervisory control scheme for coupling remote droop-regulated microgrids. *IEEE Trans. Smart Grid*, 9(5):5381–5392, 2018.

2 Xiaoyu Wu, Yin Xu, Xiangyu Wu, Jinghan He, Josep M Guerrero, Chen-Ching Liu, Kevin P Schneider, and Dan T Ton. A two-layer distributed control method for islanded networked microgrid systems. *IEEE Trans. Smart Grid*, 11(2):942–957, 2019.

3 Iraklis P Nikolakakos, Hatem H Zeineldin, Mohamed Shawky El-Moursi, and Nikos D Hatziargyriou. Stability evaluation of interconnected multi-inverter microgrids through critical clusters. *IEEE Trans. Power Syst.*, 31(4):3060–3072, 2016.

4 Liang Che, Mohammad Shahidehpour, Ahmed Alabdulwahab, and Yusuf Al-Turki. Hierarchical coordination of a community microgrid with AC and DC microgrids. *IEEE Trans. Smart Grid*, 6(6):3042–3051, 2015.

5 Zhuoli Zhao, Ping Yang, Yuewu Wang, Zhirong Xu, and Josep M Guerrero. Dynamic characteristics analysis and stabilization of PV-based multiple microgrid clusters. *IEEE Trans. Smart Grid*, 10(1):805–818, 2017.

6 Farhad Shahnia and Ali Arefi. Eigenanalysis-based small signal stability of the system of coupled sustainable microgrids. *Int. J. Electr. Power Energy Syst.*, 91:42–60, 2017.

7 Mohammad S Golsorkhi, David John Hill, and Hamid Reza Karshenas. Distributed voltage control and power management of networked microgrids. *IEEE J. Emerging Sel. Top. Power Electron.*, 6(4):1892–1902, 2018.

8 Md Jahangir Hossain, Md Apel Mahmud, Federico Milano, Seddik Bacha, and Ahmad Hably. Design of robust distributed control for interconnected microgrids. *IEEE Trans. Smart Grid*, 7(6):2724–2735, 2016.

9 Hyeong-Jun Yoo, Thai-Thanh Nguyen, and Hak-Man Kim. Multi-frequency control in a stand-alone multi-microgrid system using a back-to-back converter. *Energies*, 10(6):1–18, 2017.

10 Inam Ullah Nutkani, Poh Chiang Loh, and Frede Blaabjerg. Distributed operation of interlinked AC microgrids with dynamic active and reactive power tuning. *IEEE Trans. Ind. Appl.*, 49(5):2188–2196, 2013.

11 Ramon Zamora and Anurag K Srivastava. Multi-layer architecture for voltage and frequency control in networked microgrids. *IEEE Trans. Smart Grid*, 9(3):2076–2085, 2018.

12 Cheng-Yu Tang, Yen-Fu Chen, Yaow-Ming Chen, and Yung-Ruei Chang. DC-link voltage control strategy for three-phase back-to-back active power conditioners. *IEEE Trans. Ind. Electron.*, 62(10):6306–6316, 2015.

13 Jaewan Suh, Dong-Hee Yoon, Yoon-Sung Cho, and Gilsoo Jang. Flexible frequency operation strategy of power system with high renewable penetration. *IEEE Trans. Sustainable Energy*, 8(1):192–199, 2017.

14 Ritwik Majumder and Gargi Bag. Parallel operation of converter interfaced multiple microgrids. *Int. J. Electr. Power Energy Syst.*, 55:486–496, 2014.

15 Ritwik Majumder, Arindam Ghosh, Gerard Ledwich, and Firuz Zare. Power management and power flow control with back-to-back converters in a utility connected microgrid. *IEEE Trans. Power Syst.*, 25(2):821–834, 2010.

16 Tiankui Sun, Jian Lu, Zhimin Li, David Lee Lubkeman, and Ning Lu. Modeling combined heat and power systems for microgrid applications. *IEEE Trans. Smart Grid*, 9(5):4172–4180, 2017.

17 Joan Rocabert, Alvaro Luna, Frede Blaabjerg, and Pedro Rodriguez. Control of power converters in AC microgrids. *IEEE Trans. Power Electron.*, 27(11):4734–4749, 2012.

18 Nagaraju Pogaku, Milan Prodanovic, and Timothy C Green. Modeling, analysis and testing of autonomous operation of an inverter-based microgrid. *IEEE Trans. Power Electron.*, 22(2):613–625, 2007.

19 Francis K Tuffner, Kevin P Schneider, Jacob Hansen, and Marcelo A Elizondo. Modeling load dynamics to support resiliency-based operations in low-inertia microgrids. *IEEE Trans. Smart Grid*, 10(3):2726–2737, 2018.

20 Hassan Bevrani, Bruno François, and Toshifumi Ise. *Microgrid dynamics and control*. John Wiley & Sons, Hoboken, NJ, 2017.

21 Magdi Sadek Mahmoud, S Azher Hussain, and Mohammad Ali Abido. Modeling and control of microgrid: An overview. *J. Franklin Inst.*, 351(5):2822–2859, 2014.

22 Ling Luo and Sairaj V Dhople. Spatiotemporal model reduction of inverter-based islanded microgrids. *IEEE Trans. Energy Convers.*, 29(4):823–832, 2014.

23 Valerio Mariani, Francesco Vasca, Juan C Vásquez, and Josep M Guerrero. Model order reductions for stability analysis of islanded microgrids with droop control. *IEEE Trans. Ind. Electron.*, 62(7):4344–4354, 2014.

24 Md Rasheduzzaman, Jacob A Mueller, and Jonathan W Kimball. Reduced-order small-signal model of microgrid systems. *IEEE Trans. Sustainable Energy*, 6(4):1292–1305, 2015.

25 Petr Vorobev, Po-Hsu Huang, Mohamed Al Hosani, James L Kirtley, and Konstantin Turitsyn. High-fidelity model order reduction for microgrids stability assessment. *IEEE Trans. Power Syst.*, 33(1):874–887, 2017.

26 Iraklis P Nikolakakos, H H Zeineldin, Mohamed Shawky El-Moursi, and James L Kirtley. Reduced-order model for inter-inverter oscillations in islanded droop-controlled microgrids. *IEEE Trans. Smart Grid*, 9(5):4953–4963, 2017.

27 Nathaniel Bottrell, Milan Prodanovic, and Timothy C Green. Dynamic stability of a microgrid with an active load. *IEEE Trans. Power Electron.*, 28(11):5107–5119, 2013.

28 Alireza Kahrobaeian and Yasser Abdel-Rady I Mohamed. Analysis and mitigation of low-frequency instabilities in autonomous medium-voltage converter-based microgrids with dynamic loads. *IEEE Trans. Ind. Electron.*, 61(4):1643–1658, 2013.

29 Yousef Khayat, Mobin Naderi, Qobad Shafiee, Yazdan Batmani, Mohammad Fathi, Josep M Guerrero, and Hassan Bevrani. Decentralized optimal frequency control in autonomous microgrids. *IEEE Trans. Power Syst.*, 34(3):2345–2353, 2018. doi: 10.1109/TPWRS.2018.2889671.

30 Zhikang Shuai, Yelun Peng, Xuan Liu, Zuyi Li, Josep M Guerrero, and Z John Shen. Dynamic equivalent modeling for multi-microgrid based on structure preservation method. *IEEE Trans. Smart Grid*, 2018. doi: 10.1109/TSG.2018.2844107.

31 Farhad Shahnia. Stability and eigenanalysis of a sustainable remote area microgrid with a transforming structure. *Sustainable Energy Grids Netw.*, 8:37–50, 2016.

32 Yun Zhang, Le Xie, and Qifeng Ding. Interactive control of coupled microgrids for guaranteed system-wide small signal stability. *IEEE Trans. Smart Grid*, 7(2):1088–1096, 2016.

33 Renke Han, Michele Tucci, Andrea Martinelli, Josep M Guerrero, and Giancarlo Ferrari-Trecate. Stability analysis of primary plug-and-play and secondary leader-based controllers for DC microgrid clusters. *IEEE Trans. Power Syst.*, 34(3):1780–1800, 2018.

34 Jinghan He, Xiaoyu Wu, Xiangyu Wu, Yin Xu, and Josep M Guerrero. Small-signal stability analysis and optimal parameters design of microgrid clusters. *IEEE Access*, 7:36896–36909, 2019.

35 Ouyang Jianna, Liang Shuo, Chen Weidong, and Guo Min. Modeling and analysis of microgrid cluster simulation based on RTDS. In *2nd IEEE Conference on Energy Internet and Energy System Integration (EI2)*, pages 1–5. IEEE, 2018.

36 Zhiyi Li and Mohammad Shahidehpour. Small-signal modeling and stability analysis of hybrid AC/DC microgrids. *IEEE Trans. Smart Grid*, 10 (2):2080–2095, 2017.

37 Pan Wu, Wentao Huang, Nengling Tai, and Shuo Liang. A novel design of architecture and control for multiple microgrids with hybrid AC/DC connection. *Appl. Energy*, 210:1002–1016, 2018.

38 Mobin Naderi, Yousef Khayat, Qobad Shafiee, Tomislav Dragicevic, Hassan Bevrani, and Frede Blaabjerg. Interconnected autonomous AC microgrids via back-to-back converters—Part I: Small-signal modeling. *IEEE Trans. Power Electron.*, 35(5):4728–4740, 2019.

39 Chengshan Wang, Hao Yu, Peng Li, Jianzhong Wu, and Chengdi Ding. Model order reduction for transient simulation of active distribution networks. *IET Gener. Transm. Distrib.*, 9(5):457–467, 2015.

40 Yan Li, Peng Zhang, and Peter B Luh. Formal analysis of networked microgrids dynamics. *IEEE Trans. Power Syst.*, 33(3):3418–3427, 2018.

41 Yan Li, Peng Zhang, and Meng Yue. Networked microgrid stability through distributed formal analysis. *Appl. Energy*, 228:279–288, 2018.

42 Mehdi Monadi, Hossein Hooshyar, Luigi Vanfretti, Farhan Mahmood, Jose Ignacio Candela, and Pedro Rodriguez. Measurement-based network clustering for active distribution systems. *IEEE Trans. Smart Grid*, 10(6):6714–6723, 2019.

43 Kishan Veerashekar, Stefan Eichner, and Matthias Luther. Modelling and transient stability analysis of interconnected autonomous hybrid microgrids. In *2019 IEEE Milan PowerTech*, pages 1–6. IEEE, 2019.

44 Yun Zhang and Le Xie. Online dynamic security assessment of microgrid interconnections in smart distribution systems. *IEEE Trans. Power Syst.*, 30(6):3246–3254, 2014.

45 Yun Zhang and Le Xie. A transient stability assessment framework in power electronic-interfaced distribution systems. *IEEE Trans. Power Syst.*, 31(6):5106–5114, 2016.

46 Jingang Lai, Xiaoqing Lu, Xinghuo Yu, and Antonello Monti. Cluster-oriented distributed cooperative control for multiple AC microgrids. *IEEE Trans. Ind. Inf.*, 15(11):5906–5918, 2019.

47 Shivraman Mudaliyar, Bhanu Duggal, and Sukumar Mishra. Distributed tie-line power flow control of autonomous DC microgrid clusters. *IEEE Trans. Power Electron.*, 35(10):11250–11266, 2020.

48 Mobin Naderi, Qobad Shafiee, Hassan Bevrani, and Frede Blaabjerg. Low-frequency small-signal modeling of interconnected AC microgrids. *IEEE Trans. Power Syst.*, 36(4):2786–2797, 2020.

49 A Hasib Chowdhury and Md Asaduz-Zaman. Load frequency control of multi-microgrid using energy storage system. In *8th International Conference on Electrical and Computer Engineering*, pages 548–551, 2014.

50 Meysam Gheisarnejad and Mohammad Hassan Khooban. Secondary load frequency control for multi-microgrids: Hil real-time simulation. *Soft Comput.*, 23(14):5785–5798, 2019.

51 Deepak Kumar Lal, Ajit Kumar Barisal, and M Tripathy. Load frequency control of multi area interconnected microgrid power system using grasshopper optimization algorithm optimized fuzzy PID controller. In *2018 Recent Advances on Engineering, Technology and Computational Sciences (RAETCS)*, pages 1–6. IEEE, 2018.

52 Gary J Balas, John C Doyle, Keith Glover, Andy Packard, and Roy Smith. *μ-analysis and synthesis toolbox: For use with Matlab*. The MathWorks Inc., 1994.

53 Wei Liu, Wei Gu, Yinliang Xu, Yong Wang, and Kaifeng Zhang. General distributed secondary control for multi-microgrids with both PQ-controlled and droop-controlled distributed generators. *IET Gener. Transm. Distrib.*, 11(3):707–718, 2017.

54 Lingyu Ren, Yanyuan Qin, Yan Li, Peng Zhang, Bing Wang, Peter B Luh, Song Han, Taofeek Orekan, and Tao Gong. Enabling resilient distributed power sharing in networked microgrids through software defined networking. *Appl. Energy*, 210:1251–1265, 2018.

55 Shengxuan Weng, Yusheng Xue, Jianbo Luo, and Yanman Li. Distributed secondary control for islanded microgrids cluster based on hybrid-triggered mechanisms. *Processes*, 8(3):370, 2020.

56 Xiaoqing Lu, Jingang Lai, and Xinghuo Yu. A novel secondary power management strategy for multiple AC microgrids with cluster-oriented two-layer cooperative framework. *IEEE Trans. Ind. Inf.*, 17(2):1483–1495, 2020.

57 Yousef Khayat, Qobad Shafiee, Rasool Heydari, Mobin Naderi, Tomislav Dragičević, John W Simpson-Porco, Florian Dörfler, Mohammad Fathi, Frede Blaabjerg, Josep M Guerrero, et al. On the secondary control architectures of AC microgrids: An overview. *IEEE Trans. Power Electron.*, 35(6):6482–6500, 2019. doi: 10.1109/TPEL.2019.2951694.

58 Xiaoqiang Guo, Zhigang Lu, Baocheng Wang, Xiaofeng Sun, Lei Wang, and Josep M Guerrero. Dynamic phasors-based modeling and stability analysis of droop-controlled inverters for microgrid applications. *IEEE Trans. Smart Grid*, 5(6):2980–2987, 2014.

59 Mobin Naderi, Yousef Khayat, Qobad Shafiee, Tomislav Dragicevic, Frede Blaabjerg, and Hassan Bevrani. An emergency active and reactive power exchange solution for interconnected microgrids. *IEEE J. Emerging Sel. Top. Power Electron.*, pages 1–13, 2019.

60 Amirnaser Yazdani and Reza Iravani. *Voltage-sourced converters in power systems*, volume 34. Wiley Online Library, Hoboken, NJ, 2010.

61 Mobin Naderi, Yousef Khayat, Qobad Shafiee, Hassan Bevrani, Rasool Heydari, Tomislav Dragicevic, and Frede Blaabjerg. Modeling and stability analysis of back-to-back converters in networked microgrids. In *45th Annual Conference of the IEEE Industrial Electronics Society (IECON)*, volume 1, pages 3925–3930. IEEE, 2019.

62 Mobin Naderi, Yousef Khayat, Rasool Heydari, Mateja Novak, Qobad Shafiee, Tomislav Dragicevic, Hassan Bevrani, and Frede Blaabjerg. Model validation of power electronics-based networked micro-grids by prony analysis. In *2019 21st European Conference on Power Electronics and Applications (EPE '19 ECCE Europe)*, pages 1—10. IEEE, 2019.

63 Mobin Naderi, Yousef Khayat, Qobad Shafiee, Tomislav Dragičević, Hassan Bevrani, and Frede Blaabj. Interconnected autonomous AC microgrids via back-to-back converters-Part II: Stability analysis. *IEEE Trans. Power Electron.*, 35(11):11801–11812, 2020.

64 Hassan Bevrani. *Robust power system frequency control*. Springer, 2014.

65 Ubaid Ubaid, Stephen Daley, and S A Pope. Design of remotely located and multi-loop vibration controllers using a sequential loop closing approach. *Control Eng. Pract.*, 38:1–10, 2015.

9

Stability of Interconnected Microgrids

9.1 Introduction

Stability is a fundamental requirement of each power grid, and has always been of interest due to the huge number of interconnected components, which their interactions may cause instability. The interconnected microgrids (IMGs) stability is also taken into account due to their large scale and the presence of many distributed energy resources (DERs), different loads, and nested hierarchical control.

Stability analysis of IMG operation and its improvement are important due to possible interactions between the individual controllers within IMGs as well as between these controllers and power exchange controllers, which are placed in a central controller or distributed across MGs and interface devices. Different architectures of IMGs manifest several stability challenges. dc link voltage stability of IMGs via back-to-back converters (BTBC-IMGs), power sharing and frequency stability of IMGs via circuit breakers (CB-IMGs), and stability issues of coordination control in hybrid and multiple ac/dc IMGs are some of the IMG stability challenges. Moreover, owing to the dominant low-frequency modes of the IMGs, power and frequency oscillations are necessary to be taken into account [1]. In CB-IMGs, synchronization control is very important, which is investigated in Chapter 11.

The stability should be preserved in different operation modes, i.e. emergency and scheduled interconnections. The emergency condition in an islanded MG is defined as large and sudden disturbances such as large load change or DER trip, which change the balance between generation and consumption [2]. The emergency-based IMG formation is determined usually due to such an emergency condition in a remote MG in order to support frequency and load-shedding mechanisms [3]. In fact, this is an alternative solution instead of adding energy storage systems and extra-sized diesel generators [4] in emergency conditions. Nevertheless, the emergency operation of IMGs requires a powerful logic control to detect the emergency condition, apply the emergency control signals, and distinguish nonemergency situations.

In BTBC-IMGs, BTBCs are more flexible interfaces, which result in a separate frequency control of IMGs [1, 5–8]. Generally, the control and operation of BTBC-IMGs are similar to the multiarea power systems inter-tied through high-voltage dc links [8–10]. Both systems similarly lead to frequency stability improvement due to eliminating inter-area frequency interactions. However, the voltage strength of the two systems and the correlated stability margin are different. In each area of

Microgrids: Dynamic Modeling, Stability and Control, First Edition. Qobad Shafiee, Mobin Naderi, and Hassan Bevrani.

the conventional power system, many synchronous generators participate in the load supply and inertia support. Furthermore, common load changes are less than 5% of the total supplied load [11]. Therefore, ac voltage of each area is stiff. On the contrary, MG's DERs are mostly inertia-less voltage source converter (VSC)-based units have a considerable contribution to the load supply. In addition, load steps can be larger than 20% of the total MG load [12]. Hence, each DER outage or load change has a remarkable impact on the MG voltage/frequency, which behaves as a weak grid. Consequently, the voltage and frequency stability in these IMGs are much challenging and need to be studied.

9.2 IMG Stability Review

The small-signal and transient stabilities of CB-IMGs have been analyzed in the literature [13–21]. An eigenvalue analysis method is presented in order to determine the suitable range of control parameters for IMG's DERs and to guarantee the IMG stability employing the sensitivity analysis [13]. In [14], the impact of rating/number of DERs and loads, as well as the autonomous MG topology on the IMG stability, is investigated. A similar stability analysis is performed to study the impact of different interconnecting points within a distribution network on the IMG formation [16]. Parameter stability margins are also calculated using sensitivity analysis. In a more general method, a parametric criterion for IMG stability is achieved by applying Lyapunov stability on the simplified droop-based IMG model [15, 17]. In contrast, a detailed small-signal model of PV-based IMGs is proposed considering the dynamics of the PV controllers and dc sides [18]. Then, sensitivity analysis is employed to find acceptable parameter ranges affecting the dominant critical modes (DCMs). In [19], authors have presented a two-layer, four-level distributed control strategy for IMGs, and then its impact on the small-signal stability is analyzed. Similar work has been presented for distributed voltage control and power management of IMGs [20]. The authors have simplified the MG model with primary and secondary controllers and have studied their impacts on the IMG small-signal stability.

Stability analysis of BTBC-IMGs is evaluated in [1, 21, 22]. In [1], a specific configuration of MGs is considered in which any autonomous MG has a STATCOM/ESS unit to coordinate its generation with BTBC power exchange. A robust distributed controller is presented for such BTBC-IMGs to dampen the oscillatory modes due to the interaction of local DER controllers and make the system robust against the parametric uncertainties. A common structure of two autonomous ac MGs interconnected by a BTBC is considered to prove IMG frequency support [22]. Then, time-domain frequency stability analysis is conducted, which shows the frequency independence for BTBC-IMGs during faults. In [21], the study focuses on multiple MGs connected to a strong grid, but it does not address the main challenges of IMGs, e.g., power exchange and voltage/frequency control.

9.3 Small-Signal Stability Analysis

The small-signal stability type is analyzed based on the linear/linearized model around an operating point. These models were represented in detail for CB-IMGs and BTBC-IMGs in Chapter 8. For each model represented by the state space equations, e.g. (8.26) or (8.26), the small-signal stability can be assessed using eigenvalue analysis as

$$|\lambda I - A_{IMG}| = 0, \tag{9.1}$$

where λ shows the eigenvalues, I is the identity matrix, and $|.|$ indicates the matrix determinant operator. To achieve an asymptotic stability, all eigenvalues should have negative real values. The stability margin can also increase by moving away from the imaginary axis to its left side.

9.3.1 Eigenvalue Analysis of CB-IMGs

In this study, the two IMGs introduced in Section 8.5.3 and [23] are selected. In order to understand the impact of each power/control part of the IMGs, e.g. DER coupling line and voltage controller, on IMG stability, their detailed model is considered. Therefore, the possible impact of various parts can be seen using the eigenvalues and participation factors. The operating point information of the IMGs is shown in Table 9.1, which is obtained using a time-domain simulation of the studied IMGs in MATLAB/SimPowerSystems.

Figure 9.1 shows all eigenvalues of the state matrix obtained numerically using the interconnection method. Generally, four different ranges of real values can be observed. Very far eigenvalues from the imaginary axis are related to the DER power part, i.e. RLC filters, which are shown as Group 1. This group of eigenvalues is not important in the stability analysis due to their large stability margin. However, sensitivity analysis should be employed to understand the behavior of such eigenvalues for changing the filter parameters. It is possible for stable eigenvalues to move to the right, even crossing the imaginary axis. The eigenvalues around −600 real correlate to the MG loads, power network (coupling lines), interlinking line, and somewhat inner controllers, which are called Group 2. They also may reduce the stability margin by changing the parameters of the mentioned contributors. The eigenvalues around −200 real are generally related to the inner controllers, i.e. current and voltage controllers, which are addressed as Group 3.

The dominant modes are mostly related to the secondary and primary controllers as well as a contribution of DER coupling lines (see Group 4 in Figure 9.1). These eigenvalues and their contributing state variables are the most remarkable items in the IMG small-signal stability. Note that the precise contribution of the state variables belonging to the mentioned IMG parts can be obtained using the participation matrix. Nevertheless, changing the parameters of the control parts, e.g. primary and secondary controllers, shows the contribution clearer by sensitivity analysis outputs. Moreover, the stability margins for such important parameters can be achieved.

Table 9.1 Operating point information of the studied two IMGs obtained from a time-domain simulation in MATLAB/SimPowerSystem.

Parameter (unit)	Value	Parameter (unit)	Value	Parameter (unit)	Value	Parameter (unit)	Value
$\delta_{10}^{MG_1}$ (rad)	0	$i_{od20}^{MG_1}$	2.03	$\delta_{10}^{MG_2}$ (rad)	0.0003	$i_{od20}^{MG_2}$	2.22
$\delta_{20}^{MG_1}$ (rad)	0	$i_{oq20}^{MG_1}$	−1.07	$\delta_{20}^{MG_2}$ (rad)	0.0003	$i_{oq20}^{MG_2}$	−1
$\omega_{10}^{MG_1}$ (rad/s)	314.4	$v_{od10}^{MG_1}$ (V)	316.5	$\omega_{10}^{MG_2}$ (rad/s)	314.4	$v_{od10}^{MG_2}$ (V)	317.6
$\omega_{20}^{MG_1}$ (rad/s)	314.4	$v_{oq10}^{MG_1}$ (V)	0	$\omega_{20}^{MG_2}$ (rad/s)	314.4	$v_{oq10}^{MG_2}$ (V)	0
$i_{ld10}^{MG_1}$ (A)	1.02	$v_{od20}^{MG_1}$ (V)	316.5	$i_{ld10}^{MG_2}$ (A)	1.12	$v_{od20}^{MG_2}$ (V)	317.6
$i_{lq10}^{MG_1}$ (A)	−0.16	$v_{oq20}^{MG_1}$ (V)	0	$i_{lq10}^{MG_2}$ (A)	−0.12	$v_{oq20}^{MG_2}$ (V)	0
$i_{ld20}^{MG_1}$ (A)	2.02	$v_{pcc,d0}^{MG_1}$ (V)	310	$i_{ld20}^{MG_2}$ (A)	2.22	$v_{pcc,d0}^{MG_2}$ (V)	311.5
$i_{lq20}^{MG_1}$ (A)	−0.32	$v_{ppc,q0}^{MG_1}$ (V)	−3	$i_{lq20}^{MG_2}$ (A)	−0.24	$v_{ppc,q0}^{MG_2}$ (V)	−3
$i_{od10}^{MG_1}$ (A)	1.02	$i_{ILd0}^{MG_1}$ (A)	−0.26	$i_{od10}^{MG_2}$ (A)	1.11		
$i_{oq10}^{MG_1}$ (A)	−0.53	$i_{ILq10}^{MG_1}$ (A)	0.08	$i_{oq10}^{MG_2}$ (A)	−0.5		

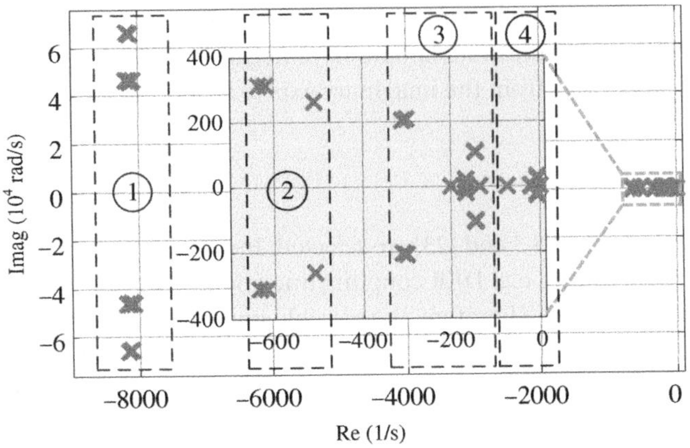

Figure 9.1 All eigenvalues of the studied two CB-IMGs.

9.3.2 Frequency Stability of CB-IMGs

The frequency stability and dynamics of three IMGs shown in Figure 8.19(a) are studied to introduce the inter-MG DLFMs and show a stable operation of the secondary controllers designed in Section 8.6.1. All the MGs are of low-voltage ac type, which have the same nominal frequency and voltage as 50 Hz and 400 V. They have a general structure shown in Figure 8.19, where MG_1 consists of four grid-forming (GFO) DERs, MG_2 has two GFO and one grid-following (GFL) DERs, and MG_3 contains three droop-based and two GFL DERs. The general data of the MG modules is presented in [24] and the parameters of the frequency response model shown in Figure 8.19(b) are found by applying the four-step method introduced in Figure 3.10 for the three MGs, which are indicated in Table 9.2. Finally, Figure 8.19(c) shows the low-order frequency model of the studied three IMGs. Note that each box indicates the low-order frequency model of each MG, e.g. the contents of the MG_1 box are shown in Figure 8.19(b) within a dashed outline. The frequency interactions of the MGs are exhibited in Figure 8.19(c).

9.3.2.1 Intermicrogrid Oscillatory Modes

The participation matrix of the state-space model of the three IMGs shown in Figure 8.19, can be seen in Figure 9.2. SOM, SC, and Int indicate the state variables of the single-order MG model, secondary PI controller, and the integrator modeling tie-line power dynamics, respectively. The contribution of each state variable in each eigenvalue is calculated and then normalized. The participation matrix results are also shown in Figure 9.3 by the qualitative interpretations added to the nine eigenvalues of the 3-IMGs. λ_1–λ_6 are the main inter-MG modes affected by all modules including the inner MG models, the secondary controllers, and the tie-line power values. According to

Table 9.2 Calculated parameters of the three IMGs in Figure 8.19.

	$D\left(\frac{\text{kW·s}}{\text{rad}}\right)$	$J\left(\frac{\text{W·s}^2}{\text{rad}}\right)$	$\beta\left(\frac{\text{kW·s}}{\text{rad}}\right)$	$T_{qn}\left(\frac{10^5\ \text{W}}{\text{rad}}\right)$	K_p^{SC}	K_i^{SC}
MG_1	5.6	1.45	7	$T_{21} = 1.4$ $T_{31} = 2.3$	0	−9.16
MG_2	1.9	0.71	1.9	$T_{12} = 1.4$ $T_{32} = 0.9$	0	−4.71
MG_3	3.8	1.4	3.8	$T_{13} = 2.3$ $T_{23} = 0.9$	0	−8.98

	$-1.74 \pm j38.26$		$-1.10 \pm j35.24$		$-5.35 \pm j8.24$		-8.48	-4.87	0
	λ_1	λ_2	λ_3	λ_4	λ_5	λ_6	λ_7	λ_8	λ_9
SOM_{MG_1}	0.26	0.26	0.03	0.03	0.17	0.17	0.02	0.01	0
SC_1	0.06	0.06	0.01	0.01	0.20	0.20	0.35	0.02	0
Int_1	0.24	0.24	0.03	0.03	0.06	0.06	0.03	0.02	0.55
SOM_{MG_2}	0.15	0.15	0.21	0.21	0.10	0.10	0	0.03	0
SC_2	0.02	0.02	0.03	0.03	0.05	0.05	0	0.84	0
Int_2	0.15	0.15	0.20	0.20	0.05	0.05	0	0.03	0.15
SOM_{MG_3}	0.05	0.05	0.23	0.23	0.16	0.16	0.03	0.01	0
SC_3	0.01	0.01	0.06	0.06	0.16	0.16	0.53	0.01	0
Int_3	0.05	0.05	0.22	0.22	0.03	0.03	0.03	0.02	0.3
	Interaction between MG_1 and MG_2		Interaction between MG_2 and MG_3		Interaction among all three MGs		Dynamic modes of secondary controllers		Integrators mode

Figure 9.2 Participation matrix of the studied three interconnected ac microgrids shown in Figure 8.19.

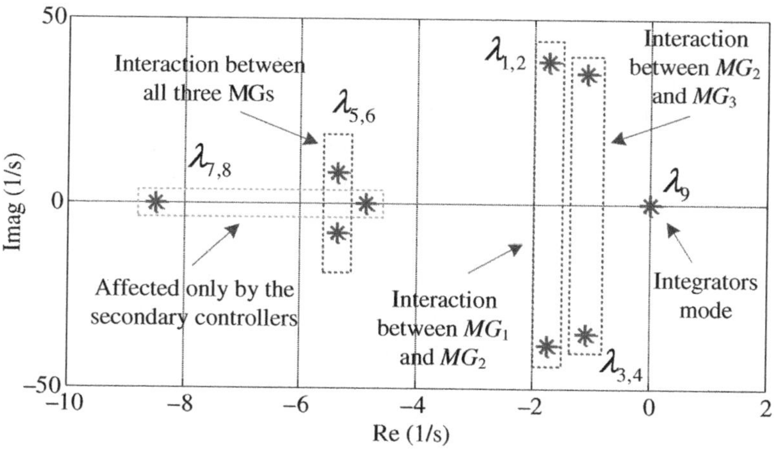

Figure 9.3 Eigenvalues of the studied three interconnected ac microgrids shown in Figure 8.19.

the larger participation factors of each dynamic mode, i.e. column data of the participation matrix shown in Figure 9.2, $\lambda_{1,2}$, $\lambda_{3,4}$, and $\lambda_{5,6}$ show the interaction between MG_1 and MG_2, the interaction between MG_2 and MG_3, and the interaction between all three MGs, respectively. $\lambda_{7,8}$ are the nonoscillatory modes affected only by the secondary controllers, and λ_9 is a noncritical zero mode due to the tie-line power integrators.

Note that the main inter-MG modes are oscillatory, which may cause frequency oscillation and instability, especially considering the low inertia of IMGs. The possible frequency instability and even frequency fluctuations can be confronted by designing the secondary controller based on the behavior of inter-MG modes.

9.3.2.2 Frequency Response

The frequency response of the MGs is shown in Figure 9.4. Three load changes +18%, −18%, and +23% of the rated controllable power for frequency control (sum of GFO DER powers) are applied

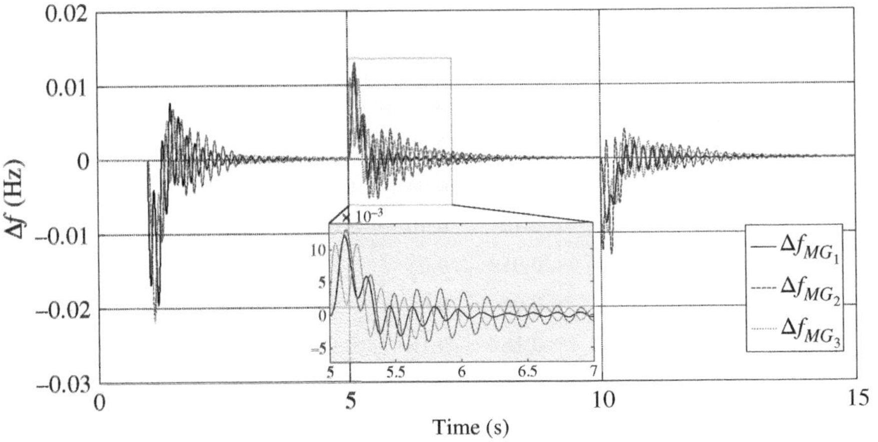

Figure 9.4 Frequency response of the studied three interconnected ac microgrids. MG_1 load increases as 18% rated controllable power at $t = 1$ s, MG_3 load decreases as 18% rated controllable power at $t = 10$ s, MG_2 load increases as 23% rated controllable power at $t = 20$ s.

as ΔP_D in MG_1, MG_2, and MG_3, respectively. Generally, the dynamics of three MG frequencies are similar due to interlinking using circuit breakers, i.e. the power section of IMGs is uniform unlike interconnection using back-to-back converters [6, 25, 26]. Moreover, the frequency oscillations can be seen, which are caused by the inter-MG modes. Nevertheless, there are differences in the transients, e.g. nadir and rate of change of frequency due to the different amounts of virtual J and D. It can be seen that the frequency stability is guaranteed against several small disturbances by the secondary PI controllers.

It is noteworthy to remember that the proposed IMG model is a small-signal type, which cannot be used to investigate the system behavior during large disturbances, e.g. DER/MG disconnection, or inner nonlinear phenomena, e.g. limiter saturation of converter controllers. Furthermore, the fast and medium speed dynamics are removed. Therefore, the corresponding variables studies cannot be done using the proposed model, e.g. the voltage stability analysis.

9.3.3 Eigenvalue Analysis of BTBC-IMGs

In order to analyze the small-signal stability of BTBC-IMGs, two case studies consisting of two interconnected MGs through one BTBC (case study 1) and three interconnected MGs through three BTBCs (case study 2) are modeled in MATLAB/Editor, as shown in Figure 9.5. The MG structure is shown in Figure 8.3 without GFL DERs. The information on the simulated BTBC-IMGs is presented in Table 9.3, except for the interconnected line (IL) impedances displayed in Figure 9.5. The state matrix A_{IMG} is calculated numerically for the detailed models of both IMGs. A_{2IMG} is an 81×81 matrix, and A_{3IMG} is a 159×159 matrix. In this section, case study 1 is analyzed, with details of the BTBC structure shown in Figure 8.8(a). The operating point information of the DERs within the MGs is similar to that represented for CB-IMGs in Table 9.1. Moreover, the operating point information of the BTBC in case study 1 is shown in Table 9.4, where it is also obtained using a time-domain simulation of the studied IMGs in MATLAB/SimPowerSystems.

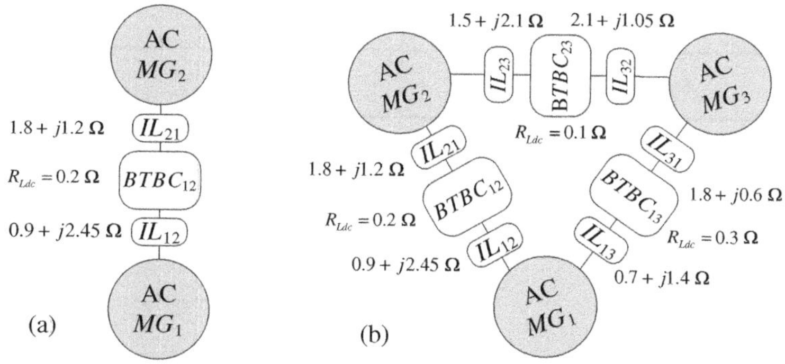

Figure 9.5 The structure of the simulated interconnected MGs: (a) case study 1: two autonomous ac MGs interconnected through one BTBC, (b) case study 2: three autonomous ac MGs interconnected through three BTBCs.

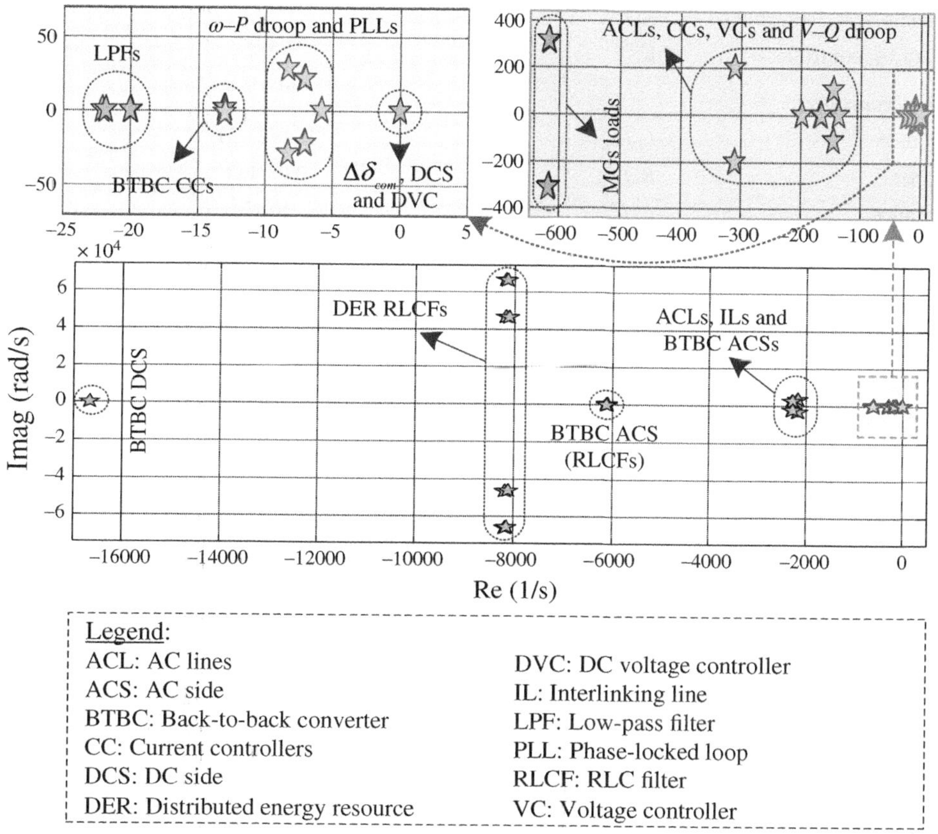

Figure 9.6 All eigenvalues of the studied two BTBC-IMGs in case study 1.

Figure 9.6 shows all eigenvalues of the studied IMGs, where the corresponding modules to each eigenvalue cluster are indicated generally using the participation matrix. The eigenvalues around −5000 real to −8000 real are mostly affected by DER and BTBC RLC filters. The eigenvalues between −2000 real and −3000 real are due to the ac lines, ILs, and BTBC ac sides. More dominant modes can be categorized into three clusters. The cluster near −600 real is due to the MG loads. The

Table 9.3 Power and control information of case studies 1 and 2.

Parameters	Symbol	Value					
General features							
Nominal phase voltage	v_n	326 V					
DC link voltage	v_{dc}	780 V					
Nominal frequency	ω_n	100π rad/s					
Back-to-back converters							
DC capacitor	C_{dc}	600 μF					
DC resistance	R_{dc}	0.1 Ω					
CC proportional gain	K_p^{CC}	20					
CC integral gain	K_p^{CC}	300					
Proportional gain of V_{dc} controller	K_p^{DVC}	5					
Integral gain of V_{dc} controller	K_i^{DVC}	1787					
RLC filter for all converters							
Series inductance	L_f/L_{fc}	1.08/2 mH					
Series resistance	R_f/R_{fc}	0.05/1 Ω					
Shunt capacitance	C_f/C_{fc}	10 μF					
Microgrids							
MG rated power	S_{MG}	3354 VA					
Microgrid number		1		2		3	
Active load (W)		1500		1800		900	
Reactive load (VA)		750		900		450	
DERs		1	2	1	2	1	2
Line ind. (mH)	L_{li}	10.6	5	5	2.5	8.5	4.2
Line res. (Ω)	R_{li}	1.6	0.8	1.2	0.6	1.5	0.7
$\omega - P$ droop gain (rad/kW·s)	m_p	1.6	0.8	1.6	0.8	1.6	0.8
$V - Q$ droop Gain (10 V/kVAr)	n_q	6.5	3.2	6.5	3.2	6.5	3.2
				K_P		K_I	
PI voltage controller				0.05		20	
PI current controller				30		500	

eigenvalues cluster between −100 real and −350 real generally indicates the dynamic participation of the ac lines, DER current controllers, DER voltage controllers, and $V - Q$ droop characteristics. The DCMs (third cluster) are related to the LPFs, BTBC current controllers, phase-locked loops (PLLs), $\omega - P$ droop characteristics, BTBC dc side, dc voltage controller, and $\Delta\delta_{com}$. The eigenvalue results show the remarkable BTBC influence on the IMG stability. In fact, most of the critical eigenvalues are in danger of instability due to the BTBC parameter changes. Complementary studies can be done using sensitivity analysis to show the most effective components in the small-signal stability of BTBC-IMGs.

Table 9.4 Operating point information of the BTBC of the studied two BTBC-IMGs obtained from a time-domain simulation in MATLAB/SimPowerSystem.

Parameter (unit)	Value	Parameter (unit)	Value	Parameter (unit)	Value	Parameter (unit)	Value
v_{dc10} (V)	780	$v_{fc,d10}$ (V)	317.5	$i_{fc,d10}$ (A)	−1.8	m_{d10}	0.82
i_{c10} (A)	1.1	$v_{fc,q10}$ (V)	5	$i_{fc,q10}$ (A)	0	m_{q10}	−0.01
ω_{B10} (rad/s)	313.7	$E_{c,d10}$ (V)	317.5	$v_{fc,d10}$ (V)	320	i_{c10} (A)	−1.11
ω_{B20} (rad/s)	314.2	$E_{c,q10}$ (V)	6	$v_{fc,q10}$ (V)	−2.5		
$i_{fc,d10}$ (A)	1.8	m_{d10}	0.81	$E_{c,d10}$ (V)	320		
$i_{fc,q10}$ (A)	0	m_{q10}	0.015	$E_{c,q10}$ (V)	−4		

9.4 Sensitivity Analysis

9.4.1 CB-IMGs

Figures 9.7 and 9.8 show the sensitivity analysis of the most dominant parameters of the CB-IMGs, i.e. primary and secondary controller parameters. In Figures 9.7(a) and 9.7(b), the $\omega - P$ droop characteristic gains of MG_1's DERs and MG_2's DERs are modified within $[0.1\ 5]\% \cdot (\omega_n/P_{max}^{DER})$, respectively. The most dominant modes affected by changing $\omega - P$ droop gain K_p are shown by λ_1 and λ_2. λ_1 is very close to the imaginary axis for a very low value of K_p. The modes change to an oscillatory mode by increasing K_p and will decrease the stability margin. For the same criteria for changing the K_p gain, $K_p^{MG_1}$ increase does not cause instability; however $K_p^{MG_2}$ increase leads

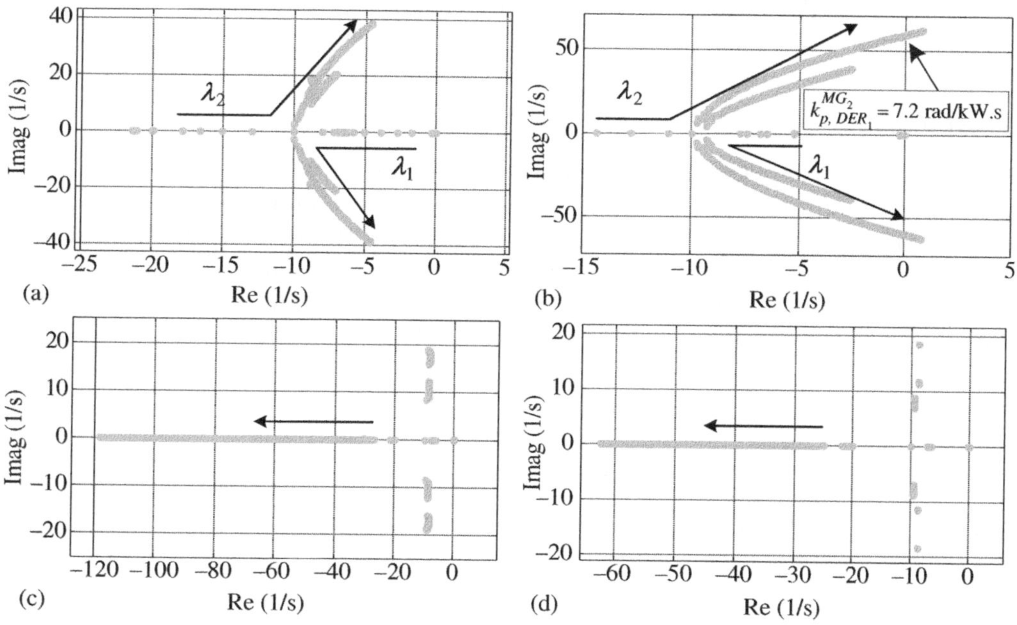

Figure 9.7 Sensitivity analysis output for the primary controller parameters: (a) $K_p^{MG_1} \in [0.1\ 5]\% \cdot (\omega_n/P_{max}^{DER})$, (b) $K_p^{MG_2} \in [0.1\ 5]\% \cdot (\omega_n/P_{max}^{DER})$, (c) $K_q^{MG_1} \in [1\ 15]\% \cdot (v_n/Q_{max}^{DER})$, and (d) $K_q^{MG_2} \in [1\ 15]\% \cdot (v_n/Q_{max}^{DER})$.

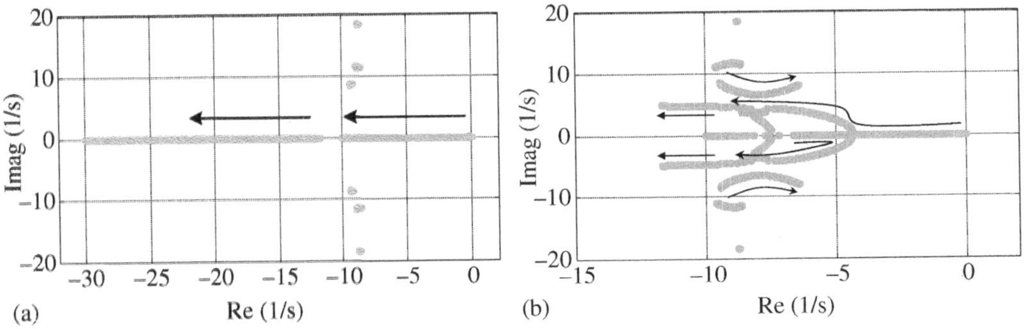

Figure 9.8 Sensitivity analysis output for the secondary controller parameters: (a) leader DERs $T_{sc}^{MG_1} \in [0.1\ 10]$ s, and (b) follower DERs $\tau_{sc}^{MG_1} \in [0.1\ 10]$ s.

to instability for $K_{p,DER1}^{MG_2} > 7.2$ rad/kW · s. In Figures 9.7(c) and 9.7(d), the $v - Q$ droop characteristic gains of MG_1's DERs and MG_2's DERs are changed within [1 15]% · (v_n/Q_{max}^{DER}), respectively. Increasing the K_q for both MGs results in the stability margin enhancement. Note that the nearest eigenvalues to the imaginary axis are not affected by the K_q gain modifications. Moreover, there exist small changes in the oscillatory modes, especially in Figure 9.7(d) for changes in the MG_2 $v - Q$ droop gains. However, in Figure 9.7(c), the frequency of the dominant oscillatory modes increases to 4 Hz.

According to Figures 9.8(a) and 9.8(b), when both the secondary controller parameters T_{sc} and τ_{sc} increase, the most dominant eigenvalue moves from the very close position to the imaginary axis to the left side, where the stability margin is improved. Other dominant real modes in Figure 9.8(a) have the same behavior for the T_{sc} change. Nevertheless, the dominant oscillatory modes have negligible movement for the T_{sc} change. On the other hand, the dominant oscillatory modes have considerable changes in their real value for modifying the τ_{sc} as shown in Figure 9.8(b).

9.4.2 BTBC-IMGs

9.4.2.1 DC Side and Voltage Controller of Back-to-Back Converter

Table 9.5 shows the participation factors for the dc side and its controller in case study 1. The nondominant mode is only impacted by the dc side parameters. However, the DCMs are related to both the power and control parts of the dc link. Figure 9.9 indicates the trajectory of the $\lambda_{70,71}$ for changing the dc voltage controller parameters as $0.1 < K_p^{DVC} < 200$ and $1 < K_i^{DVC} < 4000$. As a general result, increasing the K_p^{DVC} results in a low stability improvement, and an increase in K_i^{DVC}

Table 9.5 Participation matrix for BTBC dc side.

	Dynamic modes		
State variable	$\lambda_{23} = -16667$	$\lambda_{70} = j1.05$	$\lambda_{71} = -j1.05$
ΔX_{dc1}	0.5	0.249	0.249
ΔX_{dc2}	0.5	0.249	0.249
ΔX_{DVC}	0	0.499	0.499
Other	0	<0.003	<0.003

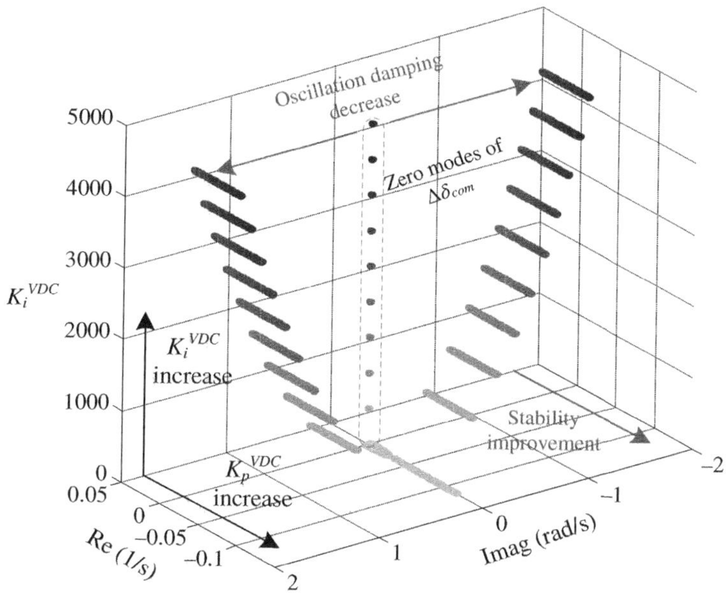

Figure 9.9 Sensitivity analysis output for the BTBC voltage controller parameters: $0.1 < K_p^{DVC} < 200$, $1 < K_i^{DVC} < 4000$, and $R_{Ldc} = 0$.

leads to reduced oscillation damping. In order to maintain stability, the K_p^{DVC} must be chosen larger in relation to a larger K_i^{DVC}. For instance, in the case of $K_i^{DVC} = 1778$, the dominant frequency modes are stable just for $K_p^{DVC} > 7.7$. Note that the DCMs are not very sensitive to the dc side elements. However, all resistive elements, e.g. R_{Ldc} improve the stability.

9.4.2.2 PLLs of Back-to-Back Converter

Figure 9.10 shows the loci of BTBC PLL-affected eigenvalues for changing parameters as $0.01 < K_p^{PLL} < 1$ and $0.1 < K_i^{PLL} < 5$ in case study 1. In accordance with the two oscillatory modes, one can conclude: (i) when the K_p^{PLL} increases, the stability boundary improves. However, it decreases

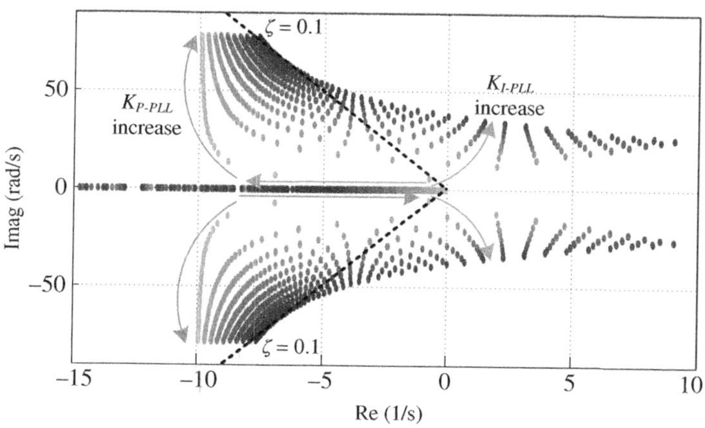

Figure 9.10 Sensitivity analysis output for phase-locked loop parameters: $0.01 < K_p^{PLL} < 1$ and $0.1 < K_i^{PLL} < 5$.

the oscillation damping for all values of K_i^{PLL}. (ii) Very low values of K_p^{PLL} cause instability, and their amounts increase by increasing K_i^{PLL}. This means that the difference between K_p^{PLL} and K_i^{PLL} values should not be too large. (iii) K_i^{PLL} increment causes instability for low K_p^{PLL} values. For the stable nonoscillatory mode, the damping improves by decreasing K_p^{PLL} and/or increasing K_i^{PLL}. In order to satisfy a degree of robust stability and a specified value of the oscillation damping ($\zeta = 0.1$ [18]), acceptable ranges of the PLL parameters can be found, e.g. for case study 1 they are $0.2 \leq K_p^{PLL} \leq 0.7$ and $0.5 \leq K_i^{PLL} \leq 2$.

9.4.2.3 $\omega - P$ Droop Characteristic

The m_p changes by DER rating (S_{DER}) variation at a constant $\Delta\omega_{max}$, which in turn is related to MG rated power (S_{MG}). Figure 9.11 shows the DCM trajectories for 335.4 VA < S_{MG_2} < 33.5 kVA and $\Delta\omega_{max} = 1.57$ rad/s. S_{MG_2} <420 VA cause a severe instability based on $\lambda_{1,2}$, and S_{MG_2} <3051 VA ($m_{p1}^{MG_2} > 1.73$ rad/kW·s or $m_{p2}^{MG_2} > 0.86$ rad/kW·s) causes a very slow instability based on $\lambda_{3,4}$. The droop gain upper limits are already demonstrated for autonomous MGs [27] and CB-IMGs [16, 18]. The loci of $\lambda_{5,6}$ shows that oscillatory DCMs convert to real DSMs by increasing S_{MG_2}. Note that the λ_6 movement decreases the system speed.

9.4.2.4 Cutoff Frequency (ω_c) of LPFs

Figure 9.12 displays the sensitivity analysis output for 1 rad/s < ω_c <50 rad/s, where ω_c is the variable for the LPFs of all DER power controllers and PLLs in one case, and it is constant for the PLLs in another case (case study 1). In the first case, the variable ω_c of PLLs causes instability in a larger interval as ω_c < 5 rad/s. Whereas, the constant ω_c of PLLs leads to the larger stable range as ω_c >1.5 rad/s. Note that the IMG instability is more affected by the LPFs of PLLs. The upper band of ω_c is limited to the required high-frequency rejection.

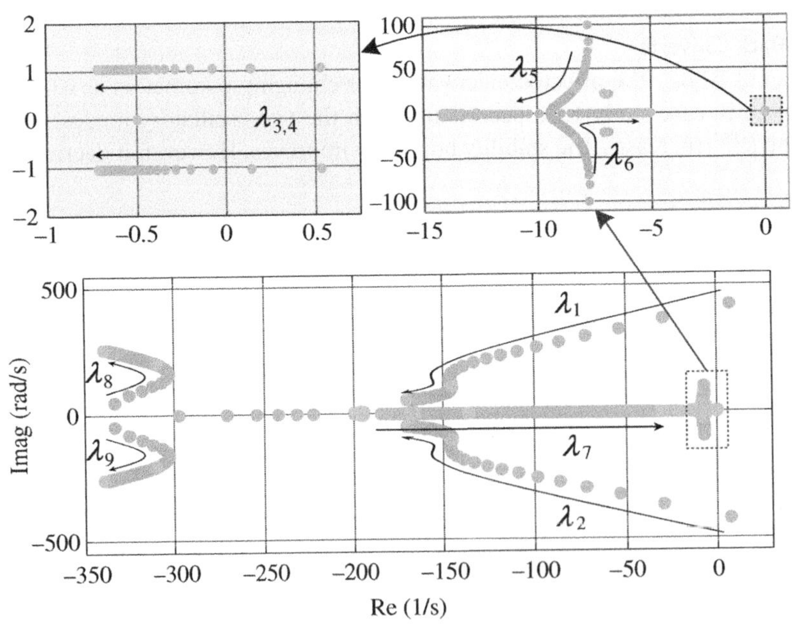

Figure 9.11 Sensitivity analysis output for MG_2 rating as 335.4 VA < S_{MG_2} < 33.5 kVA, $\Delta V_{max} = 32.6$ V and $\Delta\omega_{max} = 1.57$ rad/s.

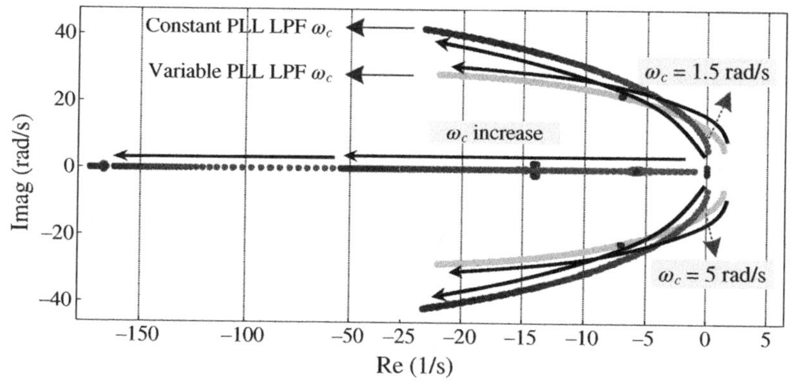

Figure 9.12 Sensitivity analysis output for the ω_c of LPFs (1 rad/s < ω_c < 50 rad/s) in cases of: (i) ω_c is variable for all DER power controller and phase-locked loop LPFs, (ii) ω_c is variable only for DER power controller LPFs.

9.4.2.5 Initial DC Voltage (V_{dc0}^2)

The V_{dc0}^2 should be provided by a supplementary controller, e.g. using a parallel battery with C_{dc2}, which will be bypassed after pre-charging. Figure 9.13 shows the sensitivity analysis outputs for 100 V < V_{dc0}^2 < 800 V in the case study 1. According to Figure 9.13(a), some medium-frequency modes λ_{16-21} improve the stability by increasing the V_{dc0}^2. In addition, improved IMG stability margin can be seen in Figure 9.13(b) in accordance with λ_{10-15} by increasing the V_{dc0}^2. Although the $\lambda_{8,9}$ decrease the system damping. Figure 9.13(c) shows the DCMs λ_{3-7}, which are maintained stable

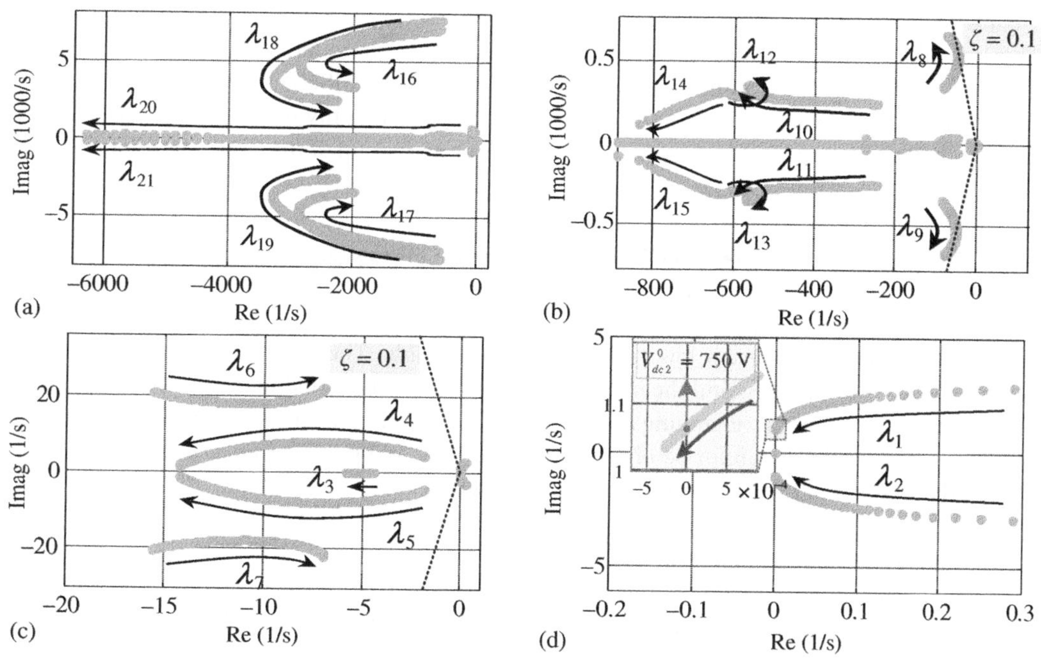

Figure 9.13 Sensitivity analysis output for 100 V < V_{dc2}^0 < 800 V in different frequency ranges: (a) large frequencies, (b) medium frequencies, (c) stable low-frequency modes, and (d) unstable low-frequency modes for V_{dc2}^0 > 750.

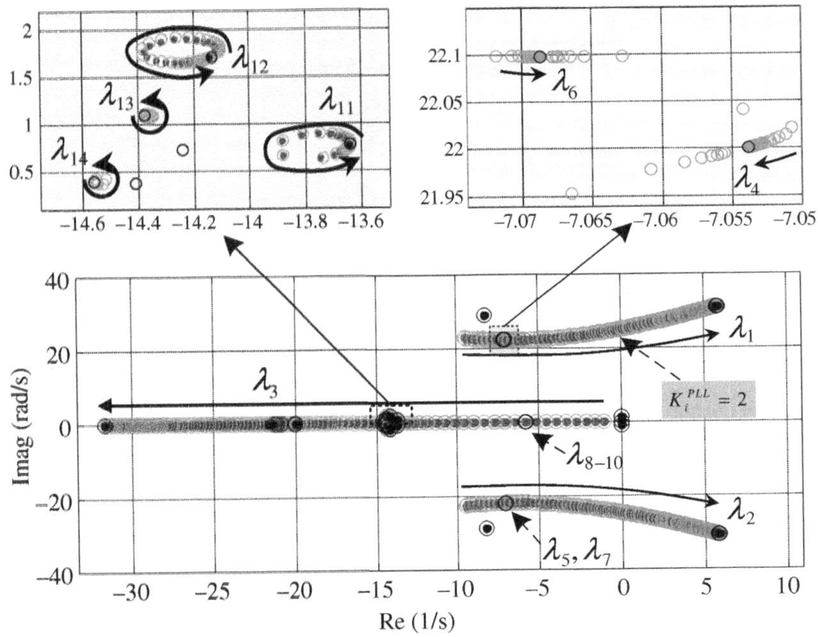

Figure 9.14 Sensitivity analysis output when the $BTBC_{12}$'s K_i^{PLL} varies from 0.1 to 5 for case study 1 (points) and case study 2 (circles).

for all values of the V_{dc0}^2. Nevertheless, the DCMs $\lambda_{1,2}$ can be stable only for $V_{dc0}^2 > 750$ V as shown in Figure 9.13(d).

This instability is due to the dc voltage controller's inability in charging the C_{dc2} before disturbing the ac sides. In fact, the minimum stabilizing dc voltage criterion (MSDVC) is exceeded. It is noteworthy that the sensitivity analysis outputs in Figure 9.13 are not only useful to show the impact of the V_{dc0}^2 but also convenient to generally express the participation of the BTBC dc voltage in the IMG stability.

9.4.2.6 Comparison Between Two and Three Interconnected Microgrids

Figure 9.14 shows the sensitivity analysis results for changing K_i^{PLL} of $BTBC_{12}$ from 0.1 to 5. In both case studies 1 and 2, $\lambda_{1,2}$ behave similarly. In case study 2, $\lambda_{4,5}$ and $\lambda_{6,7}$, related to $BTBC_{13}$ and $BTBC_{23}$, respectively, have a low tendency to be unstable. The λ_{8-10} are independent from changing the K_i^{PLL} completely, and the other, λ_{11-14} and their conjugated modes, have a negligible sensitivity to it. Therefore, by changing an effective parameter of one of the BTBCs on the DCMs, a considerable interaction cannot be observed between the BTBCs themselves and the MGs. In addition, the number of IMGs cannot affect this interaction.

9.4.2.7 Number of Interconnected Microgrids

By increasing the number of IMGs, the structure of the ac passive elements changes, and the number of on-mission control instruments increases. According to the eigenvalue analysis and participation matrix results presented in Figure 9.6, for the case of two IMGs, the ac passive elements, e.g. RLC filters, do not have a considerable impact on the DCMs. In addition, parametric changes of the control modules have approximately the same impact on the different numbers of IMGs. Figure 9.14 shows this fact for case studies 1 and 2. It can be realized simply for more IMGs using the proposed modeling method. As a pleasant result, the dominant modes are not sensitive to the number of IMGs. On the other hand, the increment of on-mission control instruments especially BTBCs leads to an increase in the number of DCMs, which in turn decreases the IMG security.

9.5 Transient Stability of BTBC-IMGs: BTBC DC Voltage

According to the results of eigenvalue analysis, participation matrix, and sensitivity analysis, BTBC dc voltage and its controller are very effective on the IMG small-signal stability. In this section, it is proven that the dc voltage impacts also the transient stability. Here, large power flow changes are considered as large-signal disturbances.

9.5.1 Energy-Based Transient Stability Analysis

During each power exchange, the dc voltage V_{dc}^{j} should be settled on the reference $V_{dc,ref}^{j}$ in order to form acceptable ac voltages on the BTBC's ac sides (see Figure 8.8). Note that tracking the voltage reference should also be in a limited time interval ΔT_{stab} in which the ac voltages are stable. It is due to the weak MGs' voltages and their negative effect on the BTBC performance, which is explained in Section 9.5.3. Otherwise, the BTBC ac voltages will be unstable due to the level difference with MGs voltages (e.g. V_{PCC}^{i} and V_{fc}^{i} for VSC$_i$ in Figure 8.8). In addition, the assumption of $V_{dc0}^{j} < V_{dc,ref}^{j}$ is usually reasonable due to partial capacitor discharges when the BTBC is off. Thus, the minimum required energy W_{dc}^{min} to stabilize the dc voltage before ac voltage instability is given as

$$W_{dc,j}^{\min} = \frac{1}{2} C_{dc}^{j} \left({V_{dc,ref}^{j}}^{2} - {V_{dc0}^{j}}^{2} \right), \tag{9.2}$$

where V_{dc0}^{j} is the initial dc voltage before power flow change, and (9.2) is based on the well-known capacitor energy relationship. According to the dc voltage controller duty, the VSC$_j$ tries to provide W_{dc}^{min} by drawing power from the MG_j in ΔT_{dc}. One can give the power–energy relationship as follows:

$$W_{dc,j} = \int_{\Delta T_{dc}} P_{dc,j}(t) dt, \tag{9.3}$$

where ΔT_{dc} is the dc capacitor charging time interval and $P_{dc,j}$ is the net charging power. As mentioned, in order to preserve stability, the ΔT_{dc} should be limited to the ΔT_{stab}, which is a small time before instability. Therefore, it is easy to recognize $\Delta T_{dc} \ll 1$ and the average required power can be calculated as

$$P_{dc,j} = \frac{W_{dc,j}}{\Delta T_{dc}}. \tag{9.4}$$

By replacing $W_{dc,j}$ with its minimum value $W_{dc,j}^{min}$ and the ΔT_{dc} with its maximum value ΔT_{stab}, (9.4) appears as follows:

$$P_{dc0,j} \geq \frac{W_{dc,j}^{min}}{\Delta T_{stab}}, \tag{9.5}$$

where 0 in subscript indicates the average power required during the ΔT_{dc}. By substituting (9.2) in (9.5), one can find

$$P_{dc0}^{j} \geq \frac{1}{2\Delta T_{stab}} C_{dc}^{j} \left({V_{dc,ref}^{j}}^{2} - {V_{dc0}^{j}}^{2} \right). \tag{9.6}$$

Then, the V_{dc0}^{j} can be achieved as follows:

$$V_{dc0}^{j} \geq \sqrt{{V_{dc,ref}^{j}}^{2} - \frac{2\Delta T_{stab} P_{dc0}^{j}}{C_{dc}^{j}}}. \tag{9.7}$$

This relationship is called minimum stabilizing dc voltage criterion (MSDVC), which shows that the BTBC and generally the IMG system can have a stable operation after enabling the BTBC if V_{dc0}^{j} is greater than or equal to a certain value related to $V_{dc,ref}^{j}$, ΔT_{stab}, C_{dc}^{j}, and P_{dc0}^{j}.

9.5.2 Minimum Stabilizing DC Voltage Criterion

Although the $V_{dc,ref}^{j}$ in MSDVC, (9.7), is a constant value, other correlated components can change the value of V_{dc0}^{j} as discussed later. This section deals with a comparison among MSDVC, critical clearing time (CCT), and power transfer limit (PTL) as common transient assessment tools.

9.5.2.1 Time Interval ΔT_{stab}

The dc-link capacitor voltage has ripples with twice the frequency of the ac side [28]. The V_{dc}^{j} should be stabilized in the first ripples after the event. Otherwise, the ac side voltage disturbance increases and causes instability. In fact, during the stable time interval ΔT_{stab}, the VSC$_j$ output voltage E_{C0}^{j} should increase to be approximately equal to the v_{fc}^{j}, which is equal to the v_{PCC}^{j} before power exchange. In other words, the stability condition is given as

$$E_{C0}^{j}\Big|_{t\leq \Delta T_{stab}} = v_{PCC}^{j}, \tag{9.8}$$

where the E_{C0}^{j} and V_{PCC}^{j} can be obtained as follows:

$$E_{C0}^{j} = \frac{1}{2} m_{0}^{j} V_{dc0}^{j}, \tag{9.9a}$$

$$v_{PCC}^{j} = v_{o}^{m} - \left(R_{li}^{m} + jL_{li}^{m}\omega\right) i_{o}^{m}, \tag{9.9b}$$

where m_{0}^{j} is the input control signal of PWM$_j$ during the ΔT_{stab}. Equation (9.9a) is true for three-phase full-bridge converters, and (9.9b) is obtained according to DER$_m$ in MG_j in steady state (see Figure 8.3). Note that the v_{PCC}^{j} is an acceptable value less than the nominal ac voltage due to the voltage drop of the DER lines' and V–Q droop coefficients.

In stability condition (9.8), m_{0}^{j} is the only variable that can be manipulated to accelerate the increment of E_{C0}^{j} to be stable. m_{0}^{j} is a function of ΔT_{stab}, which can be found easily according to the dc voltage controller block diagram in Figure 8.8. It is obvious that increasing both proportional and integral gains of both current and v_{dc}^{j} controllers can accelerate the increase of E_{C0}^{j}. However, such a gain increase is not practical. Because the gains are limited to the small-signal stability, which is shown in the sensitivity analysis results. Therefore, ΔT_{stab} is hard to be controlled due to the lack of an independent controller. In fact, it is affected by the current and v_{dc}^{j} controllers, which should be limited to other control/stability objectives. Moreover, according to (9.9b), MG voltages and currents are the disturbances for finding ΔT_{stab}. Hence, it is also hard to calculate. Finally, ΔT_{stab} cannot be selected either as transient stability criteria or as an accessible variable to easily find its impact on MSDVC.

9.5.2.2 Capacitance of the C_{dc}^{j}

The capacitance of C_{dc}^{j} is based on the ability to regulate the dc voltage under transient disturbances. Larger values of C_{dc}^{j} lead to more required energy to charge it, but the settled voltage is robust against larger disturbances. Therefore, lower values of C_{dc}^{j} are appropriate in terms of MSDVC. However, the capacitance cannot be selected just based on MSDVC.

9.5.2.3 Injected Power to the DC Link $\left(P_{dc0}^{j}\right)$

According to MSDVC, the larger P_{dc0}^{j} permits the smaller values of V_{dc0}^{j} to stabilize the dc link voltage. In other words, the P_{dc0}^{j} increment results in a larger transient stability margin for BTBC-IMGs. According to the active power balance of BTBC dc and ac sides and neglecting switching losses, P_{dc0}^{j} can be given as follows:

$$P_{dc0}^{j} = P_{dc0}^{line} - P_{ji0} = -\left(P_{dc0}^{i} + P_{ij0}\right) - P_{ji0}, \tag{9.10}$$

where P_{dc0}^{line} is the dc line power, P_{dc0}^{i} is the initial VSC$_i$ capacitor power, and P_{ij0} and P_{ji0} are the VSC$_i$ and VSC$_j$ initial active powers in the ac side, respectively, as shown in Figure 8.8, which can be calculated as follows:

$$P_{ij0} = E_{CO}^{i} i_{ILO}^{ij} = \frac{1}{2} m_0^{i} V_{dc0}^{i} i_{ILO}^{ij}, \tag{9.11a}$$

$$P_{ji0} = E_{CO}^{j} i_{ILO}^{ji} = \frac{1}{2} m_0^{j} V_{dc0}^{j} i_{ILO}^{ji}, \tag{9.11b}$$

where i_{ILO}^{ij} and i_{ILO}^{ji} are the initial three-phase currents of VSC$_i$ and VSC$_j$. Regarding (9.10), (9.11a), and (9.11b), the following points can be extracted.

i) Terms P_{dc0}^{i} and P_{ij0} in (9.10) are as the disturbances in controlling the V_{dc0}^{j}. It is reasonable to consider the same value and initial condition for both the C_{dc}^{i} and C_{dc}^{i} due to the same VSCs and same nominal ac voltages. Therefore, the C_{dc}^{i} needs same initial power, i.e. $P_{dc0}^{i} = P_{dc0}^{j}$. As a result, the C_{dc}^{i} cannot be an appropriate contributor in providing the P_{dc0}^{j}. However, the P_{ij0} can be an effective contributor with a dual behavior according to the power flow direction, which is illustrated in Section 9.5.4.
ii) Another important point is the presence of V_{dc0}^{j} in the P_{ji0} (9.11(b)). Although the m_0^{j} is only controllable signal to increase the V_{dc0}^{j}, in an instability situation, the term V_{dc0}^{j} can resonate instability.
iii) Equations (9.11a) and (9.11b), and also m_i and m_j shown in Figure 8.8, express the nonlinear behavior of the BTBC-IMGs at the beginning of power exchange. Furthermore, they show the complexity and hardness of analytical calculation due to nonlinear relationships and many contributing variables. Hence, the problem of initial BTBC dc voltage in interconnected weak MGs can be listed below the transient stability assessment, which MSDVC is presented for and considered as an appropriate analysis method. Note that enabling the BTBC to flow power between MGs is a large-signal disturbance.

9.5.2.4 MSDVC Comparison with Common Transient Stability Criteria

Generally, the transient stability analysis has been presented for fault occurrence as the common large-signal disturbance. Two common assessment criteria are CCT and PTL. For a fault and given operating condition, CCT is defined as the maximum fault duration, while the power system is still transiently stable [29]. PTL is the maximum power flow through an interface device, e.g. a BTBC, while the power system is still transiently stable. CCT is usually used for ac power flow through ac lines. However, the transient stability of dc power exchange through HVDC/BTBC is analyzed by PTL [30]. According to the third point of Section 9.5.2.3 and definitions of CCT and PTL, it is easy to find their different applications with respect to MSDVC. CCT and PTL have been used to assess the transient stability after a fault occurrence. However, MSDVC is useful to assess the transient stability after a BTBC connection to exchange power.

A relationship can exist between PTL and MSDVC. In the case of MSDVC application, the initial dc voltage of BTBC is calculated for a given acceptable level of power exchange (P^{ij}_{ref}), which is realized by P_{ij0} in (9.10) and then (9.7). One can adapt the initial PTL P_{ij0} based on a given V^{j}_{dc0}, which in turn, changes the P^{ij}_{ref}. In this case, the initial PTL will be the transient stability criterion instead of MSDVC.

In (9.10), P_{ij0} is the initial power exchange to find the initial PTL, and P_{ji0} is a combined impact of the V^{j}_{dc} controller output (u_{dc}) and $-P^{ij}_{ref}$, which is applied by m^{j}_{0}. One can find m^{j}_{0} according to the dc voltage controller block diagram shown in Figure 8.8 and calculate P_{ji0} through (9.11b). Regarding several multiplications, it is obvious that P_{ji0} is a nonlinear function of the current controller inputs u_{dc} and $-P^{ij}_{ref}$ as follows:

$$P_{ji0} = f_{NL}\left(u_{dc} - P^{ij}_{ref}\right), \tag{9.12}$$

where *NL* in the subscript of *f* indicates its nonlinearity. By substituting (9.10) and (9.12) in (9.6) and rearranging, the initial PTL can be achieved as

$$P_{ij0} \leq \frac{1}{2\Delta T_{stab}} C^{j}_{dc}\left(V^{j\,2}_{dc,ref} - V^{j\,2}_{dc0}\right) + P^{i}_{dc0} - f_{NL}\left(u_{dc} - P^{ij}_{ref}\right). \tag{9.13}$$

This relation proves an initial PTL for which the BTBC-IMGs remain transiently stable for a given V^{j}_{dc0}.

Note that both MSDVC and initial PTL try to maintain transient stability during the BTBC connection. In the case of using MSDVC, the BTBC can be connected with each given power exchange just by pre-charging the dc capacitor to the required dc voltage and then enabling the BTBC to exchange power. Whereas, by using initial PTL, the initial BTBC power reference is first set to be acceptable based on (9.13) for the given initial dc voltage. After the capacitor is full charged the reference can be modified to the desired value.

9.5.3 Grid Strength Impact

It is noteworthy that there is a considerable difference in the case of P^{j}_{dc0} value between weak VSC-based IMGs and strong conventional power grids. After connecting the VSC_j to the MG_j in order to exchange power, the charging power begins to flow from the MG_j into the VSC_j. According to Figure 8.8, this power can be calculated as

$$S^{j}_{ac0} = \left| v^{j}_{fc,0}\left(\frac{v^{j}_{fc,0} - E^{j}_{CO}}{R^{j}_{fc} + jX^{j}_{fc}}\right)^{*}\right|. \tag{9.14}$$

In the most severe case, i.e. $V^{j}_{dc0} = 0$, the S^{j}_{ac0} equals the short-circuit level of the power system at the point of common coupling (PCC) ($S^{pcc}_{SC,j}$). Nevertheless, it equals the MG_j rating (S^{j}_{n}) for weak VSC-based IMGs due to the current-limiting function employed on the DER current controllers. Since $S^{pcc}_{SC,j} \gg S^{j}_{n}$, the stabilizing initial dc voltage for conventional power systems can be much smaller than that of VSC-based IMGs according to the MSDVC. As an important result, the transient stability of initial dc voltage for conventional power systems is less challenging than weak VSC-based IMGs.

Table 9.6 Back-to-back converter power exchange impact on MSDVC.

Situation	P^j_{dc0}	V^j_{dc0}	Stability margin
Power flow from MG_j to MG_i	$S^j_n - S^i_{dc0}$	Increase	Low
Power flow from MG_i to MG_j	$S^j_n + S^i_{dc0}$	Decrease	High
dc voltage stabilization before power exchange	S^j_n	Base case	Medium

9.5.4 BTBC Power Flow Direction

Another important point is the impact of the VSC_i power direction on the MSDVC. Table 9.6 shows different situations for the BTBC power flow and their impacts on the MSDVC. In the worst case, where the power should be sent from the MG_j to the MG_i, the net injected power to the dc link is decreased to $S^j_n - S^i_{dc0}$. Thus, the stability margin of the V^j_{dc0} is decreased based on the MSDVC. In other words, the V^j_{dc0} should be more close to the $V^j_{dc,ref}$ to be stabilized after the transient. In contrast, the net injected power to the dc link is increased to $S^j_n + S^i_{dc0}$ in the case of power transfer from the MG_i to the MG_j. Hence, the stability margin is increased. A third case in Table 9.6 is related to enabling the VSC_j before the power exchange. This BTBC operation strategy leads to a medium stability margin in comparison with previous cases.

The transient stability analysis presented in the form of MSDVC is a necessary primary step for designing the dc voltage controller of BTBCs in weak BTBC-IMGs considering the initial dc voltage challenge. The designer can focus on each participant variable, including ΔT_{dc}, C^j_{dc}, and P^j_{dc0}, to improve the stability. As a simple solution, the dc voltage can be stabilized by enabling the V_{dc} controller before power exchange. Another solution is to design a robust dc voltage controller, which can be stable for a range of initial dc voltages as an uncertainty.

9.5.5 Time-Domain Simulations

In order to analyze the large-signal stability of BTBC-IMGs, the two case studies presented in Figure 9.5 consisting of two interconnected MGs through one BTBC (case study 1), and three interconnected MGs through three BTBCs (case study 2) are used. Each MG structure is considered as shown in Figure 8.3 without GFL DERs. The simulation data has previously been presented in Table 9.3. Note that real-time simulation results obtained from the OPAL-RT simulator shown in Figure 8.15 are presented in this section.

9.5.5.1 Frequency Instability of BTBC-IMGs

Figure 9.15 shows some outputs of case study 1 of BTBC-IMGs shown in Figure 9.5(a), where the MGs are isolated until $t = 0.5$ s. They are connected hereafter with $K_i^{PLL2} = 0.5$ to exchange 800 W from MG_2 to MG_1, which results in a stable operation. The K_i^{PLL2} is increased to 2.5 at $t = 2$ s. According to Figure 9.15(a), DER and PLL frequencies indicate a slow instability due to exciting correlated DCMs to the $BTBC_{12}$ PLL shown in Figure 9.10. The instability can be observed in the other measured frequencies with a smaller amplitude, although its amplitude is larger in the $f_{PLL}^{MG_2}$ itself. The impact of frequency instability on the dc link voltage can be seen in Figure 9.15(b), which in turn, spreads the instability to MG_1. Figures 9.15(c) and 9.15(d) show MG_2 support for MG_1 until $t = 2$ s, then the instability appears. As a result, the obtained

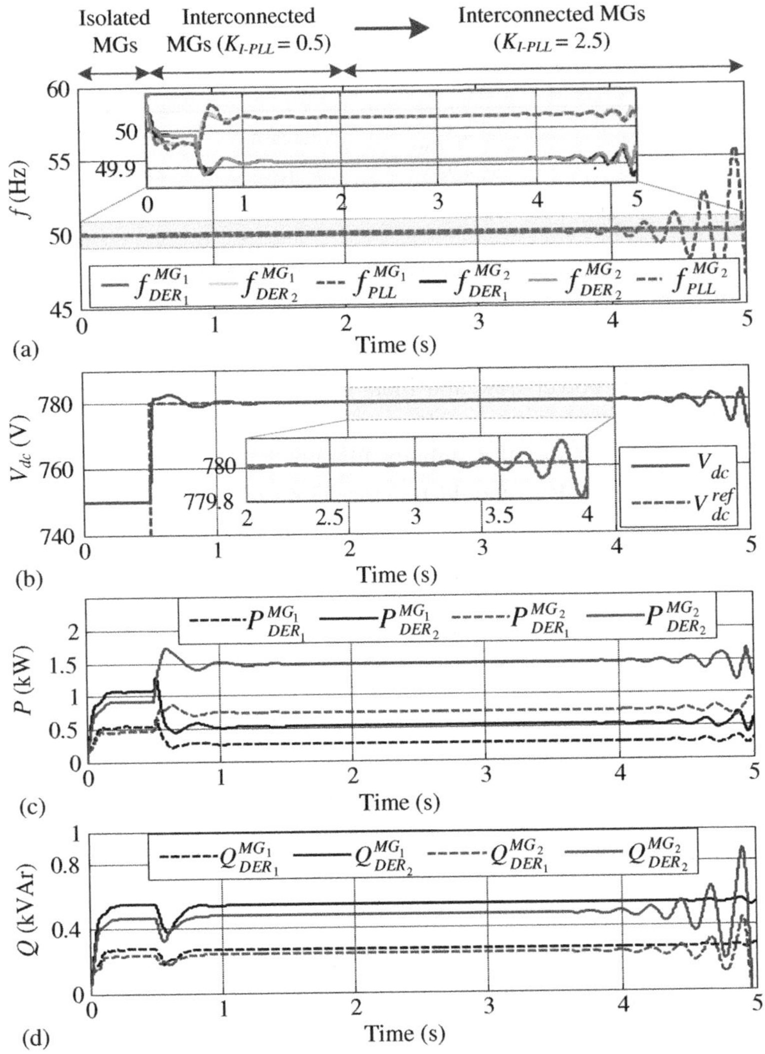

Figure 9.15 Frequency instability in case study 1: (a) DER and phase-locked loop frequencies, (b) $BTBC_{12}$ dc voltage, (c) DER's active powers, and (d) DER's reactive powers.

ranges for the parameters, e.g. K_i^{PLL2} presented in Section 9.4 are necessary to maintain IMG stability.

9.5.5.2 Voltage Instability of BTBC-IMGs

In this scenario, the MGs considered as the case study 2 in Figure 9.5(b) are isolated at the beginning. The MG_1 load increases at $t = 0.5$ s from 60% to 100% of the rated MG_1 load. MG_2 sends 600 W and 300 VAr to MG_1 at $t = 1$ s via $BTBC_{12}$ while $v_{dc0}^{12} = 750$ V. In response, MG_3 sends 600 W and 300 VAr to MG_1 at $t = 2$ s via $BTBC_{13}$ in order to fully compensate the MG_1 overload, but with $v_{dc0}^{13} = 700$ V. Since v_{dc0}^{13} is lower than the threshold of 750 V indicated in Figure 9.13(d), the v_{dc}^{13} cannot track the reference and goes to a very low value as indicated in Figure 9.16(a). The impact of the dc voltage instability on the active and reactive powers of DERs can be observed in Figures 9.16(b)

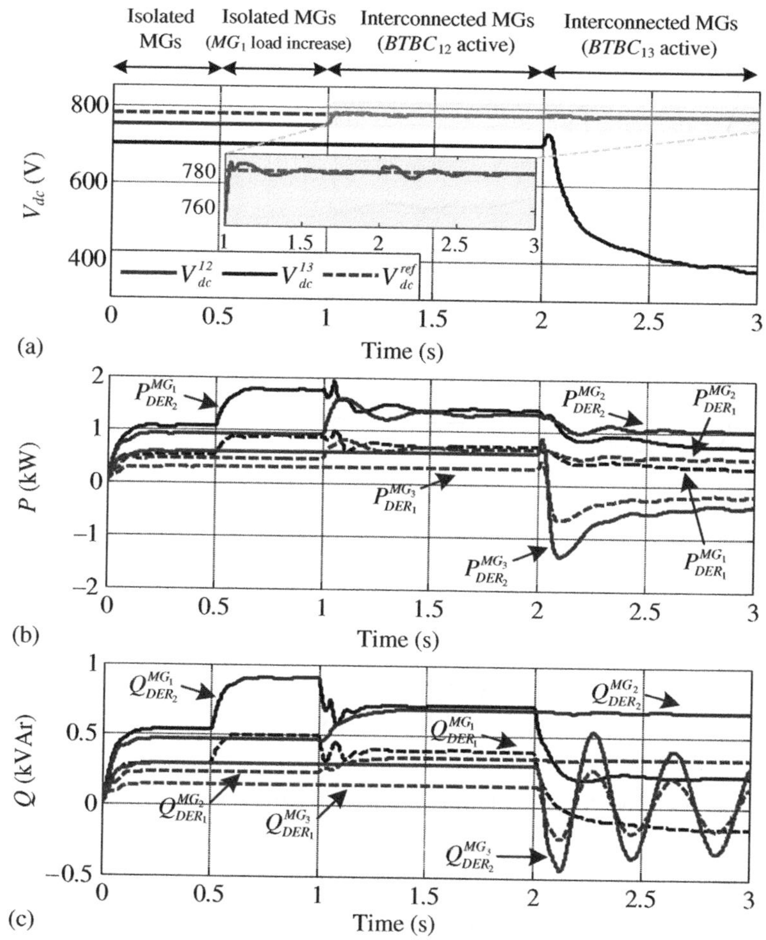

Figure 9.16 Voltage instability in case study 2: (a) dc voltage of back-to-back converters, (b) active powers of DERs, and (c) reactive powers of DERs.

and 9.16(c). Such transient voltage instability indicates the necessity of preserving the MSDVC for BTBC-IMGs.

9.5.5.3 Power Flow Direction

As mentioned in Section 9.5.4, the initial dc voltage stability or MSDVC correlates with the BTBC power flow direction. In this scenario, the same power $P_{12}^{ref} = 855$ W is considered to flow from MG_2 to MG_1 in Situation 1 and from MG_1 to MG_2 in Situation 2. In order to calculate the V_{dc0} stability margin, it is assumed that $R_{dc} = 0\ \Omega$ and $\Delta T_{dc} = 0.01$ s. Then, employing (9.7) and using information in Tables 9.3 and 9.4, $V_{dc0}^{S1} \geq 752.9$ V for Situation 1 and $V_{dc0}^{S2} \geq 733.7$ V for Situation 2. Therefore, when the power flows from MG_2 to MG_1 as 855 W, the MSDVC equals 752.9 V, while it is decreased to 733.7 V in Situation 2. For example, $V_{dc0} = 740$ V can stabilize the two IMGs in Situation 2 (see Figures 9.17(a) and 9.17(c)), however it cannot stabilize them in Situation 1 (see Figures 9.17(a) and 9.17(b)). In Figure 9.17, the two MGs are isolated until $t = 1$ s. They are interconnected in both Situations 1 and 2 at $t = 1$ s. The transient stability is satisfied for Situation 2 due to a stabilizing $V_{dc0}^{S2} = 740$ V, whereas it cannot be achieved in Situation 1 due to MSDVC

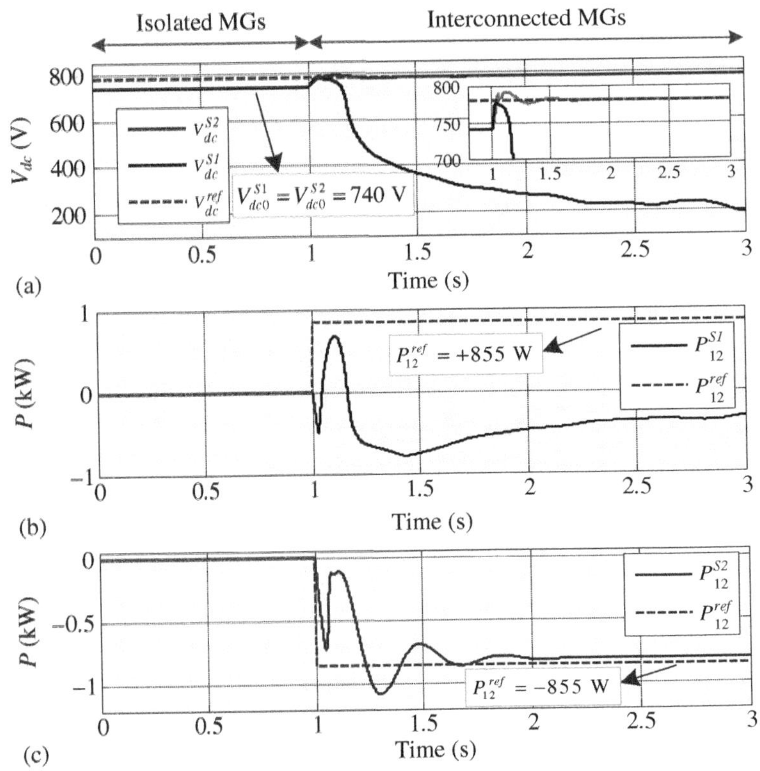

Figure 9.17 Impact of the back-to-back converter power flow direction on the transient dc voltage stability: (a) $BTBC_{12}$ dc voltage, (b) Situation 1: power flow from MG_2 to MG_1, and (c) Situation 2: power flow from MG_1 to MG_2.

violation for the $V_{dc0}^{S1} = 740$ V. Therefore, the worst power direction must be considered for calculating MSDVC in order to stabilize IMGs under each power flow direction.

9.5.5.4 Pre-charging Before Power Flow

As mentioned in Section 9.5.4, pre-charging the dc capacitor before power flow (base case in Table 9.6) can be as a simple solution for transient stability improvement in the case of enabling BTBC. Similar to Section 9.5.5.3, the same power $P_{12}^{ref} = 855$ W is considered to be flown from MG_2 to MG_1 (Situation 1) and from MG_1 to MG_2 (Situation 2) with the same $V_{dc0} = 740$ V. In contrast, the dc capacitor is charged before enabling power flow in 1 to 2 s for both situations. Then, power is flowing in 2 to 3 s. Figure 9.18 shows the transient stability for both situations in comparison to Figure 9.17, in which the dc capacitor is not initially charged. Therefore, pre-charging the dc capacitor has a considerable improvement in the transient stability. Another important observation is decoupling the dynamics of the dc capacitor charge and the power flow. In both Figures 9.18(a) and 9.18(b), the same dynamics can be seen for the capacitor charging in Situations 1 and 2 during 1 to 2 s and for the power flow symmetrically during 2 to 3 s.

9.5.5.5 Initial Power Transfer Limit

Here, the initial PTL (IPTL) is considered as the transient stability criterion instead of MSDVC. In all scenarios, which applied as power exchange from MG_2 to MG_1 in case study 1, the MGs are isolated before $t = 1$ s, and then they are interconnected with different V_{dc0} and P_{ref}^{ij}. For $V_{dc0} = 730$ V, $IPTL_1$

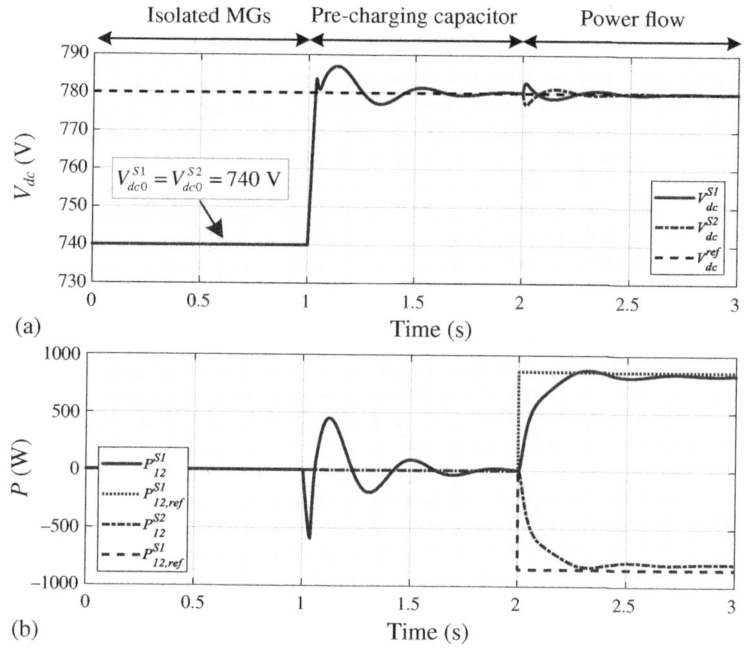

Figure 9.18 Pre-charging the BTBC dc capacitor (case study 1): (a) $BTBC_{12}$ dc voltage, (b) active power flow for both flow directions from MG_2 to MG_1 (Situation 1), and from MG_1 to MG_2 (Situation 2).

is obtained as 370 W by increasing P_{ref}^{ij} and checking for stability. In the first scenario, $P_{ref}^{ij} = 114$ W, which is less than $IPTL_1$. Hence, the dc voltage is transiently stable (Figure 9.19(a)), and the active power is settled on the reference after the oscillations (Figure 9.19(b)). In the second scenario, $P_{ref}^{ij} = 399$ W, which is a little larger than $IPTL_1$. Therefore, the dc voltage will be unstable and the power cannot be stably exchanged. For the same power reference, V_{dc0} is increased to 740 V in the third scenario. The corresponding $IPTL_2$ to this new V_{dc0} is found as 712 W. Since $P_{ref}^{ij} = 399$ W is less than $IPTL_2$, the stability is preserved in this scenario. Finally, in order to validate $IPTL_2$, $P_{ref}^{ij} = 798\text{ W} > IPTL_2$ is considered. The instability can be seen in Figure 9.19. As a result, IPTL can be another criterion of BTBC enabling transient stability, which has a close relation with the initial dc voltage or MSDVC.

9.6 Summary

This chapter presents an eigenvalue analysis for all CB-IMGs and BTBC-IMGs modeled in Chapter 8. The eigenvalues are spread in a wide range of real and imaginary values and can be classified into three or four groups, including critical dominant, dominant, non-dominant, and nonimportant groups. Using participation factors, one can obtain the participation of each state variable or each module on each eigenvalue or, generally, on IMG stability. Participation factors can show the meaningful relationship between IMG modules and their stability only for a set of parameters. For example, inter-MG oscillatory modes of CB-IMGs were studied using a participation matrix and the simplified model of CB-IMGs introduced in Chapter 8. However, each IMG includes many parameters, and each one can fluctuate within a feasible range and make

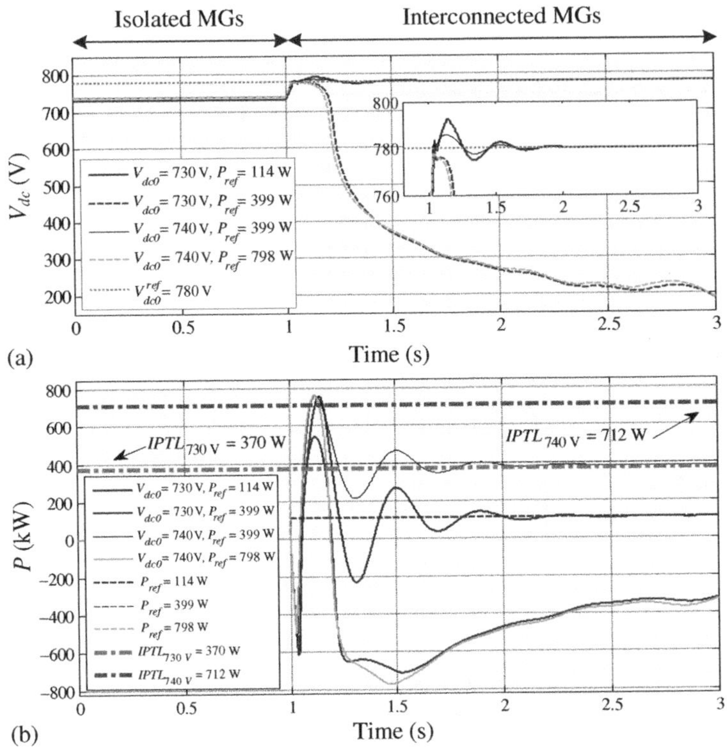

Figure 9.19 Initial power transfer limit (IPTL) for two initial dc voltages as 730 and 740 V (case study 1): (a) $BTBC_{12}$ dc voltage and (b) active power flow.

considerable changes in IMG stability. Therefore, several sensitivity analyses have been done for the most important parameters.

In CB-IMGs, primary and secondary control parameters were analyzed, where a considerable impact on the critical dominant and dominant eigenvalues was observed. Primary control parameters, including droop gains and low-pass filter cutoff frequency, are also effective on BTBC-IMG stability. Moreover, stability is highly sensitive to most BTBC parameters, such as the dc link voltage controller and PLL parameters. The initial dc voltage value of BTBCs is a critical issue in low-voltage and weak IMGs, which is taken into account in this chapter through sensitivity analysis of the small-signal model and large-signal stability assessment using time-domain real-time simulations. Grid weakness and power flow direction have a considerable impact on the initial dc link voltage and, in turn, IMG instability. Therefore, pre-charging of the dc capacitor or limiting the initial power transfer is necessary before connecting MGs via BTBCs and power flow.

References

1 Md Jahangir Hossain, Md Apel Mahmud, Federico Milano, Seddik Bacha, and Ahmad Hably. Design of robust distributed control for interconnected microgrids. *IEEE Trans. Smart Grid*, 7(6):2724–2735, 2016.

2 Hassan Bevrani, Bruno François, and Toshifumi Ise. *Microgrid dynamics and control*. John Wiley & Sons, Hoboken, NJ, 2017.

3 Ehsan Pashajavid, Arindam Ghosh, and Firuz Zare. A multimode supervisory control scheme for coupling remote droop-regulated microgrids. *IEEE Trans. Smart Grid*, 9(5):5381–5392, 2018.

4 Seyed Ali Arefifar, A-R I Mohamed Yasser, and Tarek H M El-Fouly. Optimum microgrid design for enhancing reliability and supply-security. *IEEE Trans. Smart Grid*, 4(3):1567–1575, 2013.

5 Inam Ullah Nutkani, Poh Chiang Loh, Peng Wang, Tseng King Jet, and Frede Blaabjerg. Inter-tied AC–AC microgrids with autonomous power import and export. *Int. J. Electr. Power Energy Syst.*, 65:385–393, 2015.

6 Ramon Zamora and Anurag K Srivastava. Multi-layer architecture for voltage and frequency control in networked microgrids. *IEEE Trans. Smart Grid*, 9(3):2076–2085, 2018.

7 Julius Susanto, Farhad Shahnia, Arindam Ghosh, and Sumehda Rajakaruna. Interconnected microgrids via back-to-back converters for dynamic frequency support. In *IEEE Power Engineering Conference (AUPEC)*, pages 1–6, September 2014.

8 Hyeong-Jun Yoo, Thai-Thanh Nguyen, and Hak-Man Kim. Multi-frequency control in a stand-alone multi-microgrid system using a back-to-back converter. *Energies*, 10(6):1–18, 2017.

9 Nikolas Flourentzou, Vassilios G Agelidis, and Georgios D Demetriades. VSC-based HVDC power transmission systems: An overview. *IEEE Trans. Power Electron.*, 24(3):592–602, 2009.

10 Amirnaser Yazdani and Reza Iravani. *Voltage-sourced converters in power systems*, volume 34. Wiley Online Library, Hoboken, NJ, 2010.

11 Joseph H Eto, John Undrill, Peter Mackin, Ron Daschmans, Ben Williams, Brian Haney, Randall Hunt, Jeff Ellis, Howardand Illian, Carlos Martinez, et al. Use of frequency response metrics to assess the planning and operating requirements for reliable integration of variable renewable generation. *Lawrence Berkeley National Laboratory*, 2011.

12 Kevin P Schneider, Nikitha Radhakrishnan, Yingying Tang, Francis K Tuffner, Chen-Ching Liu, Jing Xie, and Dan Ton. Improving primary frequency response to support networked microgrid operations. *IEEE Trans. Power Syst.*, 34(1):659–667, 2019.

13 Farhad Shahnia and Ali Arefi. Eigenanalysis-based small signal stability of the system of coupled sustainable microgrids. *Int. J. Electr. Power Energy Syst.*, 91:42–60, 2017.

14 Farhad Shahnia. Stability and eigenanalysis of a sustainable remote area microgrid with a transforming structure. *Sustainable Energy Grids Netw.*, 8:37–50, 2016.

15 Yun Zhang and Le Xie. Online dynamic security assessment of microgrid interconnections in smart distribution systems. *IEEE Trans. Power Syst.*, 30(6):3246–3254, 2014.

16 Iraklis P Nikolakakos, Hatem H Zeineldin, Mohamed Shawky El-Moursi, and Nikos D Hatziargyriou. Stability evaluation of interconnected multi-inverter microgrids through critical clusters. *IEEE Trans. Power Syst.*, 31(4):3060–3072, 2016.

17 Yun Zhang, Le Xie, and Qifeng Ding. Interactive control of coupled microgrids for guaranteed system-wide small signal stability. *IEEE Trans. Smart Grid*, 7(2):1088–1096, 2016.

18 Zhuoli Zhao, Ping Yang, Yuewu Wang, Zhirong Xu, and Josep M Guerrero. Dynamic characteristics analysis and stabilization of PV-based multiple microgrid clusters. *IEEE Trans. Smart Grid*, 10(1):805–818, 2017.

19 Xiaoyu Wu, Yin Xu, Xiangyu Wu, Jinghan He, Josep M Guerrero, Chen-Ching Liu, Kevin P Schneider, and Dan T Ton. A two-layer distributed control method for islanded networked microgrid systems. *IEEE Trans. Smart Grid*, 11(2):942–957, 2019.

20 Mohammad S Golsorkhi, David John Hill, and Hamid Reza Karshenas. Distributed voltage control and power management of networked microgrids. *IEEE J. Emerging Sel. Top. Power Electron.*, 6(4):1892–1902, 2018.

21 Ritwik Majumder and Gargi Bag. Parallel operation of converter interfaced multiple microgrids. *Int. J. Electr. Power Energy Syst.*, 55:486–496, 2014.

22 Mojtaba Khederzadeh, Hamed Maleki, and Vahid Asgharian. Frequency control improvement of two adjacent microgrids in autonomous mode using back to back voltage-sourced converters. *Int. J. Electr. Power Energy Syst.*, 74:126–133, 2016.

23 Mobin Naderi, Yousef Khayat, Rasool Heydari, Mateja Novak, Qobad Shafiee, Tomislav Dragicevic, Hassan Bevrani, and Frede Blaabjerg. Model validation of power electronics-based networked micro-grids by prony analysis. In *2019 21st European Conference on Power Electronics and Applications (EPE '19 ECCE Europe)*, pages 1–10. IEEE, 2019.

24 Mobin Naderi, Yousef Khayat, Qobad Shafiee, Tomislav Dragicevic, Frede Blaabjerg, and Hassan Bevrani. An emergency active and reactive power exchange solution for interconnected microgrids. *IEEE J. Emerging Sel. Top. Power Electron.*, 9(5):5206–5218, 2019.

25 Mobin Naderi, Yousef Khayat, Qobad Shafiee, Tomislav Dragičević, Hassan Bevrani, and Frede Blaabjerg. Interconnected autonomous AC microgrids via back-to-back converters-Part II: Stability analysis. *IEEE Trans. Power Electron.*, 35(11):11801–11812, 2020.

26 Mobin Naderi, Yousef Khayat, Qobad Shafiee, Tomislav Dragicevic, Hassan Bevrani, and Frede Blaabjerg. Interconnected autonomous AC microgrids via back-to-back converters—Part I: Small-signal modeling. *IEEE Trans. Power Electron.*, 35(5):4728–4740, 2019.

27 Nagaraju Pogaku, Milan Prodanovic, and Timothy C Green. Modeling, analysis and testing of autonomous operation of an inverter-based microgrid. *IEEE Trans. Power Electron.*, 22(2):613–625, 2007.

28 Mahesh K Mishra and K Karthikeyan. A fast-acting DC-link voltage controller for three-phase DSTATCOM to compensate AC and DC loads. *IEEE Trans. Power Delivery*, 24(4):2291–2299, 2009.

29 Prabha Kundur, Neal J Balu, and Mark G Lauby. *Power system stability and control.* McGraw-Hill, New York, 1994.

30 Raouia Aouini, Bogdan Marinescu, Khadija Ben Kilani, and Mohamed Elleuch. Stability improvement of the interconnection of weak AC zones by synchronverter-based HVDC link. *Electr. Power Syst. Res.*, 142:112–124, 2017.

10

Control of Interconnected Microgrids

10.1 Introduction

In Part I of this book, the control objectives of microgrids (MGs) were obviously explained in both grid-connected and islanded/autonomous operating modes. The duty of grid-connected MGs as well as single distributed energy resources (DERs), whose power generation ranges are not considerable with respect to the main generation units of the grid, is to generate a certain amount of power and maximize renewable energy generation. These entities are not responsible for frequency and voltage control/regulation. On the contrary, DERs in autonomous and grid-independent MGs have the main responsibility for frequency and voltage control. Therefore, DERs' control objectives should be switched for these two operating modes between frequency/voltage control and power control. These two general control strategies, i.e. frequency/voltage and power control strategies, are the main control goals of MGs.

In interconnected microgrids (IMGs), there exist two vital control objectives added with respect to individual MGs, which are power exchange and power sharing controls. These are main control objectives specific to IMGs that require satisfying some secondary or higher level objectives like cooperative frequency control in IMGs via circuit breakers (CB-IMGs) and power and dc voltage control aims of back-to-back converters (BTBCs) in BTBC-IMGs. In fact, the power sharing control is specific to CB-IMGs (as well as DC-IMGs through dc CBs), and power exchange control is specific to BTBC-IMGs (as well as DC-IMGs through dc–dc converters). The power sharing control, which is used within MGs for DERs, is also used among CB-IMGs, and its main control functionalities are voltage and frequency tuning as well as power sharing among MGs. On the other hand, the power exchange control is specific to BTBC-IMGs, where BTBCs are controllable devices and can determinately exchange certain power amounts among IMGs. Generally speaking, one can say CB-IMG power sharing control is similar to frequency/voltage control of autonomous MGs with the same control objectives, while taking into account some considerations. Furthermore, the power exchange control of BTBC-IMGs is similar to the power control of grid-connected MGs. Nevertheless, coordination control is required in the upper control layers to determine the allowable ranges of power exchanges for BTBCs.

Since there exist well-known and reasonable hierarchical control (HC) strategies including three control layers [1] and four control layers [2], i.e. primary, secondary, central, and global layers, IMG control should also be standardized through the existing control strategies. In other words, it should be determined that in which control layer, IMG control objectives are considered. The basic IMG control objectives, i.e. power sharing control of CB-IMGs or power exchange control of

Microgrids: Dynamic Modeling, Stability and Control, First Edition. Qobad Shafiee, Mobin Naderi, and Hassan Bevrani.

BTBC-IMGs, can be considered in different control layers of the hierarchical control. In the next section, the literature of IMG control is reviewed, and the control layer in which the IMG control objectives are considered is determined. Moreover, the most common situations in IMG control are discussed afterward.

Another important point to consider is the differing controllability of CBs and BTBCs, which influences the choice of power sharing control for CB-IMGs and power exchange control for BTBC-IMGs. Since CBs are not controllable and are just switchable, they are not appropriate for power control and certain power exchanges. In fact, in CB-IMGs, the control objective should be achieved by the existing actuators, i.e. DERs and their control units. In autonomous power grids such as IMGs, the main DERs with an acceptable power capacity should be operated for frequency and voltage control. Hence, such DERs can only participate in power sharing, where some considerations are required to coordinate intra-MG and inter-MG power sharing control objectives. On the other hand, BTBCs are relatively independent controllable devices. In BTBC-IMGs, BTBCs have the main role of power control. Therefore, power control and determining a certain value for power exchange among IMGs is possible. It is noteworthy to mention that power exchange among BTBC-IMGs is not only related to the capability of BTBCs but also in accordance with MG conditions, e.g. spinning reserve and existing energy/power. For this goal, different levels of communication are required.

In this chapter, control of both CB-IMGs and BTBC-IMGs is taken into consideration. Frequency control of the CB-IMG frequency model and power sharing control of the nonlinear model of CB-IMGs simulated in MATLAB/SymPowerSystems are addressed. Moreover, both planned and emergency operations of BTBC-IMGs are represented, with focus on power exchange control among IMGs, and frequency/voltage support of a critical MG. Most of the results are obtained by OPAL real-time simulations.

10.2 Overview on IMG Control

The IMG control can be studied from different viewpoints. According to control architecture, it can be classified into centralized, decentralized, and distributed. All three architectures can be realized in a standard hierarchical form including three [1]/four [2] control levels, i.e. primary, secondary, central, and global layers. Concepts and literature of these architectures for individual MGs are reviewed in Chapter 5 and [3]. The point that power exchange/tie-line controller is realized in which control layer, i.e. primary, secondary, central, and global, leads to another classification. Various controller types and design methods, e.g. proportional-integral, robust, and model-predictive designs are used in IMG controllers.

Table 10.1 reviews the IMG control literature with useful classifications, validation methods, and main control objectives. The IMG structure and control architecture investigated in the literature, and the control layer, i.e. primary control (PC), secondary control (SC), and tertiary control (TC) including both central and global control layers, selected to implement the special IMG control objectives, are mentioned in the second, third, and fourth columns, respectively. In the fourth column, HC means that the IMG control aims are distributed in all three layers, and BTBC and dc–dc converter (DDC) indicate that a part of IMG controllers is realized at their controllers. Control type/design methods and controller validation techniques are presented in the fifth and sixth columns, and the corresponding illustrations for the abbreviations can be found at the bottom of Table 10.1. The final column consists of main and secondary control objectives of different IMG structures investigated in the literature.

Table 10.1 Literature review of interconnected microgrids control.

References	IMG structure	Control architecture	Control layer	Control type/design method	Validation method	Main control objective
[4]	Two IMG	Decentralized	PC	PR	TDS, ER	Coupling effects among MGs
[5–7]	BTBC-MIMG	Centralized	BTBC-TC	PI	FR, TDS	Power sharing of multiple frequency/voltage IMG
[8]	ac/dc MIMG	Centralized	HC	PI, AD	TDS	IMG economic operation
[9–11]	Two IMG	—	SC	LFC, FL	LMTDS	Theory of IMG load-frequency control (LFC)
[12]	CB-MIMG	Distributed	HC	PI, CP, MAS	TDS	Network topology-based power control
[13]	SS-MIMG	Distributed	SC	PI, CP	TDS, ER	Voltage and power exchange control
[14]	dc IMG	Distributed	SC, TC	PI, VI, MAS	RTS	Coordinated power flow control
[15]	CB-MIMG	Centralized	PC, TC	Robust droop	TDS	Adaptive power flow control
[16, 17]	CB-MIMG	Distributed	HC	I, CP, PSO	TDS	IMG power sharing
[18]	BTBC-MIMG	Distributed	BTBC-TC	Robust	TDS	Transient stability improvement
[19]	BTBC-MIMG	Centralized	TC	PI, AD	TDS	Power exchange via dc lines
[20]	dc IMG	Decentralized	DAB-PC	IOL	TDS	Coordinated control of IMG energy storage
[21, 22]	BTBC-MIMG	Decentralized	BTBC-TC	PI	TDS	Frequency control improvement
[23–25]	Two IMG	Decentralized	BTBC-PC	PI, droop	ER	Active and reactive power exchange
[26]	BTBC-MIMG	Decentralized	BTBC-PC	PI, droop, LC	RTS	Emergency frequency/voltage support
[27, 28]	dc IMG	Decentralized	DDC-PC	PI	TDS	Interconnection of dc IMGs
[29, 30]	CB-MIMG	Distributed	SC	I, PC	TDS	Cooperative IMG power sharing
[31]	dc IMG	Distributed	HC	PI, FTCP	TDS	Power management enhancement
[32, 33]	CB-MIMG	—	SC	VI	LMTDS	Frequency control improvement
[34]	CB-MIMG	Distributed	TC	PI, CP, ET	TDS	Power sharing and communication mitigating
[35]	CB-MIMG	Distributed	TC	MAS	TDS	Fully distributed IMG control

(*continued*)

Table 10.1 (Continued)

References	IMG structure	Control architecture	Control layer	Control type/design method	Validation method	Main control objective
[36]	CB-MIMG	Distributed	SC	I, PCP, MAS	TDS	IMG control with grid-following/forming DERs
[37]	dc IMG	Distributed	HC	PI, PCP	RTS, ER	Economic tie-line power control
[38]	Two IMG	Centralized	PC, TC	AD, LC	TDS	Preventing power shortfall of critical MG
[39]	CB-MIMG	Distributed	SC	I, CP, ET	RTS, ER	Software-defined networking-based power sharing
[40]	dc IMG	Distributed	SC	PI, PCP	TDS	Improving dynamic performance of IMG control
[41]	CB-MIMG	Decentralized	PC	PI	TDS	Improving frequency response
[42]	dc IMG	Distributed	SC	PI, CP	RTS	Power exchange based on battery SOC
[43]	Two dc IMG	Decentralized	DDC-PC	PI	RTS, ER	Active power flow control
[44]	ac/dc MIMG	Distributed	SC	PI	TDS, ER	Asynchronous power control
[45–47]	dc IMG	Distributed	—	PI, HSS	TDS	IMG power flow control
[48]	CB-MIMG	Distributed	SC	I, PCP, ET	TDS	Reducing communication burden
[49]	CB-MIMG	Decentralized	PC	PI, VI	TDS	Virtual inertia control to improve dynamic stability
[50]	BTBC-MIMG	Decentralized	PC, SC	AD, LC	TDS	IMG control for different operating points
[51]	ac/dc MIMG	Distributed	TC	PI, CP, ET	TDS	Power sharing/support by low communication
[52]	ac/dc MIMG	Centralized	HC	PI, PR	TDS, ER	Energy pool for power exchange

Legend: PC: primary control, SC: secondary control, TC: tertiary control, HC: hierarchical control, PI: proportional integral, PR: proportional-resonance, AD: adaptive droop, FL: fuzzy logic, CP: consensus protocol, FTCP: finite-time CP, VI: virtual inertia, LMTDS: linear model TDS, IOL: input–output linearization, DDC: dc–dc converter, ET: event triggered, MAS: multiagent systems, HSS: Hamiltonian surface shaping, LC: logic control, I: integral, DAB: dual active bridge.

10.2.1 CB-IMGs

In CB-IMGs, decentralized control cases are limitedly represented in the literature [41, 49]. In [41], a simple MG structure including one grid-forming (GFO) DER and a number of grid-following (GFL) DERs is considered, where the need for coordination control (in the secondary or a higher level) is eliminated to focus on primary frequency dynamics. A modified automatic voltage

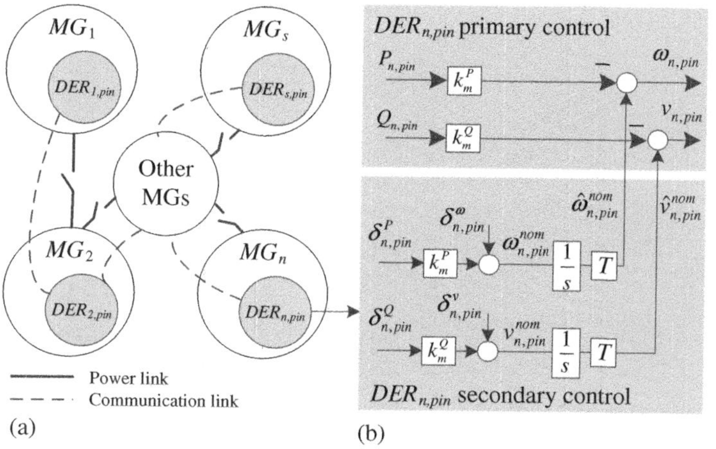

Figure 10.1 Control of CB-MIMG: (a) pinning consensus-based distributed secondary control architecture and (b) corresponding uniform control structure, where the input signals exist in Table 10.2.

regulator is proposed for the GFO DERs to improve the frequency response. The same objective is taken into account by virtual inertia control using super-capacitors without considering the control architecture [49]. Therefore, one can say CB-MIMG control just at the primary level without data communication is impossible.

In [15], a wide-area measurement and communication system including MGCCs and a global distribution system control are represented. Nevertheless, most of the CB-MIMG literature belongs to distributed control methods, which seem to be the most reasonable solution to control power exchange among such IMGs.

The distributed IMG control is realized in one of secondary, central, or global levels or by a complete hierarchical control (see Table 10.1). Some authors have tried to find a standard method like the tie-line power control of traditional power systems. Hence, the IMG control is implemented in the secondary control level (SCL) with the aims of IMG active/reactive power sharing and global frequency/voltage regulation. Generally, the pinning consensus protocol is used in which one of the DERs in each MG is selected as a leader (pinning) agent. Therefore, in a two-layer communication infrastructure, DER_{pin} is a leader for follower DERs within the MG to determine the nominal voltage/frequency in the intra-MG layer, and it is a communicating node in the inter-MG layer to share the information with the other MG leaders.

Figure 10.1(a) shows the communication type in the pinning consensus-based distributed secondary control architecture for CB-MIMGs. Figure 10.1(b) shows a uniform control structure for the leader DER in nth MG. The error signals $\delta^{\omega}_{n,pin}$, $\delta^{v}_{n,pin}$, $\delta^{P}_{n,pin}$, $\delta^{Q}_{n,pin}$, which are the inputs of the integral secondary controller are listed in Table 10.2. In [16, 17], the same method is utilized; however, the MG PCC data is the basis of the inter-MG layer communication instead of the pinning DERs data, which leads to excess measurement and thus lower reliability.

10.2.2 BTBC-IMGs

Figure 10.2 shows the control architectures of BTBC-IMGs. Decentralized control for BTBC-IMGs is realized using droop characteristics. Modified droop characteristics are used in controlling interlinking voltage source converters (VSCs) to consider MG overload and overgeneration conditions [23–26, 50], e.g. (10.1) of [50] using the state of charge (SOC) and nominal power P_n of energy

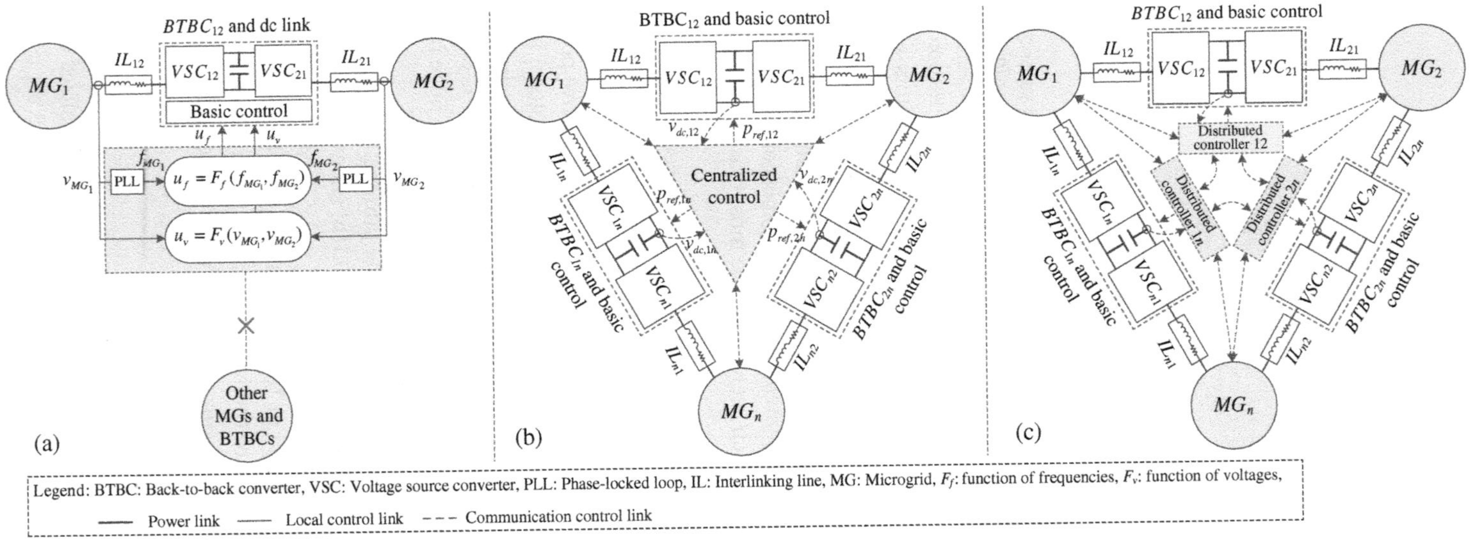

Figure 10.2 Control architectures of BTBC-IMG: (a) decentralized control, (b) centralized control, and (c) distributed control.

Table 10.2 Pinning consensus-based distributed secondary control inputs for leader DERs in CB-MIMG correlated with Figure 10.1.

References	Upper: $\delta^{\omega}_{n,pin}$, Lower: $\delta^{v}_{n,pin}$	Upper: $\delta^{P}_{n,pin}$, Lower: $\delta^{Q}_{n,pin}$
[36]	$\sum_{l=1}^{s}\tilde{a}_{nl}\left\{\left(\omega_{l,pin}-\omega_{l}^{rated}\right)-\left(\omega_{n,pin}-\omega_{n}^{rated}\right)\right\}$	$-\frac{1}{k^{P}_{n,pin}}d_{n,pin}\left(\omega_{n,pin}-\omega_{n}^{rated}\right)$
	$\sum_{l=1}^{s}\tilde{a}_{nl}\left\{\left(v_{l,pin}-v_{l}^{rated}\right)-\left(v_{n,pin}-v_{n}^{rated}\right)\right\}$	$-\frac{1}{k^{Q}_{n,pin}}d_{n,pin}\left(v_{n,pin}-v_{n}^{rated}\right)$
[39]	$\sum_{l=1}^{s}\tilde{a}_{nl}\left(\hat{\omega}_{l,pin}^{nom}-\hat{\omega}_{n,pin}^{nom}\right)$	0
	0	0
[29]	$\sum_{l=1}^{s}\tilde{a}_{nl}\left(\omega_{l,pin}-\omega_{n,pin}\right)+\tilde{a}_{n0}\left(\omega^{rated}-\omega_{n,pin}\right)$	$\frac{1}{k^{P}_{n,pin}}\sum_{l=1}^{s}\tilde{a}_{nl}\left(k^{P}_{l,pin}P_{l,pin}-k^{P}_{n,pin}P_{n,pin}\right)$
	$\sum_{l=1}^{s}\tilde{a}_{nl}\left(v_{l,pin}-v_{n,pin}\right)+\tilde{a}_{n0}\left(v^{rated}-v_{n,pin}\right)$	$\frac{1}{k^{Q}_{n,pin}}\sum_{l=1}^{s}\tilde{a}_{nl}\left(k^{Q}_{l,pin}Q_{l,pin}-k^{Q}_{n,pin}Q_{n,pin}\right)$
[48]	$\sum_{l=1}^{s}\tilde{a}_{nl}\left(\omega_{l,pin}-\omega_{n,pin}\right)+\tilde{a}_{n0}\left(\omega^{rated}-\omega_{n,pin}\right)$	$\frac{1}{k^{P}_{n,pin}}\sum_{l=1}^{s}\tilde{a}_{nl}\left(k^{P}_{l,pin}P_{l,pin}-k^{P}_{n,pin}P_{n,pin}\right)$
	0	0
[30]	$\sum_{l=1}^{s}\tilde{\varsigma}\tilde{a}_{nl}\left(\omega_{l,pin}-\omega_{n,pin}\right)+\tilde{a}_{n0}\left(\omega^{rated}-\omega_{n,pin}\right)$	$sat\left(P_{n,pin}(0)+\frac{1}{T}\int_{0}^{t}\sum_{l=1}^{s}\tilde{\varsigma}\tilde{a}_{nl}\left(M_{l,pin}(s)-M_{n,pin}(s)\right)ds\right)$
	$\sum_{l=1}^{s}\tilde{\varsigma}\tilde{a}_{nl}\left(v_{l,pin}-v_{n,pin}\right)+\tilde{a}_{n0}\left(v^{rated}-v_{n,pin}\right)$	$-\int\omega_{c}Q_{n,pin}+\int\omega_{c}Q_{n,pin}^{measured}$

Legend: $\tilde{a}_{nl}$: lmth array of the inter-MGs network adjacency matrix, $\tilde{a}_{n0}$: is 1 if the leader DER can receive the rated information, otherwise it is 0, $d_{n,pin}$: pinning control gain, M: leader DER increment cost, for $sat(.)$ and other variables see the references.

storage systems (ESSs) as GFO DERs.

$$\omega=\begin{cases}\omega_{n}-k_{p}\left(P-P_{n}\right), & \begin{array}{l}P\leq P_{n}\text{ and }SOC\geq SOC_{max}\\ P>P_{n}\text{ and }SOC\leq SOC_{min}\end{array}\\ \omega_{n}, & \begin{array}{l}P>P_{n}\text{ and }SOC<SOC_{max}\\ P\leq P_{n}\text{ and }SOC>SOC_{min}\end{array}\end{cases} \tag{10.1}$$

A generalized droop is also used to determine active and reactive power references in (13) and (19) of [26]. Frequency and/or voltage are locally being measured to detect the emergency condition by applying logic functions [21, 26, 50] (See Figure 10.2(a)). By communicating some useful information, e.g. DERs surplus powers, centralized control architecture is employed in order to increase the control precision [5, 19]. The communication is minimized and thus the reliability can be improved by replacing the distributed communication instead of the centralized one [18].

Regarding the IMG operation, an emergency or planned power exchange may occur. Though decentralized control is preferred in emergencies, centralized and distributed architectures are more applicable in a planned operation, where MGs communicate their power shortfall/surplus to the controllers and, accordingly the power set points are determined and then sent to BTBCs [6, 7, 18, 26] (see Figures 10.2(b) and (c)).

10.2.3 DC-IMGs

The decentralized control architecture is implemented using dc–dc converters [27, 28] and a dual active bridge (DAB) converter [53] as power exchange actuators. Similar to CB-IMGs, the most applicable control architecture is the distributed architecture (see Table 10.1). In [14], an averaging consensus method is used to communicate the PCC voltage and the average SOC of dc MGs; thus, the voltage coordination among MGs and the power flow control are accomplished in the secondary and global layers, respectively. An HC is employed to consider minimizing power generation cost in the power exchange control [31, 37]. A pinning consensus method is also used that approximately follows the general structure presented in Figure 10.1 in which the $v - i$ droop is used. In order to feed the secondary controller, DER_{pin} voltage error [37, 40], tie-line power error [37], and load current sharing error [40]) are selected.

10.3 Frequency Control for CB-IMGs

In Section 8.6.1, the frequency model of IMGs was presented, where the modeling concentrates on the dynamics and interactions of active power and frequency. A basic PI controller was considered there for the secondary control layer, which is a common secondary controller, and its design was presented using sequential loop closing and the PID toolbox of MATLAB/SIMULINK. Moreover, the stability of three IMGs modeled by the proposed frequency model was addressed in Section 9.3.2. The secondary controllers' proportional and integral gains obtained by the method of sequential loop closing and PID tuning toolbox can stabilize the IMG frequency for small disturbances. However, the controller gains can be modified to check for better performances, which are taken into account in this section. Furthermore, one of the most important issues related to frequency control and stability is the low inertia of inverter-dominated power systems, e.g. MGs and IMGs, which is also paid attention to.

10.3.1 Tuning of Secondary Control Gains

Table 10.3 shows the cases of the secondary PI controller gains. The base case is the output of the sequential loop closing and PID tuning toolbox obtained in Chapter 8. The case K_{P1} is the same as the base case, but with a nonzero proportional gain, i.e. −1 for all the three MGs. In the case of K_{P2}, the proportional gain magnitude increases to 4, i.e. $K_P = -4$. The cases K_{I1} and K_{I2} include smaller and larger integral gains with respect to the base case, respectively.

Figure 10.3(a) displays the frequency response of MG_1 when there is a load reduction of 23% from the rated controllable power of MG_3 at $t = 5$. The other two load changes as −18% and +18% of the rated controllable power for frequency control (sum of GFO DER powers) are applied in MG_1 and

Table 10.3 The cases of the secondary PI controller gains to improve the frequency response.

Cases	K^{sc}_{p1}	K^{sc}_{i1}	K^{sc}_{p2}	K^{sc}_{i1}	K^{sc}_{p3}	K^{sc}_{i3}
Base case	0	−9.2	0	−9	0	−4.7
K_{P1}	−1	−9.2	−1	−9	−1	−4.7
K_{P2}	−4	−9.2	−4	−9	−2	−4.7
K_{I1}	−1	−4	−1	−4	−1	−2
K_{I2}	−1	−18	−1	−18	−1	−9

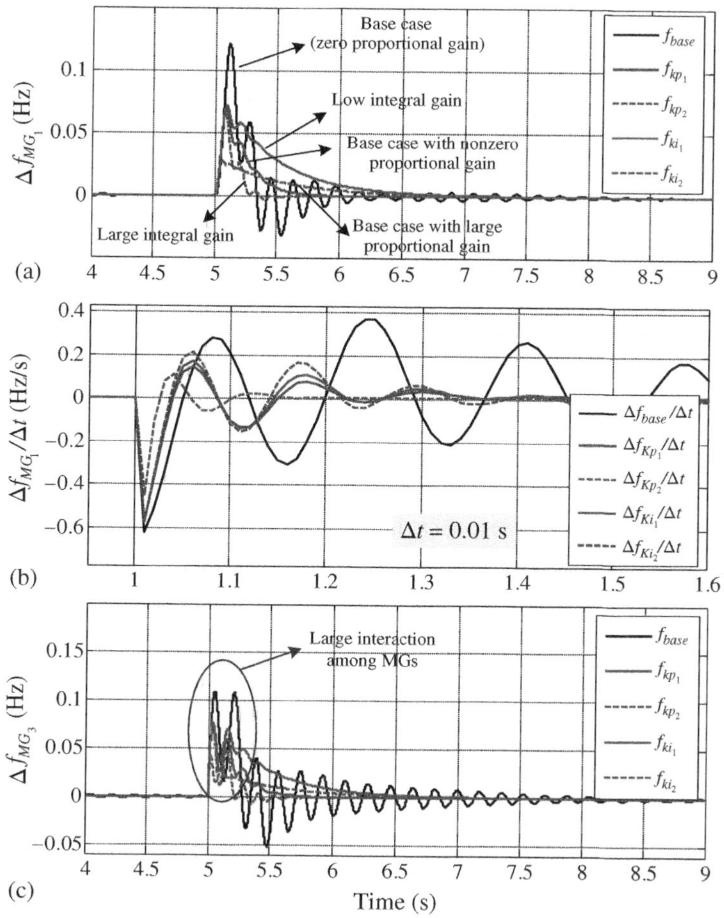

Figure 10.3 Modification of the secondary PI controller gains in the frequency model of the three CB-IMGs: (a) the frequency changes of MG_1, (b) the rate of frequency changes of MG_1, and (c) the frequency changes of MG_3.

MG_2 at $t = 1$ s and $t = 10$ s, respectively, which are not shown here. The base case has obviously the worst performance. In fact, it shows that the method of composition of sequential loop closing and PID tuning toolbox does not have optimal tuning outputs. The frequency response becomes better by considering a proportional gain as shown in the case of K_{P1}. In addition, increasing this gain magnitude results in a small overshoot and fewer oscillations as shown in the case of K_{P2}. Large integral gains lead to a smaller settling time, while smaller values cause a large settling time, but more damped oscillations. Overall, the case K_{P2} in Table 10.3 has the best transient response.

The rate of change of frequency, $\Delta f/\Delta t$, is another index for a frequency controller. Figure 10.3(b) shows $\Delta f/\Delta t$ for the Δf_{MG_1} shown in Figure 10.3(a) with $\Delta t = 10$ ms, which is a common time interval for this calculation in protection relays. One can conclude that an increase in the proportional gain has a positive effect on $\Delta f/\Delta t$. Nevertheless, the integral gain does not have a considerable impact on the rate of change of frequency. Another observation, which is shown in Figure 10.3(c), indicates a large interaction between MGs. The two continuous overshoots show this fact well. Note that general conclusions obtained for MG_1 are correct for all three IMGs, which can also be seen in Figure 10.3(c) for MG_3.

10.3.2 Virtual Inertia Control

As shown in the previous section, increasing the proportional gain of the secondary PI controllers can decrease both the frequency overshoot or undershoot/nadir, as well as the rate of change of frequency. However, much bigger proportional gains decrease the stability margins or even may cause frequency instability. A common method to improve the inertial response of inverter-based DERs and MGs is well known as virtual inertia injection [54–56]. A nonintermittent DER or a reliable ESS is usually used to apply the virtual inertia control function. Figure 10.4 shows a simple method of the virtual inertia, which is a loop from $\Delta\omega$ to ΔP. In fact, such a DER/ESS tries to produce active power with the objective of injecting virtual inertia, neither to produce a certain value of the active power nor to regulate frequency. The virtual control loop includes two subsystems, where $1/(T_{VI}s+1)$ shows a simple single-order model of the dynamics of the DER/ESS with the time constant T_{VI}, and the term $M_{VI}s$ shows the dynamics of the virtual inertia controller applied to the DER/ESS with the virtual momentum of inertia as $j_{VI} = M_{VI}/\omega_n$. By adding the virtual inertia loop, the overall inertia of each MG including the single-order model of MG (primary controller and power part), and the virtual inertia loop is expected to be increased.

Figure 10.5(a) shows the frequency response of the MG_1 for the same condition presented in Section 10.3.1, but for different values of virtual inertia, i.e. $j_{VI} = 0, 20, 50, 100\,\mathrm{W\cdot s^2/rad}$. The secondary controllers are considered the best case, i.e. the K_{P2} case. Therefore, the results are generally better than the other cases presented in Figure 10.3(b). Whereas, Figure 10.5(b) obviously shows the differences among the rate of change of frequencies for various virtual inertias. A larger virtual inertia leads to a smaller rate of change of frequency (RoCoF). This fact is better shown in Figure 10.3(c) by calculating $\Delta f/\Delta t$ during the simulation time for a low $\Delta t = 10$ ms. As a result, the virtual inertia loop can effectively improve the inertia response including the frequency overshoot/undershoot and its rate of change. Nevertheless, it is worth mentioning that increasing the virtual inertia, i.e. a larger j, results in larger oscillations, which in turn is a weakness in the transient response. Hence, a

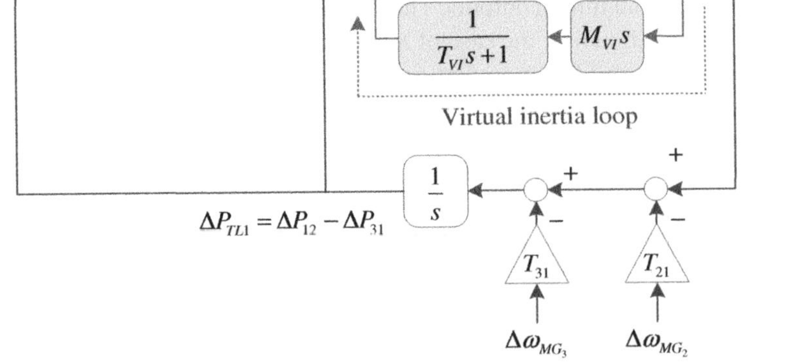

Figure 10.4 Virtual inertia loop added to the frequency model and control of the three CB-IMGs.

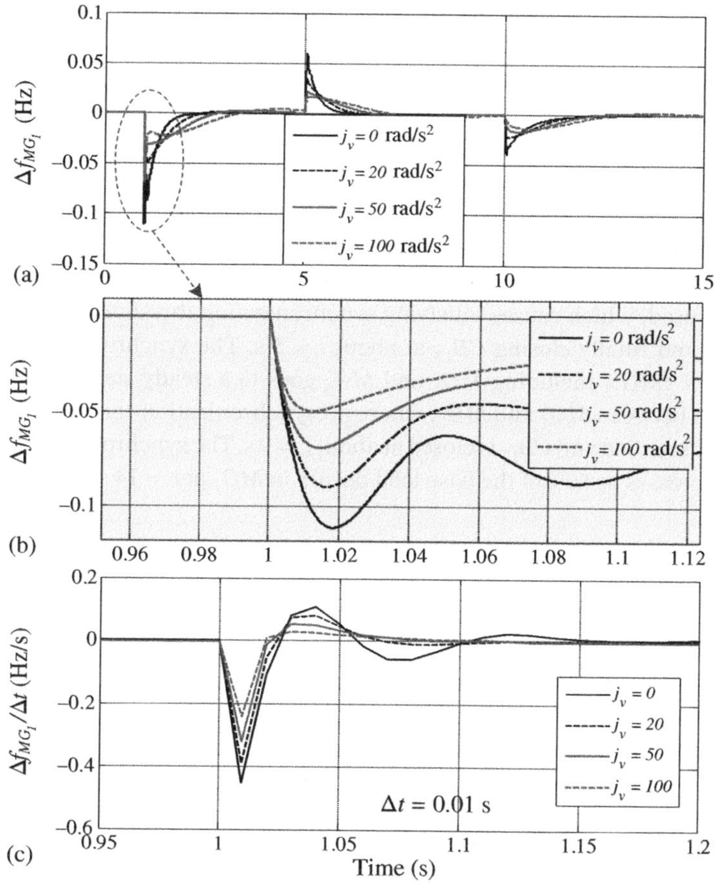

Figure 10.5 Modification of the virtual inertia value in the frequency model of the three CB-IMGs: (a) the frequency changes of MG_1 for all the three load changes, (b) zoomed frequency changes for the largest load changes at $t = 1$ s, and (c) the rate of frequency changes of MG_1.

trade-off is required for the j_{VI} value. In a more applicable case, the constraints of the rate of change of frequency and other important indices should be determined to optimize the j_{VI} value through an optimization problem.

10.4 Power Sharing Control for CB-IMGs

As mentioned earlier, the pinning consensus-based secondary control is a suitable strategy to share power among CB-IMGs, as well as voltage and frequency controls. It considers two cyber communication layers in which both intra-MG and inter-MG power sharing control objectives are embedded. The details of modeling and control loops of the pinning consensus-based secondary controllers in CB-IMGs have been presented in Section 8.3.1.1. This method is very compatible with islanded MG control functions and needs minimal control structure changes to also cope with inter-MG power sharing control. Here, the pinning consensus-based control strategy is considered at the secondary control layer of IMGs. The upper control layers, i.e. central and global are considered to do their duties well, where their dynamics are intensively slower than the studied dynamics. Therefore, there are just some static commands from the global control layer to apply on the secondary controllers, which are considered to be well done through communication links.

In this scenario, three IMGs are interconnected through two CBs and two lines in the general form of Figure 10.1(a), where MG_1 and MG_2 are connected via CB_{12} and IL_{12} with $z_{IL_{12}} = 1.6 + j0.6\ \Omega$, and MG_1 and MG_3 are connected via CB_{13} and IL_{13} with $z_{IL_{13}} = 1.6 + j0.6\ \Omega$. Each MG consists of two DERs in which the pinning-based consensus distributed secondary controller [29] is used and DER_1 is selected as the leader DER. The rated power of DER_2 is twice that of DER_1.

Figure 10.6 shows the simulation results for a sequence of events. The MGs are isolated till $t = 4$ s. The presynchronization starts from $t = 4$ s and a synchronization algorithm presented at [57] and will be explained in Chapter 11 is used, which causes satisfying synchronization thresholds, $\Delta f = 0.05$ Hz, $\Delta\varphi = 5°$, and $\Delta v = 5$ V, and finally closing CB_{13} at about $t = 5$ s. The synchronization control loops are opened at $t = 6$ s. IMG_{13} including MG_1 and MG_3 goes to a steady state in $t = (6, 8)$ s. Another interconnection is between MG_1 and MG_2, where presynchronization starts at $t = 8$ s, synchronization thresholds are satisfied and CB_{12} is closed at about $t = 9$ s. The synchronization control is off at $t = 10$ s. A load increases as 50% of the base load occurs in MG_1 at $t = 14$ s.

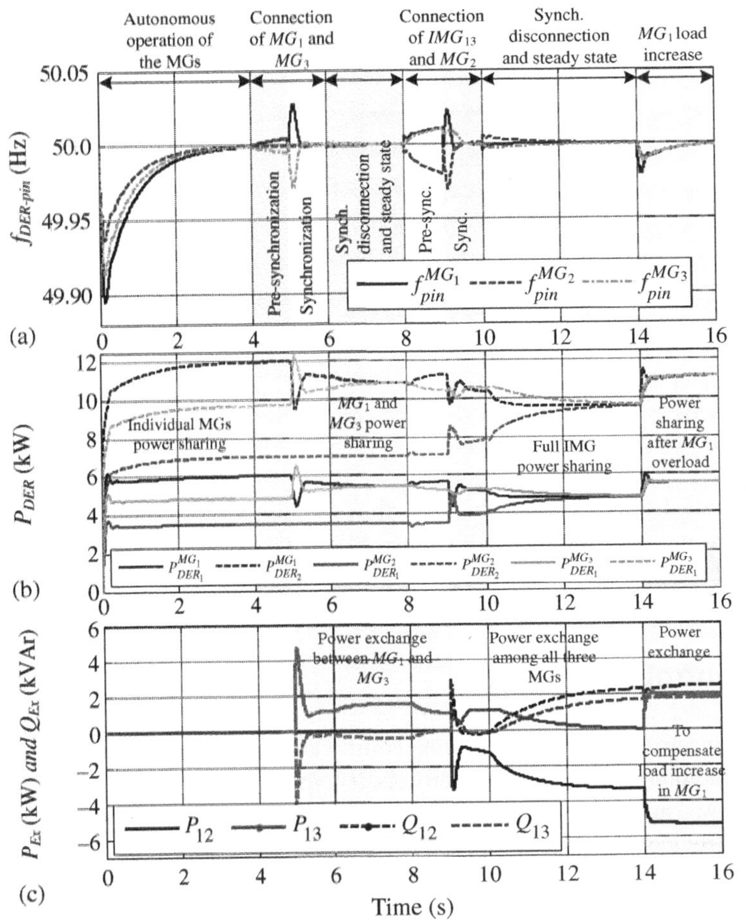

Figure 10.6 Autonomous operation, synchronization process, interconnection, and MG_1 load increase in the three CB-IMGs: (a) pinning/leader DERs frequencies, (b) active powers of all DERs, and (c) exchanging active and reactive powers.

According to Figure 10.6(a), frequency is well regulated during autonomous operation, synchronization transients, and load change. As shown in Figure 10.6(b), active power sharing is done in autonomous MGs before $t = 4$ s, in IMG_{13} within $t = (6, 8)$ s, and in the complete IMG including the three MGs in $t = (12, 14)$ s. Furthermore, active power sharing is maintained for the MG_1 load increase. As shown in Figure 10.6(c), power exchanges are zero before the IMG_{13} formation at $t = 5$ s. In IMG_{13}, MG_3 sends the active power to MG_1 and the reactive power exchange is about zero during $t = (5, 8)$ s. In the complete IMG, MG_2 has the priority to support MG_1 with respect to MG_3 due to its lower local load. The MG_1 load increase is compensated by increasing local MG_1 generation shown in Figure 10.6(b), as well as receiving power from both MG_2 and MG_3. The reactive power is also exchanged to improve bus voltages.

10.5 Power Exchange Control for BTBC-IMGs

Figure 8.1 shows a general structure of ac IMGs, including autonomous MGs, ac and dc ILs, and BTBCs as the main components. The MGs may have different structures with any number of DERs, loads, and lines, which are interconnected using BTBCs, and ac/dc ILs. The control process of BTBC-IMGs is a coordination among individual MG controllers and BTBC controllers in order to exchange certain power amounts among MGs, while they preserve their stability, and independence in inter-MG control.

10.5.1 Prerequisites of Individual Microgrid Control

A four-level hierarchical control presented in [2] is considered here to take the control responsibility of MGs in both individual and interconnected operating modes. The primary control level (PCL) comprises decentralized droop controllers implemented for each DER cascading with inner voltage and current controllers. The vital objectives of voltage/frequency stability, active/reactive power sharing, and current limiting are fulfilled in the PCL. The SCL is mainly responsible for voltage/frequency restoration and power sharing improvement. However, some ancillary objectives such as power quality improvement may be done in the SCL. The most common architectures for the SCL are centralized and distributed types, where communication links are used to share the data among DERs. Nevertheless, decentralized methods are also addressed to just restore frequency and share active power [58]. The central control level (CCL) provides the supervisory and emergency MG capabilities, e.g. grid connecting and load shedding. The high-level energy/power management is done at the global control level (GCL). The optimal power flow among BTBC-IMGs or among individual MGs and the main grid is accomplished in the GCL by communicating the production and consumption data of all participants. However, the GCL control signals and commands are usually applied to the lower control levels, mostly CCL [5–7, 18, 19, 21, 22], as well as SCL [50] and PCL [26]. Figure 10.7 shows a simple schematic of the MG hierarchical control and typical communication among the control layers.

Here, the control challenges at the PCL of MGs are taken into account, which is coordinated with BTBC control at the GCL. Therefore, the SCL and CCL dynamics are not considered. In Section 10.6, the CCL emergency function is studied and updated by the concept of IMG power exchange. In the IMG operating mode, the GCL role can be divided into two divisions: (i) an optimization process based on the communicated data and (ii) sending set points to control units including the CCLs and BTBCs as shown in Figure 8.1. These GCL duties are almost done as a static action without dynamic studies and are considered to be well established in the studied time scale. This is due to the fact

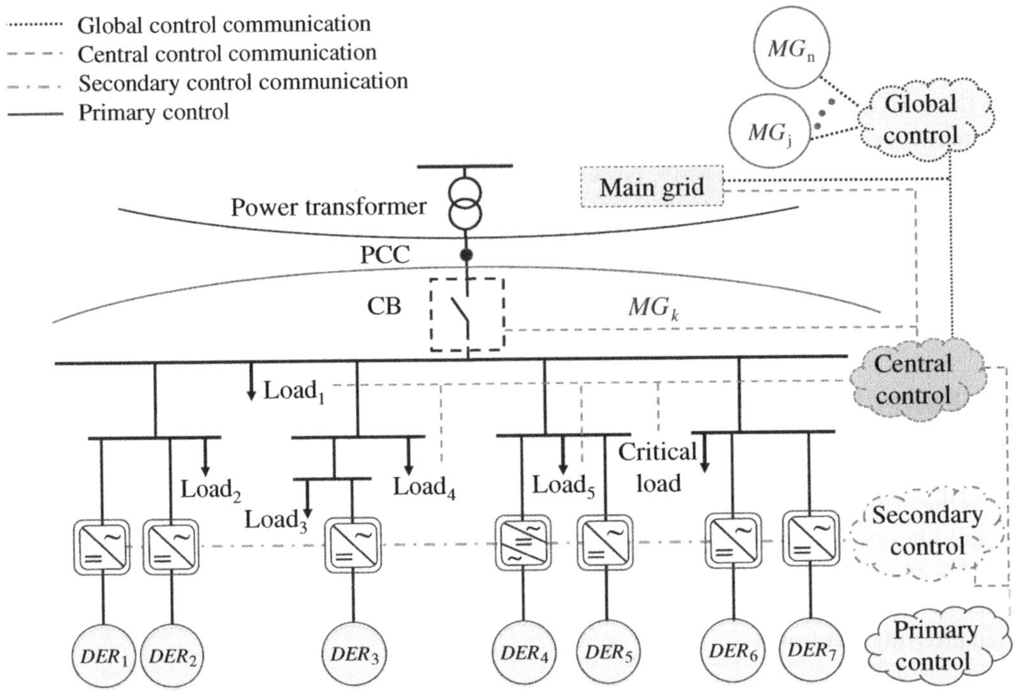

Figure 10.7 A simple schematic of the hierarchical four-level microgrid control comprising primary, secondary, central and global control levels for BTBC-IMGs concentrating on MG_k control and communications.

that the GCL time scale (minutes to an hour) and the CCL time scale (a few seconds to minutes) in sending set points are different from the desired dynamics (seconds or fraction of a second). In other words, the CCL and BTBC set points sent by the GCL and the set points sent from the CCL to the downstream controllers are approximately permanent during small-signal disturbances.

It is noteworthy to mention that the coordination among all IMGs and BTBCs is according to the GCL set points. Let us assume that a power shortage is reported to the GCL by MG_i in a specific time span. The amount of power flow to MG_i, the number of involved BTBCs and MGs as well as MG commitment are determined by the optimal power flow management at the GCL. Hence, the GCL set points sent to all control units result in a coordinated operation for IMGs. Communicating the data at all levels is considered ideal, i.e. without delay dynamics.

The PCL consists of the droop characteristics and internal loops, i.e. the current and voltage loops, as shown in Figure 10.8. The *abc* voltages and currents of any DER are measured locally, and then transformed to *dq0* equivalents. Thus, the active and reactive output powers of DER_i are calculated and filtered as follows:

$$P_i = \frac{\omega_c}{s+\omega_c}\left(v_d^i i_{od}^i + v_q^i i_{oq}^i\right), \tag{10.2a}$$

$$Q_i = \frac{\omega_c}{s+\omega_c}\left(v_d^i i_{oq}^i - v_q^i i_{od}^i\right), \tag{10.2b}$$

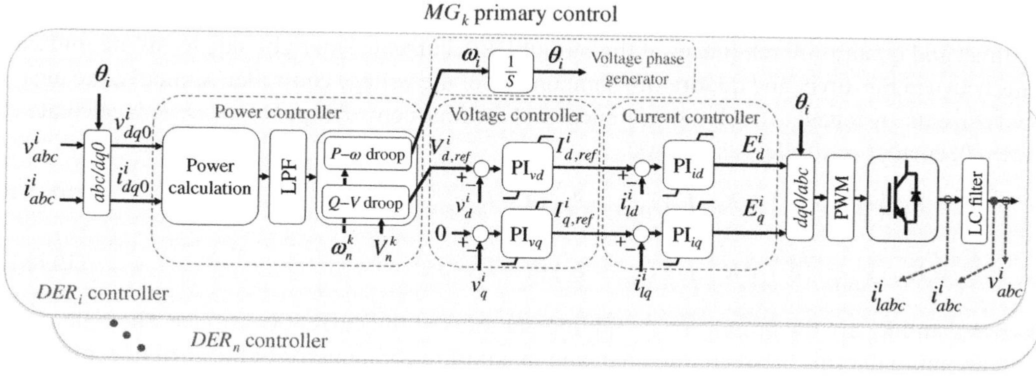

Figure 10.8 Primary control level for the coordinated operation of the typical MG_k within BTBC-IMGs structure.

where ω_c is the cutoff frequency of the low-pass filter (LPF), s is the Laplace operator, v^i_d, v^i_q, i^i_{od}, and i^i_{oq} are the direct and quadrature components of the filter capacitor's voltage and the output current of DER_i, respectively, and P_i and Q_i are the average active and reactive output powers of DER_i. The DER_i frequency, ω_i, and voltage magnitude reference, $V^i_{d,ref}$ are provided using the well-known $\omega - P$ and $v - Q$ droop characteristics as follows:

$$\omega_i = \omega^k_n - m^i_p P_i, \tag{10.3a}$$

$$V^i_{d,ref} = V^k_n - n^i_q Q_i, \tag{10.3b}$$

where ω^k_n and V^k_n are the nominal frequency and voltage amplitude of MG_k. Note that nominal values of frequency and voltage amplitude can be different for each MG in the structure of BTBC-IMGs. Then, the voltage angle of DER_i, θ_i, which is used in the abc to $dq0$ and reverse transformations, can be obtained by an integrator as shown in Figure 10.8. The parameters m^i_p and n^i_q are the $\omega - P$ and $v - Q$ droop coefficients of DER_i, which are usually calculated as follows:

$$m^i_p = \Delta\omega^k_{max}/P^i_r, \tag{10.4a}$$

$$n^i_q = \Delta V^k_{max}/Q^i_r, \tag{10.4b}$$

where $\Delta\omega^k_{max}$ and ΔV^k_{max} are the maximum acceptable deviations of the frequency and voltage amplitude in MG_k, and P^i_r and Q^i_r are the rated active and reactive powers of DER_i. The quadrature voltage reference of DER_i, $v^i_{q,ref}$, is usually considered to be zero in order to earn a zero voltage phase angle at the phase a.

In the voltage control loop, $v^i_{d,ref}$ and $v^i_{q,ref}$ are the voltage references, which are followed by the measured voltages using two PI controllers as follows:

$$i^i_{d,ref} = K_{Pv}\left(v^i_{d,ref} - v^i_d\right) + K_{Iv}\int\left(v^i_{d,ref} - v^i_d\right)dt - C^i_f\omega_n v^i_q, \tag{10.5a}$$

$$i^i_{q,ref} = K_{Pv}\left(v^i_{q,ref} - v^i_q\right) + K_{Iv}\int\left(v^i_{q,ref} - v^i_q\right)dt + C^i_f\omega_n v^i_d, \tag{10.5b}$$

where K_{Pv} and K_{Iv} are the proportional and integral gains of the voltage controller. $i^i_{d,ref}$ and $i^i_{q,ref}$ are the direct and quadrature references of the output DER_i current. The right-side terms are mutual terms between the direct and quadrature components of the voltage controller. Similarly, the measured currents are forced to track the references and the final control input of the DER_i is calculated via two PI current controllers as follows:

$$E^i_d = K_{Pi}\left(i^i_{d,ref} - i^i_d\right) + K_{Ii}\int\left(i^i_{d,ref} - i^i_d\right)dt - L^i_f\omega_n i^i_q, \tag{10.6a}$$

$$E^i_q = K_{Pi}\left(i^i_{q,ref} - i^i_q\right) + K_{Ii}\int\left(i^i_{q,ref} - i^i_q\right)dt + L^i_f\omega_n i^i_d, \tag{10.6b}$$

where K_{Pi} and K_{Ii} are the proportional and integral gains of the current controller. The right side terms are mutual terms between the direct and quadrature components of the current controller. E^i_d and E^i_q are the references for the direct and quadrature components of the output converter voltage, which are transformed to the *abc* equivalent in order to be used in the PWM unit of DER_i. Note that the main duty of the current loop is to limit the converter current through the saturation functions.

10.5.2 Interlinking Back-to-Back Converter Control

As mentioned, the power flow between two IMGs can be controlled using a BTBC. Figure 10.9 shows the BTBC including two VSCs and their basic control units. The VSC_1 is controlled as a power flow controller, while the VSC_2 is controlled as a dc voltage controller. The power controller tries to flow power between IMGs based on the active and reactive power references, and the dc voltage controller stabilizes the dc capacitor voltage during the power flow operating mode. The BTBC is not connected to MGs at the usual autonomous operation of MGs. The BTBC should be connected to the MGs when it receives a command from the GCL stating the beginning of an interconnection operating mode. Therefore, two PLLs are required to synchronize the VSCs of the BTBC to corresponding MGs during a power exchange.

The power controller receives the active and reactive power references, P^{C1}_{ref} and Q^{C1}_{ref}, from an upper control level. In the case of planned power exchanges, this upper control level is generally the global MG control as shown in Figure 10.7. The active and reactive powers can be exchanged

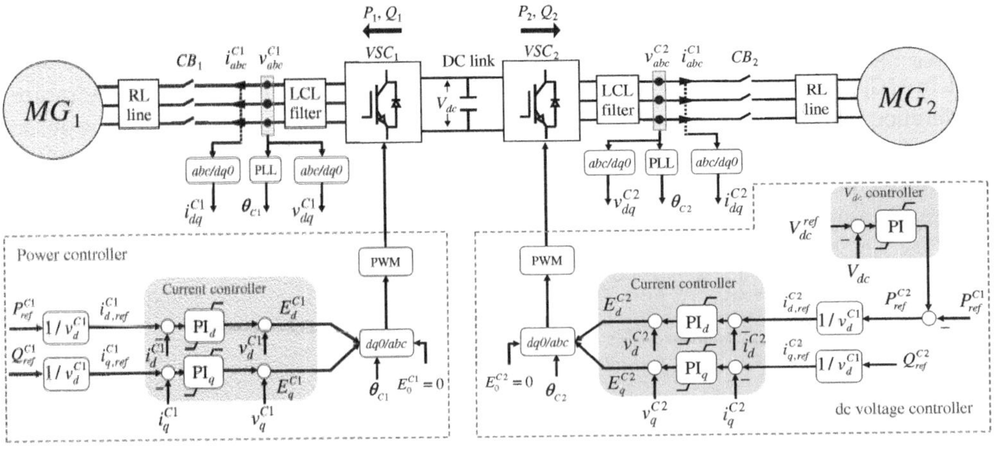

Figure 10.9 Basic control of the back-to-back converter including power and dc voltage controllers in order to exchange power between microgrids.

by tracking the references using a PI current controller. The current references, $i_{d,ref}^{C1}$ and $i_{q,ref}^{C1}$, are calculated as [59]:

$$i_{d,ref}^{C1} = 2P_{ref}^{C1}/3V_{d}^{C1}, \tag{10.7a}$$

$$i_{q,ref}^{C1} = -2Q_{ref}^{C1}/3V_{d}^{C1}, \tag{10.7b}$$

where V_d^{C1} is the direct component of the ac side voltage of the VSC_1, which can be considered as the nominal MG voltage. Therefore, a PI current controller can be employed to achieve the power control as follows:

$$E_d^{C1} = K_{PC1}e_d^{C1} + K_{IC1}\int e_d^{C1}\,dt - L_f^{C1}\omega_n i_q^{C1} + v_d^{C1}, \tag{10.8a}$$

$$E_q^{C1} = K_{PC1}e_q^{C1} + K_{IC1}\int e_d^{C1}\,dt + L_f^{C1}\omega_n i_d^{C1} + v_q^{C1}, \tag{10.8b}$$

where $e_{dq}^{C1} = i_{dq,ref}^{C1} - i_{dq}^{C1}$, i_d^{C1} and i_q^{C1} are the measured currents of VSC_1 ac side, L_f^{C1} is the filter inductance of VSC_1, E_d^{C1} and E_q^{C1} are the direct and quadrature components of the VSC_1 output voltage, K_{PC1} and K_{IC1} are the proportional and integral gains of the VSC_1 current controller, and v_d^{C1} and v_q^{C1} are direct and quadrature components of the VSC_1 voltage, which are added as feed-forward terms. Hence, by applying a control effort on the VSC_1 output voltage as (10.8), the desired powers expressed in (10.7) are flowing.

The injected/absorbed power to/from the dc link should be delivered/received to/from the ac side of VSC_2. Otherwise the power cannot be exchanged and the dc voltage will be unstable. Therefore, a similar power control should be applied on VSC_2 with a mandatory active power reference $P_{ref}^{C2} = -P_{ref}^{C1}$ and an optional reactive power reference, Q_{ref}^{C2}. In addition, the dc voltage of BTBC should be controlled by the VSC_2 controller. Therefore, another PI controller is employed, which tries to control V_{dc} as follows:

$$P_{ref}^{C2} = K_P^{dc}\left(V_{dc}^{ref} - V_{dc}\right) + K_I^{dc}\int\left(V_{dc}^{ref} - V_{dc}\right)dt - P_{ref}^{C1}. \tag{10.9}$$

A fully correct operation of the BTBC depends on two other loops including the PLL and initial dc voltage control. Both PLLs should be tuned well in order to ignore their dynamics in the main controllers. Besides, the dc capacitor should be charged to have an initial voltage near the V_{dc}^{ref}. This decreases the possibility of instability during the BTBC start up by decreasing the oscillations and overshoot of the V_{dc}. The capacitor charger is not included in this study.

10.5.3 Simulation Results for Planned BTBC-IMG Power Exchange

The neighbor MGs can be interconnected based on a prescheduled plan of the global control layer to support power/energy deficiency in some time intervals. Moreover, the energy price can be a good reason for energy exchange between two MGs. In fact, in a situation, in which renewable energies of one of the MGs are cheaper and the other requires in that time span, IMG formation and power exchange are applicable and economic.

In a planned BTBC-IMG formation and power exchange, power and energy sell/buy is the main objective, and voltage/frequency support is not taken into consideration. However, power/energy management and optimal power flow are done while voltage and frequency constraints are considered. Therefore, in the planned IMG power exchange scenarios, the goal is to study the control considerations and requirements in the lowest control layers, where the time constant and bandwidth are related to the time/frequency range studied in this book.

In order to analyze the large-signal stability of BTBC-IMGs, the two case studies have been presented in Figure 9.5. Each MG structure is as shown in Figure 8.3 without GFL DERs. The simulation data has previously been presented in Table 9.3. Note that real-time simulation results obtained from the OPAL-RT simulator shown in Figure 8.15 are presented in this section.

In this section, the two case studies are given including two MGs interconnected through one BTBC (Case study 1) and three MGs interconnected via three BTBCs (Case study 2). The MG structure is similar to that shown in Figure 8.3 without GFL DERs. The general IMG data is represented in Table 9.3. However, different power references and nominal DER powers are used here.

10.5.3.1 Two BTBC-IMGs

In this scenario, the two MGs are islanded before $t = 1$ s, and they are interconnected at this time. The MG_1 load increases at $t = 3$ s, and the BTBC active power reference is changed at $t = 5$ s to send more active power from MG_2 to MG_1 and support MG_1 power shortage according to the command of the global control layer.

Figure 10.10 shows the active and reactive powers exchanged between the two MGs, i.e. MG_1 and MG_2. The power references are also shown. Note that the reactive power exchange between MGs are not considered here, and thus the reactive power reference is set to zero. The first important point is the steady-state error, which is shown in both active and reactive power waveforms. It is due to the voltage drop at the PCCs, where the BTBC is connected to the MGs. The BTBC current references are obtained from the scheduled power references according to (10.7), where the nominal direct voltage component is used instead of the measured one due to stability considerations. Therefore, the used nominal value is always a little larger than the real value and causes smaller real powers. However, the error is less than 6% in the worst case and can be neglected.

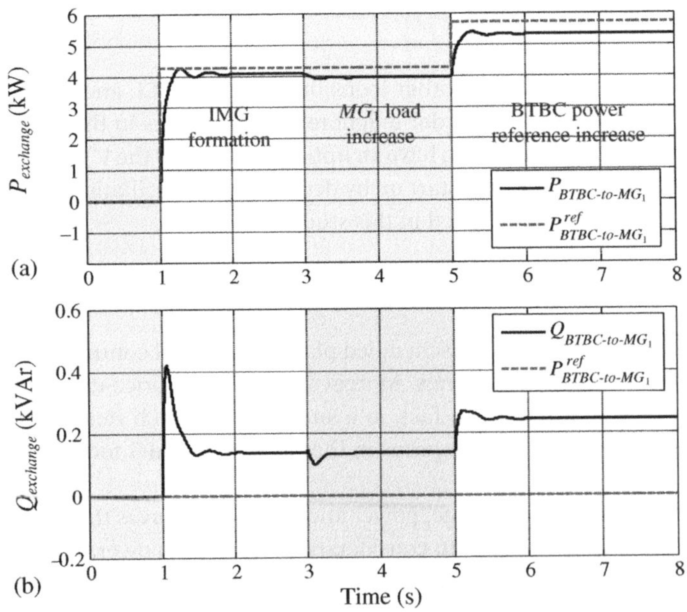

Figure 10.10 Power exchange from MG_2 to MG_1 measured at the BTBC VSC_1: (a) the measured active power and the reference and (b) the measured reactive power and the reference.

The second remarkable point, which can be seen in Figure 10.10, is the lack of considerable change in the BTBC power caused by the MG_1 load increase. There are a few variations due to the more voltage drop of the MG_1 PCC, which is negligible. Therefore, in BTBC-IMGs, the exchanging power is not affected directly by the MG power changes. In fact, the BTBC is basically controlled as GFL converters. However, the BTBC active power increases after the global control command based on increasing the active power reference.

Figure 10.11 shows the measured active and reactive powers of the DERs. Regarding the power values of each MG at different times, the power sharing function of the PCL is well performed, especially for active powers with a small error of less than 0.5%. Nevertheless, the error of the reactive power sharing is larger. Note that the SCL, which tries to improve power sharing, is not performed here. An important point in accordance with power sharing is that the MGs do not have a unified power sharing after interconnection. In BTBC-IMGs unlike CB-IMGs, the power exchanging objective is generally taken into account instead of the power sharing objective. Therefore, there is no global power sharing in opposition to CB-IMGs, where the main strategy of IMG control is based on overall power sharing.

The MG_1 load increase is well seen in Figure 10.11 after $t = 3$ s. It can also be seen that the MG_2 DER powers are not changed at all due to this load increase in the interconnected MG_1. It is due to the PQ control of the BTBC, which is not normally affected by the load/generation changes unlike v/f droop-based control. In contrast, after $t = 5$ s when the BTBC reference values are manipulated, both droop controls of MGs are independently affected. In fact, the basic control of BTBC may cause instability in IMGs if coordination considerations are not realized in the global control layer.

The dc link voltage of BTBC is mainly studied in Chapter 9, especially Section 9.5. Here, it is considered to be set on the nominal value, i.e. 780 V, before power exchange. Figure 10.12(a) shows the BTBC dc voltage and its reference. The oscillations can be seen, which mention the vital

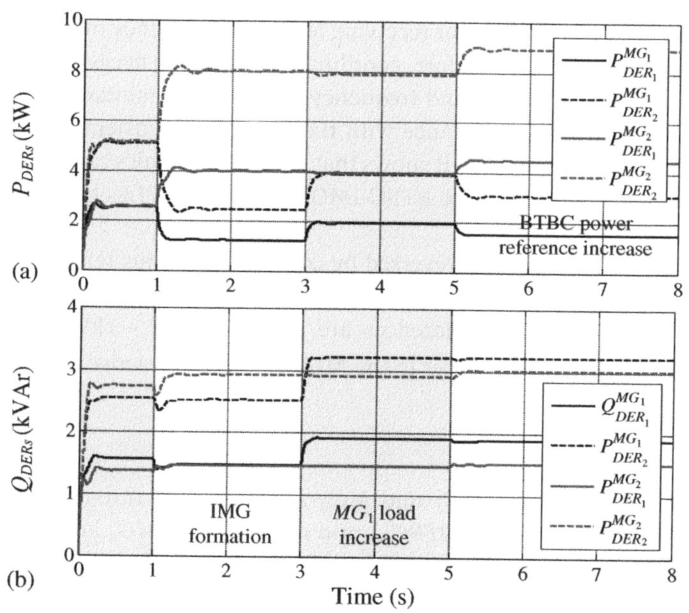

Figure 10.11 DER powers of MG_1 and MG_2: (a) the measured active powers and (b) the measured reactive powers.

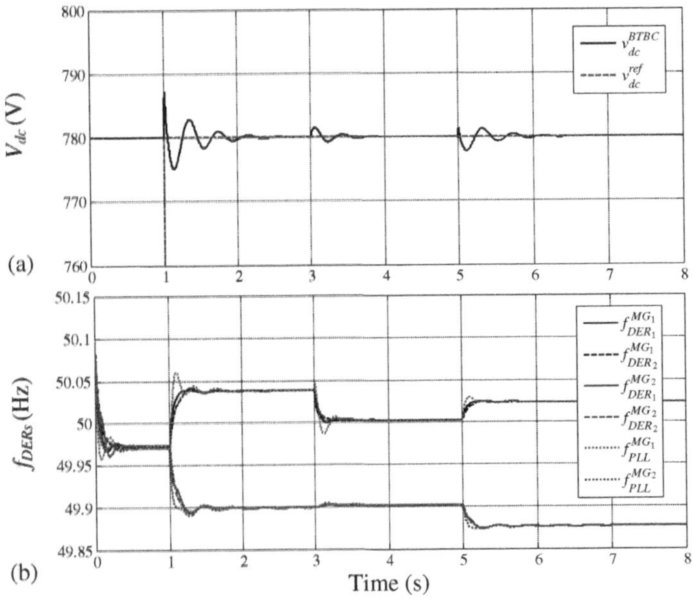

Figure 10.12 (a) dc voltage of the BTBC and (b) frequency of MGs measured at the place of DERs and BTBC ac sides during power exchange between MGs.

role of the BTBC dc voltage control. In such a non-critical operating condition, the dc voltage is controlled without any stability challenges. However, various uncertainties and disturbances may occur, which should be taken into consideration in the control of BTBC-IMGs in future research.

Figure 10.12(b) shows the MG frequencies measured at different places including DER controllers and BTBC PLLs. It is obvious that the MGs are operated at different frequencies. The frequencies change with the BTBC power exchange. Power receiving leads to a frequency increase and power injecting leads to a frequency decrease. Therefore, coordination control is necessary to manage BTBC power exchanges such that the voltage and frequency of MGs are maintained in acceptable ranges. Another important point is in accordance with the frequency transients. The frequencies measured by the PLLs have larger transients. It shows that the PLL dynamics can have a considerable impact on the stability and control of these BTBC-IMGs because the PLLs have an important role in BTBC synchronization.

Note that the power exchange direction can easily be reversed by considering minus reference values for the BTBC powers. Figure 10.13 shows the active and reactive powers of the BTBC when the IMG is formed at $t = 1$ s and the power references are set as $P_{BTBC}^{ref} = -3$ kW and $Q_{BTBC}^{ref} = +1$ kVAr. Therefore, the MG_1 sends reactive power to the MG_2 and receives active power from it.

10.5.3.2 Multiple BTBC-IMGs

In this case, three MGs are considered, where the MG_1 and MG_2 are connected through the $BTBC_{12}$, the MG_1 and MG_3 are connected through the $BTBC_{13}$, and the MG_2 and MG_3 are connected via the $BTBC_{23}$. Hence, both active and reactive powers are able to be directly exchanged from each MG to another one. The first power exchange is at $t = 1$ s in which the $BTBC_{12}$ sends

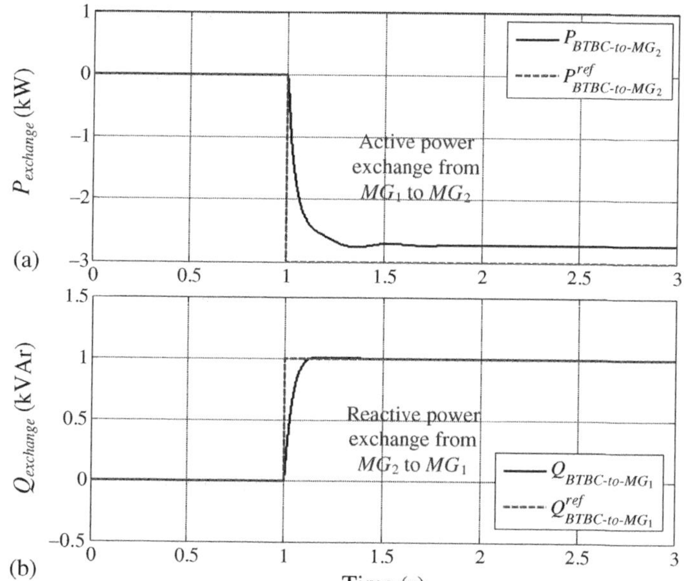

Figure 10.13 Power exchange from MG_2 to MG_1 measured at the BTBC VSC_1: (a) the measured active power and the reference, and (b) the measured reactive power and the reference.

about 7.5 kW and 3.5 kVAr from MG_2 to MG_1. This power exchange is set to come back to zero at $t = 3$ s and is completely suspended at $t = 4$ s by opening the CBs at both sides of the $BTBC_{12}$. Another power exchange is from MG_3 to MG_1 via the $BTBC_{13}$ starting at $t = 2$ s to the end of simulation time. The last power exchange is started at $t = 4$ from MG_2 to MG_3 through the $BTBC_{23}$.

Figure 10.14 shows the power exchanges among the IMGs measured at the BTBCs. Here, both active and reactive powers are flown among the BTBC-IMGs to show the high flexibility of such systems in power exchange. According to the measurement devices, the positive power exchanges are from MG_2 to MG_1 for $BTBC_{12}$, from MG_3 to MG_1 for $BTBC_{13}$, and from MG_3 to MG_2 for $BTBC_{23}$. As can be seen, the power exchanges are done in a stable and controllable situation. Exchanging both active and reactive powers in both directions at each BTBC makes BTBC-IMGs much more flexible than CB-IMGs.

The IMG global controller can recognize MGs with excess power and MGs with power deficiency and then find and establish direct or indirect routes to flow the power from the first to the second. Therefore, in BTBC-IMGs, the possibility of consuming renewable energies and also the possibility of supplying loads, especially critical loads, increase due to much more energy exchange routes with respect to islanded MGs. Moreover, the flexibility of BTBC-IMG in which power exchanges is possible, is more than the CB-IMG flexibility, where there is just power sharing among IMGs. It's worth mentioning that the initial investment and installation costs for BTBC-IMGs exceed those of CB-IMGs, primarily due to the acquisition of BTBCs. Nevertheless, their economic efficiency is often higher thanks to their flexibility, allowing them to compensate the initial costs over time.

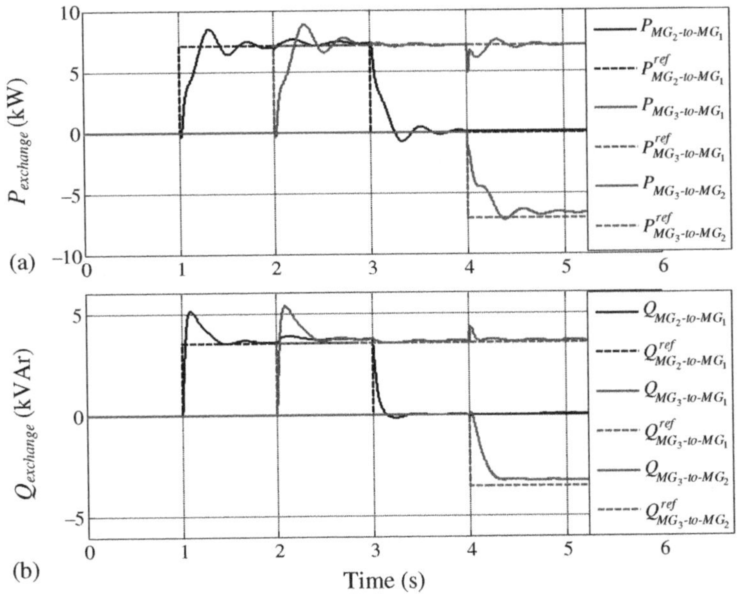

Figure 10.14 A typical power exchange among MG_1, MG_2, and MG_3 measured at the BTBCs: (a) the measured active power and the reference, and (b) the measured reactive power and the reference.

10.6 Emergency Control for BTBC-IMGs

The planned IMG formation, control, and operation are mostly related to the energy/power management level, which is done by the global control level of MGs/IMGs or the distribution system operator (market operator) [60–63]. This high-level control does not consider the transition dynamics and other low-level control characteristics such as the voltage and frequency waveforms. On the other hand, in the emergency IMG formation, the transients and their control are much more important due to the disturbance occurred before the IMG formation. The emergency condition in an islanded MG is defined as large and sudden disturbances like large load changes or a DER trip, which change the balance between generation and consumption [2]. The emergency-based IMG formation is usually due to such an emergency condition in a remote MG to support frequency and delay load-shedding mechanisms [38]. It is an alternative solution for using ESSs and extra-sized diesel generators [64], which can be more flexible.

In this section, an emergency-based IMG composition and decomposition are studied. A fully implemented application of ac BTBC-IMGs is represented in order to cope with an emergency condition of islanded MGs, which can normally activate load-shedding mechanisms. A completely automatic detection of emergency condition is proposed without using communication links. Both voltage and frequency deviations of individual MGs are measured to detect the emergency condition. The voltage and frequency differences of BTBC ac sides are a complete variables set to determine the value of active and reactive powers to be exchanged. This set is applied to a generalized droop controller (GDC) to regulate the power references for the BTBC. The GDC method shares the power robustly between IMGs against the interlinking line X/R ratio. Only local voltages and frequencies of each BTBC ac side are used in all BTBC control processes including emergency and remote detections, and determining required exchange powers. Compatibility of this emergency power exchange method with the individual primary MG controllers is guaranteed by considering the rated powers and maximum acceptable voltage and frequency deviations of individual MGs in the GDC and logical control designs.

To benefit from the surplus energy of the adjacent MGs in a critical MG, an emergency control is required to detect the emergency condition. On the other hand, it should be checked whether the adjacent MGs have enough energy to exchange without exceeding limitations or not. In such a control process, logical controllers and constraints are inevitable. Nevertheless, their nonlinear role can be reduced by a better selection of the controller and their desired variables.

In the presented emergency control, both the detection of the critical condition in the receiving MG and the excess power in the sending MG are based on the voltage magnitude and frequency monitoring. According to the droop characteristics, an overload, a DER trip, or any change in the MG production/consumption can be seen as voltage and frequency deviations. Therefore, the output DER powers or load powers do not need to be measured and communicated in order to detect the emergency/normal condition. Using only the local voltage magnitude and frequency of ac sides of the interlinking BTBC avoids the communication links and facilitates the emergency power exchange.

Figure 10.15 shows the schematic of the emergency control framework. The open-circuit voltage amplitude and frequency of ac BTBC sides before the CBs are available via voltage sensors and PLLs of the BTBC. These voltages are used in the basic BTBC control in order to synchronize the BTBC with both ac sides.

Before using the voltage magnitude and frequency of ac sides in the emergency controller, they generally should be normalized. Because the nominal frequencies and voltages of the two

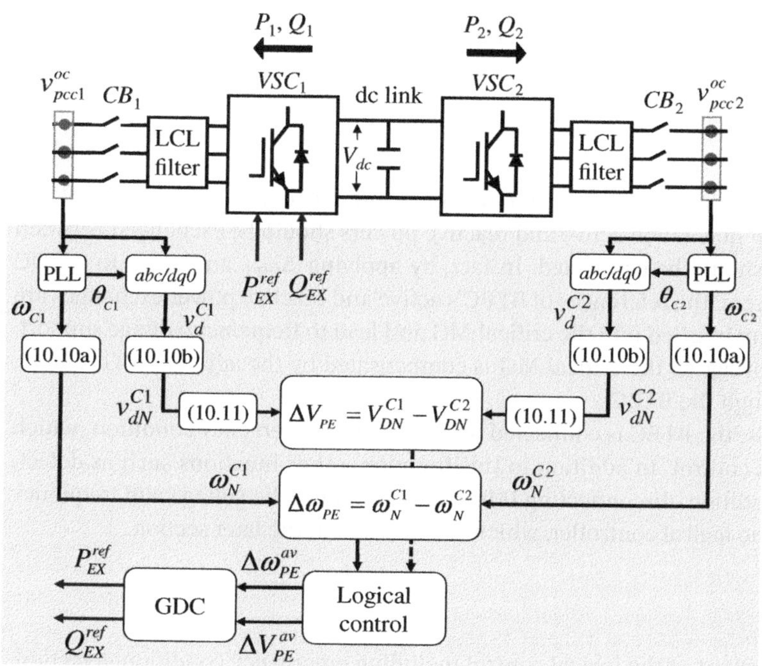

Figure 10.15 Overview of the presented emergency control for power exchange of interconnected microgrids via back-to-back converters.

MGs may be different in BTBC-IMGs. The normalization is done using the nominal values as follows:

$$\omega_N^{Ci} = \omega_{Ci}/\omega_n^k, \tag{10.10a}$$

$$v_{dN}^{Ci} = v_d^{Ci}/V_n^k, \tag{10.10b}$$

where ω_{Ci} and v_d^{Ci} are the measured frequency and voltage amplitude of VSC_i, and ω_N^{Ci} and v_{dN}^{Ci} are the normalized frequency and voltage amplitude and $i = 1, 2$. In addition, the voltage amplitudes should be transformed in a common reference frame. This is because in practice they are calculated in two different reference frames by the asynchronous PLLs. The transformation is straightforward as

$$\Delta V_{DQ} = T_s \cdot \Delta v_{dq} + T_\delta \cdot \Delta\delta, \tag{10.11}$$

where the V_{DQ} and v_{dq} express variables in common and individual reference frames, respectively, $\Delta\delta$ is the angle difference between each individual reference frame and the common reference frame, and

$$T_s = \begin{bmatrix} \cos\delta_0 & -\sin\delta_0 \\ \sin\delta_0 & \cos\delta_0 \end{bmatrix},$$

$$T_\delta = \begin{bmatrix} -\left(v_{d0}\sin\delta_0 + v_{q0}\cos\delta_0\right) \\ v_{d0}\cos\delta_0 - v_{q0}\sin\delta_0 \end{bmatrix}.$$

Note that the common reference frame can be selected from each one of individual reference frames of MGs or even another different frame.

The difference between the voltage magnitude and frequency of two ac BTBC sides after the normalization and transformation, $\Delta\omega_{PE}$ and ΔV_{PE}, are provided as follows

$$\Delta\omega_{PE} = \omega_N^{C1} - \omega_N^{C2}, \tag{10.12a}$$

$$\Delta V_{PE} = V_{DN}^{C1} - V_{DN}^{C2}. \tag{10.12b}$$

Since $\Delta\omega_{PE}$ and ΔV_{PE} are the voltage magnitude and frequency differences between the critical MG and a normal MG, they show how much active and reactive powers should be exchanged between them and when the BTBC should be connected. In fact, by applying $\Delta\omega_{PE}$ and ΔV_{PE} to a GDC after a necessary logical process, the references of BTBC's active and reactive power exchanges are determined. These powers are injected into the critical MG and lead to frequency/voltage support. In other words, a power shortage in the critical MG is compensated by the adjacent MG reserve, which should be flown through the BTBC.

In this application of IMGs, the BTBC is connected only during an emergency condition, which can be detected by a logical control. In addition to this function, other functions such as detecting the return to normal condition, disconnecting IMGs, and averaging the voltage and frequency deviations are included in the logical controller, which is explained in the later section.

10.6.1 Logical Control

Figure 10.16 shows the flowchart of the logical control including emergency condition detection, averaging the deviations, and normal condition detection after an event. The $\Delta\omega_{PE}$ and ΔV_{PE} are

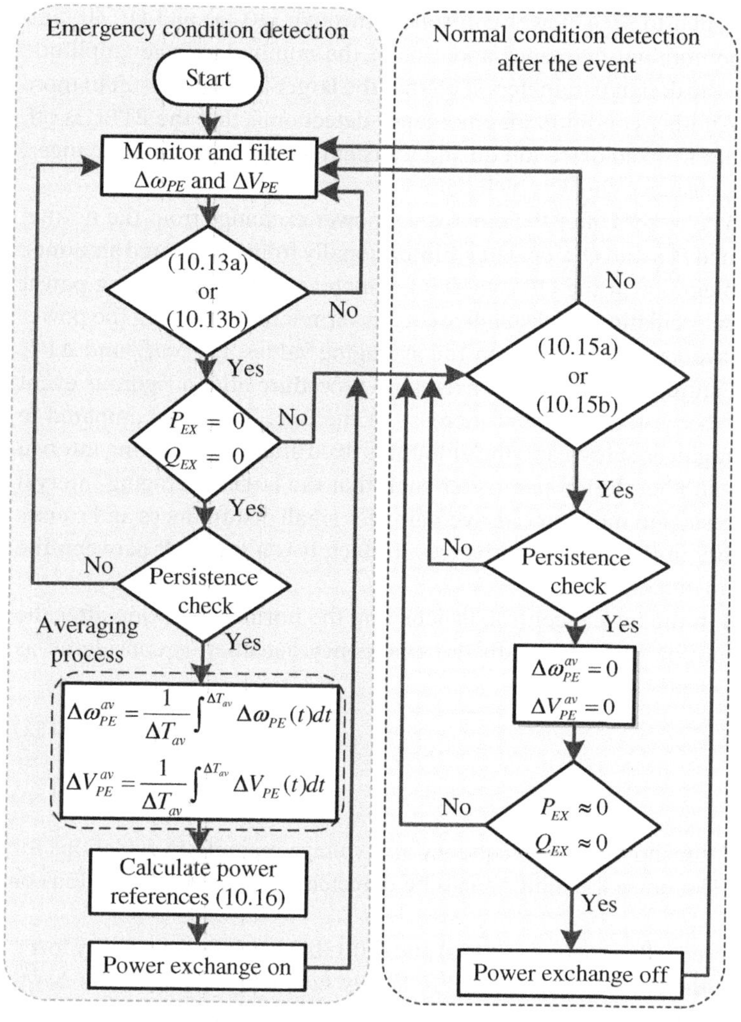

Figure 10.16 Flowchart of the logical control, used in the emergency power exchange including detection of emergency and normal conditions.

measured continuously as (10.12) and compared with their thresholds $\Delta\omega_{Tr}^{On}$ and ΔV_{Tr}^{On} as follows

$$|\Delta\omega_{PE}| > \Delta\omega_{Tr}^{On}, \tag{10.13a}$$

$$|\Delta V_{PE}| > \Delta V_{Tr}^{On}. \tag{10.13b}$$

Note that the emergency condition can be detected by each one of (10.13a) or (10.13b). The threshold values are defined such that the normal condition is not included and the emergency condition is detected before exceeding the limitations of the individual receiving MG as follows

$$\left|\Delta\omega_{NC,max}^{MG}\right| < \Delta\omega_{Tr}^{On} < \left|\Delta\omega_{max}^{k}\right|, \tag{10.14a}$$

$$\left|\Delta V_{NC,max}^{MG}\right| < \Delta V_{Tr}^{On} < \left|\Delta V_{max}^{k}\right|, \tag{10.14b}$$

where $\Delta\omega^k_{max}$ and ΔV^k_{max} are related to MG_k power controllers through (10.4a) and (10.4b) and can be calculated as 0.5% of the nominal frequency and 10% of the nominal voltage amplitude [65]. $\Delta\omega^{MG}_{NC,max}$ and $\Delta V^{MG}_{NC,max}$ are the design parameters such that the larger amounts result in more strict detection and vice versa. Another condition for emergency detection is that the BTBC is off, i.e. $P_{EX} = 0$ and $Q_{EX} = 0$. In order to avoid detection during transient voltage/frequency changes, in addition to using an LPF, a persistence check is included.

Once $\Delta\omega_{PE}$ and ΔV_{PE} start to go to zero after the emergency power exchange from the normal MG to the critical MG, their related P^{ref}_{EX} and Q^{ref}_{EX} deviate unintentionally from the desired amounts. This is due to the lack of an integrator in the GDC method, which causes an oscillating power exchange. In order to avoid these oscillations and stabilize the power references during the power exchange, $\Delta\omega_{PE}$ and ΔV_{PE} are averaged and frozen on the averaging values, i.e. $\Delta\omega^{av}_{PE}$ and ΔV^{av}_{PE} before exchanging the power. Figure 10.17 shows the averaging procedure after a rigorous event and its detection in the case of $\Delta\omega_{PE}$. If a persistent increase in the $\Delta\omega_{PE}$ or ΔV_{PE} compared to their thresholds occurs, the averaging is started and the amount is fixed after the averaging interval ΔT_{av}. The averaging interval is another design parameter such that the larger averaging interval leads to slower detection. This results in more precise averaging for small disturbances and causes poor averaging or even instability in the large disturbances. In fact, it is a trade-off between the averaging accuracy and detection quickness.

The last function, considered in the logical control, is detecting the normal condition after the event. A special constraint to not be mixed up with the emergency detection is considered as follows:

$$|\Delta\omega_{PE} + \Delta\omega^{av}_{PE}| < \Delta\omega^{Off}_{Tr}, \tag{10.15a}$$

$$|\Delta V_{PE} + \Delta V^{av}_{PE}| < \Delta V^{Off}_{Tr}, \tag{10.15b}$$

where $\Delta\omega^{Off}_{Tr}$ and ΔV^{Off}_{Tr} are the thresholds of the frequency and voltage magnitude deviations for stopping the power exchange. The normal condition can be detected by each one of (10.15a) or (10.15b).

The thresholds should be theoretically zero and (10.15a) and (10.15b) should be in equality form. Nevertheless, in practical cases, the equality conditions are hardly satisfied. Therefore, the $\Delta\omega^{Off}_{Tr}$ and ΔV^{Off}_{Tr} are considered less than 10% of $\Delta\omega^{On}_{Tr}$ and ΔV^{On}_{Tr}, respectively. The compulsory constraint before stopping the power exchange by opening the CB reduces the exchanging power to zero. This condition is included in the logical controller as $P_{EX} \approx 0$ and $Q_{EX} \approx 0$, which can be seen in Figure 10.16. A persistence check is included in this function.

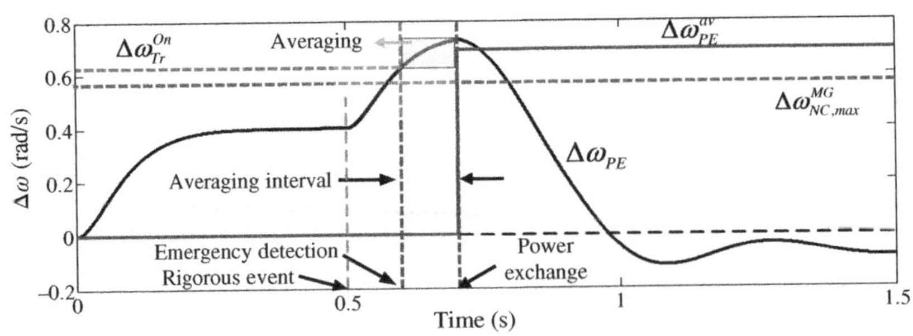

Figure 10.17 Detecting, averaging, and fixing processes of the frequency deviation after an emergency condition.

10.6.2 Generalized Droop Control

As mentioned, the $\Delta\omega_{PE}^{av}$ and ΔV_{PE}^{av} determine the references of the exchanged active and reactive powers. For this purpose, the droop characteristics can be employed. As the ac interlinking lines may have different lengths, in order to decouple the active and reactive powers, a GDC method is used to separate the impact of $\Delta\omega_{PE}^{av}$ and ΔV_{PE}^{av} on the P_{EX}^{ref} and Q_{EX}^{ref}, respectively. Therefore, the rotational power references as ancillary variables [66, 67], are obtained as

$$
\begin{aligned}
P'^{ref}_{EX} &= \frac{X_{IL}}{Z_{IL}}\left[P_{EX}^{0} - \frac{\Delta\omega_{PE}^{av}}{k_p}\right] + \frac{R_{IL}}{Z_{IL}}\left[\frac{\Delta V_{PE}^{av}}{k_p} - Q_{EX}^{0}\right],\\
Q'^{ref}_{EX} &= \frac{R_{IL}}{Z_{IL}}\left[P_{EX}^{0} - \frac{\Delta\omega_{PE}^{av}}{k_p}\right] + \frac{X_{IL}}{Z_{IL}}\left[Q_{EX}^{0} - \frac{\Delta V_{PE}^{av}}{k_p}\right],
\end{aligned}
\tag{10.16}
$$

where R_{IL}, X_{IL}, and Z_{IL} are the resistance, inductance, and impedance of the interlinking line, and P_{EX}^{0} and Q_{EX}^{0} are the nominal active and reactive exchanged powers. k_p and k_q are the coefficients of the conventional droop characteristics in the case of power exchange, i.e. $\Delta\omega_{PE} - P_{EX}$ and $\Delta V_{PE} - Q_{EX}$. Two main points should be considered to find the coefficients:

1) The frequency and voltage magnitude of each MG should not be exceeded due to exchanging the power. This can be satisfied by limiting $\Delta\omega_{PE}^{max}$ and ΔV_{PE}^{max} to the maximum deviations of the same variables from their nominal values in both sending and receiving MGs. Figure 10.18 shows the relationship between the droop characteristics of the receiving MG ($y_{MG}^{R} - x_{MG}^{R}$), its power exchange ($y_{PE}^{R} - x_{PE}^{R}$), sending MG ($y_{MG}^{S} - x_{MG}^{S}$), its power exchange ($y_{PE}^{S} - x_{PE}^{S}$), and the

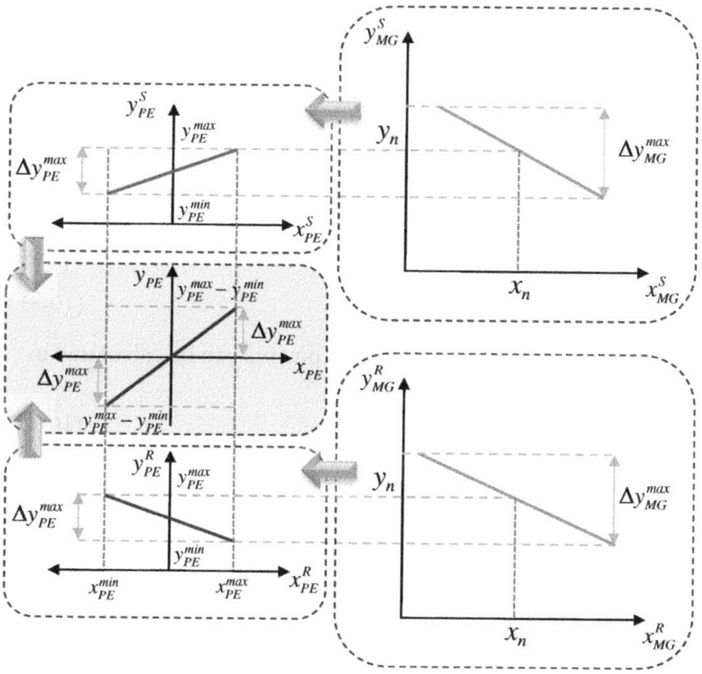

Figure 10.18 Relationship between different droop characteristics used in calculating gains of the presented GDC method including main MG droops, power exchange MG droops, and the resulted GDC-based droop.

finally resulting GDC-based emergency controller ($\Delta y_{PE} - x_{PE}$). Note that the power exchange droops are related to the VSCs of the BTBC. Without loss of generality and in order to show clearer graphs, MG_1 and MG_2 are considered as sending and receiving, MGs, respectively and the same nominal values are considered for both MGs. Besides, the output power (x) direction is considered in the power direction of VSC_1. According to the MG droop characteristics, the maximum y-axis deviation of the power exchange droop for each sending/receiving MG, Δy_{PE}^{max}, is obtained as follows:

$$\Delta y_{PE}^{max} = (1/2) \cdot \Delta y_{MG}^{max}, \tag{10.17}$$

where $y \in \left[\omega \ V\right]^T$ for $x \in \left[P \ Q\right]^T$, respectively. Δy_{MG}^{max} equals $\Delta\omega_{max}^k$ or ΔV_{max}^k of MG_k. This constraint limits the sender MG to send power only for situations that there is enough power equivalent to utmost $(1/2) \cdot \Delta y_{MG}^{max}$.

2) According to (10.12), the presented GDC-based droop characteristic is a subtraction of the power exchange droop characteristics of sending and receiving MGs as

$$\Delta y_{PE} = y_{PE}^S - y_{PE}^R. \tag{10.18}$$

Hence, the k_p and k_q are calculated according to the frequency and voltage limitations and maximum powers of GDC-based droop characteristics (10.18) as follows

$$k_p = \frac{\left(\omega_{PE}^{max} - \omega_{PE}^{min}\right) - \left(\omega_{PE}^{min} - \omega_{PE}^{max}\right)}{P_{PE}^{max} - \left(-P_{PE}^{max}\right)} = \frac{\Delta\omega_{PE}^{max}}{P_{PE}^{max}}, \tag{10.19a}$$

$$k_q = \frac{\left(V_{PE}^{max} - V_{PE}^{min}\right) - \left(V_{PE}^{min} - V_{PE}^{max}\right)}{Q_{PE}^{max} - \left(-Q_{PE}^{max}\right)} = \frac{\Delta V_{PE}^{max}}{Q_{PE}^{max}}, \tag{10.19b}$$

where $\Delta\omega_{PE}^{max}$ and $\Delta\omega_{PE}^{max}$ are calculated using (10.17), and P_{PE}^{max} and Q_{PE}^{max} are the rated active and reactive powers of the interlinking BTBC determined as

$$x_{PE}^{max} = (1/2) \cdot x_{MGs}^{max}, \tag{10.20}$$

where x_{MGs}^{max} is the active/reactive power of the sender MG, which can be recognized by the frequency/voltage deviation sign according to (10.12). This constraint limits the rated $BTBC_{ij}$ power between MG_i and MG_j as

$$x_{PEij}^{max} = (1/2) \cdot \max\{x_{MGi}^{max}, x_{MGj}^{max}\}. \tag{10.21}$$

Finally, the main references of the active and reactive exchanged powers can be calculated as

$$\begin{bmatrix} P_{EX}^{ref} \\ Q_{EX}^{ref} \end{bmatrix} = K_z \begin{bmatrix} X_{IL}/Z_{IL} & R_{IL}/Z_{IL} \\ -R_{IL}/Z_{IL} & X_{IL}/Z_{IL} \end{bmatrix} \begin{bmatrix} P_{EX}^{ref} \\ Q_{EX}^{ref} \end{bmatrix}', \tag{10.22}$$

where $K_z = \left(\left(R_{IL}/Z_{IL}\right)^2 + \left(X_{IL}/Z_{IL}\right)^2\right)^{-1}$. Note that, P_{EX}^{ref} and Q_{EX}^{ref} are used in (10.7) as the power references.

10.6.3 Coordination of BTBC Emergency Controls

The emergency control for two BTBC-IMGs illustrated previously can be generalized for more IMGs. A coordination is required among connected BTBCs to each MG, in fact among all BTBCs in IMGs. The coordination is based on the minimum number of interconnecting, which means if the first BTBC/connection compensates for the power shortage, the second BTBC/connection should be avoided. It is realized by off-line adjusting for frequency and voltage thresholds of BTBCs as well as averaging intervals as follows:

$$\Delta\omega_{Tr1}^{On} < \Delta\omega_{Tr2}^{On} < \cdots < \Delta\omega_{Trk}^{On} < \cdots < \Delta\omega_{Trn}^{On}, \tag{10.23a}$$

$$\Delta V_{Tr1}^{On} < \Delta V_{Tr2}^{On} < \cdots < \Delta V_{Trk}^{On} < \cdots < \Delta V_{Trn}^{On}, \tag{10.23b}$$

$$\Delta T_{av1} < \Delta T_{av} < \cdots < \Delta T_{avk} < \cdots < \Delta T_{Trn}^{On}, \tag{10.23c}$$

where $\Delta\omega_{Trk}^{On}$, ΔV_{Trk}^{On}, and ΔT_{avk} are, respectively, the frequency threshold, the voltage threshold and the averaging interval of k'th priority in MG interconnecting to the critical MG. Hence, after detecting the emergency condition by $BTBC_k$, $BTBC_{k+1}$ has a delay based on the threshold difference, averaging interval difference and persistence check. If $BTBC_{k+1}$ thresholds are also exceeded, it will be connected due to a large power deficiency. Note that more reliable and flexible coordination can be achieved using the communication infrastructure [18, 51], which is left for future works.

10.6.4 Real-Time Simulation Results

In order to verify the emergency power exchange control, the presented structure in Figure 10.19 is implemented in MATLAB/SimPowerSystems. A real-time simulation is provided using the OPAL-RT digital simulator shown in Figure 8.15 to indicate a larger degree of practicality for the emergency controller. The details of the BTBCs shown in Figure 10.19 can be seen in Figure 10.9. Each MG has two DERs with the primary control shown in Figure 10.8(b). Table 10.4 depicts the system information including the power parts and controllers. In the first three scenarios, only MG_1 and MG_2 interconnected through $BTBC_{12}$ are studied, but the last scenario includes all three MGs.

10.6.4.1 Bidirectional Power Flow Support

In this scenario, MG_1 and MG_2 are operated independently at the beginning until $t = 1$ s. MG_2 is overloaded at $t = 1$ s to $1.1P_{max}^{MG_2}$ and $0.6Q_{max}^{MG_2}$ and it comes back to the normal situation at $t = 3$ s. Another rigorous event occurs at $t = 5$ s, which is the MG_1 overload as $1P_{max}^{MG_1}$ and $1.1Q_{max}^{MG_1}$. The normal condition after the event is provided at $t = 7$ s.

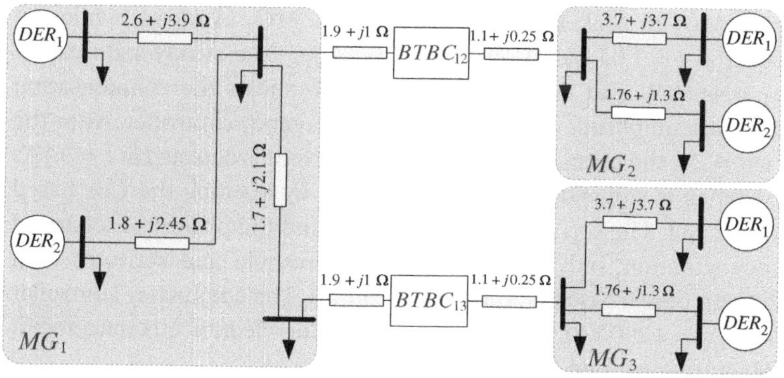

Figure 10.19 Study structure of interconnected microgrids including three autonomous microgrids and two back-to-back converters.

Table 10.4 Power and control data of the interconnected MGs.

Parameters	Value			
RMS line-line voltage (V)	400			
dc voltage and initial value (V)	780,750			
Angular frequency (rad/s)	100\pi			
dc capacitor (μF)	1200			
MGs	MG_1		MG_2, MG_3	
MG P_{max} (kW)	3		3	
MG Q_{max} (kVAr)	1.5		1.5	
Initial load	$0.5P_{max}$		$0.8P_{max}$	
	$0.3Q_{max}$		$0.4Q_{max}$	
DERs	DER_1	DER_2	DER_1	DER_2
DER ratings (S_{max}) (kVA)	1.12	2.24	1.12	2.24
Series filter inductance (mH)	1.8	1.8	1.5	1.5
Series filter resistance (Ω)	0.05	0.05	0.05	0.05
Shunt filter capacitance (μF)	3.7	7.5	5.2	10
P–f droop coefficients (10^{-3} rad/s · W)	1.6	0.8	1.6	0.8
Q–V droop coefficients (10^{-2} V/VAR)	6.5	3.2	6.5	3.2
Basic control coefficients	K_P		K_I	
PI voltage controller for all DERs	0.05		20	
PI current controller	30		500	
PI dc voltage controller	5		1778	
Emergency control	$BTBC_{12}$		$BTBC_{13}$	
Emergency detection thresholds	0.63 rad/s, 13 V		0.75 rad/s, 17 V	
Normal detection thresholds	0.35 rad/s, 7 V		0.35 rad/s, 7 V	
Averaging time interval (s)	0.1		0.2	

Figure 10.20 shows the active and reactive exchanging powers at the side of VSC_2. The power starts to flow at $t = 1.14$ s from MG_1 to MG_2 to compensate for the MG_2 overload, while the emergency condition is detected at $t = 1.04$ based on the $\Delta T_{av} = 0.1$ s. The steady-state values of exchanged powers are $P_{EX} = +650$ W and $Q_{EX} = +308$ VAr, which shows the compensation based on both frequency and voltage amplitude deviations in the emergency controller. After the overload elimination in MG_2 at $t = 3$ s, the normal condition after the event is detected at $t = 3.12$ s and the emergency power exchange is stopped completely at $t = 4$ s by opening the CBs based on the automatic controller command. The second event is detected and the power is exchanged at $t = 5.18$ s. In this emergency situation, only ΔV_{PE} exceeds its threshold and activates both voltage and frequency averaging processes based on the logical control. The emergency controller exchanges $Q_{EX} = +640$ W and $Q_{EX} = +290$ VAr from MG_2 to MG_1. The clearance is detected at $t = 7.06$ s and the BTBC and MGs are separated completely at $t = 7.67$ s.

Figures 10.21(a) and (b) show the frequency of the DERs and the PCC voltage of the MGs. The frequency and voltage of MG_2 are supported in [1 3] s as well as MG_1 frequency and voltage are

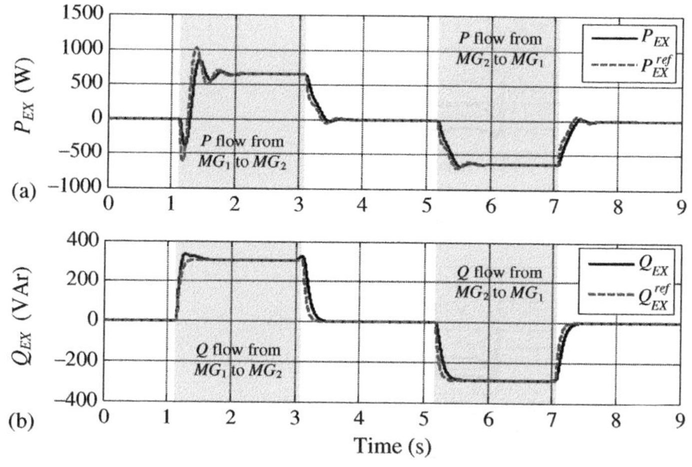

Figure 10.20 Powers exchanging from VSC_2 side: (a) the measured active power and its reference and (b) the measured reactive power and its reference.

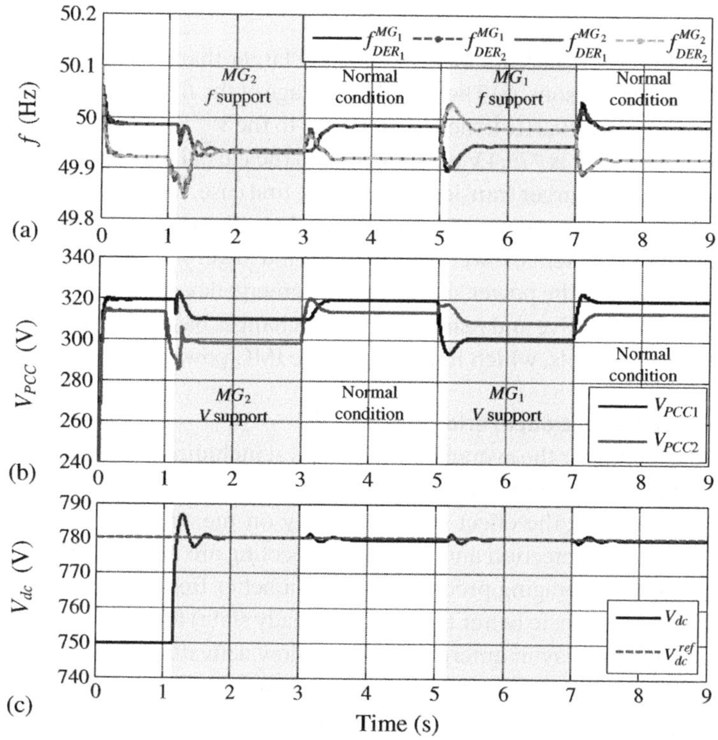

Figure 10.21 Interconnected MG features during critical condition under the emergency control performance: (a) DER frequencies, (b) MG PCC voltage amplitudes, and (c) dc link voltage of $BTBC_{12}$.

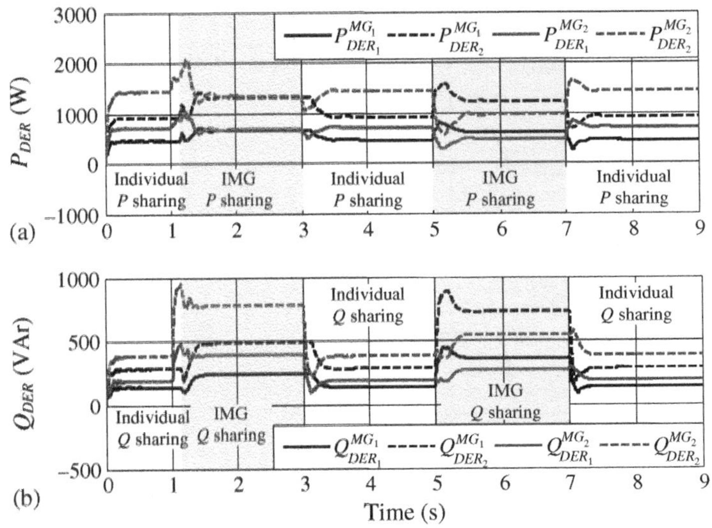

Figure 10.22 Power sharing of DERs during emergency and normal conditions: (a) active power sharing and (b) reactive power sharing.

supported in [5 7] s. The transients of power flow from MG_1 to MG_2 are larger than the transients of power flow from MG_2 to MG_1 due to two reasons. (i) The initial dc voltage of the $BTBC_{12}$ at $t = 1$ s is considered to be 750 V. As shown in Figure 10.21(c), the V_{dc} increases to the $V_{dc}^{ref} = 780$ V during power flow by full charging the dc capacitor. It is 779.4 V at $t = 5$ s due to the partial discharge. The larger difference between V_{dc} and V_{dc}^{ref} causes larger transients. (ii) In the first case, the power flows in the opposite direction of the charging capacitor power. However, they have the same direction in the second case, which leads to smoother transients, lower voltage drop, and lower frequency nadir.

Figure 10.22 depicts the power sharing and the power support by the emergency controller from the normal MG to the overloaded MG. Both active and reactive power exchanges based on the GDC method improve voltage and frequency controls, which in turn enhance IMG power sharing.

10.6.4.2 Averaging Interval Impact on the Controller Performance

In this scenario, three values are considered for the averaging time interval including 0.05, 0.1, and 0.2 s. The frequency and PCC voltage of MGs are shown in Figure 10.23 for the same condition as Section 10.6.4.1 except $V_{dc0} = 780$ V. Since the effect of ΔT_{av} is only on the activation of the emergency controller, the normal condition detection and IMG disconnecting are not shown. The larger ΔT_{av}, which gives more time to the averaging process results in a better frequency/voltage restoration. Both Figures 10.23(a) and (b) indicate better restoration (steady state) for $\Delta T_{av} = 0.2$ s. Nevertheless, the larger ΔT_{av} causes more delay in emergency power flow activating, which may lead to tripping protection devices. The delay can just be seen for the voltage in Figure 10.23(b). Because the frequency does not exceed (10.13a) and is not challenging for detection.

10.6.4.3 DER Plug-and-Play

A well-known large disturbance in MGs is DER plug-and-play, which leads to power deficiency and can be supported by the presented emergency controller. It is notable to mention that the secondary control and a synchronization method are necessary for reconnecting the DER. The distributed secondary control and the synchronization method presented in [68] and [69] are used here. DER_1 of MG_1 is disconnected at $t = 1$ s and it is reconnected at $t = 3$ s. The synchronization

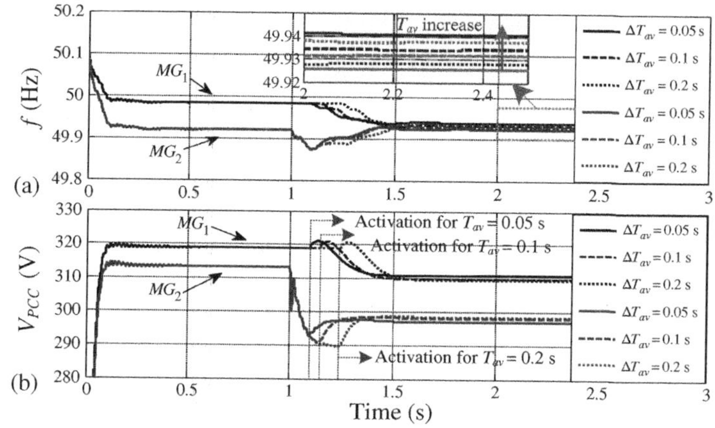

Figure 10.23 MG features for different values of the averaging time interval comprising 0.05, 0.1, and 0.2 s: (a) MG frequency and (b) PCC voltage.

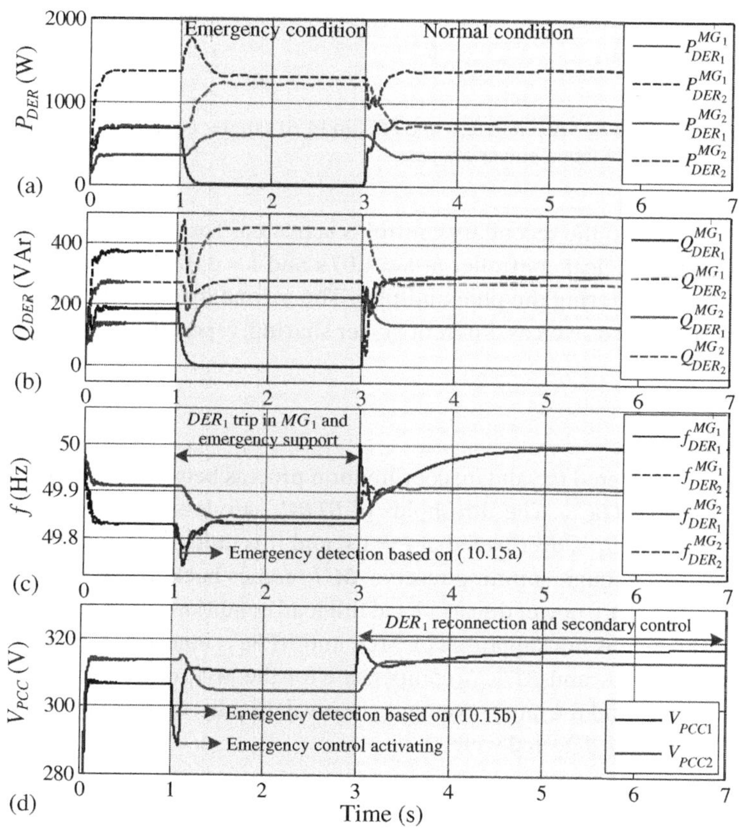

Figure 10.24 DER_1 plug-and-play scenario in MG_1 at $t = 1$ s and $t = 3$ s, respectively: (a) active power, (b) reactive power, (c) frequency, and (d) PCC voltage of MGs.

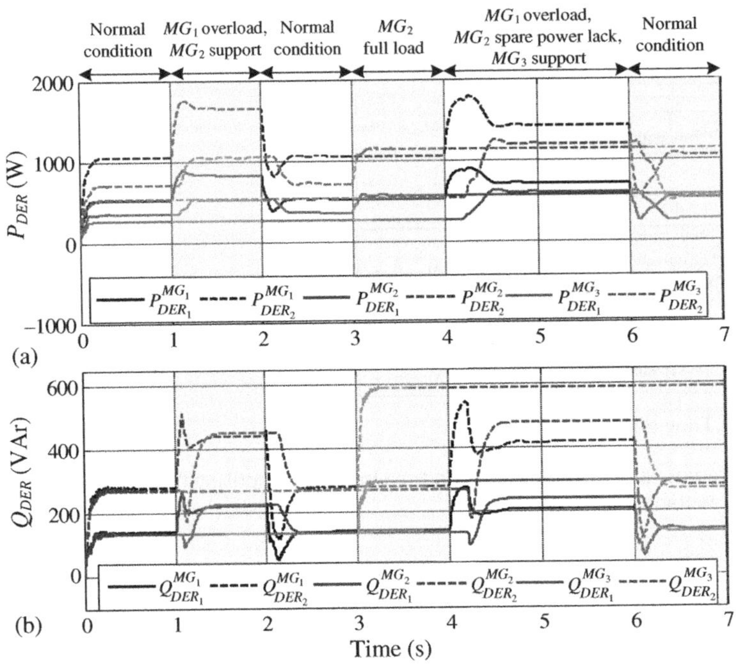

Figure 10.25 MG_1 power shortage supported by MG_2 in [1 2] s and MG_3 in [4 6] s under emergency control coordination: (a) active power of MGs and (b) reactive power of MGs.

method is activated before $t = 3$ s and then the secondary control is activated. The emergency and normal conditions are detected by the logical controller at $t = 1.07$ s and $t = 3.07$ s, respectively. Figures 10.24(a)–(d) show MG_2 support during the plug-and-play. The secondary control in MG_1 leads to frequency and voltage restoration as well as different power sharing, especially in the case of reactive power.

10.6.4.4 Three Interconnected Microgrids

In this scenario, all three MGs are considered to validate coordination process between individual emergency controllers of $BTBC_{12}$ and $BTBC_{13}$. The thresholds of $BTBC_{12}$ are lesser as shown in Table 10.4. As shown in Figure 10.25, MG_1 is overloaded at $t = 1$ s, and it is quickly supported by MG_2 due to the lesser thresholds and enough spinning reserve. MG_1 comes back to the normal condition at $t = 2$ s, and MG_2 loses the spare power due to a local full load condition at $t = 3$ s. Next similar MG_1 overload condition at $t = 4$ s is not supported by MG_2 and $BTBC_{12}$ emergency control due to lack of spare power. However, MG_3 and $BTBC_{13}$ compensate for the MG_1 overload with a larger delay due to the larger thresholds and the larger averaging interval. Finally, MG_1 comes back to the normal condition at $t = 6$ s, and it is detected well.

10.7 Summary

This chapter presents four valuable studies, including two cases of CB-IMGs and two cases of BTBC-IMGs. A simplified model of CB-IMGs is introduced in Chapter 8, which is further developed in this chapter to improve the frequency response by tuning the secondary control gains and adding a virtual inertia loop. Both methods have a positive impact on the transient frequency response. The second study of CB-IMGs focuses on power sharing between IMGs using a pinning consensus-based distributed secondary control, which is already introduced in Chapter 8. On the other hand, power exchanges among IMGs via BTBCs can be scheduled or done in an emergency situation. In the third case study, a planned power exchange among BTBC-IMGs is studied for two and three IMGs, where a stable and appropriate performance is achieved. In contrast to CB-IMGs, where power sharing is possible through the droop mechanism and leader-follower secondary control, power exchange according to exact power references is possible in BTBC-IMGs using BTBC power control and by assuring the available energy of DERs in the global control level. In the fourth case study, emergency control is developed for BTBC-IMGs using logical control and generalized droop characteristics to support power shortfalls in an adjacent MG. The proposed controller is validated for different scenarios, such as bidirectional power flow support, voltage/frequency support, and DER plug-and-play. Due to using BTBCs in BTBC-IMGs and isolating their frequencies in stable operation, synchronization of BTBC-IMGs is not a significant control challenge and can be done using existing PLLs of both converters at both BTBC sides. However, synchronization is a main challenge in CB-IMGs, where CBs connect to MGs directly, and differences in frequencies, voltage magnitudes, and phases can cause instability. This will be studied in Chapter 11 as the synchronization stability of CB-IMGs.

References

1 Josep M Guerrero, Juan C Vasquez, José Matas, Luis García de Vicuna, and Miguel Castilla. Hierarchical control of droop-controlled AC and DC microgrids—a general approach toward standardization. *IEEE Trans. Ind. Electron.*, 58(1):158–172, 2011.

2 Hassan Bevrani, Bruno François, and Toshifumi Ise. *Microgrid dynamics and control*. John Wiley & Sons, Hoboken, NJ, 2017.

3 Yousef Khayat, Qobad Shafiee, Rasool Heydari, Mobin Naderi, Tomislav Dragičević, John W Simpson-Porco, Florian Dörfler, Mohammad Fathi, Frede Blaabjerg, Josep M Guerrero, et al. On the secondary control architectures of AC microgrids: An overview. *IEEE Trans. Power Electron.*, 35(6):6482–6500, 2019. doi: 10.1109/TPEL.2019.2951694.

4 Ali Akhavan, Hamid Reza Mohammadi, Juan C Vasquez, and Josep M Guerrero. Coupling effect analysis and control for grid-connected multi-microgrid clusters. *IET Power Electron.*, 13(5):1059–1070, 2020.

5 Megha Goyal and Arindam Ghosh. Microgrids interconnection to support mutually during any contingency. *Sustainable Energy Grids Netw.*, 6:100–108, 2016.

6 Sandeep Bala and Giri Venkataramanan. Autonomous power electronic interfaces between microgrids. In *2009 IEEE Energy Conversion Congress and Exposition, ECCE'2009*, pages 3006–3013. IEEE, 2009.

7 S M Ferdous, Farhad Shahnia, and G M Shafiullah. Power sharing and control strategy for microgrid clusters. In *9th International Conference on Power and Energy Systems (ICPES)*, pages 1–5. IEEE, 2019.

8 Liang Che, Mohammad Shahidehpour, Ahmed Alabdulwahab, and Yusuf Al-Turki. Hierarchical coordination of a community microgrid with AC and DC microgrids. *IEEE Trans. Smart Grid*, 6(6):3042–3051, 2015.

9 A Hasib Chowdhury and Md Asaduz-Zaman. Load frequency control of multi-microgrid using energy storage system. In *8th International Conference on Electrical and Computer Engineering*, pages 548–551, 2014.

10 Meysam Gheisarnejad and Mohammad Hassan Khooban. Secondary load frequency control for multi-microgrids: Hil real-time simulation. *Soft Comput.*, 23(14):5785–5798, 2019.

11 Deepak Kumar Lal, Ajit Kumar Barisal, and M Tripathy. Load frequency control of multi area interconnected microgrid power system using grasshopper optimization algorithm optimized fuzzy PID controller. In *2018 Recent Advances on Engineering, Technology and Computational Sciences (RAETCS)*, pages 1–6. IEEE, 2018.

12 Chunxia Dou, Yuanzhe Li, Dong Yue, Zhanqiang Zhang, and Bo Zhang. Distributed cooperative control method based on network topology optimisation in microgrid cluster. *IET Renew. Power Gener.*, 14(5):939–947, 2020.

13 Mohammad S Golsorkhi, David John Hill, and Hamid Reza Karshenas. Distributed voltage control and power management of networked microgrids. *IEEE J. Emerging Sel. Top. Power Electron.*, 6(4):1892–1902, 2018.

14 Ying Han, Yuchen Pu, Qi Li, Wangxuan Fu, Weirong Chen, Zhiyu You, Hong Liu. Coordinated power control with virtual inertia for fuel cell-based DC microgrids cluster. *Int. J. Hydrogen Energy*, 44(46):25207–25220, 2019.

15 Ran Hao, Qian Ai, Ziqing Jiang, and Yuchao Zhu. A novel adaptive control strategy of interconnected microgrids for delay-dependent stability enhancement. *Int. J. Electr. Power Energy Syst.*, 99:566–576, 2018.

16 Jinghan He, Xiaoyu Wu, Xiangyu Wu, Yin Xu, and Josep M Guerrero. Small-signal stability analysis and optimal parameters design of microgrid clusters. *IEEE Access*, 7:36896–36909, 2019.

17 Xiaoyu Wu, Yin Xu, Xiangyu Wu, Jinghan He, Josep M Guerrero, Chen-Ching Liu, Kevin P Schneider, and Dan T Ton. A two-layer distributed control method for islanded networked microgrid systems. *IEEE Trans. Smart Grid*, 11(2):942–957, 2019.

18 Md Jahangir Hossain, Md Apel Mahmud, Federico Milano, Seddik Bacha, and Ahmad Hably. Design of robust distributed control for interconnected microgrids. *IEEE Trans. Smart Grid*, 7(6):2724–2735, 2016.

19 Blessy John, Arindam Ghosh, Megha Goyal, and Firuz Zare. A DC power exchange highway based power flow management for interconnected microgrid clusters. *IEEE Syst. J.*, 13(3):3347–3357, 2019.

20 Md Tanvir Arafat Khan, Aranya Chakrabortty, Iqbal Husain, and Rafael Cisneros. Coordinated control of energy storage in networked microgrids under unpredicted load demands. In *2019 American Control Conference (ACC)*, pages 1573–1578. IEEE, 2019.

21 Mojtaba Khederzadeh, Hamed Maleki, and Vahid Asgharian. Frequency control improvement of two adjacent microgrids in autonomous mode using back to back voltage-sourced converters. *Int. J. Electr. Power Energy Syst.*, 74:126–133, 2016.

22 Hyeong-Jun Yoo, Thai-Thanh Nguyen, and Hak-Man Kim. Multi-frequency control in a stand-alone multi-microgrid system using a back-to-back converter. *Energies*, 10(6):1–18, 2017.

23 Inam Ullah Nutkani, Poh Chiang Loh, and Frede Blaabjerg. Distributed operation of interlinked AC microgrids with dynamic active and reactive power tuning. *IEEE Trans. Ind. Appl.*, 49(5):2188–2196, 2013.

24 Inam Ullah Nutkani, Poh Chiang Loh, and Frede Blaabjerg. Power flow control of intertied AC microgrids. *IET Power Electron.*, 6(7):1329–1338, 2013.

25 Inam Ullah Nutkani, Poh Chiang Loh, Peng Wang, Tseng King Jet, and Frede Blaabjerg. Intertied AC–AC microgrids with autonomous power import and export. *Int. J. Electr. Power Energy Syst.*, 65:385–393, 2015.

26 Mobin Naderi, Yousef Khayat, Qobad Shafiee, Tomislav Dragicevic, Frede Blaabjerg, and Hassan Bevrani. An emergency active and reactive power exchange solution for interconnected microgrids. *IEEE J. Emerging Sel. Top. Power Electron.*, 9(5):5206–5218, 2019.

27 Srayashi Konar and Arindam Ghosh. Interconnection of islanded DC microgrids. In *2015 IEEE PES Asia-Pacific Power and Energy Engineering Conference (APPEEC)*, pages 1–5. IEEE, 2015.

28 Mahesh Kumar, Suresh Chandra Srivastava, Sri Niwas Singh, and Mylavarapu Ramamoorty. Development of a control strategy for interconnection of islanded direct current microgrids. *IET Renew. Power Gener.*, 9(3):284–296, 2014.

29 Jingang Lai, Xiaoqing Lu, Xinghuo Yu, and Antonello Monti. Cluster-oriented distributed cooperative control for multiple AC microgrids. *IEEE Trans. Ind. Inf.*, 15(11):5906–5918, 2019.

30 Xiaoqing Lu, Jingang Lai, and Xinghuo Yu. A novel secondary power management strategy for multiple AC microgrids with cluster-oriented two-layer cooperative framework. *IEEE Trans. Ind. Inf.*, 17(2):1483–1495, 2020.

31 Yilin Li, Ping Dong, Mingbo Liu, and Guokang Yang. A distributed coordination control based on finite-time consensus algorithm for a cluster of DC microgrids. *IEEE Trans. Power Syst.*, 34(3):2205–2215, 2018.

32 Yuko Hirase, Yuki Ohara, and Hassan Bevrani. Virtual synchronous generator based frequency control in interconnected microgrids. *Energy Rep.*, 6:97–103, 2020.

33 Mobin Naderi, Qobad Shafiee, Hassan Bevrani, and Frede Blaabjerg. Low-frequency small-signal modeling of interconnected AC microgrids. *IEEE Trans. Power Syst.*, 36(4):2786–2797, 2020.

34 Yaran Li, Ke Meng, and Zhao Yang Dong. Distributed consensus control with event-triggered communication for multi-microgrid cluster. In *2019 IEEE Milan PowerTech*, pages 1–6. IEEE, 2019.

35 Qiang Li, Mengkai Gao, Houfei Lin, Ziyu Chen, and Minyou Chen. MAS-based distributed control method for multi-microgrids with high-penetration renewable energy. *Energy*, 171:284–295, 2019.

36 Wei Liu, Wei Gu, Yinliang Xu, Yong Wang, and Kaifeng Zhang. General distributed secondary control for multi-microgrids with both PQ-controlled and droop-controlled distributed generators. *IET Gener. Transm. Distrib.*, 11(3):707–718, 2017.

37 Shivraman Mudaliyar, Bhanu Duggal, and Sukumar Mishra. Distributed tie-line power flow control of autonomous DC microgrid clusters. *IEEE Trans. Power Electron.*, 35(10):11250–11266, 2020.

38 Ehsan Pashajavid, Arindam Ghosh, and Firuz Zare. A multimode supervisory control scheme for coupling remote droop-regulated microgrids. *IEEE Trans. Smart Grid*, 9(5):5381–5392, 2018.

39 Lingyu Ren, Yanyuan Qin, Yan Li, Peng Zhang, Bing Wang, Peter B Luh, Song Han, Taofeek Orekan, and Tao Gong. Enabling resilient distributed power sharing in networked microgrids through software defined networking. *Appl. Energy*, 210:1251–1265, 2018.

40 Subham Sahoo, Sukumar Mishra, Seyed Mahdi Fazeli, Furong Li, and Tomislav Dragičević. A distributed fixed-time secondary controller for DC microgrid clusters. *IEEE Trans. Energy Convers.*, 34(4):1997–2007, 2019.

41 Kevin P Schneider, Nikitha Radhakrishnan, Yingying Tang, Francis K Tuffner, Chen-Ching Liu, Jing Xie, and Dan Ton. Improving primary frequency response to support networked microgrid operations. *IEEE Trans. Power Syst.*, 34(1):659–667, 2018.

42 Qobad Shafiee, Tomislav Dragičević, Juan C Vasquez, and Josep M Guerrero. Hierarchical control for multiple DC-microgrids clusters. *IEEE Trans. Energy Convers.*, 29(4):922–933, 2014. doi: 10.1109/TEC.2014.2362191.

43 Umamaheswararao Vuyyuru, Suman Maiti, and Chandan Chakraborty. Active power flow control between DC microgrids. *IEEE Trans. Smart Grid*, 10(5):5712–5723, 2019.

44 Zhaojian Wang, Laijun Chen, Feng Liu, Peng Yi, Ming Cao, Sicheng Deng, and Shengwei Mei. Asynchronous distributed power control of multi-microgrid systems. *IEEE Trans. Control Netw. Syst.*, 7(4):1960–1973, 2020.

45 Wayne W Weaver, Rush Robinett, Gordon G Parker, and David G Wilson. Distributed control and energy storage requirements of networked DC microgrids. *Control Eng. Pract.*, 44:10–19, 2015.

46 David G Wilson, Wayne W Weaver, Rush D Robinett, and Steve F Glover. Nonlinear power flow control design for networked AC/DC based microgrid systems. In *2018 American Control Conference (ACC)*, pages 5698–5705. IEEE, 2018.

47 David G Wilson, Rush D Robinett III, Wayne W Weaver, and Steven F Glover. Nonlinear power flow control for networked AC/DC microgrids, May 26 2020. US Patent 10,666,054.

48 Shengxuan Weng, Yusheng Xue, Jianbo Luo, and Yanman Li. Distributed secondary control for islanded microgrids cluster based on hybrid-triggered mechanisms. *Processes*, 8(3):370, 2020.

49 Li Yang, Zhijian Hu, Shiwei Xie, Shunfei Kong, and Weiwei Lin. Adjustable virtual inertia control of supercapacitors in PV-based AC microgrid cluster. *Electr. Power Syst. Res.*, 173:71–85, 2019.

50 Ramon Zamora and Anurag K Srivastava. Multi-layer architecture for voltage and frequency control in networked microgrids. *IEEE Trans. Smart Grid*, 9(3):2076–2085, 2018.

51 Jianguo Zhou, Huaguang Zhang, Qiuye Sun, Dazhong Ma, and Bonan Huang. Event-based distributed active power sharing control for interconnected AC and DC microgrids. *IEEE Trans. Smart Grid*, 5(6):6815–6828, 2018.

52 Xiaoping Zhou, Leming Zhou, Yandong Chen, Josep M Guerrero, An Luo, Wenhua Wu, and Ling Yang. A microgrid cluster structure and its autonomous coordination control strategy. *Int. J. Electr. Power Energy Syst.*, 100:69–80, 2018.

53 Atif Zaman Khan and Farrukh Shahzad. A PC based software package for the equal area criterion of power system transient stability. *IEEE Trans. Power Syst.*, 13(1):21–26, 1998.

54 Abdolwahhab Fathi, Qobad Shafiee, and Hassan Bevrani. Robust frequency control of microgrids using an extended virtual synchronous generator. *IEEE Trans. Power Syst.*, 33(6):6289–6297, 2018. doi: 10.1109/TPWRS.2018.2850880.

55 Thongchart Kerdphol, Fathin Saifur Rahman, Yasunori Mitani, Masayuki Watanabe, and Sinan Küfeo Küfeolu. Robust virtual inertia control of an islanded microgrid considering high penetration of renewable energy. *IEEE Access*, 6:625–636, 2017.

56 Thongchart Kerdphol, Fathin Saifur Rahman, Masayuki Watanabe, Yasunori Mitani, Dirk Turschner, and Hans-Peter Beck. Enhanced virtual inertia control based on derivative technique to emulate simultaneous inertia and damping properties for microgrid frequency regulation. *IEEE Access*, 7:14422–14433, 2019.

57 Mobin Naderi, Qobad Shafiee, Frede Blaabjerg, and Hassan Bevrani. Synchronization stability of interconnected microgrids with fully inverter-based distributed energy resources. *Journal of Modern Power Systems and Clean Energy*, 1–12, 2023. doi: 10.35833/MPCE.2022.000594.

58 Yousef Khayat, Mobin Naderi, Qobad Shafiee, Yazdan Batmani, Mohammad Fathi, Josep M Guerrero, and Hassan Bevrani. Decentralized optimal frequency control in autonomous microgrids. *IEEE Trans. Power Syst.*, 34(3):2345–2353, 2018. doi: 10.1109/TPWRS.2018.2889671.

59 Amirnaser Yazdani and Reza Iravani. *Voltage-sourced converters in power systems*, volume 34. Wiley Online Library, Hoboken, NJ, 2010.

60 Mohammad Fathi and Hassan Bevrani. Statistical cooperative power dispatching in interconnected microgrids. *IEEE Trans. Sustainable Energy*, 4(3):586–593, 2013.

61 Hao Wang and Jianwei Huang. Incentivizing energy trading for interconnected microgrids. *IEEE Trans. Smart Grid*, 9(4):2647–2657, 2018. doi: 10.1109/TSG.2016.2614988.

62 Bingying Zhang, Qiqiang Li, Luhao Wang, and Wei Feng. Robust optimization for energy transactions in multi-microgrids under uncertainty. *Appl. Energy*, 217:346–360, 2018.

63 Eddy H Trinklein, Gordon G Parker, Rush D Robinett, and Wayne W Weaver. Toward online optimal power flow of a networked DC microgrid system. *IEEE J. Emerging Sel. Top. Power Electron.*, 5(3):949–959, 2017.

64 Seyed Ali Arefifar, A-RI Mohamed Yasser, and Tarek HM El-Fouly. Optimum microgrid design for enhancing reliability and supply-security. *IEEE Trans. Smart Grid*, 4(3):1567–1575, 2013.

65 British Standard. Voltage characteristics of electricity supplied by public distribution networks. *British Standard*, pages 1505–1513, 2007.

66 Hassan Bevrani and Shoresh Shokoohi. An intelligent droop control for simultaneous voltage and frequency regulation in islanded microgrids. *IEEE Trans. Smart Grid*, 4(3):1505–1513, 2013.

67 Joan Rocabert, Alvaro Luna, Frede Blaabjerg, and Pedro Rodriguez. Control of power converters in AC microgrids. *IEEE Trans. Power Electron.*, 27(11):4734–4749, 2012.

68 Qobad Shafiee, Josep M Guerrero, and Juan C Vasquez. Distributed secondary control for islanded microgrids—a novel approach. *IEEE Trans. Power Electron.*, 29(2):1018–1031, 2014.

69 Juan C Vasquez, Josep M Guerrero, Mehdi Savaghebi, Joaquin Eloy-Garcia, and Remus Teodorescu. Modeling, analysis, and design of stationary-reference-frame droop-controlled parallel three-phase voltage source inverters. *IEEE Trans. Ind. Electron.*, 60(4):1271–1280, 2013.

11

Synchronization in Interconnected Microgrids

11.1 Introduction

AC interconnected microgrids (IMGs) can be operated synchronously with the same frequency or asynchronously with different frequencies. Synchronous IMGs can be constructed using circuit breakers (CBs) [1–5] or static switches (SSs) [6], and asynchronous IMGs are possible using back-to-back converters (BTBCs) [7–9] as interlinking devices. Synchronization is not a prominent challenge in asynchronous IMGs due to the dc links among each two IMGs. In fact, each ac side of BTBCs should just be synchronized with the MG of its side using a phase-locked loop (PLL) and feeding forward the MG point of common coupling (PCC) voltage components to the corresponding ac side controller of the BTBC. Nevertheless, synchronization is a vital issue in synchronous IMGs, where all the MGs are interconnected without any dc interface. Especially in IMGs with fully inverter-based distributed energy resources (DERs) unlike the conventional power systems with a large number of rotary machines, there exists no strong and high-inertia grid-forming DER or MG to provide a stiff grid for other grids to be synchronized with.

The synchronization control has been presented for the seamless transition of single MGs from islanded to grid-connected mode [10–17], where there is a strong grid, which is supposed to be an ideal voltage source, i.e. an infinite bus. In fact, the total power of the MG is much less than the grid synchronous power produced by the massive grid and cannot destabilize it. However, the MG itself may experience large disturbances due to weak synchronization processes. Single DER synchronization [10] has been enhanced to multiple/all DER synchronization based on distributed protocols [11, 16]. In [11, 13, 16], the small-signal stability and control of MG transition modes are analyzed however, the synchronization-related parameters are not taken into account in the stability assessment. Robust and model predictive controllers are employed to improve the seamless transition including both islanding and grid synchronizing [17]. The synchronization process is developed for multi-bus MGs using a distributed leader–follower control strategy [16] and multi-master and multi-slave MGs [18]. The synchronization controller gains are optimized to decrease the inrush current [12], where the threshold values of the voltage, phase, and frequency differences of the MG PCCs and the grid are considered as 3%, 1°, and 0.05 Hz, respectively. Note that IEEE Std. 1547-2003 recommends the synchronization criteria as voltage, phase, and frequency differences as ±10%, 20°, and 0.3 Hz, respectively, in order to synchronize DERs and MGs for less than 500 kVA rated capacity to the utility grid [19].

Microgrids: Dynamic Modeling, Stability and Control, First Edition. Qobad Shafiee, Mobin Naderi, and Hassan Bevrani.

In the IMG research, [20] have focused on the transition mode of forming IMGs and have presented the logic control required for the synchronization and interconnection by static switches. Moreover, a transient stability criterion is proposed as the required time of a successful synchronization for different frequency and phase differences on IMG PCCs. The seamless transition of IMGs is discussed in [21]. A fuzzy-based droop controller [22] and an adaptive control scheme [23] are represented to enhance the transient response. The synchronization modeling and stability are focused in [24], where both small-signal and transient stability analyses have been taken into consideration.

In this chapter, a simple method of synchronization for CB-IMGs is presented, which is adopted from the synchronization control of single MGs [11, 12, 16]. The stability of the synchronization process of weak IMGs with fully inverter-based DERs is analyzed. At first, the inrush power at the beginning of the interconnection instance is modeled and analyzed, which is the main source of disturbance in synchronization studies. The most effective power and control parameters in the inrush power amount are investigated, and the transition from the inrush power situation to the scheduled power after the disturbances is discussed. Small-signal model of the synchronization control unit is proposed, and small-signal stability of IMGs focusing on the parameters affecting the synchronization process, i.e. synchronization and secondary controller parameters, is assessed. Finally, the transient stability of IMG synchronization is analyzed using time-domain simulations and considering the synchronization constraints thresholds and disturbance length impact to investigate the challenges of a complete synchronization process.

11.2 Synchronization Control Requirements

Figure 11.1(a) shows a simple diagram of synchronous IMGs connected through CBs and interlinking lines. They are called synchronous because they are operating with the same frequency after closing the CBs and passing synchronization transients. A general structure of MGs shown in Figure 11.1(b) is studied in this paper. Each DER can generate power up to its rated power, which is flown after filtering the voltage and current by an LC filter via a coupling line to the MG PCC. Both local and remote loads can be supplied, where they are considered as a lumped load at the PCC. The interconnections are established among MGs at PCCs. The MGs can either be operated independently or be interconnected using a coordination control to share the power among them.

11.2.1 Basic Control of CB-IMGs

Figure 11.2 shows a complete view of the control architecture including inner, primary, and secondary control layers, where the control loops of leader and follower DERs are separately represented. Inner and primary control structures are the same for both DER types. Inner voltage and current controllers regulate the voltage and current, respectively, and provide the PWM unit with the inverter voltage reference to work as a voltage source converter (VSC). The primary control unit comprises $\omega - P$ and $v - Q$ droop characteristics, where the frequency and voltage amplitude references are provided for the DERs.

A pinning consensus-based secondary control is selected in which the control architecture is different for leader and follower DERs [3, 4, 25–29]. MGs are generally considered as single-leader DER and multiple follower DERs. Only the leader DERs have rated voltage and frequency values. They have two main duties. The first is communicating power, frequency, and voltage data

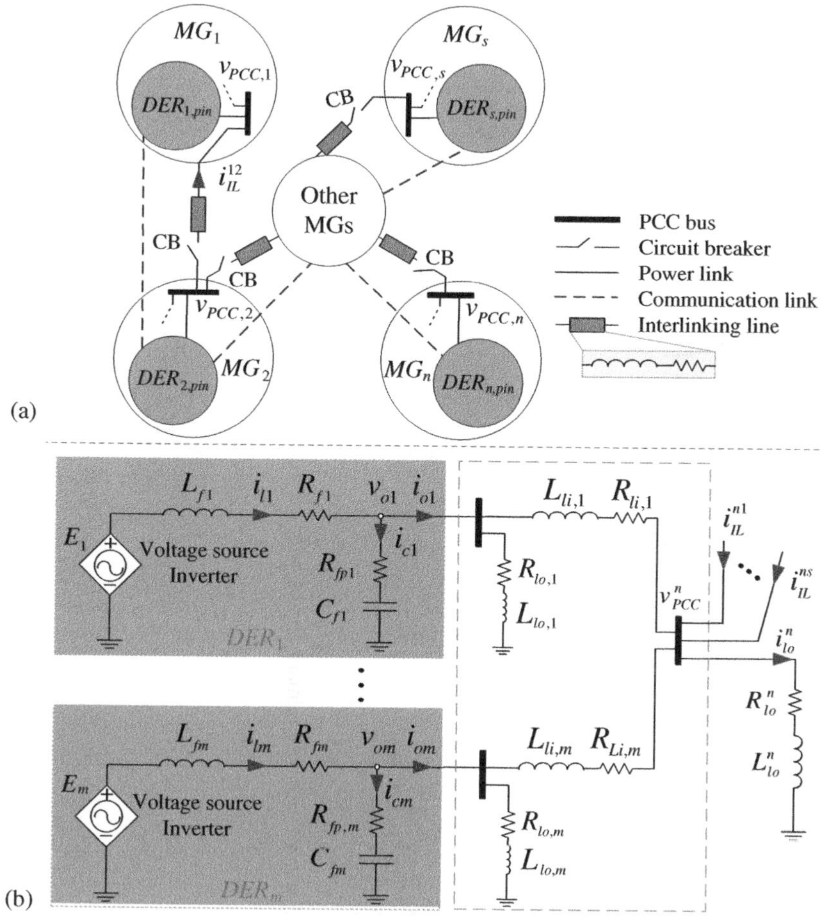

Figure 11.1 A simple diagram of synchronous interconnected MGs: (a) MGs connections through CBs and interlinking lines, (b) MG power parts including voltage source inverters, LC filters, coupling lines, and loads.

with other leader DERs in the neighbor MGs ($a_{mi} = 1$) to control the overall IMG power sharing during the interconnected operation mode. The second is communicating the rated voltage and frequency instantaneously with its follower DERs ($a_{mi} = 1$) inside the MG to achieve stable intra-MG power sharing and restore voltage and frequency to the rated values. There is another communication among neighbor follower DERs ($a_{mj} = 1$) to improve power sharing and voltage/frequency restoration.

11.2.2 Synchronization Control of CB-IMGs

In order to connect two MGs within a set of IMGs, synchronizing MGs at PCCs is necessary. The synchronization process is done by matching frequencies, PCC voltage phases, and magnitudes. The difference of frequencies can be considered to be limited to zero due to frequency restoration done by the secondary frequency loops and having the same rated frequencies. Nevertheless, phase and magnitude differences of the PCC voltages should be controlled to be decreased as much as possible [30]. The lower differences lead to the lower inrush power at the time of interconnection,

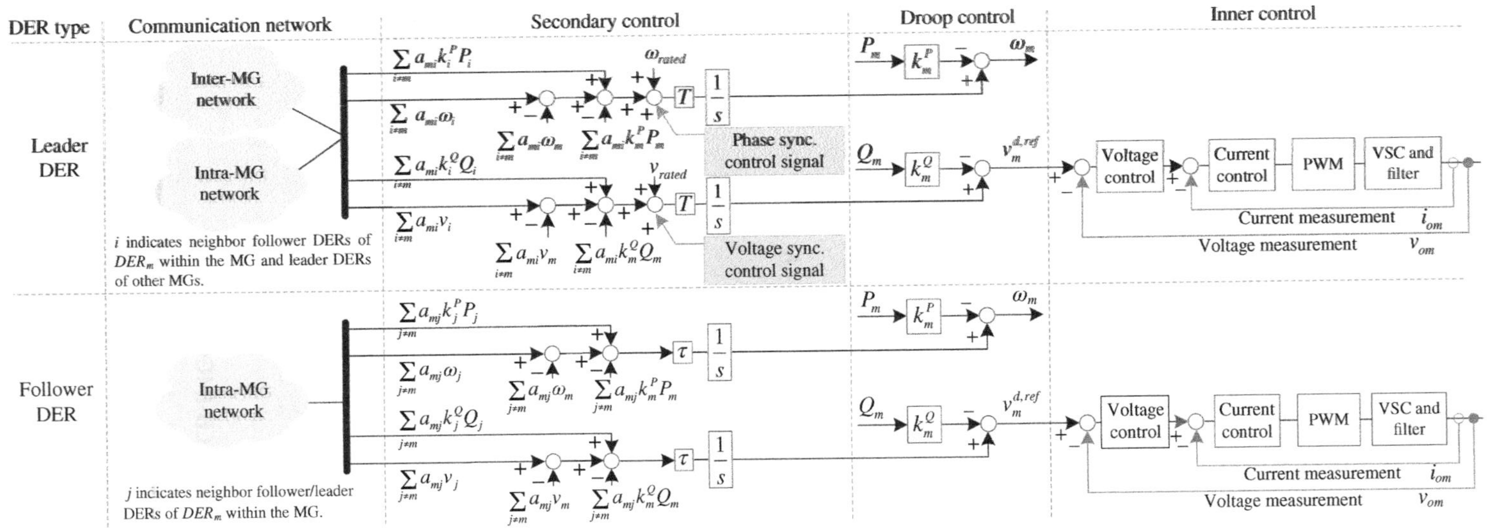

Figure 11.2 A schematic of the basic control of interconnected microgrids including primary, and secondary controllers as well as showing the points, where synchronization control signals are added.

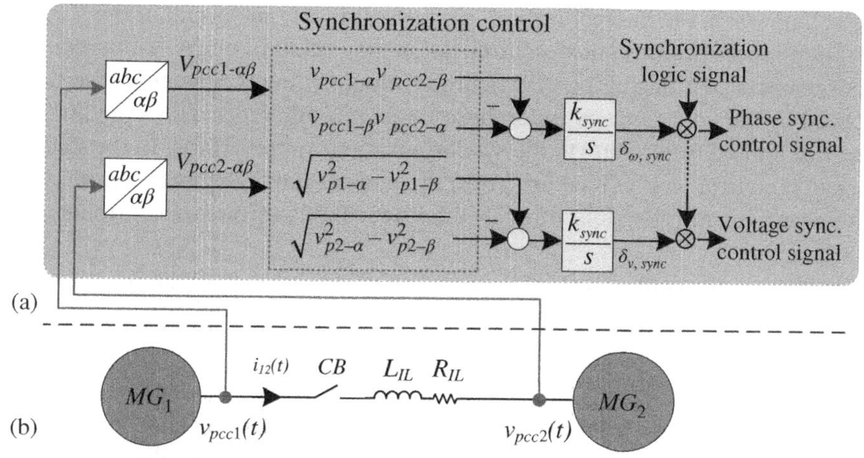

Figure 11.3 The IMG parts involved in synchronization process: (a) synchronization control of interconnected microgrids, (b) PCCs of MG_1 and MG_2 and an interlinking line between them.

The PCC voltage phase and magnitude differences of two typical MGs, MG_1, and MG_2, in $\alpha\beta$ frame are given as follows [11, 12, 16]:

$$\Delta\phi_{pcc} = \phi_{pcc1} - \phi_{pcc2} \propto v_{pcc1-\alpha}v_{pcc2-\beta} - v_{pcc1-\beta}v_{pcc2-\alpha}, \tag{11.1a}$$

$$\begin{aligned}\Delta v_{pcc} &= \left|v_{pcc1}\right| - \left|v_{pcc2}\right| \\ &= \sqrt{v^2_{p1-\alpha} - v^2_{p1-\beta}} - \sqrt{v^2_{p2-\alpha} - v^2_{p2-\beta}},\end{aligned} \tag{11.1b}$$

where ϕ_{pcc1}, $\left|v_{pcc1}\right|$, ϕ_{pcc2}, and $\left|v_{pcc2}\right|$ are the PCC_1 and PCC_2 voltage phases and magnitudes, respectively. According to (11.1a) and (11.1b), the difference signals shown in Figure 11.3(a) are used in the synchronization controllers. The controller is selected as an integral type with a small gain k_{sync} to keep the system stable. Such a common synchronization control structure is used to focus on the stability analysis.

As shown in Figure 11.2, the phase synchronization control signal is added to the $\omega - P$ error signal of the secondary controller in the leader DER control units. Similarly, the voltage magnitude synchronization control signal is added to the $v - Q$ error signal. Note that the phase difference control efforts disturb the frequency and lead to a steady-state error during presynchronization and synchronization. However, applying it is necessary for achieving smaller inrush power.

The Synchronization logic signal determines when the synchronization control unit should be impacted by the IMG control. In fact, this signal changes from 0 to 1 a while before the interconnection instance, which is the presynchronization period. After the synchronization process, the signal goes back to 0. Presynchronization is necessary to make enough time to decrease the phase and voltage differences.

11.3 Inrush Power Analysis

11.3.1 Modeling of Inrush Power

Figure 11.3(b) shows MG_1 and MG_2 and an interlinking line as well as a CB between them. There is no synchronization process to see the pure impact of the voltage magnitude/phase/frequency difference on the inrush power at the interconnection instance. It is also important at the beginning of a synchronization process because presynchronization may result in nonzero differences. Therefore, the inrush power at the beginning of synchronization, i.e. interconnection instance, is intensively influenced by the voltage magnitude/phase/frequency difference. However, after a while, the differences are limited to zero due to the synchronization control efforts. In the steady state, the synchronization control loop will be open and the power exchanging between MGs is fully controlled by the central control system through the leader DERs secondary controllers to share IMG loads.

Considering a single-phase equivalent circuit of the interconnection shown in Figure 11.3(b) and using KVL, one can find:

$$R_{IL}i_{12}(t) + L_{12}\frac{di_{12}}{dt} = v_{pcc1}(t) - v_{pcc2}(t), \tag{11.2}$$

where $i_{12}(t)$ is the single-phase current from the MG_1 to the MG_2. By assuming sine voltages as $v_{pcc1}(t) = V_{p1}\sin(\omega_1 t + \phi_1)$ and $v_{pcc2}(t) = V_{p2}\sin(\omega_2 t + \phi_2)$ and considering particular solutions, respectively, as $i_{12,pr1}(t) = I_{pr1}\sin(\omega_1 t + \theta_1)$ and $i_{12,pr2}(t) = I_{pr2}\sin(\omega_2 t + \theta_2)$ one can solve (11.2) and find the complete solution as follows:

$$\begin{aligned} i_{12}(t) &= (I_{pr2}\sin(\theta_2) - I_{pr1}\sin(\theta_1))e^{-(R/L)t} \\ &\quad + i_{12,pr1}(t) - i_{12,pr2}(t), \end{aligned} \tag{11.3}$$

where I_{pr1}, I_{pr2}, θ_1, and θ_2 are obtained using superposition principle, and replacing $i_{12,pr1}(t)$ and $i_{12,pr2}(t)$ in (11.2), separately as follows:

$$I_{pr1(pr2)} = \frac{V_{p1(p2)}}{\sqrt{R_{IL}^2 - (\omega_{1(2)}L_{IL})^2}}, \tag{11.4a}$$

$$\theta_{1(2)} = \phi_{1(2)} - tg^{-1}\left(\frac{\omega_{1(2)}L_{IL}}{R_{IL}}\right). \tag{11.4b}$$

The current shown in (11.2) is only true at the beginning times after CB closing, which is named the inrush current. Because, in real situations, the controllers change the PCC voltages after passing their time constants. Then, the exchanging current will be changed. The controller effects are not modeled here. According to the exponential form of the current in (11.3), there is no extremum point and thus the maximum current value equals the steady-state amount as

$$I_{inrush} = i_{12,pr1}(t) - i_{12,pr2}(t)\Big|_{t\geq 4L_{IL}/R_{IL}}. \tag{11.5}$$

Finally, by generalizing the voltage and current in a balanced three-phase form, the inrush active and reactive powers at PCC_1 are obtained as follows:

$$P_{inrush} = v_{p1a}(t)i_{12a}(t) + v_{p1b}(t)i_{12b}(t) + v_{p1c}(t)i_{12c}(t), \tag{11.6a}$$

$$\begin{aligned} Q_{inrush} &= \frac{1}{\sqrt{3}}[(v_{p1b}(t) - v_{p1c}(t))i_{12a}(t) + (v_{p1c}(t) \\ &\quad - v_{p1a}(t))i_{12b}(t) + (v_{p1a}(t) - v_{p1b}(t))i_{12c}(t)]. \end{aligned} \tag{11.6b}$$

11.3.2 Impact of PCC Voltage Parameters on the Inrush Power

Figure 11.4 shows the inrush power values at the interconnection instance when the voltage magnitude, phase, and frequency of the PCCs change in probable ranges. Two scenarios are considered,

Table 11.1 Values of influencing variables in the inrush power.

Variable (unit)	Scenario 1 (Figures. 11.4(a) and 11.4(b))	Scenario 2 (Figures. 11.4(c) and 11.4(d))
f_1 (Hz)	[49, 51]	50
V_{pcc1} (V)	[280, 330]	320
ϕ_{pcc1} (°)	5	5
f_2 (Hz)	50	50
V_{pcc2} (V)	320	[280, 330]
ϕ_{pcc2} (°)	0	[−20, 20]

where the parameters are given in Table 11.1. In both scenarios, the time after the interconnection is as $t = 0.15$ s, $R_{IL} = 1.6\,\Omega$, and $L_{IL} = 1.9$ mH. As shown in Figure 11.4(a), the active inrush power increases when V_{pcc1} and f_1 increase. However, changing the frequency is more effective on the inrush power value. There exists also a minimum point with respect to the frequency at $f_1 = 50$ Hz. The reactive inrush power has an inverse behavior with f_1 changes as shown in Figure 11.4(b). A symmetry can be seen by f_1 changes, where Q_{inrush} decreases when f_1 increases. In the second scenario, decreasing V_{pcc2} and increasing ϕ_{pcc2} cause an increase in P_{inrush} as shown in Figure 11.4(c). Q_{inrush} also increases intensively by increasing ϕ_{pcc2} as shown in Figure 11.4(d). In fact, for $\Delta\phi_{pcc} > 0$, the reactive inrush power flows from PCC_2 to PCC_1 and vice versa. Generally, the effectiveness level in the inrush power value belongs, respectively, to frequency, phase, and then voltage magnitude differences.

In autonomous IMGs, there exists no strong grid, unlike the conventional power systems to provide a stiff and robust frequency for other grids to be synchronized with. Therefore, each MG in a synchronization process suffers from a weak and insecure frequency, where the frequency may highly vary around the nominal value due to disturbances, e.g. sudden load changes. Moreover, PCCs may far from the VSC DERs enough that experience large voltage sags and phase shifting, hence leading to different PCC voltage magnitudes and phases.

11.3.3 Impact of X/R Ratio and Impedance Value on the Inrush Power

Figures 11.5 and 11.6 show the inrush powers for different X/R ratios as [0.15] and different impedance values as $[0.1, 2.6]\Omega$ of the interlinking line. Three frequency difference values Δf are considered as −0.1, 0, and 0.1 Hz; however, the voltage magnitude and phase differences are constant as $\Delta V_{pcc} = 5$ V, and $\Delta\phi_{pcc} = 5°$. Another important parameter change in Figure 11.5 is decreasing impedance values by increasing the X/R ratio, which is based on the typical line data for low, medium, and high voltages [21]. The line impedance is reduced from 0.85 to 0.25 Ω when the X/R ratio increases from 0.1 to 5, i.e. from a low-voltage to a high-voltage line.

According to Figures 11.5(a) and 11.5(b), both the active and reactive inrush powers increase by increasing the X/R ratio. However, for $\Delta f = -0.1$ Hz, the changing of P_{inrush} is negligible. It is due to the inverse impact of this Δf with respect to the ΔV_{pcc} and $\Delta\phi_{pcc}$. The Q_{inrush} has also the lowest changes in Scenario 1. Note that the curves are also shown for constant impedance in Figure 11.5. Another important point is the inverse directions of P_{inrush} and Q_{inrush} for the situation of the same sign of Δf, ΔV_{pcc}, and $\Delta\phi_{pcc}$. Finally, for IMGs, which are generally implemented in $X/R < 2$, P_{inrush} has the lowest values, but Q_{inrush} shows the highest values.

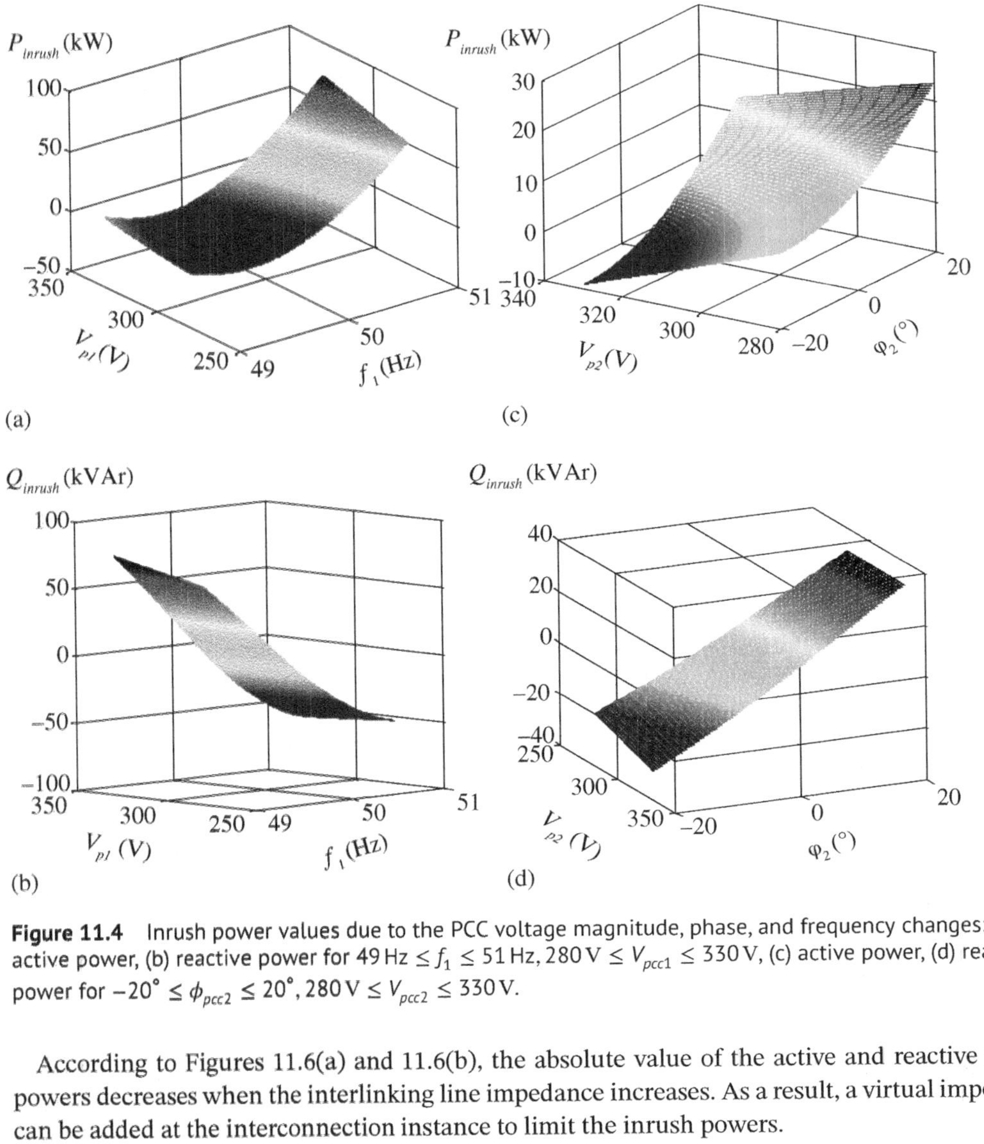

Figure 11.4 Inrush power values due to the PCC voltage magnitude, phase, and frequency changes: (a) active power, (b) reactive power for $49\,\text{Hz} \leq f_1 \leq 51\,\text{Hz}, 280\,\text{V} \leq V_{pcc1} \leq 330\,\text{V}$, (c) active power, (d) reactive power for $-20° \leq \phi_{pcc2} \leq 20°, 280\,\text{V} \leq V_{pcc2} \leq 330\,\text{V}$.

According to Figures 11.6(a) and 11.6(b), the absolute value of the active and reactive inrush powers decreases when the interlinking line impedance increases. As a result, a virtual impedance can be added at the interconnection instance to limit the inrush powers.

11.4 Small-Signal Modeling and Stability Analysis

A small-signal model of the IMGs including the synchronization control loops is derived, where the interconnection method given in [1, 7] is used.

11.4.1 Small-Signal Modeling of IMGs

11.4.1.1 Modeling of Sub-Systems

A complete process of small-signal modeling of IMGs interconnected through BTBCs [7], and interconnected through CBs (CB-IMGs) [1] are already reported by the authors. A list of all subsystems

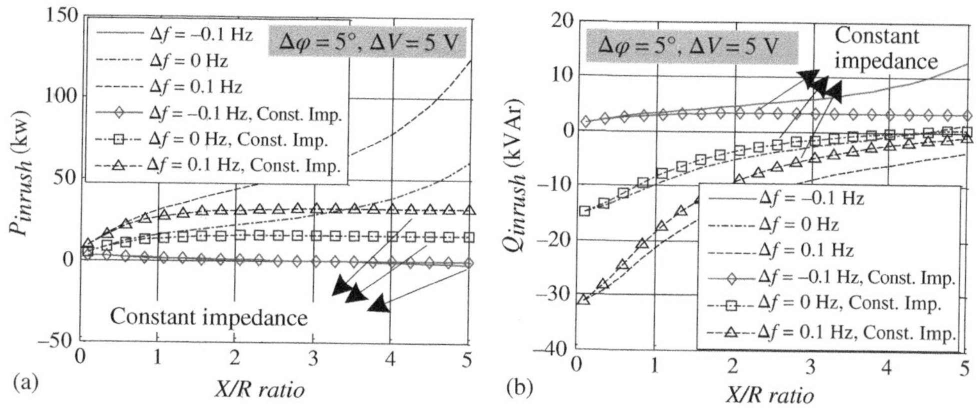

Figure 11.5 Inrush power values for different X/R ratios as [0.1, 5] when $\Delta V_{pcc} = 5$ V and $\Delta \phi_{pcc} = 5°$: (a) active power, (b) reactive power.

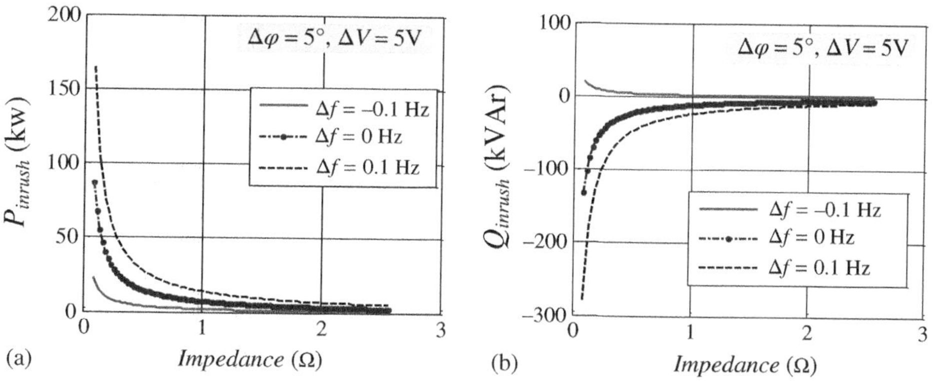

Figure 11.6 Inrush power values for different interlinking line impedance values as [0.1, 2.6] Ω when $\Delta V_{pcc} = 5$ V and $\Delta \phi_{pcc} = 5°$: (a) active power, (b) reactive power.

including interlinking lines, MGs, and interlinking devices to be modeled should be available. In the case of CB-IMGs, the important subsystems participating in the dominant IMG dynamics are the interlinking lines and MGs, which in turn, MGs consist of loads, DER coupling lines, LC filters, and current, voltage, primary, and secondary controllers. Each subsystem is modeled as an independent module considering inputs and outputs from/to other subsystems using a state-space representation [7] as follows:

$$\dot{X}_M(t) = A_M X_M(t) + B_M U_M(t), \tag{11.7a}$$

$$Y_M(t) = C_M X_M(t) + D_M U_M(t), \tag{11.7b}$$

where $X_M(t)$, $U_M(t)$, and $Y_M(t)$ are the module state vector, input vector of all the subsystem input signals, and output vector of all the subsystem output signals. A_M, B_M, C_M, and D_M are the model matrices, which should be found for all sub-systems. For instance, for a DER coupling line shown in Figure 8.1, the inductor currents in the dq frame are the state variables, which are also considered as model outputs, and voltages of the LC filter and the PCC in the dq frame are as inputs. The matrices can be easily calculated, which are presented in [7] for most of the subsystems.

Table 11.2 Parameters a_x, b_x, c_x, and d_x of k_x in (11.10).

	a_x	b_x	c_x	d_x
k_{d1}	V_{d20}	V_{q20}	1	−1
k_{q1}	V_{q20}	V_{d20}	1	1
k_{d2}	V_{d10}	V_{q10}	1	1
k_{q2}	V_{q10}	V_{d10}	1	−1
$k_{\omega 1}$	$\omega_{10}(V_{q10}V_{d20} - V_{d10}V_{q20})$	$\omega_{10}(V_{d10}V_{d20} + V_{q10}V_{q20})$	0	−1
$k_{\omega 2}$	$\omega_{20}(V_{d10}V_{q20} - V_{q10}V_{d20})$	$\omega_{20}(V_{d10}V_{d20} + V_{q10}V_{q20})$	0	1

11.4.1.2 Modeling of Synchronization Control Unit

The only subsystem of studied IMGs, which already is not modeled, is the synchronization controller, including two integrators. The integrators' outputs are considered as the state variables (see Figure 11.3). Hence, the state space representation of each synchronization controller is calculated after a linearization process as

$$\dot{X}_{sync}(t) = A_{sync}X_{sync}(t) + B_{sync}U_{sync}(t), \tag{11.8a}$$

$$Y_{sync}(t) = C_{sync}X_{sync}(t) + D_{sync}U_{sync}(t), \tag{11.8b}$$

where $U_{sync} = [\Delta v_{p1,dq}\ \Delta v_{p2,dq}\ \Delta\omega_{MG_1}\ \Delta\omega_{MG_2}]^T$, $X_{sync} = [\Delta\delta_{\omega,sync}\ \Delta\delta_{v,sync}]^T$, and the matrices are as $A_{sync} = 0_{2\times 2}$, $C_{sync} = I_{2\times 2}$, $D_{sync} = 0_{2\times 6}$, and

$$B_{sync} = k_{sync}\begin{bmatrix} k_{p1-\omega} & k_{p2-\omega} & k_{\omega 1} & k_{\omega 2} \\ k_{p1-v} & k_{p1-v} & 0 & 0 \end{bmatrix}, \tag{11.9}$$

where $k_{pi-\omega} = [k_{di-\omega}, k_{qi-\omega}]$, $k_{pi-v} = [k_{di-v}, k_{qi-v}]$ for $i = 1, 2$, and generally k_x, $x \in \{di-\omega, qi-\omega, \omega i \,|\, i = 1, 2\}$ is obtained as follows:

$$k_x = \sqrt{a_x^2 + b_x^2}\sin(\Delta\omega_{pcc0}t_0 + c_x\pi + d_x tg^{-1}(b_x/a_x)), \tag{11.10}$$

where $\Delta\omega_{pcc0} = \omega_{MG_1} - \omega_{MG_2}$ is the frequency difference at the operating point, t_0 is the time of inrush power occurrence, and a_x, b_x, c_x, and d_x are shown in Table 11.2.

11.4.1.3 Modeling of Overall IMGs

In order to calculate the overall model of IMGs, the subsystem models should be interconnected, which with the aim of small-signal stability analysis, leads to a free-motion state-space model as follows:

$$\dot{X}_{IMG}(t) = A_{IMG}X_{IMG}(t), \tag{11.11}$$

where $X_{IMG}(t)$ comprises all state variables of subsystems. A_{IMG} is a large state matrix, where the main diagonal arrays include the state matrices of the subsystems, and they can be calculated numerically as described in [7].

11.4.2 Small-Signal Stability Analysis

In addition to the synchronization controller parameter, the secondary control parameters are also important in the stability analysis of the synchronization control as the control signals are applied through the secondary controllers.

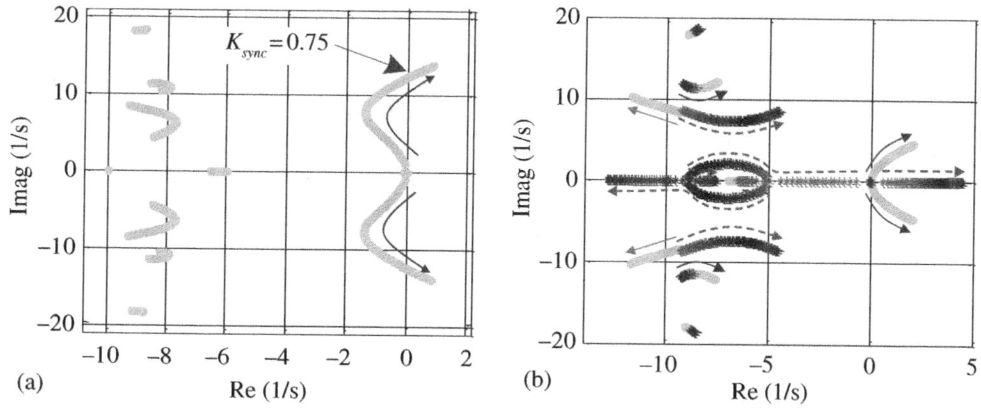

Figure 11.7 The dominant eigenvalues loci of studied IMGs for k_{sync} changing within [0.01, 1]: (a) all PCC voltage and frequency differences are zero, (b) $\Delta V_{pcc,d} = -5$ V for circles with solid lines, and $\Delta V_{pcc,d} = 5$ V for stars with dashed lines.

11.4.2.1 Synchronization Control Parameter

Figure 11.7 shows the loci of the studied two IMGs, where each MG has one leader DER and one follower DER. The MG structure is shown in Figure 8.1(b) and the control requirements are fully considered as shown in Figures 11.2 and 11.3. IMG parameters are given in Table II of [8]. The synchronization control parameter k_{sync} is varied in [0.01, 1]. In general, the arrows show the increasing direction of the parameter. In Figure 11.7(a), all PCC voltage and frequency differences are zero. In fact, it is considered that the presynchronization is done well. Two eigenvalues near the imaginary axis are so critical, and they will be unstable for $k_{sync} \geq 0.75$. Very low amounts of k_{sync} are not desired. Therefore, intermediate values are appropriate to achieve a better small-signal stability margin. Further analysis can be done to meet the performance objectives such as response speed and oscillation damping. Figure 11.7(b) shows the same loci when $\Delta V_{pcc,d} = -5$ V for circles with solid lines and $\Delta V_{pcc,d} = 5$ V for stars with dashed lines. In both situations, the IMGs cannot be stable for any value of k_{sync}. It also shows the importance of a successful synchronization.

11.4.2.2 Secondary Control Parameters

Figure 11.8 shows the loci of the most dominant eigenvalues of the studied IMGs when the secondary control parameter of the leader $DER_{1,1}$, T, changes in [0.1, 20]. In Figure 11.8(a), the PCC voltage, phase, and frequency differences are zero. Therefore, the stabilizing range of T can be found, which is as [5.1, 13.6] s. Nevertheless, in Figure 11.8(b), there is no stabilizing value for $\Delta V_{pcc,d} = -5$ V (circles with solid line) and a small stabilizing range for $\Delta V_{pcc,d} = 5$ V (stars with dashed line).

A similar condition to that in Figure 11.8 is considered in Figure 11.9 for changing the secondary control parameter of the follower $DER_{1,2}$, τ, in [0.1, 5]. According to Figure 11.9(a), the IMGs are asymptotically stable for $0.1 \leq \tau \leq 2.4$ s, and according to Figure 11.9(b), they cannot be stable for any value of τ in the range due to nonzero PCC voltage differences, i.e. $\Delta V_{pcc,d} = \pm 5$ V.

11.5 Transient Stability Assessment

For the sake of transient stability assessment, two nonlinear models are used. In Section 11.5.1, a simplified nonlinear model of exchanging power presented in (11.6) is used to show the power

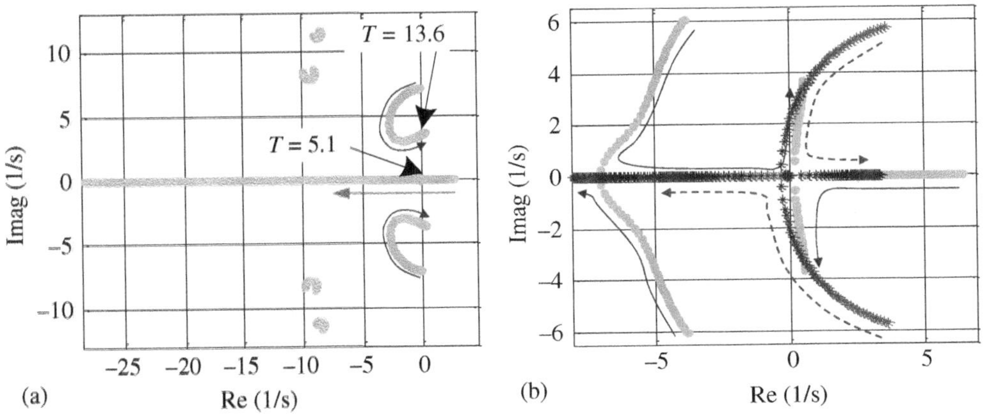

Figure 11.8 The dominant eigenvalues loci of studied IMGs for the leader secondary control parameter T changing within [0.1, 20] s: (a) all PCC voltage and frequency differences are zero, (b) $\Delta V_{pcc,d} = -5$ V for circles with solid lines, and $\Delta V_{pcc,d} = 5$ V for stars with dashed lines.

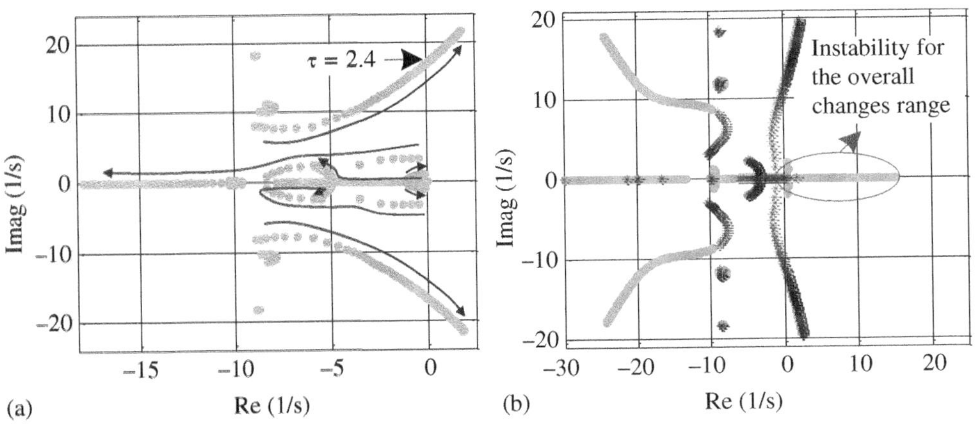

Figure 11.9 The dominant eigenvalues loci of studied IMGs for the follower secondary control parameter τ changing within [0.1, 5] s: (a) all PCC voltage and frequency differences are zero, (b) $\Delta V_{pcc,d} = -5$ V for circles, and $\Delta V_{pcc,d} = 5$ V for stars.

transitions during synchronization. In Section 11.5.2, the detailed nonlinear model of the studied IMGs is used in SimPowerSystems/MATLAB to study the synchronization stability challenges in the time domain.

11.5.1 Transition During Synchronization

In Section 11.3, the behavior of inrush power for different parameter changes is investigated. Here, the whole transition from presynchronization to the disconnection of synchronization control is taken into account. Figure 11.10 shows the active power exchange between the studied two IMGs versus the PCC phase difference during synchronization and after that for the PCC voltage and frequency differences. The system operating points before and after synchronization are obtained from a simulation of a successful synchronization using the control loops explained in Section 11.2.2. During presynchronization, the synchronization controller tries to reduce $\Delta\phi$ to

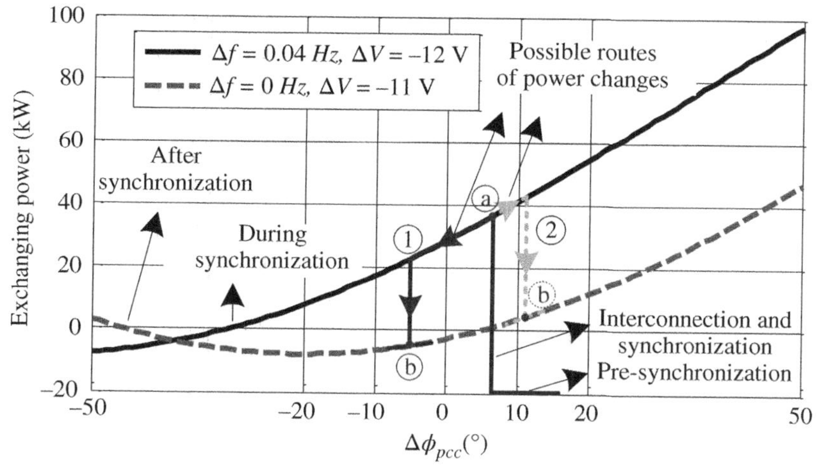

Figure 11.10 Active power exchanging between the IMGs versus phase difference changing for a transition during synchronization.

a predefined value or based on a certain constraint, while the exchanging power is as the inrush power (black solid line). During the controlled process, the inrush power may experience different routes of modification to be reduced to the scheduled power after the synchronization (dashed line). The route depends on the new operating point, i.e. the scheduled power amount. Route (1) belongs to a power exchange from MG_2 to MG_1, while route (2) shows a power exchange from MG_1 to MG_2. The routes are not necessarily for fixed voltage and frequency differences as shown as the control loops are continuously trying to change characteristics to meet the new operating point. Therefore, different routes from (a) to (b) may occur according to the control parameters and the length of inrush power. Finally, the disconnection of synchronization control leads to a change in the power to just satisfy the IMG power sharing objective.

11.5.2 Time-Domain Simulations

11.5.2.1 Trade-Off Between Synchronization Control Objectives

The studied IMGs are autonomously operated before $t = 5$ s. The presynchronization is started from $t = 3$ s, i.e. two seconds before interconnection time $t = 5$ s. The synchronization control is fully disconnected, and the MGs are interconnected at $t = 7$ s. Figures 11.11(a) and 11.11(b) show the PCC phase and voltage magnitude differences. The phase difference is considerable. However, the voltage magnitude difference is insignificant due to the MGs' operation condition. Although the presynchronization decreases the phase difference to 3° before synchronization, it causes a difference between the frequencies shown in Figure 11.11(c). It is due to applying both phase and frequency errors on the frequency control loop. Therefore, a trade-off between phase and frequency difference decreasing is necessary.

Figures 11.12(a) and 11.12(b) show the active and reactive power exchange between the IMGs. The synchronization is stable, and the transients are not large. However, they can be smaller using synchronization algorithms [12, 16, 20]. The steady-state values are tracked after synchronization control disconnection. Figures 11.13(a) and 11.13(b) show the DERs' active and reactive powers, i.e. power sharing before, during, and after the synchronization. Although the power sharing is affected by the synchronization control, it is still stable and the active power sharing is well done after the synchronization disconnection.

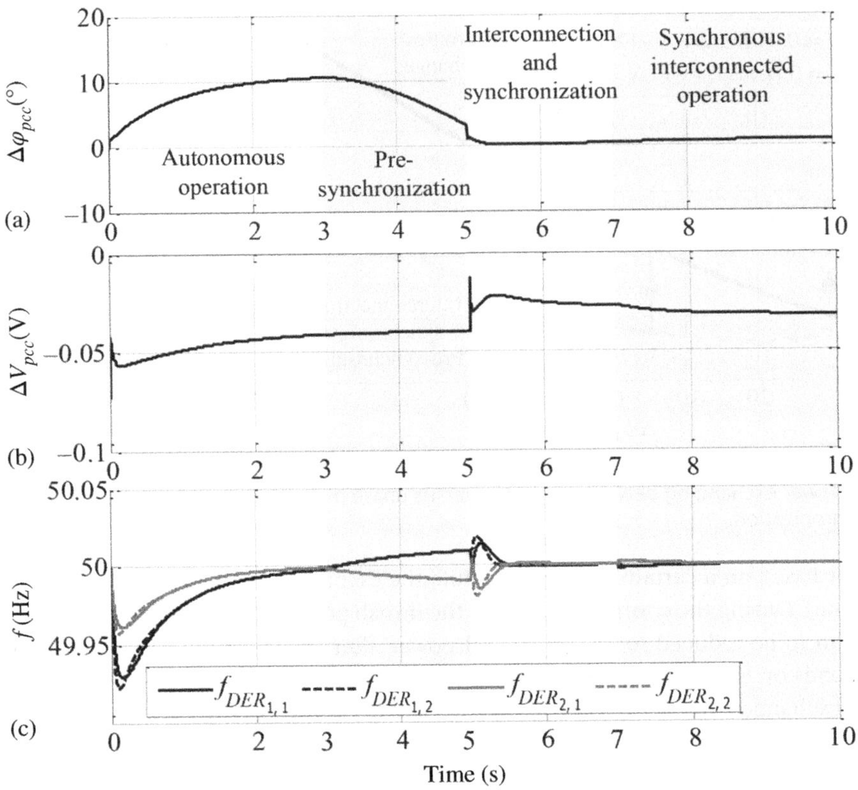

Figure 11.11 Synchronization signals before, during, and after the process: (a) phase, (b) voltage magnitude, and (c) frequency differences of the PCCs.

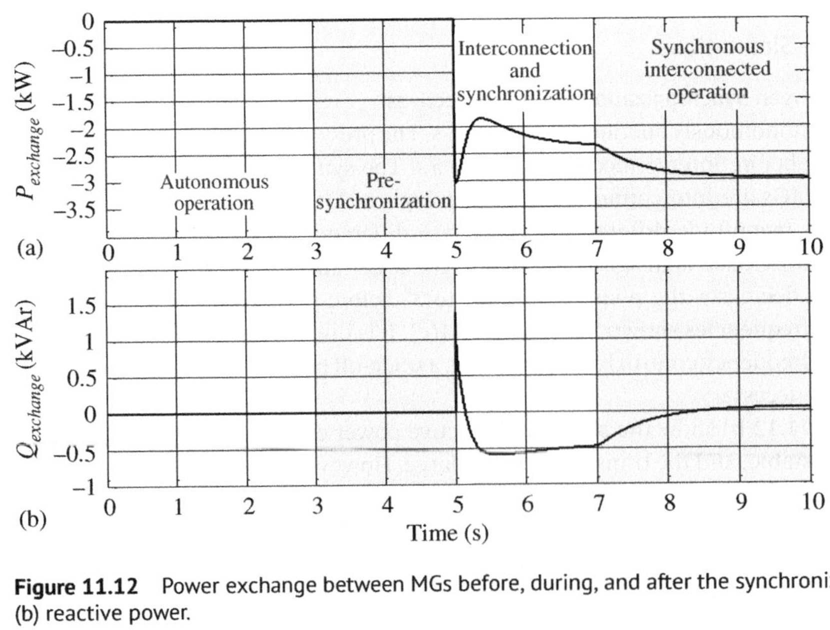

Figure 11.12 Power exchange between MGs before, during, and after the synchronization: (a) active power, (b) reactive power.

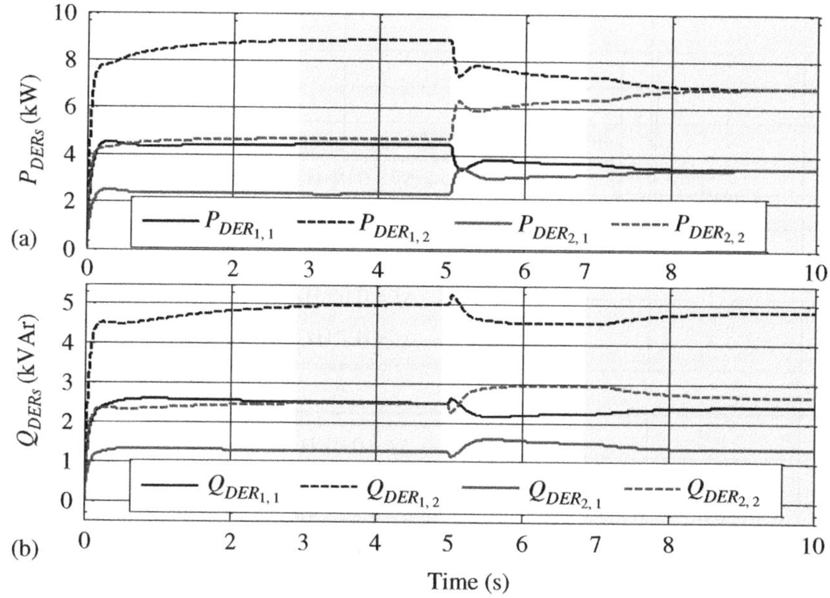

Figure 11.13 DERs' power sharing in isolated and interconnected MGs before, during, and after the synchronization: (a) active powers, (b) reactive powers.

Table 11.3 Threshold values for scenarios in Section 11.5.2.2.

Scenario	Δf_{Thr} Hz	$\Delta\phi_{Thr}$ °
①	0.01	7
②	0.02	2
③	0.03	5
④	0.05	3
⑤	0.1	10

11.5.2.2 Constraints Impact on the Synchronization Transients

As discussed in Sections 11.3 and 11.5.2.1, the PCC frequency and phase differences are more effective on the synchronization transients, and a trade-off is required in their minimization. Here, a simple algorithm is considered to apply the required constraints of synchronization, i.e. simultaneously $|\Delta f| \leq \Delta f_{Thr}$, $|\Delta\phi_{pcc}| \leq \Delta\phi_{Thr}$, and $|\Delta v_{pcc}| \leq \Delta v_{Thr}$. Table 11.3 shows the threshold values for five different scenarios. Scenarios 1, 3, and 4 are considered based on the trade-off between PCC frequency and phase differences. Scenario 2 is assumed as the best constraint with both low phase and frequency thresholds. Scenario 5 is the easiest constraint to be satisfied having large phase and frequency thresholds.

Figures 11.14(a) and 11.14(b) show the active and reactive powers exchange between the MGs. Scenario 2 has the lowest transients during the synchronization. The algorithm has been able to satisfy both thresholds; however, they are both small. Small thresholds may not be satisfied with a hard condition, e.g. high-load states.

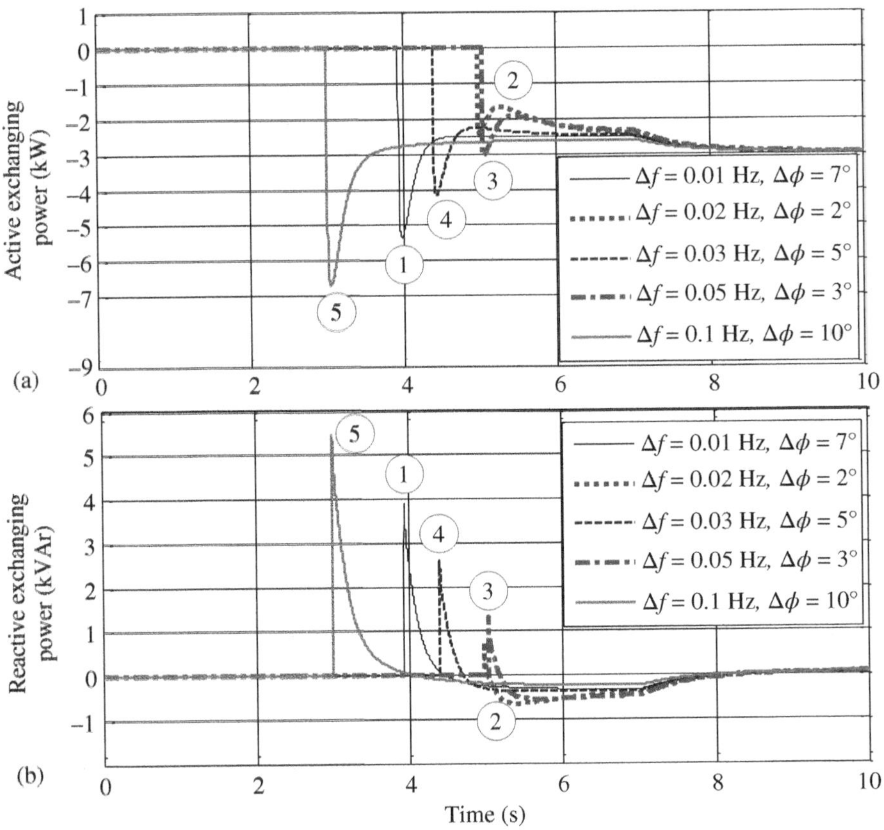

Figure 11.14 Exchanges between MGs for different synchronization constraints with the situations specified in Table 11.3: (a) active powers, (b) reactive powers.

11.5.2.3 Synchronization During High-Load MGs

In this condition, MGs' loads increase 2 times the basic load, i.e. a high-load condition occurs. The presynchronization is started at $t = 3$ s, and two different constraints are considered as $\Delta f = 0.1$ Hz, $\Delta\phi_{pcc} = 10°$, $\Delta f = 0.02$ Hz, and $\Delta\phi_{pcc} = 2°$. Figure 11.15(a) shows the frequencies of the leader DERs for both constraints, where there exists high transients, especially for the first constraint. Although both scenarios of constraints will be stable in the steady state, the rate of change of frequency is high during transients. Therefore, the operation may be interrupted by the protection devices. Figure 11.15(b) shows the active power exchange between MGs, where large transients can also be seen. The MGs' voltage/frequency will be severely affected by such transients of exchanging powers.

11.6 Summary

This chapter investigates the synchronization stability of CB-IMGs through small-signal analysis and transient stability assessment. Inrush power plays a vital role in synchronization stability, which is very sensitive to the frequency and phase differences of the MG's interconnection points, i.e. the ends of the interlinking line. Additionally, low-voltage MGs with low X/R ratios

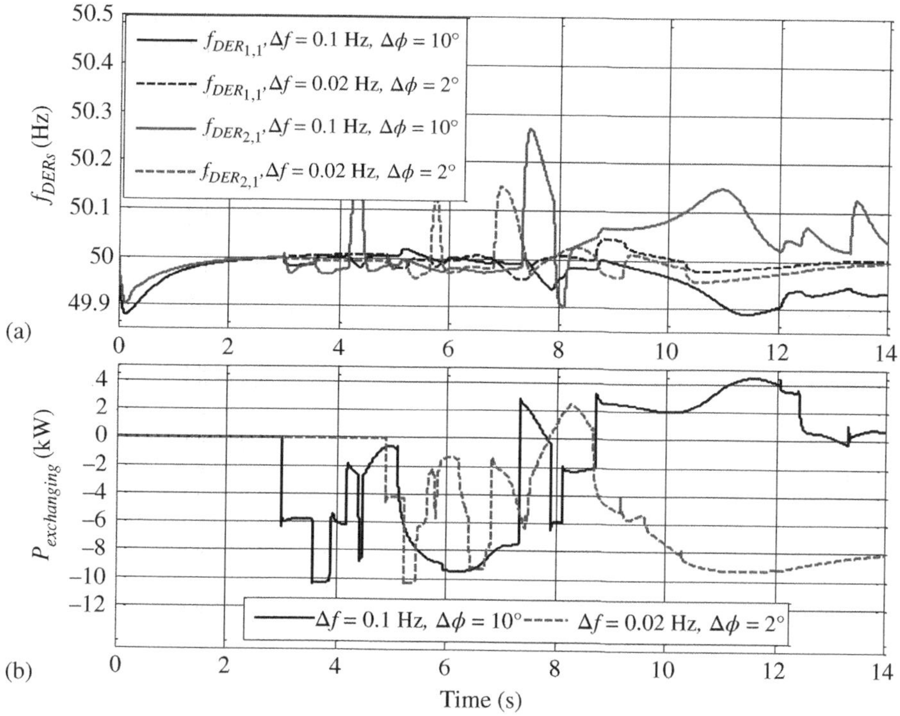

Figure 11.15 Synchronization during high-load conditions: (a) frequencies of the leader DERs, (b) active power exchanging between MGs.

of interlinking lines have lower active but greater reactive inrush power. However, a lower line impedance leads to greater active and reactive inrush power. During the transition from inrush to scheduled power, the amplitudes of the transient states represent a random variable that depends on the inrush and scheduled power values and controller gains. Nevertheless, achieving lower inrush power is generally a reasonable objective since the usual values of the scheduled power are less than the usual values of the inrush power. Nonzero voltage/phase/frequency differences may lead to small-signal instability at the primary operation point, and synchronization can only be small-signal stable for zero differences.

The stabilizing ranges of the control gains are calculated for synchronization, and secondary controllers are significant for synchronization stability. Studies show that lower thresholds of the synchronization controller result in lower transients. Transient stability is satisfied for low-load conditions and different thresholds of synchronization constraints. However, high-load operations lead to synchronization with higher transient states, particularly for large thresholds that are still less than those of the IEEE Standard 1547 criteria, which can also cause instability. Accordingly, this standard should be revised for synchronization of weak MGs during interconnections, i.e. forming CB-IMGs.

References

1 Mobin Naderi, Qobad Shafiee, Hassan Bevrani, and Frede Blaabjerg. Low-frequency small-signal modeling of interconnected AC microgrids. *IEEE Trans. Power Syst.*, 36(4):2786–2797, 2020.

2 Xiangyu Wu, Yin Xu, Jinghan He, Xiaojun Wang, Juan C Vasquez, Josep M Guerrero. Pinning-based hierarchical and distributed cooperative control for AC microgrid clusters. *IEEE Trans. Power Electron.*, 35(9):9867–9887, 2020.

3 Shengxuan Weng, Yusheng Xue, Jianbo Luo, and Yanman Li. Distributed secondary control for islanded microgrids cluster based on hybrid-triggered mechanisms. *Processes*, 8(3):370, 2020.

4 Xiaoqing Lu, Jingang Lai, and Xinghuo Yu. A novel secondary power management strategy for multiple AC microgrids with cluster-oriented two-layer cooperative framework. *IEEE Trans. Ind. Inf.*, 17(2):1483–1495, 2020.

5 Iraklis P Nikolakakos, Hatem H Zeineldin, Mohamed Shawky El-Moursi, and Nikos D Hatziargyriou. Stability evaluation of interconnected multi-inverter microgrids through critical clusters. *IEEE Trans. Power Syst.*, 31(4):3060–3072, 2016.

6 Munira Batool, Farhad Shahnia, and Syed M Islam. Multi-level supervisory emergency control for operation of remote area microgrid clusters. *J. Mod. Power Syst. Clean Energy*, 7(5):1210–1228, 2019.

7 Mobin Naderi, Yousef Khayat, Qobad Shafiee, Tomislav Dragicevic, Hassan Bevrani, and Frede Blaabjerg. Interconnected autonomous AC microgrids via back-to-back converters—Part I: Small-signal modeling. *IEEE Trans. Power Electron.*, 35(5):4728–4740, 2019.

8 Mobin Naderi, Yousef Khayat, Qobad Shafiee, Tomislav Dragičević, Hassan Bevrani, and Frede Blaabjerg. Interconnected autonomous AC microgrids via back-to-back converters—Part II: Stability analysis. *IEEE Trans. Power Electron.*, 35(11):11801–11812, 2020.

9 Blessy John, Arindam Ghosh, Megha Goyal, and Firuz Zare. A DC power exchange highway based power flow management for interconnected microgrid clusters. *IEEE Syst. J.*, 13(3):3347–3357, 2019.

10 Tine L Vandoorn, Bart Meersman, Jeroen D M De Kooning, and Lieven Vandevelde. Transition from islanded to grid-connected mode of microgrids with voltage-based droop control. *IEEE Trans. Power Syst.*, 28(3):2545–2553, 2013.

11 Xiaochao Hou, Yao Sun, Jinghang Lu, Xin Zhang, Leong Hai Koh, Mei Su, and Josep M Guerrero. Distributed hierarchical control of AC microgrid operating in grid-connected, islanded and their transition modes. *IEEE Access*, 6:77388–77401, 2018.

12 Mohamed Ahmed Ebrahim, Beshoy Abdou Aziz Ayoub, Maged Naguib Fahmy Nashed, and Fawzy Ahmed Mohamed Osman. A novel hybrid-HHOPSO algorithm based optimal compensators of four-layer cascaded control for a new structurally modified AC microgrid. *IEEE Access*, 9:4008–4037, 2020.

13 Fen Tang, Josep M Guerrero, Juan C Vasquez, Dan Wu, and Lexuan Meng. Distributed active synchronization strategy for microgrid seamless reconnection to the grid under unbalance and harmonic distortion. *IEEE Trans. Smart Grid*, 6(6):2757–2769, 2015.

14 Di Shi, Xi Chen, Zhiwei Wang, Xiaohu Zhang, Zhe Yu, Xinan Wang, and Desong Bian. A distributed cooperative control framework for synchronized reconnection of a multi-bus microgrid. *IEEE Trans. Smart Grid*, 9 (6):6646–6655, 2017.

15 Girish G Talapur, Hiralal M Suryawanshi, Lie Xu, and Amardeep B Shitole. A reliable microgrid with seamless transition between grid connected and islanded mode for residential community with enhanced power quality. *IEEE Trans. Ind. Appl.*, 54(5):5246–5255, 2018.

16 Yao Sun, Chaolu Zhong, Xiaochao Hou, Jian Yang, Hua Han, and Josep M Guerrero. Distributed cooperative synchronization strategy for multi-bus microgrids. *Int. J. Electr. Power Energy Syst.*, 86:18–28, 2017.

17 Bishoy E Sedhom, Magdi M El-Saadawi, Ahmed Y Hatata, and El-Hosaini E Abd-Raboh. H-Infinity versus model predictive control methods for seamless transition between islanded-and grid-connected modes of microgrids. *IET Renew. Power Gener.*, 14(5):856–870, 2020.

18 Liu Xia and Liu Hai. A seamless transfer strategy based on multi-master and multi-slave microgrid. In *9th IEEE International Symposium on Power Electronics for Distributed Generation Systems (PEDG)*, pages 1–5. IEEE, 2018.

19 IEEE application guide for IEEE STD 1547™, IEEE standard for interconnecting distributed resources with electric power systems, 2009.

20 Farhad Shahnia and Soheil Bourbour. A practical and intelligent technique for coupling multiple neighboring microgrids at the synchronization stage. *Sustainable Energy Grids Netw.*, 11:13–25, 2017.

21 N Bilidis. Seamless transitions of multiple micro-grids between the back-bone interconnected and the islanded operational modes. Delft University of Technology, 2016.

22 Seungyeon Kim. Experimental verification for seamless mode transitions of multiple microgrids using fuzzy-based droop controller. Delft University of Technology, 2017.

23 Aisha Pasha, Mohamed Shawky El Moursi, Hatem H Zeineldin, Vinod Khadkikar, Ehab F El-Saadany, Mohamed Al Hosani, Ahmed F Bendary. Enhanced transient response and seamless interconnection of multi-microgrids based on an adaptive control scheme. *IET Renew. Power Gener.*, 15(11):2452–2467, 2021.

24 Mobin Naderi, Yousef Khayat, Qobad Shafiee, Frede Blaabjerg, and Hassan Bevrani. Robust synchronization of multiple converter-based weak microgrids. *Submitted to IEEE Trans. Power Syst.*, 2022. doi: 10.1109/TPWRS.2023.3303206.

25 Wei Liu, Wei Gu, Yinliang Xu, Yong Wang, and Kaifeng Zhang. General distributed secondary control for multi-microgrids with both PQ-controlled and droop-controlled distributed generators. *IET Gener. Transm. Distrib.*, 11(3):707–718, 2017.

26 Lingyu Ren, Yanyuan Qin, Yan Li, Peng Zhang, Bing Wang, Peter B Luh, Song Han, Taofeek Orekan, and Tao Gong. Enabling resilient distributed power sharing in networked microgrids through software defined networking. *Appl. Energy*, 210:1251–1265, 2018.

27 Jingang Lai, Xiaoqing Lu, Xinghuo Yu, and Antonello Monti. Cluster-oriented distributed cooperative control for multiple AC microgrids. *IEEE Trans. Ind. Inf.*, 15(11):5906–5918, 2019.

28 Xiaoyu Wu, Yin Xu, Xiangyu Wu, Jinghan He, Josep M Guerrero, Chen-Ching Liu, Kevin P Schneider, and Dan T Ton. A two-layer distributed control method for islanded networked microgrid systems. *IEEE Trans. Smart Grid*, 11(2):942–957, 2019.

29 Jinghan He, Xiaoyu Wu, Xiangyu Wu, Yin Xu, and Josep M Guerrero. Small-signal stability analysis and optimal parameters design of microgrid clusters. *IEEE Access*, 7:36896–36909, 2019.

30 Mobin Naderi, Qobad Shafiee, Frede Blaabjerg, and Hassan Bevrani. Synchronization stability of interconnected microgrids with fully inverter-based distributed energy resources. *J. Mod. Power Syst. Clean Energy*, 1–12, 2023. doi: 10.35833/MPCE.2022.000594

Index

e

f

g

h

m

q

r

t

u

v

W

Z

Printed and bound by CPI Group (UK) Ltd, Croydon, CR0 4YY
17/09/2026
14964259-0004